PAPERS OF THE MICHIGAN ACADEMY OF SCIENCE, ARTS, AND LETTERS

VOLUME XLVII

I. *Natural Science*
II. *Social Science*
III. *Philosophy, Literature, and Art*

Papers of the
MICHIGAN ACADEMY
OF SCIENCE, ARTS, AND LETTERS

VOLUME XLVII

Editor

HUBERT M. ENGLISH, JR.

Published for the Michigan Academy of Science, Arts, and Letters by
THE UNIVERSITY OF MICHIGAN PRESS
ANN ARBOR
1962

LIBRARY OF CONGRESS CATALOG CARD NO. 23–11028

Manufactured in the United States of America

Paperback ISBN: 978-0-472-75129-7

OFFICERS OF THE ACADEMY, 1960–61

PRESIDENT, Conrad R. Lam, Henry Ford Hospital, Detroit

PRESIDENT-ELECT, Justin W. Leonard, Michigan Department of Conservation, Lansing

VICE-PRESIDENT, Charles P. Loomis, Michigan State University

SECRETARY, F. Clever Bald, The University of Michigan, Ann Arbor

TREASURER, Volney H. Jones, The University of Michigan

EDITOR, Hubert M. English, Jr., The University of Michigan

LIBRARIAN, Frederick H. Wagman, The University of Michigan

COORDINATOR OF JUNIOR ACADEMY, Lincoln C. Pettit, Michigan State University

COORDINATOR OF JUNIOR ACADEMY FOR NORTHERN PENINSULA, Clarence Bjork, Northern Michigan College

SECTION CHAIRMEN

AMERICAN STUDIES: Russel B. Nye, Michigan State University

ANTHROPOLOGY: Leonard W. Moss, Wayne State University

BOTANY: Warren Stoutamire, Cranbrook Institute of Science

ECONOMICS: Howard Bourne, Wayne State University

FINE ARTS: G. Alden Smith, Wayne State University

FISHERIES-WILDLIFE: Richard O. Anderson, Wolf Lake Hatchery, Mattawan

FOLKLORE: Bruno Nettl, 1991 Ewald Circle, Detroit 38

FORESTRY: Charles Allen, Scott Paper Co., Detroit

GEOGRAPHY: L. A. Peter Gosling, The University of Michigan

GEOLOGY-MINERALOGY: Andrew Mozola, Wayne State University

HISTORY-POLITICAL SCIENCE: Elton Ham, Kalamazoo College

LANDSCAPE ARCHITECTURE: Norbert Gorwic, Wayne State University

LANGUAGE-LITERATURE: Sister Mary Hilda, Marygrove College

MATHEMATICS: Earl D. Rainville, The University of Michigan

MEDICAL SCIENCES: Otto W. Neuhaus, Wayne State University

PHILOSOPHY: Edmund Gettier, Wayne State University

PSYCHOLOGY: Bertram D. Cohen, Lafayette Clinic, Detroit

RUSSIAN STUDIES: Arthur E. Adams, Michigan State University

SOCIOLOGY: Paul Horton, Western Michigan University

ZOOLOGY: Gordon D. Gill, Northern Michigan College

CONTENTS

Part I. Natural Science

MEDICINE

GEOLOGY

PALEONTOLOGY

BOTANY

FORESTRY

B. Recreational Geography

C. The Geography of Asia

Part III. Philosophy, Literature, and Art

Literature

PART I

NATURAL SCIENCE

PAPERS OF THE MICHIGAN ACADEMY OF SCIENCE, ARTS, AND LETTERS
VOL. XLVII, 1962 (1961 MEETING)

MEDICINE

SURGERY'S LAST FRONTIER: THE HEART*

CONRAD R. LAM

FOR practical purposes, it can be said that the science of surgery is less than 100 years old. Although the discovery of anesthesia in 1846 made patients more willing to accept the services of the surgeon, the lack of knowledge of bacteria made the surgical wards of that time a sad collection of individuals with infected wounds, both accidentally and surgically acquired. This year we observe the centennial of the Civil War, and some of us will be reminded of the battle which both sides lost—the battle of their surgeons against wound infection. Ironically, at that exact time, the great Frenchman Pasteur was proving the existence of micro-organisms, and in 1865 Lister began his experiments in antisepsis in surgery, which were published in 1867. The boiling of instruments, skin sterilization, the "surgical scrub," and rubber gloves came later. Therefore, I think Dr. Sherman, at the American Medical Association meeting in 1902, was a little unfair to surgeons when he said, "The road to the heart is only two or three centimeters in a direct line, but it has taken surgery nearly 2400 years to travel it" (19). (This statement assumes that surgical history dates back to the days of Hippocrates.)

Furthermore, I think this speaker was premature in calling the surgical route to the heart a "road." It was hardly a footpath, and it had been very little traveled. Nine years before, the most famous surgeon of his time, the great Billroth of Vienna, had said that the surgeon who would attempt to suture a wound of the heart would lose the respect of his colleagues (2). A book entitled *Surgery of the Chest* published in England in 1896, contained the statement, "Surgery of the heart has probably reached the limits set by nature to all surgery; no new method and no new discovery can overcome the natural difficulties which attend a wound of the heart" (17). That same year, however, as if in rebuttal, a surgeon who spoke the same

* The Presidential Address delivered at the Academy Dinner Meeting, March 24, 1961. Dr. Lam is surgeon-in-charge of the Division of Thoracic Surgery, Henry Ford Hospital, Detroit, Michigan.

language as Billroth—Rehn of Frankfurt, Germany—successfully sutured a bleeding wound of the right ventricle (18). Six years later, the American Medical Association speaker to whom we have referred was able to announce triumphantly that the world literature contained reports of thirty-four instances of the suture of wounds of the heart (19).

For the next thirty-five years, during which time abdominal surgery and even brain surgery made remarkable advances, the heart remained essentially untouched. I have just consulted the textbook of surgery which I used as a medical student, published in 1931 (10). Nothing had been added to surgery of the heart except to state that several surgeons had been daring enough to cut away some of the pericardium when it was constricting the heart. There was an account of a half dozen attempts to open stenotic valves, all unsuccessful. It is no wonder that in the class of 1932 of the Yale School of Medicine there was no young doctor who said, "I want to become a heart surgeon." Furthermore, there is no record of a member of any graduating class of that year making such a statement. In the class of 1931 of the Harvard Medical School, however, there was a young doctor who was to perform an operation which would revive interest in heart surgery. Dr. Robert Gross was to ligate the ductus arteriosus in 1938 (7).

The ligation of the patent ductus arteriosus is not really a heart operation, since it involves vessels which are actually outside the pericardial cavity. But the open ductus arteriosus puts strain on the heart by allowing blood to leak from the aorta with its higher pressure into the pulmonary artery (Fig. 1), and this strain is relieved by obliteration of the shunt. It is understandable, therefore, that in our operating room statistics the division of the ductus is classified as a heart operation. Six years later two other operations involving the great vessels near the heart were added to the list. The first was the correction of coarctation of the aorta, which is a severe congenital constriction of that large vessel just after the arteries to the head and arms have been given off. The operation was first done by the Swedish surgeon Clarence Crafoord (6), but in a few weeks the same Robert Gross had done it also (8). The other operation was the famous blue baby operation of Blalock and Taussig of the Johns Hopkins Hospital (3). In this procedure an artery to the arm is borrowed to bring more blood to the lungs for oxygenation.

Two years later direct operations on the heart valves were being done. The London surgeon Brock opened the pulmonic valve with a spoon-shaped knife passed through an opening in the right ventricle (5). Bailey of Philadelphia (1) and Harken of Boston (9) opened the stenosed mitral valve by introducing the finger and various cutting instruments through the left auricle.

But the surgeon was still at a disadvantage because he could not see what he was doing. A partial answer was available in hypothermia—refrigeration. If the body temperature is lowered sufficiently, say to 85° F., the venous return to the heart can be interrupted for 5 to 10 minutes, during which time simple operations such as the closure of atrial septal defects and pulmonary valvotomy (20)

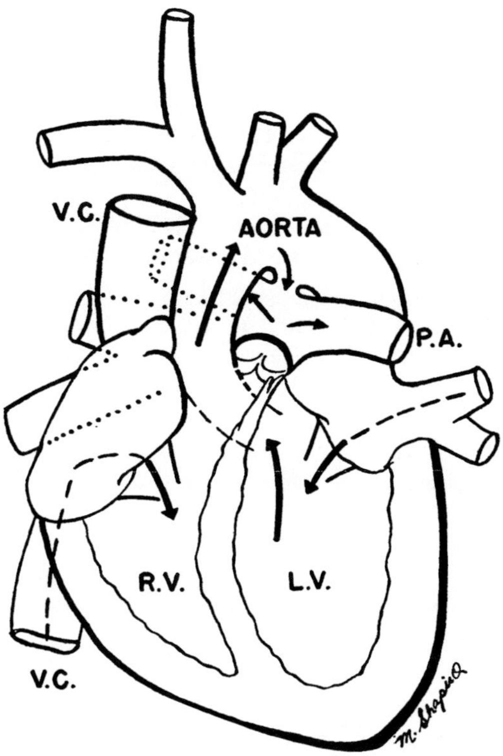

FIG. 1. Diagram of heart and great vessels, showing patent ductus arteriosus. V.C., vena cava; R.V., right ventricle; L.V., left ventricle; P.A., pulmonary artery. The arrow under the word aorta indicates the flow of blood from that vessel into the pulmonary artery.

can be carried out. But there is not time to do repairs which require careful suturing such as is necessary for the repair of a ventricular septal defect.

On April 20, 1954, a historic operation took place (13). Dr. Walton Lillehei and his team at the University of Minnesota connected an artery and vein of a four-year-old boy to an artery and vein of his father. With a pump controlling the amount of cross circulation, the lungs of the father served as the child's lungs while the heart was open for twenty-six and one-half minutes during the repair of a ventricular septal defect. For the first time a surgeon had worked inside the heart in an unhurried manner. This was a "break-through" of the greatest importance. After 40 such operations the Minneapolis group was able to announce in 1955 that a human donor was no longer needed, since a simple oxygenator of disposable plastic (Fig. 2) had been made (14). The field was now wide open for many types of heart surgery. The barriers at the last frontier had been breached.

The materials for the Lillehei-DeWall pump-oxygenator were brought to Detroit in March 1956, and in April a ventricular septal defect was closed successfully. A new principle was utilized, that of "induced cardiac arrest," (11, 12). While the sutures were being placed inside the heart, the beating of the heart was stopped by the injection of a drug, acetylcholine, into the aorta below a clamp (Fig. 3). The drug was forced into the coronary arteries and thence to the heart muscle, resulting in cessation of cardiac activity (Fig. 4). When it was desired to start the heart, the clamp on the aorta was removed and the acetylcholine was washed out of the heart by the blood coming from the pump-oxygenator. Acetylcholine for this purpose is thought to be superior to the drug potassium citrate, which was recommended by Melrose of London (15).

There is yet another good way of stopping the heart now. A heat exchanger, consisting of stainless steel tubes surrounded by a water bath, is placed between the oxygenator and the patient. The blood is passed through the tubes surrounded by cold water until the heart stops. The pump-oxygenator must continue to operate to supply the brain and other parts of the body with oxygen, although if the temperature of the whole body is also depressed the pump may be slowed down or even cut off for considerable periods of time. When it is time to resuscitate the heart the water in the heat exchanger is warmed, and the incoming warm blood warms both the heart and the rest of

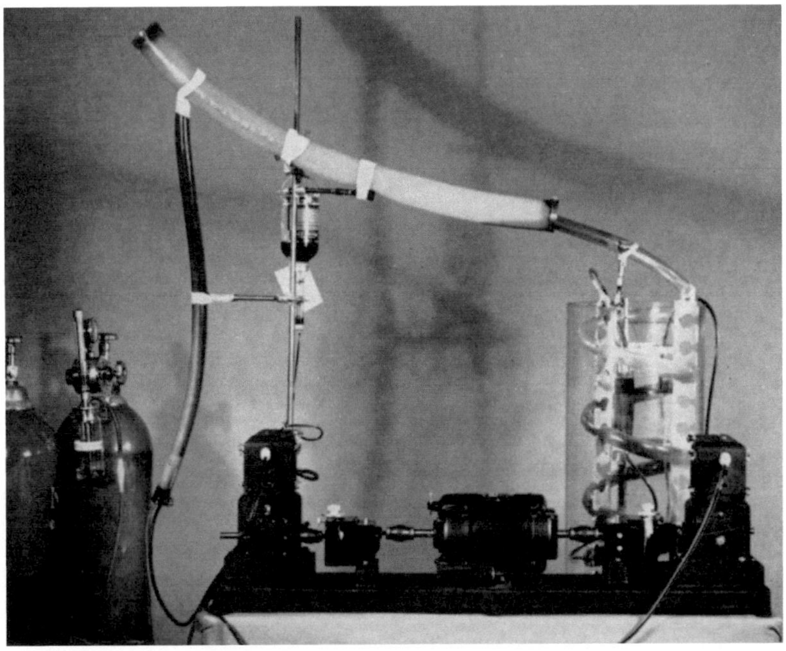

FIG. 2. Pump-oxygenator of the "bubble" type. Oxygen is bubbled through the blood in the vertical tube on the left, the bubbles are removed in the larger inclined tube, and the oxygenated blood collects in the helix in the water bath for rewarming. The box to the left of the motor is the "right ventricle" and that to the right the "left ventricle."

the body. Usually the heart responds with ventricular fibrillation, which is easily converted to a coördinated beat with a counter shock.

Extracorporeal oxygenation has made possible a great variety of heart operations. The list includes closure of auricular and ventricular septal defects, corrections of stenosis of the valves, repair of insufficiency of the mitral and aortic valves, removal of tumors from within the heart, the correction of many different kinds of abnormalities of venous drainage into the heart, the complete correction of the tetralogy of Fallot (the blue baby), and the excision of left ventricular aneurysms which may follow myocardial infarction.

I shall go into detail about the diagnosis and surgical treatment of a certain anomaly of venous drainage into the heart in which the left

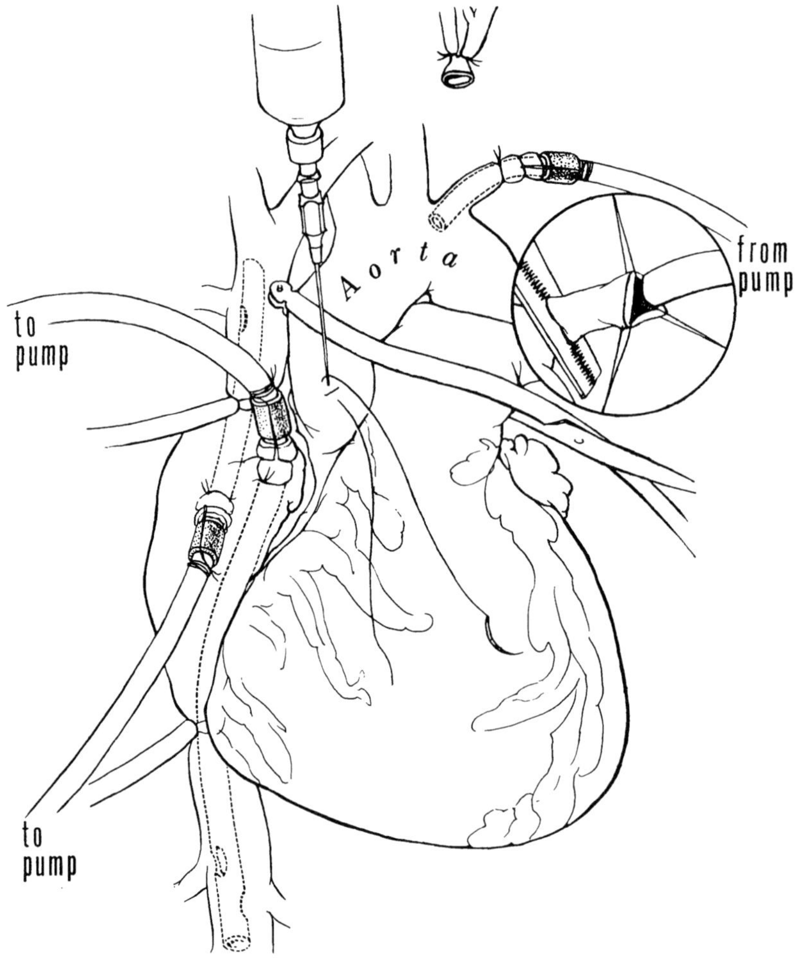

FIG. 3. Diagram of cannulations of the heart during cardio-pulmonary bypass with the pump-oxygenator. The femoral artery is now used in preference to the subclavian. A cardioplegic drug (acetylcholine) is being injected into the clamped aorta to induce cardiac arrest.

auricle is divided by a septum, so that it may be said to be a "triatrial heart." The surgical literature was reviewed by Niwayama (16), who found reports of 29 cases of a certain type of triatrial heart and added 4 cases of his own. In this type there was a membrane dividing the

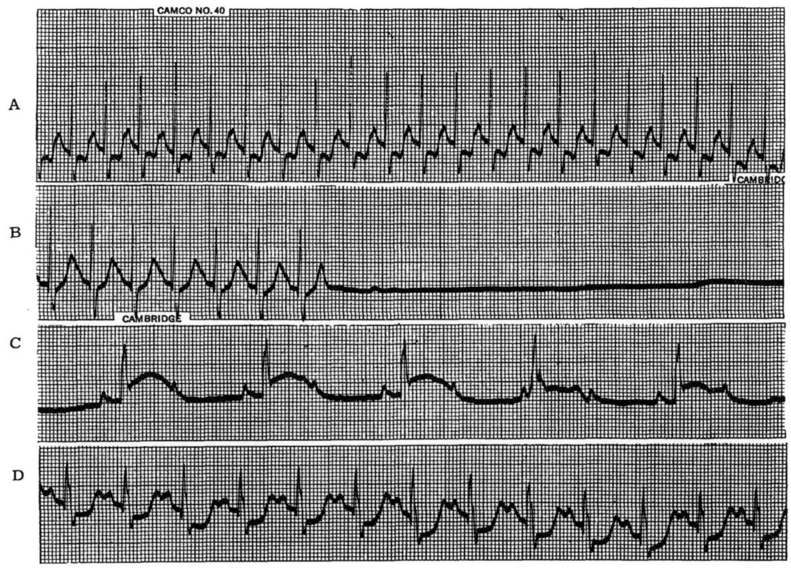

FIG. 4. Electrocardiographic records during the closure of a ventricular septal defect in a 9-year old girl. (*a*) Tracing before cardiotomy; (*b*) tracing showing instantaneous arrest of the heart with the injection of acetylcholine; (*c*) tracing during recovery phase after removal of aortic clamp; and (*d*) resumption of normal rate of rhythm. (Courtesy J. Thoracic Surg.)

left auricle into two chambers (Fig. 5, Type A). The veins bringing oxygenated blood from the lungs emptied into the upper of these chambers, but the blood could not easily reach the mitral valve in the lower chamber because it had to pass through a tiny aperture in the partition. In addition to having appeared at autopsies, this deformity had been encountered unexpectedly by two surgeons during operations.

But when we encountered a triatrial heart in a patient scheduled for the closure of an auricular septal defect, under hypothermia, it was not that kind of a triatrial heart. There were two left atria, but the partition between them was complete, and to get from the pulmonary veins to the mitral valve the blood had to pass through an atrial septal defect into the right auricle and thence through another defect into the other auricular chamber, at the bottom of which was the mitral valve (Fig. 5, Type B). The obvious corrective procedure was to re-

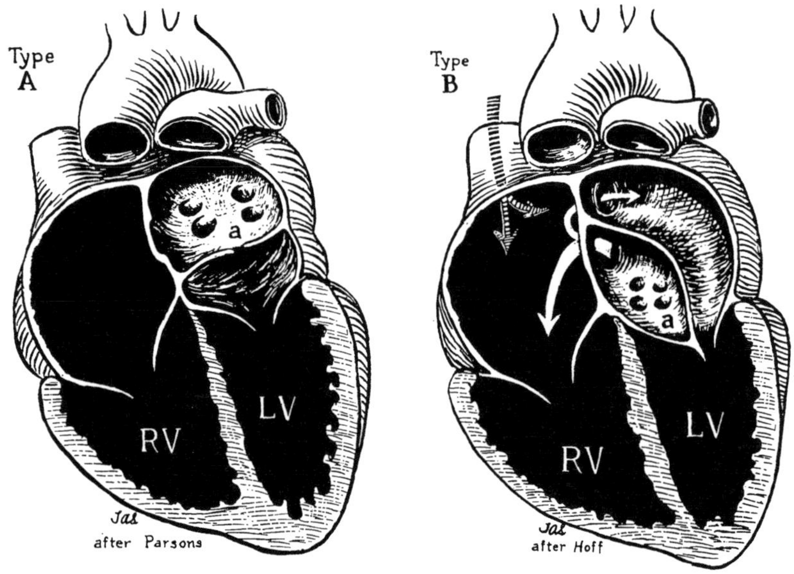

FIG. 5. Two types of triatrial heart. Left: Type A, "stenosis of the common pulmonary vein." Right: Type B. See text.

move the partition and close the resulting large defect between the right and left auricles (Fig. 6). This had to be done hastily in this patient because we were using hypothermia. Since the pump-oxygenator became available we have been able to correct four such deformities in a more leisurely fashion. We have called the type of triatrial heart which we repaired Type B, as opposed to Type A, already described in the literature.

After encountering six cases of triatrial heart Type B, we wondered how long it would be before we saw what was presumably more common in others' experience, Type A. We did not have long to wait. Dr. Edward Green, pediatric cardiologist, was studying a six-year-old boy with cardiac symptoms who showed some of the signs of mitral valve stenosis but not all of them. There was evidence of pulmonary hypertension and an enlarged left atrium, but no murmur of mitral stenosis. When radio-opaque dye was injected into the heart and followed through the lungs with rapid film exposures, there was a suggestion that the flow of blood to the mitral valve was retarded by a structure which could be the membrane of a triatrial heart, Type A.

With this information, we scheduled the boy for an operation to correct such a defect. It was gratifying to find exactly what the radiologist and cardiologist had predicted, and the defect was easily corrected while the left auricle was opened during cardio-pulmonary bypass (Fig. 7).

This is a good example of the adventures which the surgeon may encounter as he maneuvers about in surgery's last anatomic frontier. It will be noted that I now qualify the term "last frontier" as anatomic. I do not intend to get into the trap Sir James Paget put himself in when he made the remark I have already quoted: "Surgery of the heart has probably reached the limits set by nature . . ." (17). Anatomically, we can now go anywhere in the heart, but too often we do not know what to do when we get there. Our degree of sophistication might be compared to that of the pioneers on the American continent; the conquistador Balboa, for example, reached the limit of a geographic frontier at the Pacific coast line, but did not know what to do next.

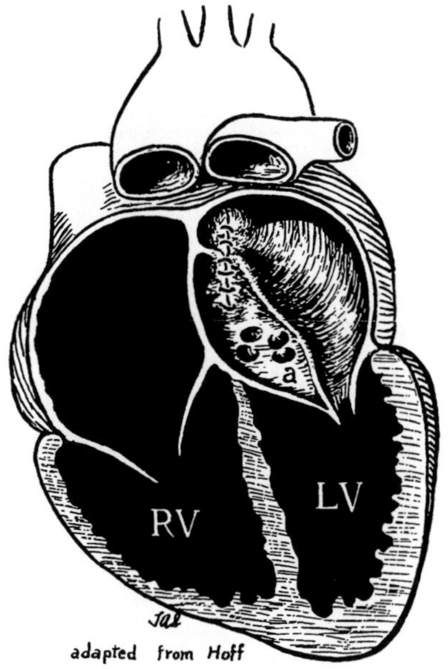

FIG. 6. Diagram showing repair of triatrial heart, Type B. The abnormal septum in the left auricle has been cut away, and the resulting large defect in the interatrial septum has been closed with sutures.

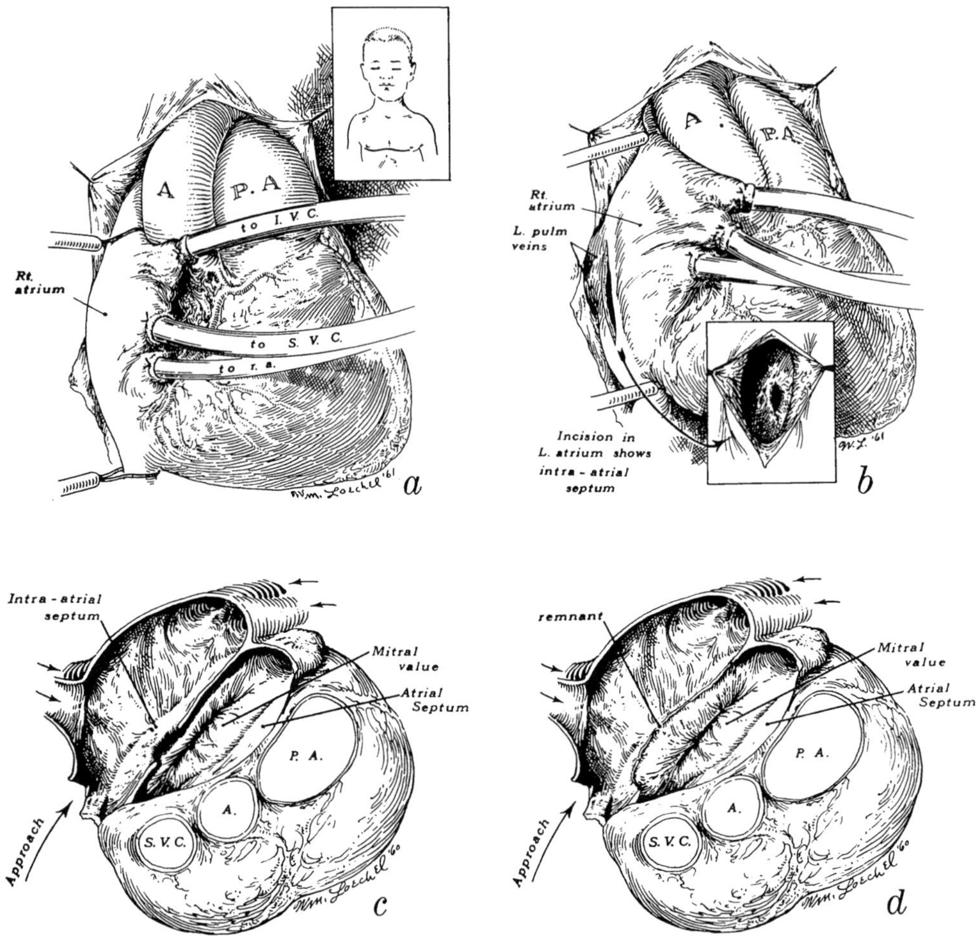

FIG. 7. Four diagrams showing operative steps in correction of triatrial heart, Type A. (*a*) Cannulations for pump-oxygenator; (*b*) incision into dry left atrium on the right side of the heart to reveal the abnormal septum in the left auricle with the small aperture for the passage of blood (inset); (*c*) horizontal section of heart to show the relationship of the abnormal septum in the left atrium; and (*d*) same as *c*, with abnormal septum removed.

Many problems in heart surgery remain to be solved. Only a few of them can even be defined at this time. One of the most obvious is the problem of coronary artery disease. It is very likely that when anatomic localization of the arterial block is more precise and operative technique more refined, the surgeon will be able to do a great amount of good before the hardworking investigators are able to tell us how to prevent atherosclerosis. Durable plastic parts for the heart are practically with us. Transplantation of the heart may become physiologically possible when the secrets of antibody formation are known, but by that time we shall probably have an implantable plastic pump with built-in atomic power. The artificial electrical excitation of the heart with "block" is already possible in a crude way, but small pacemaker units that can be recharged by induction are almost ready for placing in human cases. Much remains to be learned, however, about the electrical control of the heart. Several congenital anomalies have defied our efforts at correction. Some of these are transposition of the great vessels, tricuspid atresia, and truncus arteriosus.

I have been reading a new book containing some of the wise sayings of the late Charles F. Kettering, famous for the invention of the self-starter (4). Typical of his philosophy was this: "In my opinion, we have to date chipped away only a few fragments from the Mountain of knowledge—fragments that have changed our entire way of life. But looming ahead of us, practically intact, lies a huge mass of fundamental facts, any one of which, if uncovered, could change our civilization." This statement clearly expresses a challenge to cardiac surgeons of 1961 and to those who have indicated their interest in progress by associating themselves with the Michigan Academy of Science, Arts, and Letters.

LITERATURE CITED

1. Bailey, C. P., R. P. Glover, and T. J. E. O'Neill. 1950. The Surgery of Mitral Stenosis. J. Thoracic Surg., 19: 16.
2. Billroth, T. H. 1893. Ein Handbuch für Studierende und Aerzte. Berlin: G. Reimer.
3. Blalock, Alfred, and Helen B. Taussig. 1945. The Surgical Treatment of Malformations of the Heart in Which There Is Pulmonary Stenosis or Pulmonary Atresia. J.A.M.A., 128: 189.
4. Boyd, T. A. 1961. Prophet of Progress: Selections from the Speeches of Charles F. Kettering. New York: E. P. Dutton & Co.

5. Brock, R. C. 1948. Pulmonary Valvulotomy for Relief of Congenital Pulmonary Stenosis. Brit. M. J., 1: 1121.
6. Crafoord, Clarence, and G. Nylin. 1945. Congenital Coarctation of the Aorta and Its Surgical Treatment. J. Thoracic Surg. 14: 347.
7. Gross, R. E., and J. P. Hubbard. 1939. Surgical Ligation of a Patent Ductus Arteriosus: Report of First Successful Case. J.A.M.A. 112: 729.
8. ——, and C. A. Hufnagel. 1945. Coarctation of the Aorta: Experimental Studies Regarding Its Surgical Correction. New Eng. J. Med., 233: 287.
9. Harken, D. E., L. B. Ellis, and L. R. Norman. 1950. The Surgical Treatment of Mitral Stenosis: II Progress in Developing a Controlled Valvuloplastic Technique. J. Thoracic Surg., 19: 1.
10. Homans, John. 1931. A Textbook of Surgery. Springfield, Illinois: Charles C. Thomas.
11. Lam, C. R., T. Geoghegan, and A. Lepore. 1955. Induced Cardiac Arrest for Intracardiac Surgical Procedures. J. Thoracic Surg., 30: 620.
12. ——, T. Gahagen, C. K. Sergeant, and E. Green. 1957. Experiences in the Use of Cardioplegia (Induced Cardiac Arrest) in the Repair of Ventricular Septal Defects. J. Thoracic Surg., 34: 509.
13. Lillehei, C. W., Morley Cohen, H. E. Warden, N. R. Ziegler, and R. L. Varco. 1955. The Results of Direct Vision Closure of Ventricular Septal Defects in Eight Patients by Means of Controlled Cross Circulation. Surg., Gynec. and Obst., 101: 446.
14. ——, R. A. DeWall, R. C. Read, H. E. Warden, and R. L. Varco. 1956. Direct Visual Intracardiac Surgery in Man Using a Simple, Disposable Artificial Oxygenator. Dis. Chest, 29: 1.
15. Melrose, D. G., B. Dreyer, H. H. Bentall, and J. B. E. Baker. 1955. Elective Cardiac Arrest. Lancet, 2: 21.
16. Niwayama, Gen. 1960. Cor Triatriatum. Am. Heart J., 59: 291.
17. Paget, Steven. 1896. The Surgery of the Chest. Bristol: John Wright and Sons.
18. Rehn, L. 1897. Ueber penetrirende Herzwunden und Herznaht. Arch. Klin. Chir., 55: 315.
19. Sherman, H. M. 1902. Suture of Heart Wounds. Boston M. & S. J., 146: 653.
20. Swan, H., H. C. Cleveland, H. Muller, and S. G. Blount, Jr. 1954. Pulmonic Valvular Stenosis. J. Thoracic Surg., 28: 504.

Papers of the Michigan Academy of Science, Arts, and Letters
Vol. XLVII, 1962 (1961 Meeting)

GEOLOGY

CAVES OF THE ACROPOLIS OF ATHENS

CHARLES G. HIGGINS

University of California, Davis, California

The Athenian Acropolis, with its easily defensible, sheer, rocky slopes and smooth, gently sloping summit, has served the inhabitants of Athens as fortress and refuge for more than 4,500 years. Excavations have verified the existence of a Neolithic settlement, 2800 to 2500 B.C., on the summit, and have uncovered traces of still older Neolithic dwellings and graves on the lower slopes (Oekonomides, 1958, p. 5). According both to traditional accounts and to archaeological discoveries, these early permanent settlements, as well as those which followed, were concentrated on the north slope of the Acropolis (Broneer, 1941). This location must have been dictated not only by the need to retreat to the Acropolis in time of attack, but by the need for domestic water supplies. Until the construction, in the sixth century B.C., of the Peisistratean aqueduct and water-distribution system, parts of which have recently been unearthed (Thompson, 1956), inhabitants of the Acropolis area must have had to rely primarily on local springs and wells.

The largest spring, called Empedo (Fig. 1, *K*), originally issued from a small cave on the northwest slope. During the Kimonian period, after the Persian Wars, this spring was converted into a well and renamed Klepsydra. Its long and complex history has been worked out and described by Parsons (1942). Overflow from this spring into the scree and alluvial deposits below the north slope doubtless furnished much of the water in ancient, Neolithic wells, twenty or thirty of which have been excavated and dated by means of the pottery they contained (Hill, 1953, p. 3).

Several smaller springs on the Acropolis slopes also supplied water for domestic or religious use in ancient times. One (Fig. 1, *Z*) on the south slope was enlarged in the sixth century B.C. and enclosed in a rectangular shaft of polygonal Peisistratean masonry. In post-Roman time the enclosure was further enlarged and covered with a barrel vault. Farther east a second spring flows from a small, shallow cave (Fig. 1, *Y*). In the fourth or fifth century B.C. this cave was hollowed out to form a circular, domed room about 15 feet in diameter behind the Doric Stoa of the Asklepieion. Still later it was converted into a Christian shrine and holy spring, and is still used as such. The most complete study of this spring appears to be that by Travlos (1941); both springs are discussed in Keramopoullos' (1935) comprehensive study of the south slope.

The source of water for these springs and wells must be the Acropolis itself. The hill is approximately 900 feet long, 450 feet wide, and 150 feet high, and is composed chiefly of irregularly jointed, well-cemented, cataclastic limestone breccia. If one assumes that groundwater storage in the hill is restricted to joints in the limestone, it is difficult to believe that the Acropolis springs could have supplied sufficient water for settlements of the size suggested by historical accounts and archaeological discoveries, or even for the population of the fortress itself during wartime sieges, when the villagers retreated to the summit for protection. Either there must have been some kind of aqueduct system to augment the water supplies of the local springs as early as Mycenean, or even Neolithic, time, or else the groundwater storage capacity of the hill is greater than it would appear to be. The latter conclusion is the more likely and is supported by conclusions regarding the grottoes and caves in the Acropolis limestone.

Any natural caverns that may have existed on the west slope have been obliterated by foundations, walls, stairs, and other constructions. However, excavations have disclosed several small natural niches in the bedrock behind the bastion which supports the temple of Athena Nike (Fig. 1, *N*), and it is reasonable to assume that other parts of the low west face may also have been somewhat cavernous.

The most westerly cave on the north slope is that of the Klepsydra (Fig. 1, *K*). Most of this cave collapsed in the first century A.D. (Parsons, 1942). Like most of the Acropolis caves, it appears to have been quite shallow—a mere grotto. Above and slightly east of the Klepsydra is another shallow grotto, called cave Α. Farther east are Β, Γ, and Δ (Fig. 1). Caves Β and Γ are larger grottoes than Α, 20 to 30 feet high and 10 to 20 feet deep, but Δ is reported to be a deep, tunnel-like cavern with two branches to the east. In Classical times and before, these three caves served as religious sanctuaries associated with early

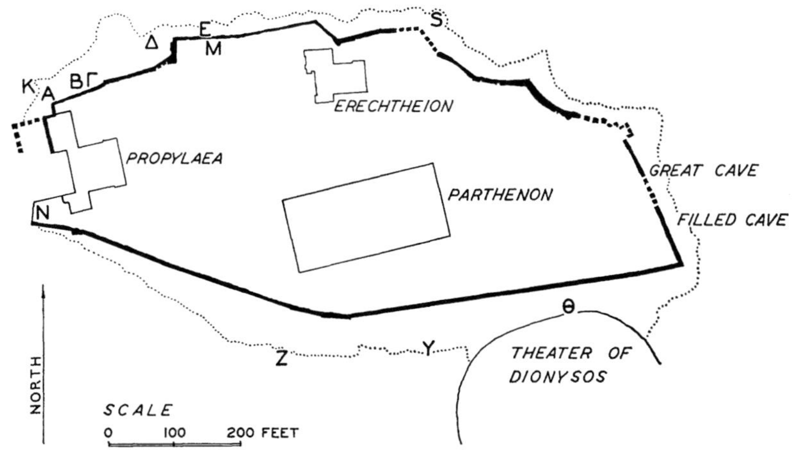

FIG. 1. Generalized Map of the Acropolis. Approximate base of the steep, bedrock slope represented by dotted line; Classical wall around the summit represented by heavy line, post-Classical wall by broken line. Only landmark buildings are indicated. Localities referred to in the text are designated as follows: *N*, temple of Athena Nike; *K*, Empedo, or Klepsydra, spring; A, B, Γ, Δ, and E, caves of the northwest slope; *M*, Mycenean well shaft; *S*, sanctuary of Eros and Aphrodite; Θ, cave behind the monument of Thrasyllos; *Y*, cave of the Asklepieion spring; *Z*, enclosed spring.

cults: B—Apollo Pythios; Γ—Ge, and perhaps Zeus Olympios or Astrapaios; Δ—Pan, or Dionysos (Wycherley, 1959). Keramopoullos (1929) has written what appears to be the most complete description of these four caves.

Cave E, the so-called "cave of Agraulos," is east of Δ, and seems to have been associated with several nature-spirit cults. It faces east and was accessible from the Acropolis by a stairway built into a deep cleft on the north face (Broneer, 1938, p. 448). Farther east along the north slope many small, irregular, natural caves and grottoes have been reported. One of the largest of these is said to be just east of the ancient sanctuary of Eros and Aphrodite (Fig. 1, *S*). This is described as a low, narrow passage, some 12 feet long and open at both ends. The central part had collapsed and been filled with modern debris some time before the cave was excavated in 1931–32 (Broneer, 1933, pp. 337–38, pl. XI). It is impossible to tell from Broneer's description whether this was a solutional cavern or simply an open space between large scree blocks, but the cave's NE-SW trend parallels the trend of the chief local joints, and hence suggests that it was formed by solution along a major joint.

The south slope has only two prominent natural openings. Both are shallow grottoes. The cave of the Asklepieion spring was noted above. East of it, above the highest rock-cut seats of the Theater of Dionysos, is another highly modified cave (Fig. 1, Θ). In 320 B.C. its entrance was enlarged and fronted by a commemorative monument dedicated by Thrasyllos. Most of this structure has since disappeared, and the interior of the grotto, approximately 20 feet square, is now occupied by the chapel of Panagia Spiliotissa. Hill (1953, p. 3) reported that Neolithic pottery had been found in another small cave "or rock-shelter" above and west of the Thrasyllos cave.

The east slope contains the largest cave in the city. It is approximately 40 feet wide, 15 to 30 feet high, and 60 feet deep. Like the other Acropolis caves which I was able to see, its ceiling is deeply furrowed along flaring, enlarged joints separated by rough pendant masses of limestone. When Broneer (1936) began excavation, the cave entrance was almost closed by rubble dumped from the Acropolis excavations of the 1890's. Broneer (1936, p. 250) reported that this material also covered the cave floor to a depth of 6 feet. His excavation showed that the clay floor had been leveled in post-Classical time and that part of the south wall had been blasted away by powder, possibly by the Venetians in 1687.

Broneer identified the original floor material of this cave as "the soft rock and clay which is found beneath the Acropolis rock on all sides of the hill" (1936, p. 250). Elsewhere, he similarly identified the material in the bottom 25 feet of the Mycenean well shaft (Fig. 1, *M*) beneath the court of the Arrhephori: "the soft clayey rock on which the limestone rock of the Acropolis rests" (Broneer, 1938, p. 448). However, in the former case, limestone bedrock exposed in the east face extends many feet below the level of the cave floor on both sides of the cave entrance, and, in the latter, it seems unlikely that a well bottomed 25 feet below the top of a continuous clayey substratum could have produced enough water to supply the Mycenean fortress during siege. A much more likely explanation for the clay in both places is that it is cave-fill material.

In June 1959, I found what appears to be a filled solutional cave or group of caves in the east slope below and south of the great cave.

This cave extends more than 30 feet along the slope, but its height and depth are unknown. It is completely filled, not with clay but with well-sorted, well-bedded, unconsolidated coarse sand, the bedding of which dips about 25° west—into the hill. Only the upper few feet of this sand is exposed in the cliff, above the top of the bordering scree slope.

The sand clearly is older than the scree and rubble which overlap it. Moreover, its good sorting and stratification suggest fluvial or marine deposition. One explanation for the origin of this sand is that a surface stream, perhaps the ancestral Ilyssus, while enlarging and deepening its valley, breached the cave and filled it with stream sediments. Another explanation is that the cave was breached by marine wave attack and became filled with beach sand. This explanation would require the Acropolis to have been a hum on a Tertiary karst landscape which was subsequently submerged and buried beneath marine sediments. Serious consideration of this hypothesis must await comparison of the sand in the cave with Tertiary sediments exposed south of the Acropolis. Either explanation would account for a steep west dip in unconsolidated, nondeformed sand.

In any case, the sand-filled cave suggests that caves were formed in the Acropolis limestone at some distant time. The relationship of the sand to the clay that Broneer discovered in the great east cave and in the Mycenean well shaft is conjectural. I have not seen this clay but suggest that it may represent an original phreatic fill like that observed in many limestone caves in the central United States (Bretz, 1942). In my brief examination of the Acropolis caves accessible to me in 1959 I found no traces of original, undisturbed phreatic clay fill. On the other hand, I saw no evidence that these caves are being enlarged by ground-water solution in the present cycle. Indeed, the enlarged ceiling joints and pendant masses observed in several caves suggest that most of the solution occurred long ago, when the caves were in the phreatic zone. Hence, these caves may once have been partly or completely filled with clay, which has since been removed by vadose percolation or perhaps, in caves such as the great east cave, reworked by the same external agent that filled the lower cave with sand.

Further study of these caves is clearly needed, but the evidence already at hand suggests that phreatic solution in the Acropolis limestone, probably in Tertiary time, formed a poorly integrated spongework of widely separated, irregular, small chambers and enlarged joints. Later, after the surface valleys were eroded around the Acropolis and the water table was lowered, some of the openings were exposed, the original clay fill, if any, was removed or reworked, and some of the caves were filled with sediment washed in from the surface. If this reconstruction of the solutional history of the Acropolis is correct, then the visible caves are not merely surface features, but belong to a cavern system that extends beneath the entire hill. Either empty or sediment-filled, such a system would help to account for the hill's large ground-water storage capacity, which seems otherwise inexplicable.

LITERATURE CITED

BRETZ, J. H. 1942. Vadose and Phreatic Features of Limestone Caverns. Jour. Geol., 50: 675–811.

BRONEER, OSCAR. 1933. Excavations on the North Slope of the Acropolis in Athens, 1931–1932. Hesperia, 2: 329–417, pl. XI.

——. 1936. The Cave on the East Slope of the Acropolis. Hesperia, 5: 247–253.

——. 1938. Discoveries on the North Slope of the Acropolis, 1938. Am. Jour. Arch., 42: 445–450.

——. 1941. Plato's Description of Early Athens in the Light of Archaeological Research. Am. Jour. Arch., 45: 92.

HILL, I. T. 1953. The Ancient City of Athens. Cambridge: Harvard Univ. Press. 258 pp.

KERAMOPOULLOS, A. D. 1929. Ὑπὸ τὰ Προπύλαια τῆς Ἀκροπόλεως. Ἀρχ. Δελτ., pp. 71–101.

——. 1935. τὸ Πελαργικὸν, τὸ Ἀσκληπιεῖον, αἱ Ὀδοὶ αἱ Ἀνάγουσαι πρὸς τα Προπύλαια. Ἀρχ. Ἐφ., 1934–1935, pp. 85–116.

OEKONOMIDES, A. N. 1958?. The Acropolis of Athens. Athens: Gouvoussis. 24 pp. + fold. map.

PARSONS, A. W. 1942. Klepsydra and the Paved Court of the Pythion. Ph.D. dissertation, Johns Hopkins Univ., 78 pp. Review by A. E. Raubitschek, Am. Jour. Arch., 49 (1945): 187–189.

THOMPSON, H. A. 1956. Activities in the Athenian Agora: 1955. Hesperia, 25: 46–68.

TRAVLOS, J. N. 1941. Ἡ Παλαιοχριστιανικὴ Βασιλικὴ τοῦ Ἀσκληπιείου τῶν Ἀθηνῶν. Ἀρχ. Ἐφ., 1939–1941, pp. 35–68.

WYCHERLEY, R. E. 1959. Two Athenian Shrines. Am. Jour. Arch., 63: 67–72.

Papers of the Michigan Academy of Science, Arts, and Letters
Vol. XLVII, 1962 (1961 Meeting)

THE BEDROCK TOPOGRAPHY OF WAYNE COUNTY, MICHIGAN

ANDREW J. MOZOLA

Wayne State University

Introduction

Since details of the bedrock surface of southeastern Michigan are not known, data on bedrock depths are being collected for eight counties within the area, namely Wayne, Monroe, Lenawee, Washtenaw, Oakland, Livingston, Macomb, and St. Clair. The objectives of the project are twofold: to determine the preglacial drainage pattern of the area and the changes that have been imposed upon this pattern by repeated glaciation during the Pleistocene. This paper is concerned primarily with the bedrock configuration of Wayne County, and the map submitted as part of this report is based on data collected from several sources over a period of years. Within the limits of the city of Detroit, bedrock elevations were obtained essentially from test-boring records. Oil, gas, and water well records provided most of the bedrock control for the rest of the county. A few bedrock elevations in Sumpter Township were determined by the seismic method.

General Considerations

Before describing the bedrock topography of Wayne County, I should like to comment briefly upon the major features of the bedrock surface of Michigan and to review the early ideas that have been proposed concerning the preglacial drainage pattern of the Great Lakes region. Any treatment of the bedrock surface of Michigan involves the Niagara and Marshall cuestas. The former cuesta, when traced on the geologic map of North America, makes a semicircular scarp around Lakes Huron and Michigan with the dip slope toward the center of the basin. The latter cuesta, developed on the Marshall sandstone and also having an inward dip slope, is within the Lower Peninsula of the state. The Marshall cuesta is nearly circular in plan except where it has been breached in several places by former streams; the most significant breaches are at Saginaw Bay in the east and in the vicinity of Muskegon in the west. Both breaches are now connected by a wide, elongated lowland that crosses the state roughly from east to west. Between the two cuestas a somewhat circular lowland, developed on less resistant Devonian and lower Mississippian strata, surrounds the Lower Peninsula of the state. Thus, beneath the present drift cover, the major physiographic features of the bedrock surface consist essentially of uplands and lowlands (Fig. 1), the former at elevations of approximately 600 to more than 1,000 feet in some areas and the latter at elevations below 600 feet.

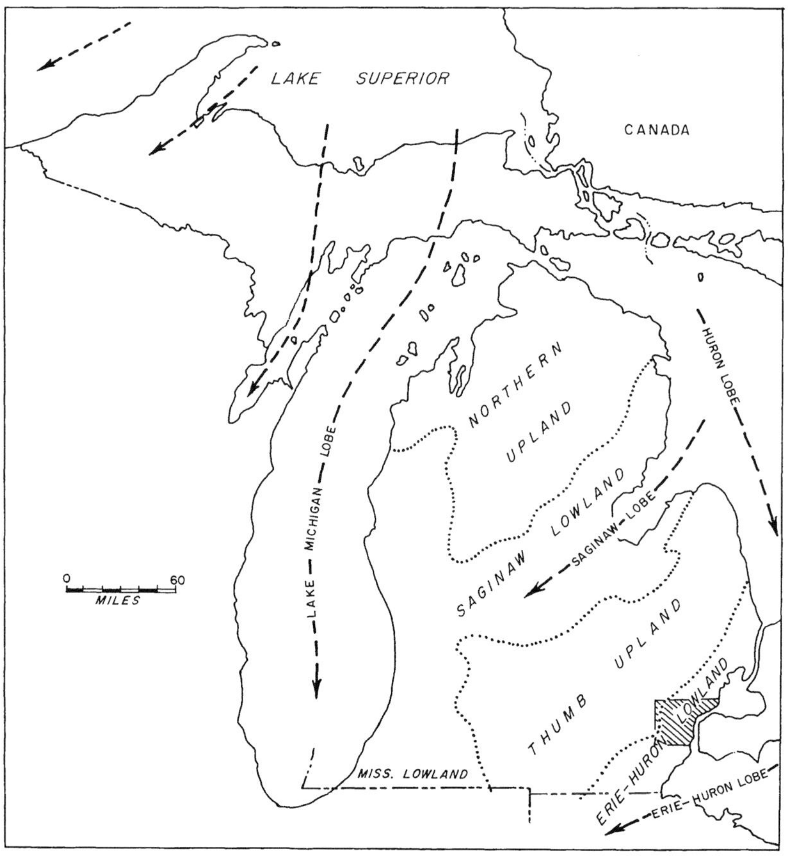

FIG. 1

With respect to preglacial drainage, it is now generally accepted that the major stream valleys were developed on belts of weak rocks presently occupied by the Great Lakes. It has also been presumed by many that the basins belonged to a single drainage system. Thus the basic question is whether the drainage of Michigan was principally tributary to the St. Lawrence or to the Mississippi.

Two hypotheses regarding the drainage systems of the Great Lakes region have been proposed, one by Spencer and another by Grabau, both of which are briefly mentioned by Fenneman (1938) in his *Physiography of Eastern United States.* The first postulates a drainage to the east which eventually became tributary to the St. Lawrence River (Fig. 2), and the second a drainage system consisting of several nearly parallel southwestwardly flowing streams (Fig. 3) which presumably discharged into the Mississippi, perhaps by way of the preglacial Teays. It is interesting to note that both hypotheses take advantage of the known breaches in the Niagara cuesta.

As indicated by morainic features produced by Wisconsin glaciation, three distinct lobes, the Lake Michigan, Saginaw, and Erie-Huron, advanced over the Lower Peninsula of the state by way of the lowland areas (Fig. 1). Of the tills older than the Wisconsin, only the Illinoian has been recognized in a few scattered exposures and as a hardpan encountered in test borings in the Detroit area just above

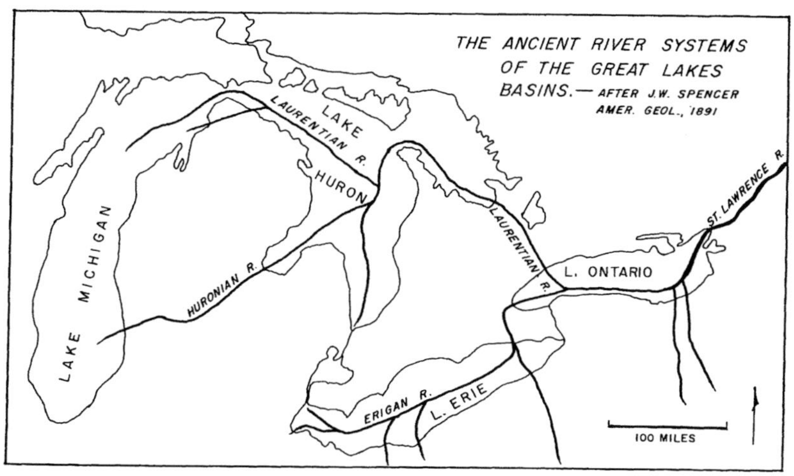

FIG. 2

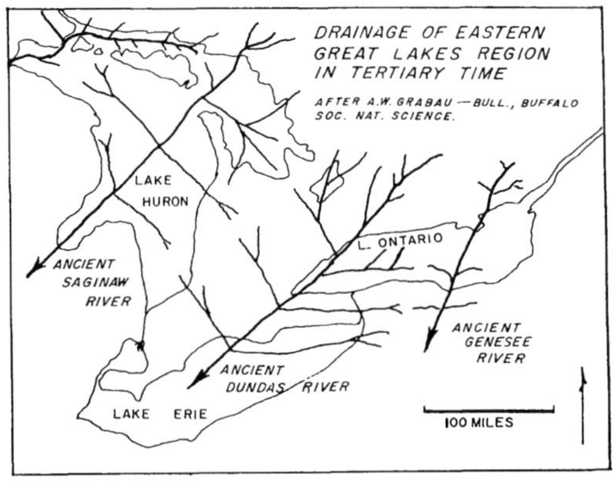

FIG. 3

bedrock. The possible presence of older tills should cause one to consider the nature of earlier glaciation in Michigan. On the basis of the gross features of pre-Pleistocene topography, it does not seem unreasonable to assume that earlier ice advances would have had a movement and lobation similar to that of Wisconsin ice.

The probable nature of the preglacial bedrock surface and proposed drainage patterns together with the manner of ice movement, particularly in Michigan, suggests certain inferences. With respect to Spencer's (1891) drainage pattern, the movement of ice over the area would result in the blocking of drainage, and the net effect would, or should, be in the development of proglacial lakes during the advancing phase of the ice. In addition, changes should be imposed upon the preglacial stream pattern: development of cross channels, possible reversals in drainage, as well as changes in the level of the proglacial lakes as outlets are affected by the advancing ice. Evidence for earlier Pleistocene lakes is suggested by deeply buried, clean, pebble-free clays revealed in a number of test-boring samples and well records in Wayne County. On the other hand, the movement of ice over a preglacial drainage system as suggested by Grabau should not seriously disrupt drainage, since it was initially to the southwest. The effects produced under these conditions should result in the further development of the valleys involved, i.e., deepening of valleys if base level was not yet attained or, if it was attained, then valley widening or deposition. Whether either one of the proposed drainage systems is essentially correct remains to be demonstrated. Modifications in either case, or perhaps even new concepts, are still possible.

Bedrock Topography

Except for the westernmost tier of townships, Wayne County is situated on the bedrock lowland of the Erie-Huron plain. The contour map of the bedrock surface (Fig. 4) shows that this lowland has an elevation between 500 and 600 feet except where it is incised by bedrock valleys. The 600-foot contour line might be tentatively used to mark the boundary of the lowland from the eastern edge of the Thumb Upland surface. In Wayne County, the lowland is developed on formations older than the Berea sandstone.

In view of the small area reported on in this paper, it may be somewhat hazardous as well as premature to assign ages and names to the bedrock valleys that are shown on the map. Such references as may be made are to be considered only tentative until mapping has been extended over a much greater area. However, several features are clearly revealed. The first of these is the preglacial valley that roughly approximates the present course of the Detroit River (which now flows south but whose bedrock gradient is to the northeast). It appears that the head of this valley terminates in the vicinity of Grosse Ile, the island being a part of an east-west divide in the southern part of the county. The absence of additional bedrock control both along the Detroit River and immediately to the east in Ontario makes it rather difficult to define this valley more precisely in terms of depth and width. The position of the valley with respect to the movement of the ice mass shows that it was undoubtedly modified by glacial scour. Borings shown on cross sections along the center lines of the Detroit-Windsor auto tunnel, the railroad tunnel, and the Ambassador Bridge failed to reveal where bedrock was encountered except near the shores, and these were not of significant depth. From the position of the divide at Grosse Ile, another valley is evident whose slope is southward, and, on the basis of present but limited data in Monroe County to the south, was tributary to a valley in Ohio having a northeast trend toward Maumee Bay. This could be a part of the drainage into Spencer's Erigan River (Fig. 2).

A well-defined preglacial valley, joining the valley just described in the vicinity of Belle Isle, is now traceable from this junction northwestwardly for twenty-five miles to the city of Pontiac in Oakland County. The valley passes through or near the cities of Highland Park, Huntington Woods, Royal Oak, Birmingham, and Bloomfield Hills (Mozola, 1954). In the Detroit area the valley has a minimum depth of 75 feet and a width of nearly two miles at its junction near Belle Isle. It decreases northwardly to a width of one mile at the Wayne-Oakland County line. On the map this Pontiac-Detroit valley has the appearance of a well-developed, mature valley with a clearly defined flood plain at an elevation slightly under 425 feet. This appearance may be misleading, however, since the very nature of the mapping project is such that the desired distribution of bedrock control is seldom obtained when relying entirely upon water well and test-boring data. It may simply mean that chance has not provided sufficient control along the valley. A few unusual elevations have been obtained and are indicated on the map, but these simply raise questions rather than answer them. Near Belle Isle, a test boring reveals the rock floor of the valley to be at 420 feet; an oil and gas record in Highland Park indicates bedrock at 403 feet; and in Section 6 of Royal Oak Township, Oakland County, a water well test hole places the bedrock floor along the same valley at 395 feet above sea level. This decrease in the floor of the rock valley northwestwardly is suggestive of a reversal in slope that could have happened as the result of blocked drainage as the ice advanced by way of the lowlands. The question then involved is the whereabouts of the outlet or outlets for the ponded waters which surely must have formed. Another possible explanation for the few unusual depths that have been indicated may be in the presence of an inner valley carved into the floor of the Pontiac-Detroit valley. Again, details of this feature are lacking for want of additional control. Winslow (1957), in his study of several areas in northeastern Ohio, reports four distinct buried valley systems: the preglacial Teays age streams, the Aftonian Deep Stage streams, the Yarmouth-Brecksville Stage streams, and streams that drained the areas in Sangamon time. According to Winslow, the Teays age drainage is indicated by a wide strath that is developed on the rock surface at an altitude of about 450 feet in Cuyahoga County. Deep Stage drainage system, in turn, is represented by a deep inner gorge that is cut in the rock floor of the Teays system. The Yarmouth-Breckville State is represented where it crosses the valley course of the two earlier drainage systems in Cuya-

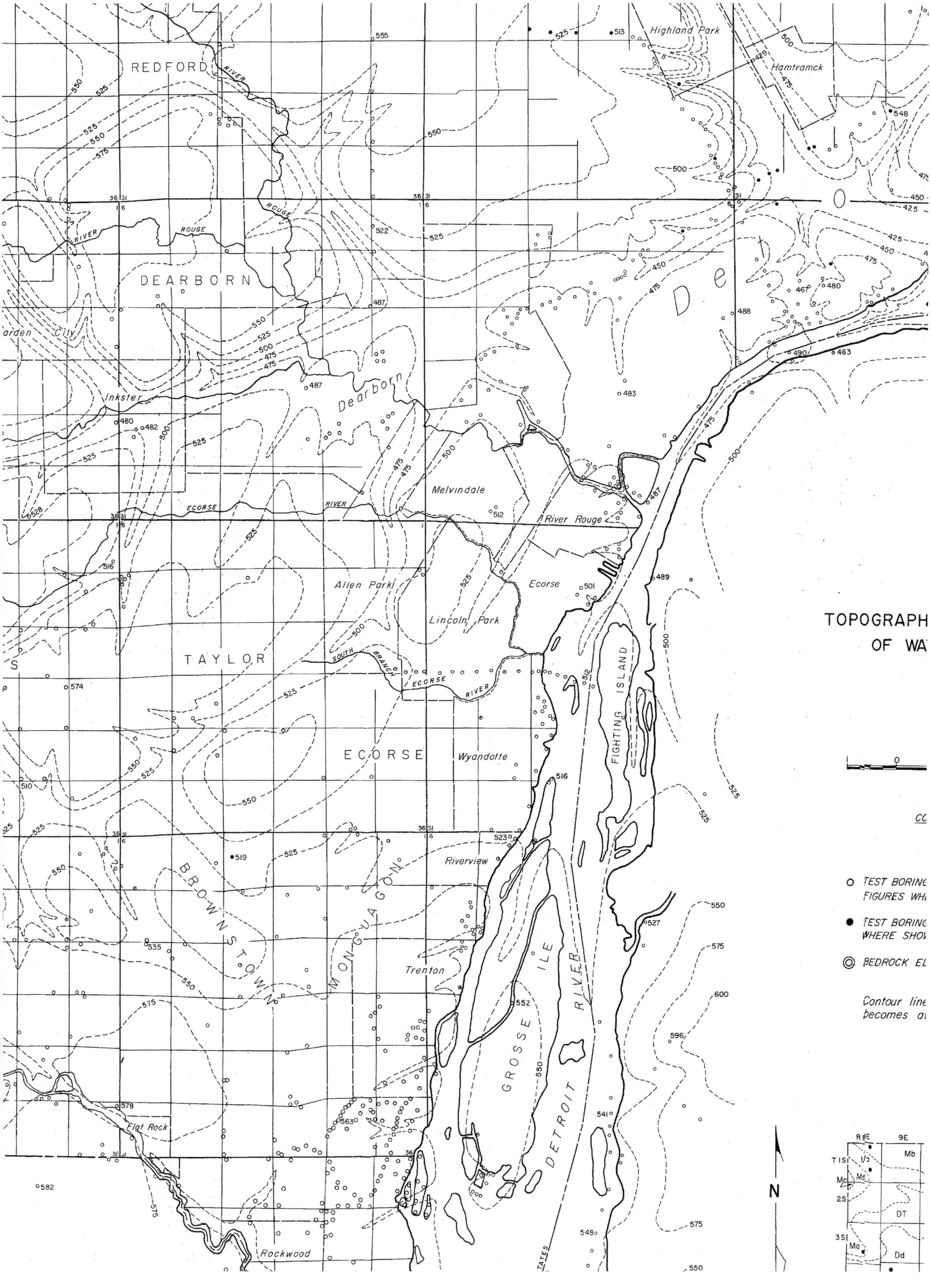

REDFORD
Highland Park
Hamtramck
DEARBORN
Garden City
Inkster
Dearborn
Melvindale
River Rouge
Ecorse
Allen Park
Lincoln Park
TAYLOR
ECORSE
Wyandotte
BROWNSTOWN
MONGUAGON
Riverview
Trenton
GROSSE ILE
FIGHTING ISLAND
DETROIT RIVER
Flat Rock
Rockwood
ROUGE RIVER
ECORSE RIVER
SOUTH BRANCH ECORSE RIVER
TOPOGRAPH
OF WA
TEST BORING
FIGURES WH
TEST BORING
WHERE SHOV
BEDROCK EL
Contour line
becomes a
N

hoga County. Sangamon drainage is less known except that it occupied parts of the earlier systems. Whether these relationships exist in southeastern Michigan is at best only suggested, and remains to be seen as details of the buried valley systems are extended over a much larger area. Unfortunately, the absence of closely spaced borings in a direction across the Pontiac-Detroit bedrock valley leaves the matter of a deeper inner gorge still in doubt. Geophysical testing to establish this feature in the Detroit area may be difficult because of dense urbanization, but perhaps it can be attempted in the more rural areas of southeastern Michigan as the study progresses.

Several other well-defined rock valleys of preglacial age that pass through or near the cities or villages of Redford, Northville, Plymouth, Belleville, and New Boston are also shown on the map (Fig. 4). The valleys are 50 to 75 feet in depth and are most conspicuously developed where their courses are incised along the eastern slope of the Thumb Upland surface. All of these ultimately converge and join the Pontiac-Detroit valley. In turn, all of these valleys are cut conspicuously by a northeast-southwest cross channel of Pleistocene age that is situated approximately along the boundary between the Thumb Upland and the Erie-Huron Lowland. The cross channel extends from the northeast corner of Redford Township (T. 1 S., R. 10 E.) and trends southwestwardly for several miles before swinging to a more southerly course through the municipalities of Wayne and New Boston. From the vicinity of New Boston it swings southwestwardly again through Sumpter Township (T. 4 S., R. 8 E.) where it leaves the county. Continuation of this channel is suggested in the southeastern part of Oakland County to the north, and a preliminary bedrock contour map of Monroe County shows its continuation to the south. In addition to this major cross channel there are others of smaller magnitude indicated: one just east of Romulus and another west of the city of Plymouth.

Some comments are probably in order with respect to the influence of bedrock topography upon the courses of the present streams. From what is indicated on the map, the present courses of the River Rouge and its tributary, the Middle River Rouge, have not been influenced by the bedrock topography. In Wayne County, at least, both stream courses appear to be independent of any bedrock influence, and this may be due to the thicker cover of Pleistocene materials. The present course of the Lower River Rouge appears nearly coincident with a bedrock valley from the vicinity of Sheldon Village eastwardly to the city of Dearborn, where the present stream also becomes a tributary to the River Rouge. Somewhat unusual is the present course of the Huron River. Its course coincides with segments of bedrock valleys of both preglacial and glacial age. It can be seen on the map that the Huron River coincides with the preglacial valley in the vicinity of Belleville, then with the cross channel just northwest of New Boston, and finally coincides again with valleys of preglacial age where it discharges into the Detroit River at Lake Erie.

Summary

The present study of the bedrock surface of Wayne County indicates that the drainage of the region was principally to the northeast except for a small area in the southeastern part of the county, where it was to the south. Preliminary mapping of the rock surface in Monroe County tends to support this southward drainage. In order to avoid later confusion, the assignment of geographic names, the dating, and the correlation of the buried rock valleys in Wayne County with valleys elsewhere has been purposely avoided until a much larger area of southeastern Michigan has been mapped.

Acknowledgments

I am extremely grateful to the Raymond Concrete Pile Company, the Michigan State Highway Department, and the Wayne County Road Commission for the privilege of searching their files of test borings for usable data in connection with this study. Numerous logs of water, oil, and gas wells were generously provided by Olen Marshall, O. O. Corsaut, Jr., Edward Kittle, Harvey Tracy, Joseph Cauffiel, and the Geological Survey Division, Michigan Department of Conservation. I also wish to acknowledge a grant received from the University Research Committee of Wayne State University for the summer of 1960, which made it possible for me to compile the present map and to make a significant start in the mapping of Monroe County. The work of Hugo Mandelbaum, for some seismic bedrock determinations, is also acknowledged.

LITERATURE CITED

FENNEMAN, N. M. 1938. Physiography of Eastern United States. McGraw-Hill Book Co., pp. 468–472.

MOZOLA, A. J. 1954. A Survey of Ground Water Resources in Oakland County, Michigan. Mich. Geol. Surv., Pub. 48: 128.

SPENCER, J. W. 1891. Origin of the Great Lakes Basins of America. Am. Geol., 7: 87.

WINSLOW, J. D. 1957. Buried Valley Systems in Parts of Northeastern Ohio. Abstract, Program A.A.A.S. Meetings, Indianapolis, Ind., p. 21.

Papers of the Michigan Academy of Science, Arts, and Letters
Vol. XLVII, 1962 (1961 Meeting)

GEOLOGY OF THE HEPBURN'S MESA BASALT, PARK COUNTY, MONTANA

RICHARD F. WARD, *Wayne State University*
KAREN E. WIER, *Bryn Mawr College*

Hepburn's Mesa is situated in the Yellowstone River valley twenty miles north of the north entrance to Yellowstone National Park. The present U. S. Route 89 passes along the base of the mesa, but a new alignment under construction will soon afford travelers to the Park a still better view of this impressive section. The scenic and educational value of the section and some interesting features of the basalt flow which caps the mesa led to this study.

The old 30-minute quadrangle in which this section lies was mapped as part of Folio no. 1 (Iddings and Weed, 1894). The geologic map is, necessarily, highly generalized and few petrographic data are given. In a geomorphic study of Yellowstone Valley, Horberg (1940) presents an illustration and discussion of the mesa for the bearing which its constituent units have on the physiographic history of the valley; but again the lithologic descriptions are brief. The Devil's Slide area a few miles to the south (upstream) of the mesa was mapped by Wilson (1934), who delimited and described the Gardiner overthrust, which locally marks the boundary of Beartooth-Snowy Mountain uplift. Small outcrops of basalt occur near the overthrust.

Howard's (1937) work on the petrology and stratigraphy of the basalts at the head of the Yellowstone River, about forty miles upstream from Hepburn's Mesa, provides the best information on the character of the volcanic activity in the region. He gives passing consideration to all the basalts of the region including the flow presently being studied.

Seager (1944) studied the Jardine mining district which lies at the edge of the Snowy Mountain block above the Yellowstone Valley about 10 miles south of Hepburn's Mesa. Seager's report includes clear and detailed petrographic descriptions of the rock types of his area, including previously undescribed basalts.

Regional Geology

The surrounding Snowy Mountain block consists of a wide variety of Precambrian schists, gneisses, and plutonic igneous rocks overlain by Tertiary and younger volcanic rocks. The surface upon which the volcanics lie has a relief of at least 4,000 feet. This relief is due in part to prevolcanic erosion and in part to faulting: a north-striking fault runs along the east side of Yellowstone Valley a few hundred yards east of the mesa. The valley lies in the downthrown block. There is evidence of relatively recent movement along the fault.

Hepburn's Mesa comprises a Tertiary tuff of uncertain stratigraphic position, a thick boulder bed, and the geologically youngest volcanic rock in the region, a basalt.

The mesa occupies seven square miles of the Yellowstone River valley. It is bounded on the west by the present Yellowstone River at elevation 4,960 feet and on the east by a former course of the River at 5,240 feet. The west-facing escarpment of the mesa rises about 280 feet to elevation 5,280 feet. All of the units present in the mesa are somewhat more widely distributed, but nowhere else is the entire section exposed.

Tuff.—This unit is buff to white, well indurated, with occasional green layers about one foot thick. It is usually fine-grained but occasional layers of fine conglomerate are present. A few leaf fossils were seen. Vertebrate remains are very common: in a few hours we found several rib fragments, a herbivore tooth, and a nearly

TABLE I

Stratigraphic Section

Unit	Exposed Thickness	Age	Comment
Erratic boulders	—	Late Wisconsin	Scattered granite boulders lying on striated basalt surface
unconformity			
Basalt	80′	Pleistocene (?)	See text
unconformity			
Gravel	80′	Pliocene (?)	See text
angular unconformity			
Tuff	200′	Mio-Pliocene (?)	See text

complete small rodent skull. Residents of the area told us that a now-deceased local collector had removed turtle and horse remains and a more or less complete camel. We have unfortunately been unable to confirm this. Frick and Jepsen (fide, Horberg, 1940) assign a Miocene-Pliocene age on the basis of vertebrates which they examined.

The tuff dips to the northeast, i.e., into the major fault, at approximately 7°. Outcrops of tuffaceous conglomerate on the west side of the Yellowstone Valley 1,700 yards southwest of the mesa appear to represent the base of this formation locally. If this relationship is valid, the total thickness, base to unconformity, is 400 feet. Additional small exposures of the tuff occur along the river for several miles north of the mesa.

In thin section the tuff is seen to be made up largely of shard-shaped glass particles set in a fine-grained (clay?) matrix. The glass shards are altered to a birefringent material around the edges but are unaltered within. Several angular grains of quartz and fresh feldspar are present. Individual glass shards and mineral grains average .1 mm.

Approximately 80 feet of coarse gravel lies unconformably above the tuff. It is well rounded and rather well sorted in a meager sandy matrix, with practically no cementation. Individual boulders range up to a foot in diameter but the median size is something less than half that. Horberg (1940) gives the composition of the gravel as 67 per cent volcanics, 24 per cent crystallines, and 9 per cent quartzite. Such materials occur upstream (south) along the Yellowstone in and near the National Park. Gravels at Tower Junction, about forty miles upstream, are so similar in composition that correlation seems reasonable (Horberg, 1940). Howard (1937) has assigned the Tower Falls gravels to a position near the Pliocene-Pleistocene border.

Basalt.—The mesa is capped with a basalt flow having an average thickness of about 30 feet. At the edges of the mesa large blocks of this basalt have slumped down and yet remain in a horizontal position. To the casual observer, therefore, there appear to be several flows. Careful examination shows that there is in fact only one. Good exposures, including a large quarry, occur on the west side of the river. Columnar joining is universal; individual columns range up to six feet in diameter in the quarry. That the columns formed normal to the isothermal surfaces is clearly displayed in exposures of irregularities in the prebasalt surface: here the columns are divergent, yet normal to the lower contact. They then curve gently upward to an approximately vertical attitude (i.e., normal to the upper contact).

The basalt is vesicular, having small crystal-lined cavities, and porphyritic, with plagioclase phenocrysts up to two inches in diameter and olivine phenocrysts up to one-half inch. Glomeroporphs of fresh olivine and plagioclase are present.

The feature of main interest, which provoked us to further study, was a conspicuous horizontal element resembling coarse bedding. The individual bands are approximately eight inches thick. On slightly weathered surfaces it can be seen that each band is fine-grained at the bottom and becomes progressively coarser toward the top. The contact with the next higher band is always quite sharp and planar. The upper portion of a band is neither glassy nor vesicular. These surfaces, then, cannot have been subaerial, so that the bands are not successive small flows.

The layers appear to record the flow characteristics of the original lava flow, the magma having been a hot, somewhat viscous fluid moving by laminar flow; the individual bands represent single, essentially independent laminar flow cells or layers. That the flow was laminar rather than turbulent is indicated by the absence of contortions or sharp irregularities of the bands and by the flat, smooth nature of their mutual contacts. This hypothesis is supported by the behavior of the layers in the neighborhood of the irregularities in the prebasalt surface. The bands retain their identity in these places and bend smoothly up and over such bedrock knobs. Successively higher layers are less and less deflected until, some ten feet above the bottom, the layers are again horizontal.

Petrography of the basalt was done on thin sections chosen to show the more conspicuous characters. Especial attention was given to a pair taken from the top and bottom of an individual flow layer. Modal analyses were made using the equipment and methods described by Chayes (1949); minerals were identified on curves given by Winchell (1951) and Poldervaart (1950), using measurements made on a 4-axis universal stage.

Texturally the basalt is holocrystalline with good ophitic texture. Olivine occurs only as phenocrysts, pyroxene is practically restricted

TABLE II

MINERALOGY OF THE BASALT

Mineral	Characteristics	Composition
Plagioclase	Composition based on extinction angles X to 010 in sections oriented normal	
Phenocrysts	to a	An_{80-64}
Groundmass		ca. An_{49}
Pyroxene	(+)2V 52, Z to c 34	ca. $En_{50}Fs_8Wo_{42}$
Olivine	(±)2V 90	ca. Fo_{87}

to the groundmass. Plagioclase is present in two rather well-defined generations. The characteristics of the minerals present are summarized in Table II.

Modes of samples taken at the top and bottom of an apparent flow cell are given in Table III. These show a small but distinct concentration of phenocrysts in the lower part of the layer as might be expected during flow. However, on the basis of the samples taken these figures have limited statistical significance.

Despite very careful search for textural variations the only petrographic feature related to the macroscopic banding is the difference in phenocryst: groundmass ratio. Routine measurements of microlite

TABLE III

MODES OF BASALT IN VOLUME PER CENT

	Top of Flow Layer		Bottom of Flow Layer	
Phenocrysts	4		8	
Olivine		2.0		0.9
Pyroxene		0.0		1.2
Plagioclase		2.1		5.7
Ground Mass	96		92	
Olivine		2.8		4.3
Pyroxene		37.9		34.2
Plagioclase		51.8		50.1
Magnetite		3.6		3.3

size and orientation and of the size of enclosing augite plates showed no difference between the top and bottom samples from the layer collected.

Till.—The uppermost unit in the mesa is a group of scattered glacial boulders lying on the basalt. These are predominantly granite with some basalt. Glacial striae on the surface of the basalt flow are preserved only where the basalt has been protected by these erratics. These trend approximately north-south, indicating that the piedmont glacier which filled the valley had its major movement in a northerly direction (i.e., away from the Yellowstone Plateau).

Discussion of Basalt

Three problems concerning the basalt have theoretical and regional significance: (1) the nature and significance of the layering; (2) its absolute age; and (3) its origin and correlation with the basalts of the upper Yellowstone Valley.

The field characteristics of the layering leave little doubt that these are flow cells. Petrographically, the evidence of crystal settling within a layer tends to support this. However, we suspect that additional variations, textural or mineralogic, within a cell are present, but that we were unable to detect them. Petrofabric analyses and more detailed mineral analyses will, we hope, shed light on the actual flow mechanism of this lava. We plan to investigate this further when funds are available for additional collecting.

The absolute ages of many of the volcanic rocks of the region, and of the basalts in particular, are matters of almost pure speculation. Scanty evidence from plant and vertebrate fossils and regional geomorphic evidence led Horberg (1940) to call the basalt Early Pleistocene and Howard (1937) to suggest Late Pliocene. Until better correlation has been established all that can be said with certainty is that the basalt is post-Miocene and pre-Wisconsin.

Every previous worker has hazarded correlations of the basalts or at least provided new data which contribute to knowledge of their distribution and petrography.

As evidence becomes accumulated, it appears that even though some correlations may be correct, the evidence for them is untrustworthy. Howard correlates the basalts on their petrographic character, particularly on the nature of their alteration. This, of course, is a function of the accident of location and has no meaning for correlation. His statement that the Tower Junction basalt and the basalt of this paper are correlative because of their lithologic similarity and the similarity of the gravels which underlie each appears to be valid, but his idea that they are older than the rhyolites of the Yellowstone Plateau because the gravels lack rhyolite particles is open to question because of lack of knowledge of the former position of drainage divides.

Wilson states that the local basalts issued from the Gardiner thrust plane. Although there is evidence neither for nor against this speculation, it appears more likely that the higher level basalts mapped by Seager poured down into the Yellowstone Valley and that the basalts below, near the overthrust, and the one at Hepburn's Mesa are remnants of a larger lava flow which originated on higher ground near Jardine.

LITERATURE CITED

CHAYES, F. 1949. A Simple Point Counter for Thin Section Analysis. Am. Mineralogist, 34: 1–11.

HORBERG, C. L. 1940. Geomorphic Problems and Glacial Geology of the Yellowstone Valley, Park County, Montana. Jour. Geol., 48: 215–303.

HOWARD, D. 1937. History of the Grand Canyon of the Yellowstone. Geol. Soc. Am. Spec. Pap. 6: 159 pp.

IDDINGS, J. P., AND W. H. WEED. 1894. Livingston, Montana *in* Geol. Atlas of the U. S., folio No. 1.

POLDERVAART, A. 1950. Correlation of Physical Properties and Chemical Composition in the Plagioclase, Olivine and Orthopyroxene Series. Am. Mineralogist, 35: 1067–1079.

SEAGER, G. F. 1944. Gold, Arsenic and Tungsten Deposits of the Jardine-Crevass Mountain District, Park County, Montana. Montana Bur. Mines and Geol. Memoir 23: 111 pp.

WILSON, C. 1934. Geology of the Thrust Fault near Gardiner, Montana. Jour. Geol., 42: 649–663.

PAPERS OF THE MICHIGAN ACADEMY OF SCIENCE, ARTS, AND LETTERS
VOL. XLVII, 1962 (1961 MEETING)

PALEONTOLOGY

UPPER CRETACEOUS MOLLUSCA FROM NIOBRARA COUNTY, WYOMING

LEWIS B. KELLUM

The University of Michigan

INTRODUCTION

THE Cannonball member of the Lance formation in the Dakotas is the youngest marine deposit recognized in the Rocky Mountain region. Preliminary to a proposed reinvestigation of its molluscan remains by a graduate student at The University of Michigan, I examined the late Mesozoic fossils in the University's Museum of Paleontology. Two small collections from Wyoming were labeled "Triceratops Beds." Since this is an older name for Lance, I studied these collections to determine if possible whether they had indeed come from the Lance formation and therefore might represent an extension of the Cannonball member into Wyoming.

These two small collections of invertebrate fossils were made in the summer of 1924 by the vertebrate paleontologist E. C. Case, the vertebrate preparator W. H. Buettner, and a graduate student, W. S. Warthin, from The University of Michigan Museum of Paleontology, and were incidental to their search for vertebrate remains in the Tertiary deposits of Wyoming.

The data with these collections and in Professor Case's field notes are:

1924/K-6. Museum of Paleontology accession number. Field label No. 89, Old Woman Creek, Niobrara County. One lot of invertebrates from nodules in Triceratops brown sandstone. Collectors: E.C.C., W.S.W., and W.H.B., July 18, 1924.

Professor Case's field notes for July 18, 1924, say: "Out to Triceratops beds near second bridge on Old Woman Creek. Invertebrates."

1924/K-7. Museum of Paleontology accession number. Field label No. 92, Old Woman Creek, Niobrara County, Wyoming. Invertebrates. Collector: W.H.B., July 19, 1924.

Professor Case's field notes for July 19, 1924, say: "Down Old Woman Creek, a few miles across upper clay to within sight of Triceratops beds. Mammal teeth and bones in outlying Triceratops beds."

Determination of the age and stratigraphic position of these invertebrate fossils from the "Triceratops beds" seemed especially significant in view of the controversy in regard to the age of the Cannonball marine member of the Lance formation.

The name "Triceratops beds" refers to deposits of the Western states that contain the remains of the horned dinosaurs belonging to the order Ornithischia, suborder Ceratopsia, and family Ceratopsidae. O. C. Marsh (1889, p. 501) called them the "Ceratops beds." Both names have been rejected for use in the classification of the U. S. Geological Survey and have been replaced by the geographic name Lance formation. (Wilmarth, 1938, pp. 390, 1142, 2183).

J. B. Hatcher (1903) first used the name "Lance Creek (Ceratops) beds" in Wyoming, from the principal stream in the region where they are best represented in Converse County, Wyoming. The U. S. Geological Survey adopted the name "Lance formation" (Stanton, 1910, p. 172) from exposures on Lance Creek, Niobrara County, which was formerly a part of Converse County.

The Lance formation in the Cannonball River Lignite field in the southwestern part of North Dakota was divided into two members (Lloyd, 1914, pp. 247–248). The lower 400 feet, of fresh-water origin, consists of shale and sandstone with thin lignite beds. The upper 250 or 300 feet of shale and shaly sandstone is of marine origin and was designated the Cannonball marine member. They were tentatively classified as probably of early Tertiary age.

Subsequently Stanton (1920) studied and described the invertebrate fauna of the Cannonball member, which comprised two species of Foraminifera, 6 of corals, and 60 of mollusca. Of the 60 molluscan forms he described 33 as new and identified 27 with previously known forms. Of the latter, 18 occur also in the Fox Hills sandstone and 12 of these range downward into the Pierre shale. Those described as new are related more or less closely to Cretaceous species. However, the ammonite genera Sphenodiscus and Scaphites which characterize the Fox Hills fauna and prove it to be Cretaceous have not been found in the Cannonball. Stanton (1920, p. 12) referred to the Cannonball fauna as a modified Fox Hills fauna and explained its distinctive characters as due to a habitat in the open sea and to modifications that occurred after Fox Hills time. Such modifications were the extinction or emigration of the ammonoids and a few other Mesozoic types and the introduction of a considerable number of new types. He concluded that the Cannonball marine member and consequently the whole of the Lance formation is of Cretaceous age.

In 1942 Fox and Ross announced the discovery of an abundant foraminiferal fauna in the upper part and middle of the type Cannonball of southern North Dakota. Their analysis of the 64 species recognized indicated that 13 were new, 8 were not determinable specifically, and 43 were identified with species of known geologic range in the Gulf Coast area. Of the latter, 21 species have long geologic ranges, 38 occur in the Paleocene Midway stage, of which 13 are restricted to the Midway, 4 to the Paleocene and late Cretaceous, and only 1 to the late Cretaceous. Since 30.2 per cent of the 43 Cannonball foraminifera of known geologic range are restricted to the Midway, they concluded that the Cannonball marine beds are Midway (Paleocene in age). At the same time they elevated the Cannonball beds from the rank of a member to that of a formation and placed the Tertiary-Cretaceous boundary at the base of the Cannonball formation.

On the correlation chart of the Western Interior of the United States (Cobban and Reeside, 1952), the marine Fox Hills sandstone of the type area in the Great Plains passes westward into the nonmarine Hell Creek and Lance formations. The Fox Hills fauna corresponds to the upper Maestrichtian fauna of the European sequence. Successive faunal zones in the Maestrichtian of the Western Interior are indicated from bottom to top by *Baculites baculus, B. grandis, Discoscaphites nicolleti, D. nebrascensis,* and *Triceratops.* In a later paper on the Powder River Basin of Wyoming and Montana, Cobban (1958, p. 117) lists *B. clinolobatus* and *Sphenodiscus* cf. *lenticularis* as zone fossils above the zone of *B. grandis.* The ammonite *Sphenodiscus,* however, seems to be associated with sandstone deposits and to occur at lower levels westward than in the typical Fox Hills area. *Sphenodiscus* cf. *S. lenticularis* (Owen) marks the youngest of the marine Cretaceous zones in the Powder River Basin (Cobban, 1958, p. 114).

Jeletzky (1952, p. 1027) pointed out that the numerous forms of *Discoscaphites* of the group *conradi* are characteristic of rocks younger than the zone of *Baculites baculus* Meek and Hayden, which he would place provisionally at the base of the Maestrichtian. He considers the so-called Fox Hills fauna to represent only the higher part of Lower Maestrichtian or some part of Upper Maestrichtian. He concludes:

"Considering that Sphenodiscus spp., which are rather characteristic of the Fox Hills fauna and occur much less commonly in the older Maestrichtian zones of the Western Interior, are essentially characteristic of the Upper Maestrichtiene in Europe, it seems by no means unlikely that at least the highest zone with the Fox Hills fauna might be of Late Maestrichtian age."

In studying the collections of fossils from Old Woman Creek three lines of investigation are followed: (1) examination of the geologic setting in which they were found; (2) identification of the species present; and (3) determination of their stratigraphic position.

Geology of the Old Woman Creek Area

Old Woman Creek rises in the southern part of Niobrara County (Fig. 1) in the Pliocene Ogallala formation at the drainage divide between the southward flowing tributaries of the Niobrara River and the northward flowing tributaries of the Cheyenne River. Within a few miles of its source Old Woman Creek enters the Oligocene White River beds and continues northward across them for about a third of its total course. Farther north it crosses an area of structural uplift and its valley is cut in successively older formations of the Cretaceous series.

For most of its course across the area of structural uplift Old Woman Creek follows the syncline between the Ant Hills anticline on the west and Old Woman anticline on the east. It meanders eastward across the north end of Old Woman anticline, then reverses its course, recrosses the axis of the anticline, and flows in a northwest direction across successively younger formations exposed on the west flank. In the lower part its valley is in highest Cretaceous formations, Fox Hills and Lance, for a distance of five or six miles to its junction with Lance Creek.

The geologic map and legend of Fig. 2 show the succession and distribution of formations in an area of more than six townships

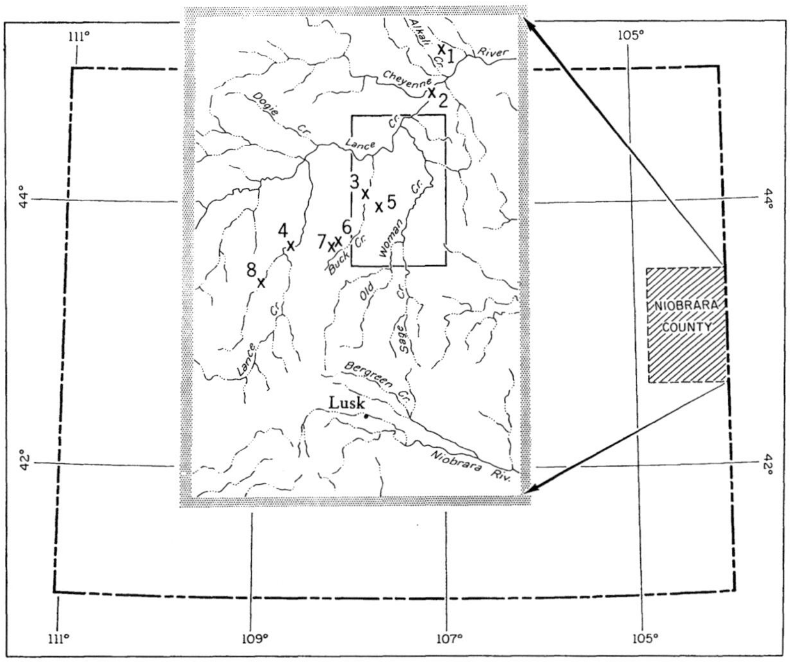

OUTLINE MAP OF WYOMING SHOWING NIOBRARA CO.

FIG. 1.

along the lower part of Old Woman Creek. The oldest formation exposed is the Jurassic Sundance formation, which crops out in a small area at the crest of the Old Woman anticline. Successively younger Jurassic and Cretaceous formations are exposed along the crest and flanks. The structurally lower Ant Hills anticline has the Pierre shale and Fox Hills formations on the surface along its crest. West of it in the township (T.36N, R.63W) at the southwest corner of Fig. 2 is the Lance Creek syncline, also developed in the Pierre shale and Fox Hills.

The "upper clay within sight of the Triceratops beds" mentioned in Professor Case's field notes undoubtedly refers to the Pierre shale, which forms a valley between the more resistant Fox Hills sandstone on the west and the steeply dipping older formations on the east.

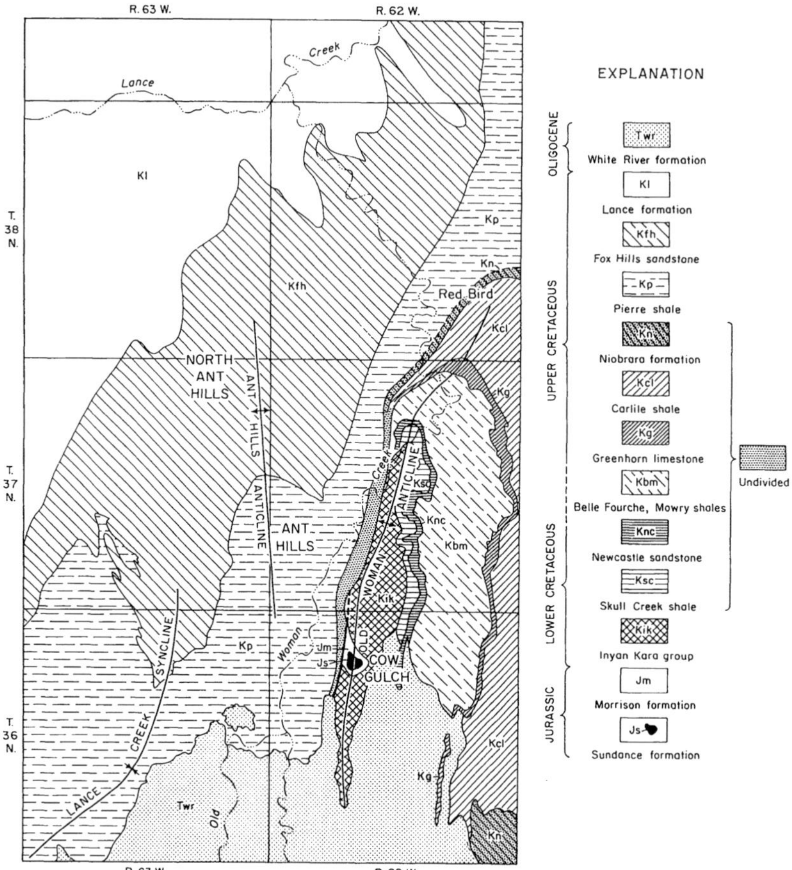

From U.S.G.S. Oil and Gas Investigations Map OM 185

GEOLOGIC MAP OF THE OLD WOMAN CREEK AREA, NIOBRARA CO., WYOMING

Scale
0 1 2 3 4 5
Miles

FIG. 2.

LATE UPPER CRETACEOUS STRATIGRAPHY IN NIOBRARA COUNTY

The lithologic succession and faunal sequence of the late Upper Cretaceous strata in Niobrara County are shown in the four stratigraphic sections on pages 46–49, and represented graphically in Fig. 3. These sections, as shown on the index map, Fig. 1, lie along a

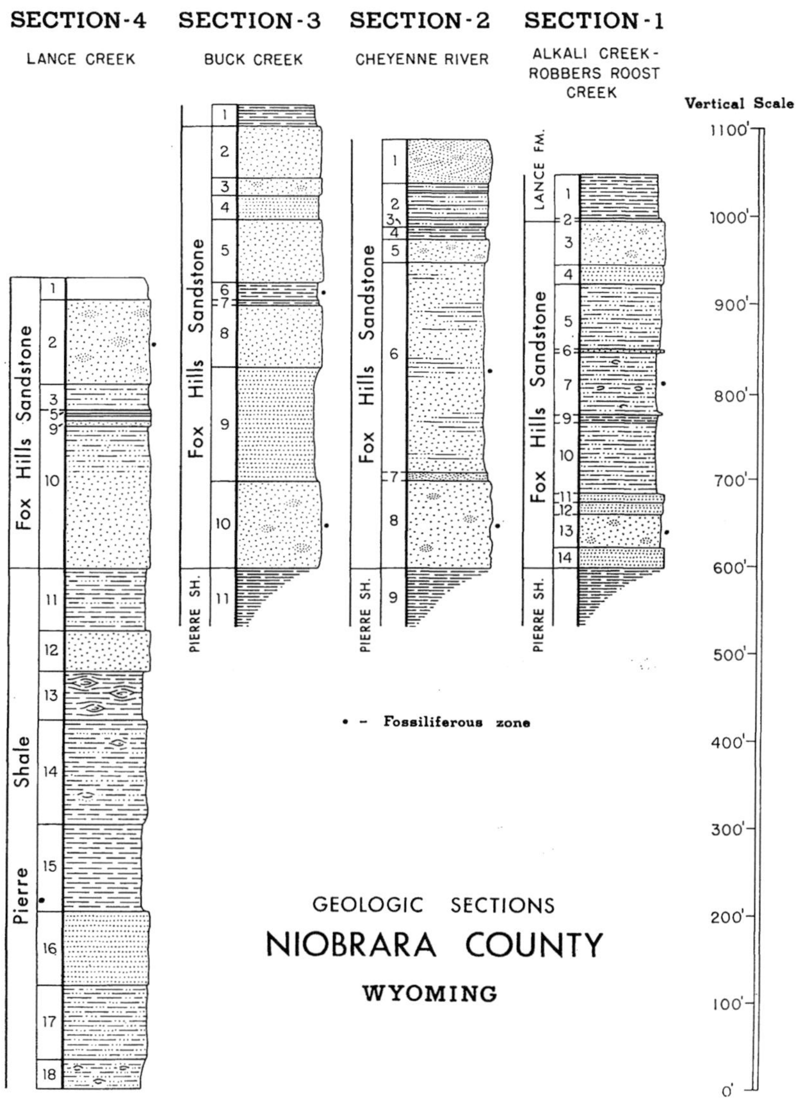
SECTION-4
LANCE CREEK
SECTION-3
BUCK CREEK
SECTION-2
CHEYENNE RIVER
SECTION-1
ALKALI CREEK-ROBBERS ROOST CREEK
Vertical Scale
1100'
1000'
900'
800'
700'
600'
500'
400'
300'
200'
100'
0'
Fox Hills Sandstone
Pierre Shale
PIERRE SH.
LANCE FM.
• - Fossiliferous zone
GEOLOGIC SECTIONS
NIOBRARA COUNTY
WYOMING

FIG. 3.

northeast-southwest trend, about five miles northwest of Old Woman Creek.

In the section between Alkali Creek and Robbers Roost Creek, at the northeast, the Fox Hills sandstone is 357 feet thick. It is predominantly sandstone, but has some 160 feet of shaly beds (units 7 to 10 incl.) near the middle. A concretionary shale and sandstone (unit 7), some 150 to 220 feet below the top, yields a varied pelecypod fauna. A concretionary sandstone interval (unit 13), 40 feet thick, situated 23 to 63 feet above the base of the formation, contains a diversified assemblage typical of the Fox Hills formation, including *Sphenodiscus lenticularis, Protocardia subquadrata,* and *Veniella humilis.*

Five miles to the south on the Cheyenne River near the mouth of Lance Creek 401 feet of Fox Hills sandstone was measured. A shaly zone (units 2 and 3), 66 feet thick, occurs 50 feet below the top. About 140 feet below the top of the section and some 25 feet below the shaly zone, there is a sandstone and sandy shale interval (unit 6), 150 feet thick, that carries a pelecypod fauna characteristic of the Fox Hills sandstone. At the base of the section a massive sandstone bearing large concretions (unit 8) 100 feet thick has yielded a rich and varied fauna which includes *Sphenodiscus lenticularis, Scaphites conradi* var. *intermedius, Protocardia subquadrata, Veniella humilis,* and many other species that make up the so-called Fox Hills fauna.

Buck Creek, a name applied to the lower reaches of Crazy Woman Creek, is another northward flowing tributary of Lance Creek, lying about seven miles west of Old Woman Creek. In the section exposed along Buck Creek the Fox Hills sandstone has a thickness of slightly more than 500 feet. It is made up almost entirely of sandstones but there is a 30-foot shaly zone (unit 6), about 180 feet below the top of the formation, which carries a pelecypod assemblage of Fox Hills species. The basal 100 feet of the formation is massive sandstone with concretions and with characteristic fossils of the Fox Hills.

About eleven miles southwest of the Buck Creek section, on the upper reaches of Lance Creek, the Fox Hills sandstone is 332 feet thick and overlies the Pierre shale, which is nearly 600 feet thick. The contact is gradational. Although there is some sandy shale interbedded with sandstone in the Fox Hills formation, there is no conspicuous shaly unit in the section. It is a succession of sandstone of varying characteristics. Twenty-five feet below the top a soft massive sandstone (unit 2), 96 feet thick, has a few sandstone concretions distributed through it. The lowest 162 feet of strata included in the Fox Hills formation in this section does not have the sandstone concretions found near the base of the other sections lying to the northeast.

The stratigraphic relationships between the sections are summarized below:

	Sections 4	3	2	1
Interval between top of Fox Hills sandstone and top of upper fossiliferous concretionary zone		180	140	151
Thickness of upper fossiliferous concretionary zone		30	150	70
Interval between the two fossiliferous concretionary zones		195	10	113
Thickness of lower fossiliferous concretionary zone		100	100	40
Interval between base of lower fossiliferous concretionary zone and base of Fox Hills sandstone		0	0	0
Thickness of Fox Hills sandstone		505	401	397
Interval between top of Fox Hills sandstone and top shaly zone		180	50	151
Thickness of shaly zone	0	30	51	161
Interval between base of shaly zone and base of Fox Hills sandstone		295	300	85

The Fox Hills sandstone as a whole thickens to the southwest. The shaly zone near its center thins to the southwest from 161 feet in the Robbers Roost area to 30 feet on Buck Creek and was not recognized on the upper part of Lance Creek. There are two rather persistent fossiliferous zones. The upper one bears concretions only in the Robbers Roost section, but the lower one has concretions in all three of the fossiliferous sections.

Stratigraphic Sections

1. Sec. 18, T.40 N., R.61 W., and Secs. 11 and 13, T.40 N., R.62 W., Wyom., between Alkali Creek and Robbers Roost Creek. (Dobbin and Reeside, 1930, pp. 19–20)

Lance formation (part):	Feet
(1) Shale, somber colored; contains abundant remains of turtles and dinosaurs	50
(2) Shale, carbonaceous, and coal	3
Fox Hills formation:	
(3) Sandstone, brown, concretionary; contains abundant remains of dinosaurs, crocodiles, and turtles near top. V. H. Barnett and D. E. Winchester reported marine fossils at a horizon 35 feet below top of this unit. The present writer did not succeed in verifying his report. This unit may belong in the Lance	50
(4) Sandstone, gray, filled with shells of *Corbicula* cf. *C. subelliptica* Meek and Hayden; a brackish-water deposit that may belong to the Lance	23
(5) Sandstone, gray, and soft sandy shale, inseparable in grass-covered slopes	74
(6) Sandstone, yellowish, hard, capped by ripple-marked fissile concretions; contains *Yolida evansi* Meek and Hayden, *Crenella?* sp., and *Haminea minor* Meek in basal part (U.S.G.S. locality 12586)	4
(7) Shale, dark gray, and sandstone, soft, light gray, interlaminated; contains botryoidal masses of limonite and concretions of platy brown sandstone; mass weathers into slopes like the Lance "somber beds"; contains *Nucula* sp. indet., *Modiola attenuata* Meek and Hayden, *Cardium speciosum* Meek and Hayden, *Tellina* sp. indet., 25 feet below top (U.S.G.S. locality 12585)	70
(8) Sandstone, platy, hard, brown	1
(9) Clay, somber-colored	10

(10) Concealed; probably interlaminated gray and brown sandy shale and sandstone much like third unit above ... 80
(11) Sandstone, gray, platy; makes bench ... 10
(12) Sandstone, brown, limonitic, hard and platy; makes the cap rock on the lowest Fox Hills escarpment ... 12
(13) Sandstone, yellow, massive, soft; contains large hard brown calcareous concretions, which yielded (U.S.G.S. locality 12584) *Membranipora* n. sp., *Gervillia subtortuosa* Meek and Hayden. *Inoceramus* (*Actinoceramus*) *fibrosus* Meek and Hayden, *Veniella humilis* Meek and Hayden, *Protocardia subquadrata* Evans and Shumard, *Tellina equilateralis* Meek and Hayden, *Dentalium gracile* Meek and Hayden, *Fasciolaria cheyennensis* Meek and Hayden, *Fasciolaria galpinianum* Meek and Hayden, *Cinulia concinna* Hall and Meek, *Discoscaphites nicolleti* (Morton), *Sphenodiscus lenticularis* (Owen), *Hypsodon* sp., and *Centrarchites?* sp. ... 40
(14) Sandstone, gray, shaly but hard enough to make a prominent bench ... 23

Total, Fox Hills ... 397

Pierre shale:

2. South side of Cheyenne River at the mouth of Lance Creek, and extending up the creek a mile and a half or two miles. (Stanton, 1910, pp. 184–185)
   (1) White cross-bedded sandstone with irregular brown indurated bands, masses, and concretions ... 50
   (2) Soft sandy shale with bands of lignitic shale. Fragments of dinosaur bone were found on the surface here ... 50
   (3) Sandy shale full of *Corbicula cytheriformis?* and *C. subelliptica* var. *moreauensis* ... ½–1
   (4) More or less carbonaceous shale ... 15

(5) Soft massive gray sandstone with many brown concretions 25
(6) Gray sandstone and sandy shale with bands of sandstone containing Fox Hills fossils, about 150
(7) Cross-bedded, ripple-marked, reddish brown sandstone with irregular base .... 8 to 10
(8) Massive soft buff sandstone with many large concretions and indurated masses and an abundant Fox Hills fauna 100
(9) Pierre shale with only the top exposed ....

3. Buck Creek, Niobrara County, Wyoming. Stanton (1910, p. 186)
(1) Sandy shale with thin beds of coal 25

Top of Fox Hills:

(2) Massive white sandstone with *Halymenites major* 60
(3) Yellowish massive sandstone with brown concretions 20
(4) More thinly bedded brown sandstone with *Halymenites* 25
(5) Massive white sandstone 75
(6) Soft somewhat sandy shales with thin sandstone bands containing marine Fox Hills shells 30
(7) Brown shaly sandstone 5
(8) Massive white sandstone 60
(9) Thin-bedded brown and gray sandstone 130
(10) Yellowish massive sandstone with concretions containing Fox Hills fauna 100
(11) Pierre shale ....

4. Along the East Bank of Lance Creek, NE¼, Sec. 28, T.36 N., R.65 W. (Hancock, 1920, p. 102).

Fox Hills sandstone:

(1) Sandstone, very light colored and very soft, not well exposed 25
(2) Sandstone, light yellow, soft and massive; contains a few concretions and rather uniformly distrib-

	uted beds of hard reddish-brown sandstone from 6 to 18 inches thick	96
(3)	Soft sandstone and sandy shale in thin beds; contains three thin beds of hard brown shelving sandstone showing numerous ripple marks	30
(4)	Sandstone, light yellow, fine grained	1
(5)	Sandstone and sandy shale in alternating layers of light yellow and dark color	3′ 6″
(6)	Sandstone, massive, yellowish	3′
(7)	Sandstone, reddish brown, slabby; dip 24°	7″
(8)	Sandstone, containing thin seams of sandy shale	4
(9)	Sandstone, very massive, light yellow	7
(10)	Sandstone, soft, containing thin beds of dark sandy shale and a few thin beds of hard sandstone, especially in the upper portion; strike east, dip 27° N.	162

Pierre shale:

(11)	Shale, very dark, containing thin layers of reddish-brown sandstone; becomes much lighter colored near the top as it merges into the Fox Hills formation; strike east, dip 27° N.	72
(12)	Sandstone; weathers to rounded surfaces; cut off abruptly by a shear zone	36
(13)	Shale, dark; contains 10 separate layers of reddish-brown concretions. *Baculites* fairly abundant. Strike east, dip at top 27½°	67
(14)	Shale, dark, slightly sandy; contains very few concretions and is not very fossiliferous	120
(15)	Interval, not well exposed but in all probability composed of very soft beds	100
(16)	Belt composed essentially of soft sand but includes many thin seams of hard yellowish-brown sandstone	84
(17)	Shale, very sandy	85
(18)	Shale, grayish brown, sandy; weathers yellowish gray. Has many calcareous concretions which contain *Inoceramus barabini* Morton and *Anchura* sp. Dip 23° N. 40° W.	34

Other Fossil Localities

In the Fox Hills formation of Niobrara County, Wyoming (Hancock, 1920, p. 104).

5. SE¼, Sec. 33, T.37 N., R.63 W., from concretions in the sandstones near the top of the formation.
6. NW¼, NW¼, Sec. 22, T.36 N., R.64 W., from concretions in the sandstones near the top of the formation.
7. NE¼, SE¼, Sec. 20, T.36 N., R.64 W., from a sandstone about 100 feet above the base of the formation.
8. SE¼, SE¼, Sec. 12, T.35 N., R.66 W., from thin beds of sandstone exposed in the east bank of Little Lightning Creek.

Upper Cretaceous Invertebrates from Old Woman Creek

R. P. Whitfield (1880) in his *Paleontology of the Black Hills of Dakota* records the following fossils from Old Woman Fork:

Pierre shale (Division No. 4)
Pteria sublevis Whitfield
Inoceramus sagensis Owen
I. simpsoni Meek
I. vanuxemi M. & H.
Endocostea typica Whitfield
Anesomyon alucolus M. & H.
Ptychoceras crassum Whitfield
Haminea subcylindrica Meek & Hayden

Pierre shale or Fox Hills formation (Division No. 4 or No. 5)
Yoldia evansi M. & H.
Fasciolaria (*Cryptorhytis*) *contorta* Meek

Fox Hills formation (Division No. 5)
Endocostea typica Whitfield
Idonearea shumardi M. & H.
Yoldia evansi M. & H.

From the Fox Hills formation (Division No. 5) on Cheyenne River, near Old Woman Fork.
Veniella humilis M. & H.
Sphaeriola transversa Whitfield

The fossils collected by Professor Case and his associates on Old Woman Creek are, with two exceptions, in calcareous nodules, crowded with organic remains. The matrix of some of the nodules is micaceous sandstone of quartz and abundant dark mineral grains. Fish scales, bone fragments, and phosphatic nodules are present. The fossils occur mainly as molds and casts of exteriors and interiors. In many, however, the original shell material is present but can rarely be preserved when the nodules are broken open and the matrix removed. Most of the specimens are small and must be studied under a binocular microscope.

Identification of the fossils from the two localities indicates a varied assemblage. Many of the organisms present are too poorly preserved for specific determination. Those identified, however, are sufficient to indicate the stratigraphic position of the beds from which they were obtained.

1924/K-6; field label 89:

Bryozoa

"Websteria" cretacea Meek

Pelecypoda

Anomia sp.

Inoceramus radiatus Kellum

Lucina? sp.

Nemodon sulcatina (Evans and Shumard)

Protocardia subquadrata (Evans and Shumard)

Protocardia? sp.

Pteria nebrascana (Evans and Shumard)

Pteria? sp. cfr. P. haydeni (Hall and Meek)

Pteria? sp.

Sphaeriola? cordata Meek and Hayden

Tellina sp.

Veniella humilis Meek and Hayden

Gastropoda

Cylichna scitula Meek and Hayden

Fasciolaria (*Piestocheilus*) *culbertsoni* (Meek and Hayden)

Fasciolaria sp.

Fusus? sp.

Cephalopoda

Discoscaphites conradi (Morton) var. *intermedius* Meek

Discoscaphites sp. (immature specimen)

Sphenodiscus lenticularis (Owen)

1924/K7; field label 92:

Pelecypoda

Protocardia subquadrata (Evans and Shumard)
Sphaeriola? cordata Meek and Hayden
Thracia prouti Meek and Hayden
Veniella humilis, Meek and Hayden

Cephalopoda

Discoscaphites sp. aff. D. conradi (Morton)

The two species in these collections that are not in nodules are *Sphenodiscus lenticularis* (Owen) and *Veniella humilis* Meek and Hayden. A single large, well-preserved specimen of *S. lenticularis* is in the collection with field label 89. *V. humilis* is the most common species in both collections.

Analysis of the Collections from Old Woman Creek

Table I shows the vertical and horizontal distribution of late Upper Cretaceous invertebrates reported from Niobrara County. Of the 68 species recorded only 3 are known to occur elsewhere in the Cannonball. The Cannonball member of the Lance formation is the only known marine deposit younger than the Fox Hills sandstone in the Rocky Mountain region. Its distribution is in North and South Dakota 170 to 300 miles northeast of Niobrara County, Wyoming. Therefore the occurrence of marine fossils in the Lance formation along Old Woman Creek would greatly extend the area of the sea which spread over the northern Great Plains at the close of the Cretaceous or beginning of Tertiary time.

In the two collections of fossils made in 1924 on Old Woman Creek I have recognized 24 distinct forms. Ten of these are identified only generically, 1 is compared to a species originally described from the Pierre shale, 1 is related to a species common in the Fox Hills sandstone, 1 is described as new, and 13 are identified with previously known species.

Of the 13 identified with previously known species, all occur elsewhere in the Fox Hills sandstone, 4 range down into the Pierre shale, and 9 have been reported only from the Fox Hills sandstone. Eight have been previously recorded from the Fox Hills sandstone in Niobrara County. Two occur also in the Cannonball member of the Lance.

Since all of the previously described species in the 1924 collections occur elsewhere in the Fox Hills sandstone and 69 per cent of them are known only from that formation, while only 2 specimens,

TABLE I

Faunal Chart
Fox Hills Sandstone
Niobrara County, Wyoming

	Fox Hills Sandstone																		
Locality	Section 1				Section 2			Section 3				5	6	7	8	1924/K-6	1924/K-7	Cannonball
Unit	4	6	7	13	3	6	8	2	4	6	10							
Calcareous Algae																		
Halymenites major Lesquereux								x	x									
Bryozoa																		
Membranipora sp.				x														
"*Websteria*" *cretacea* Meek																x		
Mollusca																		
Burrows of a boring mollusk													x					
Scaphopoda																		
Dentalium gracile Meek & Hayden				x								x						
Pelecypoda																		
Anomia sp.																x		
Avicula nebrascana Evans & Shumard										x								
Callista sp.												x						
Cardium speciosum Meek & Hayden			x			x												
Cinulia concinna Hall & Meek				x														
Corbicula cytheriformis? Meek & Hayden					x													x
C. subelliptica var. *moreauensis* Meek & Hayden						x												
C. cf. *C. subelliptica* Meek & Hayden	x																	
Crenella elegantula Meek & Hayden												x						
Crenella? sp.		x																
Cuspidaria moreauensis (Meek & Hayden)						x												
Entalis paupercula Meek & Hayden							x											
Gervillia subtortuoso Meek & Hayden				x							x							
Inoceramus radiatus Kellum																x		
Inoceramus sp.														x				
Leda (*Yoldia*) sp.							x											
Leda? sp.										x								
Linearia? formosa Meek & Hayden							x											
L. subundata Hall & Meek															x			
Lucina? sp.																x		

TABLE I—*Continued*

Locality	Fox Hills Sandstone Section 1				Section 2			Section 3				5	6	7	8	1924/K-6	1924/K-7	Cannonball
Unit	4	6	7	13	3	6	8	2	4	6	10							
Mactra warrenana Meek & Hayden						x												
Modiola attenuata Meek & Hayden			x															
Nemodon sulcatina (Evans & Shumard)																x		
Nucula cancellata Meek & Hayden													x					
N. planimarginata Meek & Hayden						x				x								
Nucula sp.													x					
Nucula sp. indet.			x															
Protocardia subquadrata (Evans & Shumard)				x			x					x	x			x	x	
Protocardia? sp.																x		
Pteria nebrascana (Evans & Shumard)																x		
P.? sp. cfr. *P. haydeni* (Hall & Meek)																x		
Pteria? sp.																x		
Sphaeriola? cordata Meek & Hayden							x				x					x	x	
Tellina equilateralis Meek & Hayden				x														
T. scitula Meek & Hayden							x			x			x					
Tellina sp.																x		
Tellina sp. indet.			x															
Thracia prouti Meek & Hayden																	x	
Veniella humilis Meek & Hayden				x			x				x	x	x	x	x	x	x	
Yoldia evansi Meek & Hayden		x													x			x
Gastropoda																		
Anchura sp.							x						x					
Cylichna scitula Meek & Hayden												x				x		x
Fasciolaria cheyennensis Meek & Hayden				x														
F. (*Piestocheilus*) *culbertsoni* Meek & Hayden							x									x		
F. galpinianum Meek & Hayden				x														
Fasciolaria sp.																x		
Fusus? sp.																x		
Haminea minor Meek		x																

TABLE I—*Concluded*

Locality	Fox Hills Sandstone																	
	Section 1				Section 2			Section 3				5	6	7	8	1924/K-6	1924/K-7	Cannonball
Unit	4	6	7	13	3	6	8	2	4	6	10							
Haminea sp.							x											
Lunatia concinna (Hall & Meek)?															x			
L. occidentalis Meek & Hayden							x											
Cephalopoda																		
Baculites ovatus Say															x			
Discoscaphites abyssinus (Morton)							x											
D. conradi (Morton)							x								x			
D. conradi (Morton)?												x						
D. conradi (Morton) var. *intermedius* Meek							x									x		
D. conradi (Morton) var. *intermedius* Meek?														x				
Discoscaphites sp. aff. *D. conradi* (Morton)																	x	
D. nicolleti (Morton)				x														
Discoscaphites sp. (immature specimen)																x		
Sphenodiscus lenticularis (Owen)				x			x									x		

or 15 per cent, occur in the Cannonball, the faunal evidence strongly indicates that the fossils are of Fox Hills age and came from the Fox Hills sandstone on Old Woman Creek and not from the overlying Lance formation or "Triceratops beds" as reported.

In trying to locate the stratigraphic position of the 1924 collections more precisely within the Fox Hills sandstone, the lithologic character and faunal associations recorded in the sections previously studied in Niobrara County are important. The two persistent fossiliferous zones have been noted, and the fact that they also bear concretions in most places is significant. The upper zone is characterized by a varied pelecypod fauna of well-known Fox Hills species but is concretionary only in the Robbers Roost section. The lower zone contains a much larger and more varied fauna of ammonites, pelecypods, and gastropods. Also, it contains concretions in all three of the fossiliferous sections. The fossiliferous concretions collected by Professor Case and his companions have the following species in common with the lower concretionary member of the Fox Hills sandstone in Niobrara County:

Protocardia subquadrata (E. & S.)
Sphaeriola? cordata M. & H.
Veniella humilis M. & H.
Fasciolaria culbertsoni M. & H.
Discoscaphites conradi (Morton) var. *intermedius* Meek
Sphenodiscus lenticularis (Owen)

I therefore conclude that the collections made in 1924 on Old Woman Creek are from the lower fossiliferous concretion-bearing unit of the Fox Hills sandstone.

Acknowledgments

I am indebted to Carroll E. Dobbin of the U. S. Geological Survey, a lifetime student of Rocky Mountain geology, for his comments on the geology of the Old Woman Creek area and for suggestions in regard to pertinent references in the literature. The geologic map of the Old Woman Creek area, text Fig. 2, is taken from a map prepared by Dobbin and others of the southeastern part of the Powder River Basin.

The fossils illustrated in this paper were photographed by Herbert W. Wienert, preparator in the Museum of Paleontology. The maps and sections of the text figures were drawn by Derwin L. Bell of the Geology Department, University of Michigan. The rubber casts of the natural molds of fossils in the rock, employed in studying the fauna and subsequently photographed for illustration, were made by E. G. Driscoll, assistant curator in the Museum of Paleontology, University of Michigan.

Notes on the Fossils from Old Woman Creek

Phylum Bryozoa Ehrenberg 1831

"*Websteria*" *cretacea* Meek

(Pl. I, Fig. 1; Pl. II, Fig. 20)

Websteria cretacea Meek, U. S. Geological Surv. of the Territories, vol. IX, 1876, pp. 3–4, pl. 28, figs. 3a, b, c.

Meek referred this fossil with doubt to the Eocene genus *Websteria* Edwards and Haime, a coral of the family Gorgoniidae, subclass Alcyonaria.

Treatise on Invertebrate Paleontology, Part F—Coelenterata 1956, p. F212, says: "*Websteria* M. Edw.-H., 1850 was assigned tentatively by its authors to the Gorgoniidae, but probably it does not belong here." On p. 88 under the heading "Generic names referred to Hydrozoa but belonging to other groups," it lists *Websteria* M. Edw.-H., 1850 (bryozoan), Eoc., Eng. Treatise, Part G—Bryozoa does not mention *Websteria*.

Figured specimen.—UMMP 37944.

Locality and stratigraphic position.—1924/K-6. Fox Hills sandstone.

Phylum Mollusca

Class Pelecypoda Goldfuss

Anomia sp.

(Pl. II, Fig. 17)

Only one specimen of this distinctive shell was recognized in the collections. It is a moderately convex upper valve with the surface ornamented by fine, regular, commonly deflected, radiating striae and widely spaced, low concentric ridges. The ornamentation and depth of the valve are similar to those of *A. micronema*.

Dimensions.—Length .34 inch; height .26 inch.

Figured specimen.—UMMP 37929.

Locality and stratigraphic position.—1924/K-6; Fox Hills sandstone.

Inoceramus radiatus n. sp., Kellum

(Pl. III, Fig. 17)

There are several specimens of the genus *Inoceramus* in our collections. The figured specimen resembles *I. fibrosa* (M. & H.) in having a radiating element of the ornamentation. Meek and Hayden placed the species *fibrosa* in the genus *Pteria* but Dobbin and Reeside (1929, p. 21) refer it to *Inoceramus*. Our specimen has coarse, regular, rounded, concentric undulations over most of the internal mold, becoming faint on the wide dorsal slope and disappearing on the umbone and adjacent steep anterior slope, which are smooth. A few broad low radial waves cross the coarse concentric undulations on the posterior and ventral slopes. The shell material characteristic of the genus *Inoceramus* has a transverse, fibrous, crystalline structure perpendicular to the surface of the valve.

The figure on Pl. III is oriented with the dorsal margin at the right.

Dimensions.—Length of dorsal margin (posterior to beak) .62 inch; greatest length from umbo to posterior-ventral angle .9 inch; height .6 inch; greatest depth of valve at umbo .18 inch.

Figured specimen.—UMMP 37933.

Locality and stratigraphic position.—1924/K-6; Fox Hills sandstone.

Lucina? sp.

(Pl. II, Fig. 9)

This tiny mold of the exterior of a left valve is doubtfully referred to the genus *Lucina.* It is more convex and the umbonal ridge more prominent than in most species in this genus. The beak is elevated slightly above the hinge line. The surface of the shell has fine concentric growth lines. Anterior to the beak the dorsal margin is convex and passes into the rounded anterior margin without angulation. Posterior to the beak the dorsal margin is straight and the growth lines terminate in a faint ridge along it. The posterior margin is broadly convex, being longer and less rounded than the anterior margin. There is a faint groove on the posterior side of the umbone, paralleling the umbonal ridge. The figure does not show the true character of the basal margin of the shell, which is broken off of the mold.

Dimensions.—Length .21 inch; height (incomplete) .15 inch; depth of valve .06 inch.

Figured specimen.—UMMP 37926.

Locality and stratigraphic position.—1924/K-6; Fox Hills sandstone.

Nemodon sulcatina (Evans and Shumard)

(Pl. III, Fig. 16)

Nemodon sulcatinus (Evans and Shumard) Meek, 1876, pp. 82–83, pl. 15, fig. 6a–b.

This species is represented by a single tiny mold of the exterior of a fragmentary left valve. Its shape and pronounced ornamentation are so characteristic that its identification is made with some assurance.

Coarse radial ribs are crossed by well-defined, more or less regularly spaced, concentric growth ridges, forming a cancellate ornamentation of the surface and giving the ribs a nodose character. Fine, thread-like secondary ribs are intercalated between some of the primary ribs toward the margins of the valve. The hinge line is long and straight; the posterior margin of the shell concave; the basal margin is straight and parallels the hinge line, forming a rounded angle of slightly less than 90° with the posterior margin. The anterior margin is missing on the cast but, judging from the growth lines that paralleled it, was convex and broadly rounded, meeting the basal margin in a sweeping curve. The umbo is elevated slightly above the hinge line. The beak is situated about a third the length of the hinge line from the anterior margin. A broad, well-defined sulcus extends from the beak to the middle of the basal margin.

Dimensions.—Length of hinge line .28 inch; length of shell (incomplete) .34 inch; height .28 inch; depth of valve .1 inch.

Figured specimen.—UMMP 37932.

Locality and stratigraphic position.—1924/K-6; Fox Hills sandstone.

Protocardia subquadrata (Evans and Shumard)

(Pl. II, Fig. 13; Pl. III, Figs. 4, 7, and 8)

Cardium subquadratum, Evans and Shumard, 1857, Trans. Acad. Sci. St. Louis, I, 39.

Protocardia (*Leptocardia*) *subquadrata* Evans and Shumard, Meek 1876, pl. 29, figs. 8a–e.

This tiny shell is fairly common in our collections. Its quadrate form, distinct, rather coarse crenulations along the posterior margin only, reflecting a few faint radiating ribs on the posterior slope, distinguish this species from others of very similar shape that occur with it.

Dimensions.—Length .21 inch; height .256 inch; thickness (two valves) .18 inch.

Figured specimens.—UMMP 37896, 37931, 37935.

Other specimens.—UMMP 37931, 38107, 43892, 43893, 43894.

Locality and stratigraphic position.—1924/K-6 and 1924/K-7; Fox Hills sandstone.

Protocardia? sp.

(Pl. III, Fig. 3)

The figured specimen is one of several small pelecypods in our collections that are very similar in external form, whose dentition is unknown, and whose systematic assignment is therefore tentative. In discussing *Protocardia subquadrata* Evans and Shumard, Meek (1876, p. 175) speaks of the obscure radiating costae of the posterior side and notes that in some individuals they are obsolete. Our figured specimen may fall in this category. It has no radiating costae or marginal crenulations but otherwise looks very much like *P. subquadrata.*

Dimensions.—Length .28 inch; height .26 inch; depth of valve .07 inch.

Figured specimen.—UMMP 37924.

Locality and stratigraphic position.—1924/K-6; Fox Hills sandstone.

Pteria nebrascana (Evans and Shumard) Meek

(Pl. II, Figs. 7, 8, 10, 11)

Avicula nebrascana Evans and Shumard, 1857, p. 38.

Pteria nebrascana Meek, 1864, p. 9.

P. (*Oxytoma*) *nebrascana* (Evans and Shumard), Meek, 1876, p. 34–36, pl. 16, figs. 3a, b; and pl. 28, fig. 11.

Several incomplete molds of the left valve are in our collection. They are characteristically convex and oblique in the umbonal area, flattening toward the margins. The beak rises slightly above the hinge line. The posterior wing decidedly longer than the anterior and defined by a pronounced, broadly rounded sinus. Surface ornamentation of fine radiating ribs and finer concentric growth lines, visible under the microscope, except on the umbonal area, which is practically smooth.

The three specimens in our collection referred to this species are natural molds of the exterior of the left valve. They are much smaller than the figure given by Whitfield. Meek gives the length of the hinge line in a medium-sized specimen as .32 inch and the greatest length from the end of the anterior wing obliquely to the posterior basal margin as .45 inch. Our specimens all have the characteristic shape, convexity, and sculpture of Evans and Shumard's species.

Dimensions.—UMMP 37925, hinge line not fully exposed is .14 inch long; height .18 inch; depth .04 inch; No. 37922, hinge line .2 inch long, height .15 inch; No. 37928, hinge line incomplete .19 inch; height incomplete .14 inch.

Figured specimens.—UMMP 37922, 37925, 37928, 37930.

Locality and stratigraphic position.—1924/K-6; Fox Hills sandstone.

Pteria? sp. cfr. *P. haydeni* (Hall and Meek)

(Pl. III, Fig. 20)

Our specimen is an internal mold of a fragmentary right valve with some of the inner layer of shell material attached and at one point a small piece showing the external ornamentation. The mold shows faint concentric undulations and radial ribs. The inner layer of shell material shows fine concentric lines of growth. The external ornamentation on the anterior slope, about a third of the distance from the anterior margin to the center of the valve, consists of thin radial ridges alternating with broad shallow interspaces. Several poorly defined, irregular, shallow depressions on the mold distributed across the central part of the disk may reflect the soft parts of the animal.

Dimensions.—Length (incomplete) .675 inch; height .7 inch; depth of valve .2 inch.

Figured specimen.—UMMP 37902.

Locality and stratigraphic position.—1924/K-6; Fox Hills sandstone.

Pteria? sp.

(Pl. II, Fig. 12)

The tiny fragment figured, an interior of the hinge area of a left valve, is tentatively referred to this genus. It has a straight hinge line with a deep posterior notch, below which extends a thickened, rounded ridge which is broken off at a distance of .08 inch from the hinge but evidently formed the posterior margin of the shell. Below and parallel to it a sulcus radiates from the beak for a distance of .2 inch. The beak, judging from the interior of the valve, was posterior of the center of the hinge line and was not elevated above it. A shallow sulcus radiates anteriorly from the beak parallel to, and just below, the dorsal margin. The interior of the valve is smooth. Anterior, ventral, and most of posterior margins are missing.

Length of hinge.—.24 inch.

Figured specimen.—UMMP 43896.

Locality and stratigraphic position.—1924/K-6; Fox Hills sandstone.

Sphaeriola? cordata Meek and Hayden

(Pl. III, Figs. 1 and 2)

Cyprina cordata Meek and Hayden (1857) p. 143.

Sphaeriola? cordata Meek and Hayden, Meek (1876) p. 137, pl. 29, fig. 3a–c.

There are three specimens of this species in our collections. All are internal molds with small parts of the exfoliated shell adhering. The figured specimen is a left valve; the others are right valves.

The mold is gibbous with rounded margins and prominent beak, elevated well above the dorsal margin. A large, rounded muscle scar at the posterior end of the hinge area forms a prominent platform on the mold in each valve reflecting depressed scars on the interior of the shell. The anterior muscle scars are not evident on our specimens. On the lower half of the mold there is a single prominent anterior radial ridge and a similar posterior radial ridge, along which the growth lines of the shell are deflected. The surface ornamentation, shown on a small fragment of shell adhering near the center of the basal margin, consists of coarse, concentric ridges with broadly rounded crests alternating with broad, shallow channels. The shell material is .2 inch thick.

Dimensions of figured specimen.—Length 1.73 inches; height 1.6 inches; depth of valve .73 inch.

Figured specimen.—UMMP 37889.

Other specimens.—UMMP 43895, 44453.

Locality and stratigraphic position.—1924/K-6 and 1924/K-7; Fox Hills sandstone.

Tellina sp.

(Pl. III, Figs. 9 and 10)

Although there is only one specimen of this species in our collection, it is in some ways better preserved than most of the other large pelecypods. The figured specimen is a natural mold of the interior, but we have also the two valves of the specimen in which it formed. Only the interior of the valves is exposed. The shell material is somewhat exfoliated and part of the margin is missing, but the hinge plate and muscle scars are preserved. The shell is excavated beneath the hinge plate and a pronounced ridge radiates from the beak to the posterior muscle scar. A small deeply impressed scar is present beneath the hinge plate above the radiation ridge and posterior to the beak.

Dimensions.—Length 1.7 inches; height 1.2 inches.

Figured specimen.—UMMP 37894.

Locality and stratigraphic position.—1924/K-6; Fox Hills sandstone.

Veniella humilis Meek

(Pl. II, Figs. 14–16, 18, 19; Pl. III, Figs. 11–15, 19)

Cyprina humilis Meek and Hayden, 1861, p. 179.

Veniella (*Venilicardia?*) *humilis* Meek and Hayden, 1876, pp. 155–156, pl. 30, fig. 5a, b, c.

Veniella humilis Whitfield, 1880, p. 414, pl. 10, figs. 7–13.

This species is reported by Whitfield from the Fox Hills sandstone on Old Woman Fork. It is the most common of the larger shells in our 1924 collections from Old Woman Creek. All of our specimens are preserved as internal molds with some of the shell material adhering. The anterior muscle scar lies directly beneath the beak, is elevated along its dorsal side on the mold but passes smoothly in the marginal area below the palial line. The umbonal ridge is well defined, passing posteriorly from the beak to the posterior-ventral angle. Another prominent ridge on the mold extends posteriorly from the beak, defining the dorsal area. This latter ridge scarcely shows on the exterior of the valve, and the umbonal ridge also is more rounded than on the mold.

Dimensions of mold (UMMP 43898).—Length 1.83 inches; height 1.3 inches; thickness 1.04 inches.

Figured specimens.—UMMP 37890, 37898, 37903, 37906, 37937, 37938, 37945.

Other cataloged specimens.—UMMP 37895, 43853.

Locality and horizon.—1924/K-6 and 1924/K-7; Fox Hills sandstone.

Thracia prouti Meek and Hayden

(Pl. III, Figs. 5 and 6)

Tellina prouti Meek and Hayden, 1857, p. 83.

Thracia prouti Meek and Hayden, 1860, p. 425.

Thracia? prouti Meek and Hayden, Meek, 1876, pp. 225–226, pl. 37, figs. 6a, b.

One specimen in our collections is referred to this species. It is an internal mold with much of the original shell material of one valve preserved. The shell is thin with concentric lines of growth on the exterior. A ridge or angulation radiating from the beak to the posterior-ventral angle is rather well defined on the mold but obscure on the shell. The sculpture on the shell between this angulation and the dorsal margin consists of arcuate growth lines convex toward the beak and perpendicular to the dorsal margin. The mold is broadly rounded anteriorly and tapers posteriorly, with the beaks about centrally located. In cross section the shell is lenticular with the right valve slightly more convex than the left.

Dimensions.—Length 1.4 inches; height 1 inch; thickness .48 inch.

Figured specimen.—UMMP 37893.

Locality and stratigraphic position.—1924/K-7; Fox Hills sandstone.

Class Gastropoda

Fusus? sp.

(Pl. II, Figs. 5 and 6)

Our single specimen of this species is an internal mold with fragments of the shell adhering but shows none of the external sculpture. In general proportions it resembles Whitfield's figure of *F. shumardi* Hall and Meek from the Benton group in the Black Hills of South Dakota. Whitfield (1880, p. 424, pl. XII, figs. 7 and 8), W. R. Calvert, and others (1914, p. 14) list *F. dakotensis* Meek and Hayden var. from the lower half of the Fox Hills formation in the Standing Rock and Cheyenne River Indian reservations of North and South Dakota, but our specimen is unrelated to that species. Dane, Pierce, and Reeside (1937, p. 232) list *Fusus?* sp. from the Fox Hills formation in eastern Colorado but have not figured the specimen.

Figured specimen.—UMMP 37901.

Locality and stratigraphic position.—1924/K-6; Fox Hills sandstone.

Cylichna scitula Meek and Hayden

(Pl. II, Fig. 1)

Cylichna scitula, Meek and Hayden, 1861, p. 178.

Cylichna scitula, Meek and Hayden, 1876, pp. 276–277, pl. 31, figs. 3, a, b.

This distinctive species, represented by at least two specimens in our collection, is characterized by its cancellate sculpture of impressed, revolving lines separated by wider interspaces and crossed by fine growth lines. Shell subcylindrical and completely involute, the aperture being longer than the height of the remainder of the outer whorl. The widest part of the shell is just below the middle.

Dimensions (incomplete).—Height .37 inch; width .22 inch.

Figured specimen.—UMMP 37934.

Other cataloged specimen.—UMMP 43911.

Locality and stratigraphic position.—1924/K-6; Fox Hills sandstone.

Fasciolaria (Piestocheilus) culbertsoni (Meek and Hayden) Whitfield

(Pl. II, Figs. 2 and 3)

Fusus culbertsoni Meek and Hayden, 1857, p. 66.

Fasciolaria (Piestocheilus) culbertsoni (Meek and Hayden), 1876, pp. 360–362, pl. 32, fig. 1, a, b, c, d, e, f.

F. (Piestocheilus) culbertsoni Whitfield, 1880, pp. 423–424, pl. 12, fig. 11.

Whitfield's figure of this species has the sutures more distinctly appressed and the whorls somewhat more inflated than the figures of Meek and Hayden indicate. In these and other characters the fragmentary specimens from Old Woman Creek resemble Whitfield's figure more closely. The elevated revolving ridges are crossed by growth lines of irregular size which slant posteriorly from the suture across the upper, appressed portion of the whorl and then curve downward to parallel the aperture.

Dimensions.—Height of figured fragment, No. 37920, .7 inch.

Figured specimens.—UMMP 37920 and 37921.

Locality and stratigraphic position.—1924/K-6; Fox Hills sandstone.

Fasciolaria sp.

(Pl. II, Fig. 4)

An internal mold of a body whorl with a fragment of the shell adhering shows an indication of the appressed suture. Six broad, faint vertical ridges or angulations in the mold probably reflect folds in the ornamentation of the shell. At the top of the whorl the mold is excavated like a shallow funnel to accommodate the base of the adjacent whorl.

This may be a fragment of the mold of *F. culbertsoni* but is too poorly preserved for identification.

Dimensions.—Height of body whorl (incomplete), .64 inch; width .5 inch.

Figured specimen.—UMMP 37904.

Locality and stratigraphic position.—1924/K-6; Fox Hills sandstone.

Phylum Cephalopoda

Discoscaphites conradi (Morton) var. *intermedius* Meek

(Pl. IV, Figs. 1 and 2; Pl. VI, Figs. 3–7)

Scaphites conradi var. *intermedius* Meek, 1876, pl. 34, figs. 3 a, b, c.

Discoscaphites conradi (Morton) var. *intermedius* (Meek), Reeside 1927, pp. 27 and 29.

The type specimen of this variety (USNM 408), of which we have a plaster cast (UMMP 16310) in The University of Michigan Museum of Paleontology, shows three distinct growth stages in its ornamentation. Our figured specimens illustrate these growth stages. Several smaller specimens in our collections have only the early ornamentation. The iridescent shell material is preserved on all except the largest figured specimen (UMMP 37888).

Dimensions (UMMP 38106).—Diameter 1.68 inches; thickness at aperture .71 inch.

Figured specimens.—UMMP 37888, 37900, 38106.

Other cataloged specimens.—UMMP 43899.

Locality and stratigraphic position.—1924/K-6 and 1924/K-7; Fox Hills sandstone.

Discoscaphites sp. aff. *D. conradi* (Morton)

(Pl. V, Figs. 4 and 5)

Our figured specimen is smaller and more compressed than the smaller specimens figured by Meek as *D. conradi* (Morton) from the Upper Missouri country. It is partly crushed but appears to be of normal thickness anteriorly. Much of the original shell material is preserved. The suture line has a greater similarity to that of the variety *intermedius* but differs in detail.

Dimensions.—Diameter 1.23 inches; thickness .39 inch.

Figured specimen.—UMMP 37897.

Locality and stratigraphic position.—1924/K-7; Fox Hills sandstone.

Discoscaphites sp. (immature specimen)

(Pl. III, Fig. 18, and Pl. V, Figs. 2 and 3)

This small specimen has the diameter proportionally wider than in *D. conradi.* In the anterior part it has a row of nodes on either side of the venter and another on the flank above the middle. These disappear on the earlier part of the whorl. The anterior primary ribs tend to bifurcate at the flank node forming two nearly equal ribs that cross the venter, which is only faintly flattened.

Dimensions.—Diameter .7 inch; thickness .34 inch.

Figured specimen.—UMMP 37899.

Locality and stratigraphic position.—1924/K-6; Fox Hills sandstone.

Sphenodiscus lenticularis (Owen)

(Pl. IV, Figs. 3 and 4; Pl. V, Fig. 1; Pl. VI, Figs. 1 and 2)

Ammonites lenticularis, Owen (1852), p. 579, tab. viii, fig. 5.

Placenticeras lenticulare (Owen), Meek (1876), pp. 473–476, pl. 34, figs. 1a, b, c.

Sphenodiscus lenticularis var. *splendens* Hyatt (1903), p. 75, pl. VIII, figs. 3–7.

One fine specimen of this species is in our 1924 collection. It has about half of the outer whorl, comprising most of the living chamber, and the inner whorls, only one of which is visible because of its involution. The lenticular shape, acute venter, and suture are characteristic of the species. The whorl preceding the living one has a faint indication of low radial waves on the flanks, which are broadly convex in cross-section sloping both toward the venter and the umbilicus. The greatest thickness is at about one-third the distance from the umbilicus to the venter.

Dimensions.—Height of outer whorl (at aperture) 5.7 inches; diameter of inner whorl 3.9 inches; thickness of outer whorl (at aperture) 2 inches; thickness of inner whorl .95 inches.

Figured specimen.—UMMP 37891.

Locality and stratigraphic position.—1924/K-6; Fox Hills sandstone.

LITERATURE CITED

BOYLE, CORNELIUS BRECKENRIDGE. 1893. A Catalogue and Bibliography of North American Mesozoic Invertebrata. U. S. Geol. Surv. Bull. No. 102, 315 pp.

COBBAN, WM. A. 1958. Late Cretaceous Fossil Zones of the Powder River Basin, Wyoming and Montana. Wyoming Geol. Assn. Guidebook, 13th Ann. Field Conf., pp. 114–119.

——, AND J. B. REESIDE, JR. 1952. Correlation of the Cretaceous Formations of the Western Interior of the United States. Bull. Geol. Soc. Am., 63: 1011–1044.

DOBBIN, C. E., AND J. B. REESIDE, JR. 1930. The Contact of the Fox Hills and Lance Formations. U. S. Geol. Surv. Prof. Pap. 158(B), pp. 9–25, 2 pls., 4 text figs.

EVANS, JOHN, AND B. F. SHUMARD. 1857. On Some New Species of Fossils from the Cretaceous Formation of Nebraska Territory. Trans. Acad. Sci. St. Louis. Vol. 1, 1856–1860, pp. 38–42.

FOX, S. K., JR., AND R. J. ROSS, JR. 1942. Foraminiferal Evidence for the Midway (Paleocene) Age of the Cannonball Formation in North Dakota. Jour. Paleontol., 16(5):600–673, 5 text figs.

HALL, JAMES, AND F. B. MEEK. 1855. Descriptions of New Species of Fossils, from the Cretaceous Formations of Nebraska, with Observations upon *Baculites ovatus* and *B. compressus*, and the Progressive Development of the Septa in Baculites, Ammonites, and Scaphites. Memoirs Am. Acad. Arts and Sci. N.S., V (Pt. II, Art. XVII), 379–411, 8 pls.

HANCOCK, E. T. 1920. The Lance Creek Oil and Gas Field, Niobrara County, Wyoming. U. S. Geol. Surv. Bull., 716 (Pt. 2, E), 91–122, 2 pls.

HATCHER, J. B. 1903. Relative Age of the Lance (Ceratops) Beds of Converse County, Wyoming, the Judith River Beds of Montana, and the Belly River Beds of Canada. Am. Geologist, 31: 369–375.

——. 1893. The Ceratops Beds of Converse County, Wyoming. Am. Jour. Sci., 3d Ser., 45: 138.

HYATT, ALPHEUS. 1903. Pseudoceratites of the Cretaceous. U. S. Geol. Surv. Monograph 44, 351 pp.

JELETZKY, J. A. 1952. Youngest Marine Rocks in Western Interior of North America and the Age of the Triceratops-Beds, with Remarks on Comparable Dinosaur-Bearing Beds Outside North America. Geological Survey of Canada Reprint 21, 1960. XXI Internat. Geol. Congress, Copenhagen. Part V, Proc. Sec. 5, pp. 25–40.

LLOYD, E. RUSSELL. 1914. The Cannonball River Lignite Field, Morton, Adams, and Hettinger Counties, North Dakota. U. S. Geol. Surv. Bull., 541: 243–291, 2 pls., 1 text fig.

MARSH, OTHNIEL CHARLES. 1889. The Skull of the Gigantic Ceratopsidae. Am. Jour. Sci., 38(3): 501–506.

MEEK, F. B. 1864. Remarks on the Family Pteriidae (= Aviculidae) with Descriptions of Some New Genera. Am. Jour. Sci., 37(2): 212–220.

——. 1876. A Report on the Invertebrate Cretaceous and Tertiary Fossils of the Upper Missouri Country. U. S. Geol. Surv. of the Territories, IX: 629 pp., 44 pls.

——, AND F. V. HAYDEN. 1857. Descriptions of New Species and Genera of Fossils Collected by Dr. F. V. Hayden in Nebraska Territory; with Some Remarks on the Tertiary and Cretaceous Formations of the West. Acad. Nat. Sci., Phila. Proc., pp. 117–148.

——, AND ——. 1861. Descriptions of New Organic Remains from the Tertiary, Cretaceous, and Jurassic Rocks of Nebraska. Proc. Acad. Nat. Sci. Phila., 1860 (1861), pp. 175–188.

REESIDE, JOHN B., JR. 1927. The Scaphites, an Upper Cretaceous Ammonite Group. U. S. Geol. Surv. Prof. Pap. 150, pp. 21–40, 3 pls.

STANTON, TIMOTHY W. 1910. Fox Hills Sandstone and Lance Formation ("Ceratops Beds") in South Dakota, North Dakota, and Eastern Wyoming. Am. Jour. Sci., 4th Ser., 30(177), Art. 16.

——. 1920. The Fauna of the Cannonball Marine Member of the Lance Formation. U. S. Geol. Surv. Prof. Pap. 128(A): 1–160, 9 pls., 2 text figs., 1 tab.

Treatise on Invertebrate Paleontology. 1956. Part F, Coelenterata. 498 pp.

WHITFIELD, R. P. 1880. Paleontology of the Black Hills of Dakota. Pp. 325–468, 16 pls. U. S. Geog. and Geol. Surv. of the Rocky Mountain Region. Rep. on the Geol. and Resources of the Black Hills of Dakota. By Henry Newton and Walter P. Jenny. 1880. 566 pp.

WILMARTH, M. G. 1938. Lexicon of Geologic Names of the United States. U. S. Geol. Surv. Bull. 896, 2 pts., 2396 pp.

EXPLANATION OF PLATE I

(magnified × 5)

Page

Websteria cretacea Meek 56

FIG. 1. Slab showing abundance of this species in a concretion of the Fox Hills sandstone. At the left, in the lower half of the plate, the interiors of two valves of *Nemodon* are shown. UMMP 37944. Locality: 1924/K-6.

PLATE I

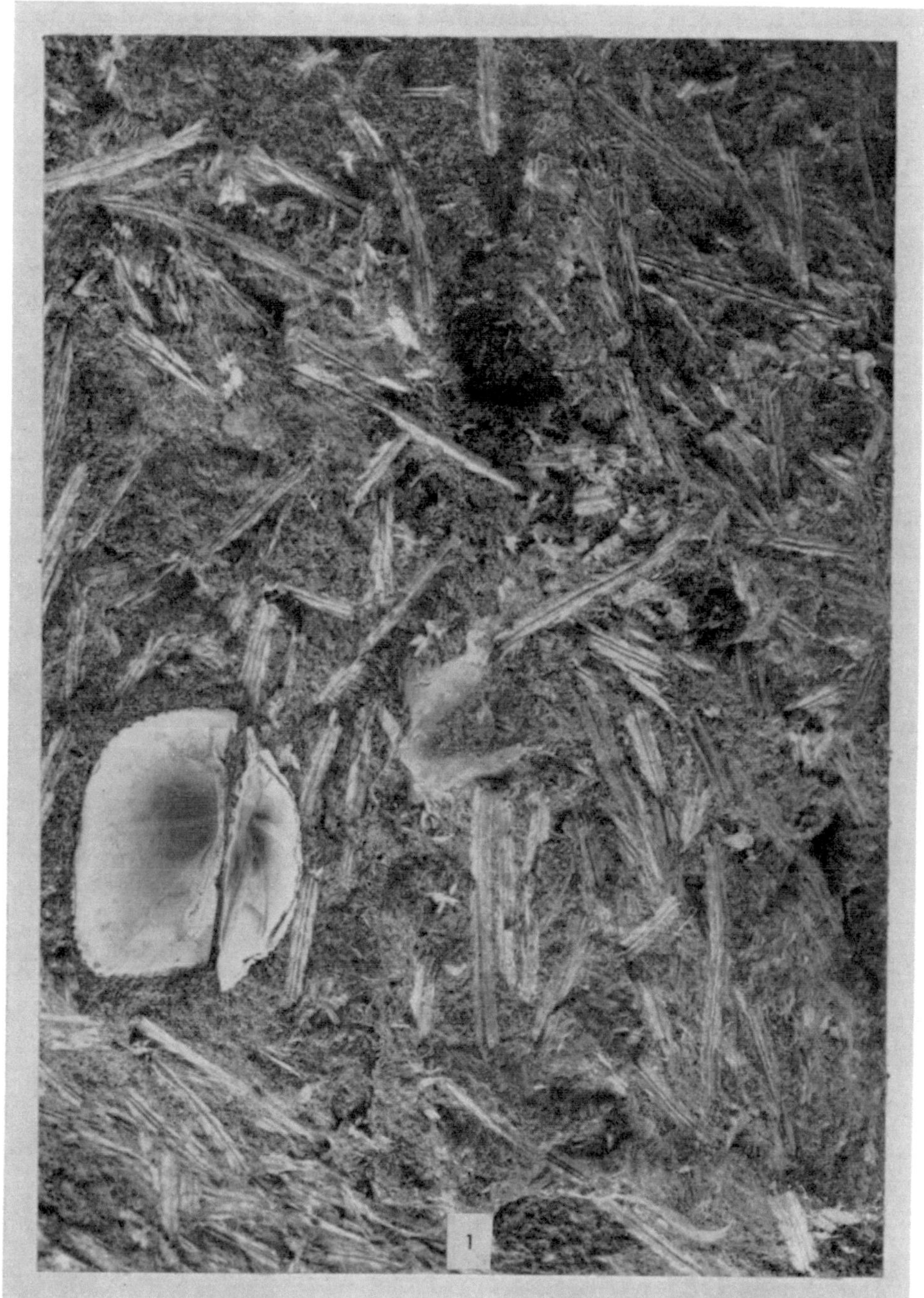

EXPLANATION OF PLATE II

(All specimens natural size except as indicated)

Page

Cylichna scitula Meek and Hayden 65

FIG. 1. Rear side view of internal mold. UMMP 37934. Locality 1924/K-6. × 4

Fasciolaria (Piestocheilus) culbertsoni (Meek & Hayden), Whitfield. 66

FIG. 2. Rear side view. Rubber cast of natural mold of exterior. UMMP 37920. Locality 1924/K-6. × 2

FIG. 3. Rear side view. Rubber cast of natural mold of exterior. UMMP 37921. Locality 1924/K-6. × 2

Fasciolaria sp. 66

FIG. 4. Posterior side view of outer whorl of internal mold. UMMP 37904. Locality 1924/K-7.

Fusus? sp. 65

FIGS. 5 AND 6. Apertural view and rear side view of internal mold. UMMP 37901. Locality 1924/K-6.

Pteria nebrascana (Evans and Shumard) Meek 60

FIG. 7. Side view of left valve. Rubber cast of natural mold of exterior. UMMP 37925. Locality 1924/K-6. × 3

FIG. 8. Side view of left valve. Rubber cast of natural mold of exterior. UMMP 37930. Locality 1924/K-6. × 3

FIG. 10. Side view of left valve. Rubber cast of natural mold of exterior. UMMP 37928. Locality 1924/K-6. × 3

FIG. 11. Side view of left valve. Rubber cast of natural mold of exterior. UMMP 37922. Locality 1924/K-6. × 3

Lucina? sp. 58

FIG. 9. Side view of incomplete left valve. Rubber cast of natural mold of exterior. UMMP 37926. Locality 1924/K-6. × 3

Pteria? sp. 61

FIG. 12. Side view of dorsal part of left valve. Rubber mold of interior. UMMP 43896. Locality 1924/K-6. × 3

Protocardia subquadrata (Evans and Shumard) 59

FIG. 13. Side view of internal mold of left valve. UMMP 37935. Locality 1924/K-6. × 4

Veniella humilis Meek and Hayden 63

FIGS. 14 AND 15. Dorsal view of both valves and oblique view of dorsal half of right valve. Rubber cast of natural mold of exterior of parts of both valves. UMMP 37938. Locality 1924/K-6

FIG. 16. Side view of anterior part of left valve. Rubber cast of natural mold of exterior. UMMP 37945. Locality 1924/K-6.

FIGS. 18 AND 19. Side view of anterior part of right valve and anterior view of both valves. Rubber cast of natural mold of exterior. UMMP 37937. Locality 1924/K-6.

Anomia sp. 57

FIG. 17. Side view of upper valve. Rubber cast of natural mold of exterior. UMMP 37929. Locality 1924/K-6. × 3

"*Websteria*" *cretacea* Meek 56

FIG. 20. Weathered fragment of frond showing possible apertures of zoecia. UMMP 37944. Locality 1924/K-6. × 15

PLATE II

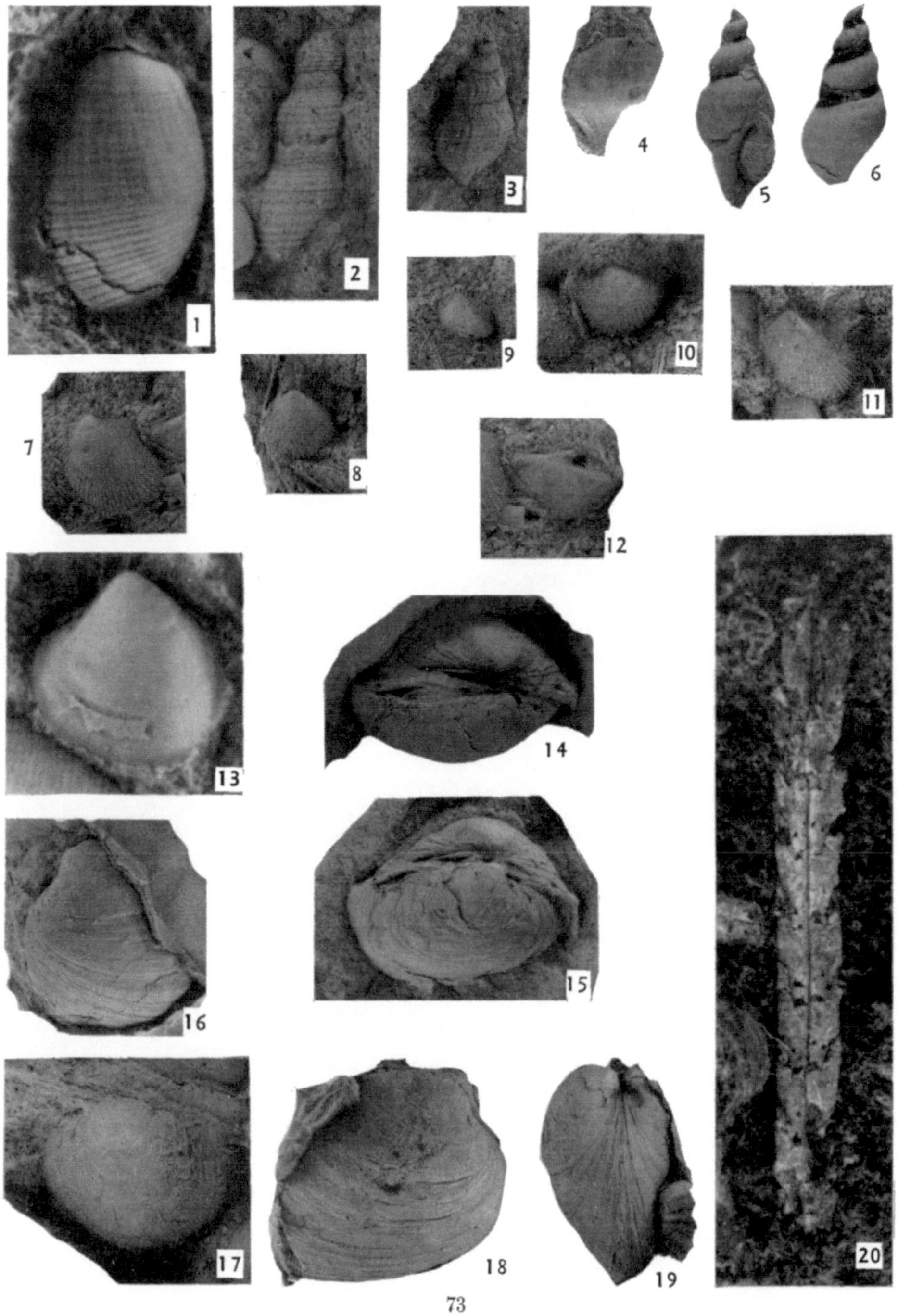

EXPLANATION OF PLATE III

(All figures natural size except as indicated)

Page

Sphaeriola? cordata Meek and Hayden 62

Figs. 1 and 2. Dorsal and side views of left valve. Internal mold. UMMP 37889. Locality 1924/K-6.

Protocardia? sp. 60

Fig. 3. Side view of left valve. Rubber cast of natural mold of interior. UMMP 37924. Locality 1924/K-6. × 3

Protocardia subquadrata (Evans and Shumard) 59

Fig. 4. Side view of posterior part of right valve. Rubber cast of natural mold of interior. UMMP 37931. Locality 1924/K-6. × 2

Thracia prouti Meek and Hayden 64

Figs. 5 and 6. Dorsal view and side view of exterior of right valve. UMMP 37893. Locality 1924/K-7

Protocardia subquadrata (Evans and Shumard) 59

Figs. 7 and 8. Side view of right valve and dorsal view of specimen. UMMP 37896. Locality 1924/K-6. × 4

Tellina sp. 63

Figs. 9 and 10. Dorsal view of specimen and side view of left valve. Natural mold of interior. UMMP 37894. Locality 1924/K-6.

Veniella humilis (Meek and Hayden) 63

Figs. 11 and 12. Dorsal view and side view of right valve. UMMP 37890. Locality 1924/K-6.

Figs. 13 and 14. Dorsal view and side view of left valve. UMMP 37906. Locality 1924/K-6.

Fig. 15. Side view of exfoliated left valve. UMMP 37898. Locality 1924/K-6.

Fig. 19. Oblique side view of right valve. UMMP 37903. Locality 1924/K-6.

Nemodon sulcatina (Evans and Shumard) 58

Fig. 16. Side view of left valve. Rubber cast of natural mold of exterior. UMMP 37932. Locality 1924/K-6. × 4.

Inoceramus radiatus Kellum 57

Fig. 17. Side view of left valve; natural mold of exterior. UMMP 37933. Locality 1924/K-6.

Discocaphites sp. (immature specimen) 68

Fig. 18. Suture line, incomplete. UMMP 37899. Locality 1924/K-6. × 7.5

Pteria? sp. cfr. *P. haydeni* (Hall and Meek) 61

Fig. 20. Side view of left valve. Natural mold of interior, incomplete. UMMP 37902. Locality 1924/K-6.

PLATE III

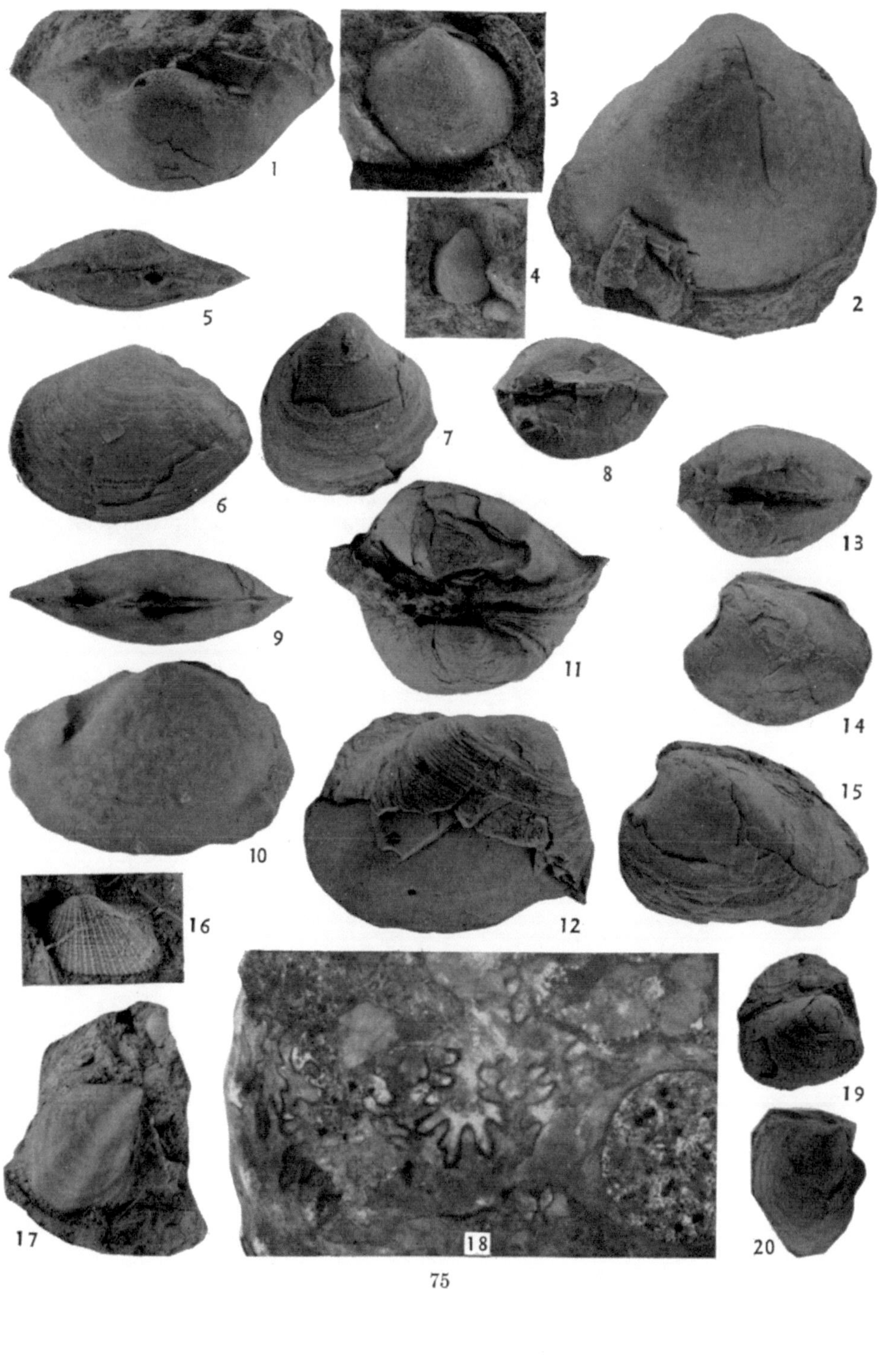

EXPLANATION OF PLATE IV

(All figures natural size except as indicated)

Page

Discoscaphites conradi (Morton) var. *intermedius* Meek 67

FIGS. 1 AND 2. Umbilical and ventral views. UMMP 38106. Locality 1924/K-6.

Sphenodiscus lenticularis (Owen) 68

FIGS. 3 AND 4. Umbilical view of an inner whorl and enlargement (× 4) of small area showing suture lines. UMMP 37891. Locality 1924/K-6.

PLATE IV

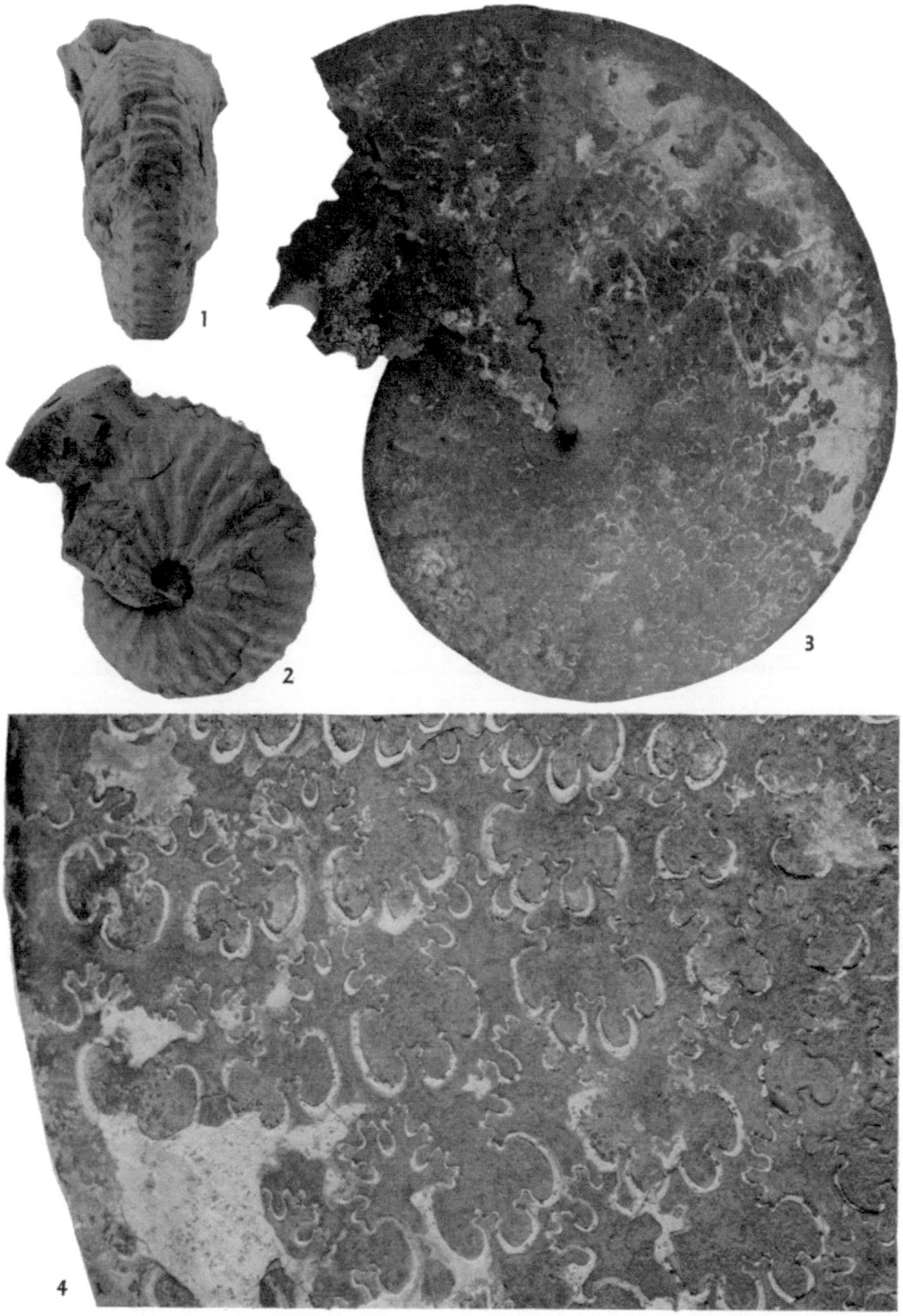

EXPLANATION OF PLATE V

(All figures natural size except as indicated)

Page

Sphenodiscus lenticularis (Owen) 68

FIG. 1. Side view of living chamber. UMMP 37891. Locality 1924/K-6. $\times \frac{7}{9}$

Discoscaphites sp. (immature specimen) 68

FIGS. 2 AND 3. Ventral and umbilical views. UMMP 37899. Locality 1924/K-6. $\times 2$

Discoscaphites sp. aff. *D. conradi* (Morton) 67

FIGS. 4 AND 5. Ventral and umbilical views. UMMP 37897. Locality 1924/K-7.

PLATE V

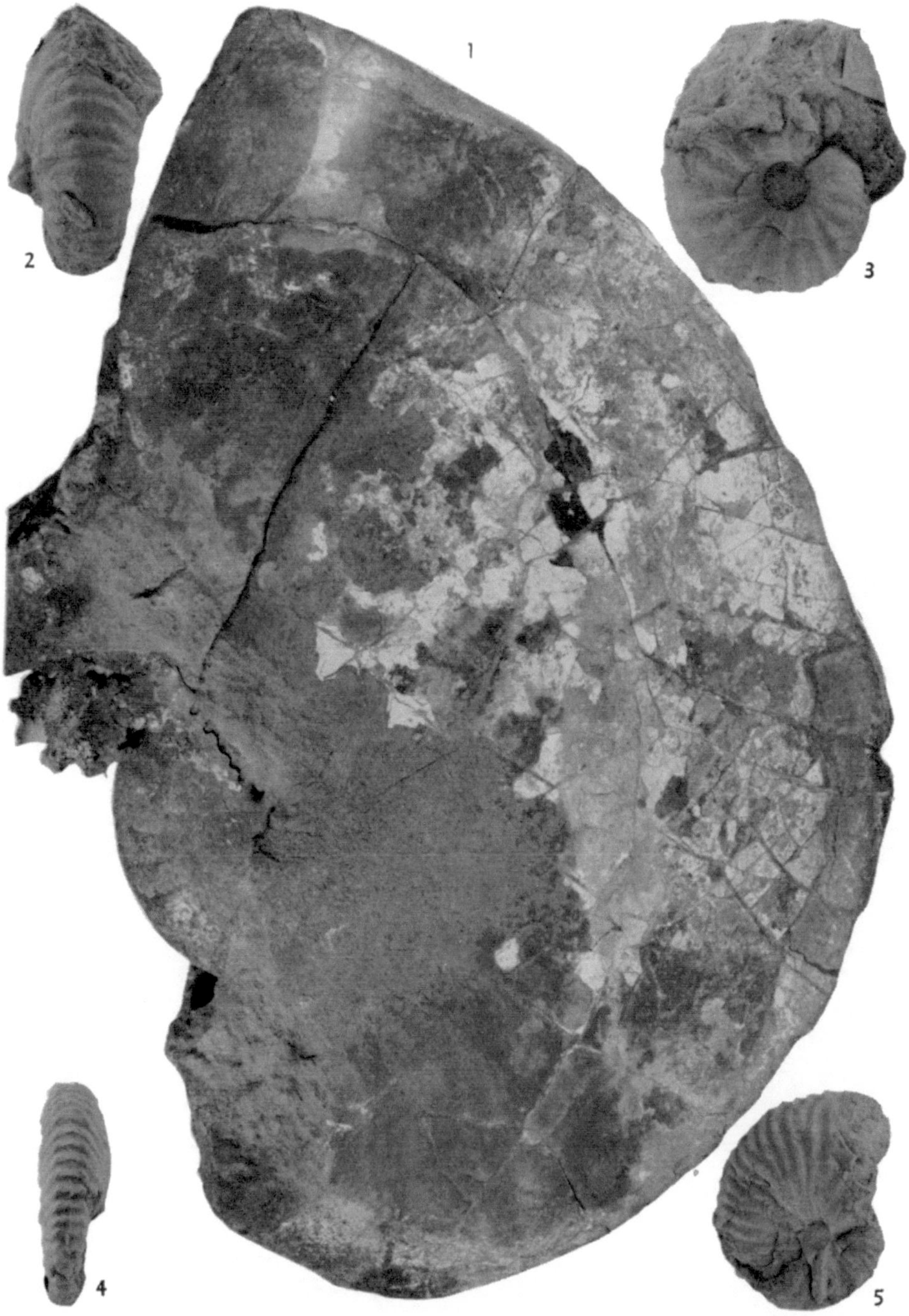

EXPLANATION OF PLATE VI

(All figures natural size except as indicated)

Page

Sphenodiscus lenticularis (Owen)................................ 68

FIGS. 1 AND 2. Apertural and ventral views. UMMP 37891. Locality 1924/K-6. $\times\frac{3}{4}$

Discoscaphites conradi (Morton) var. *intermedius* Meek............ 67

FIGS. 3 AND 5. Ventral and side views of anterior portion of outer whorl of mature individual. UMMP 37888. Locality 1924/K-6.

FIGS. 4, 6, 7. Oblique ventral, side, and apertural views of fragment near aperture of young individual. UMMP 37900. Locality 1924/K-6.

PLATE VI

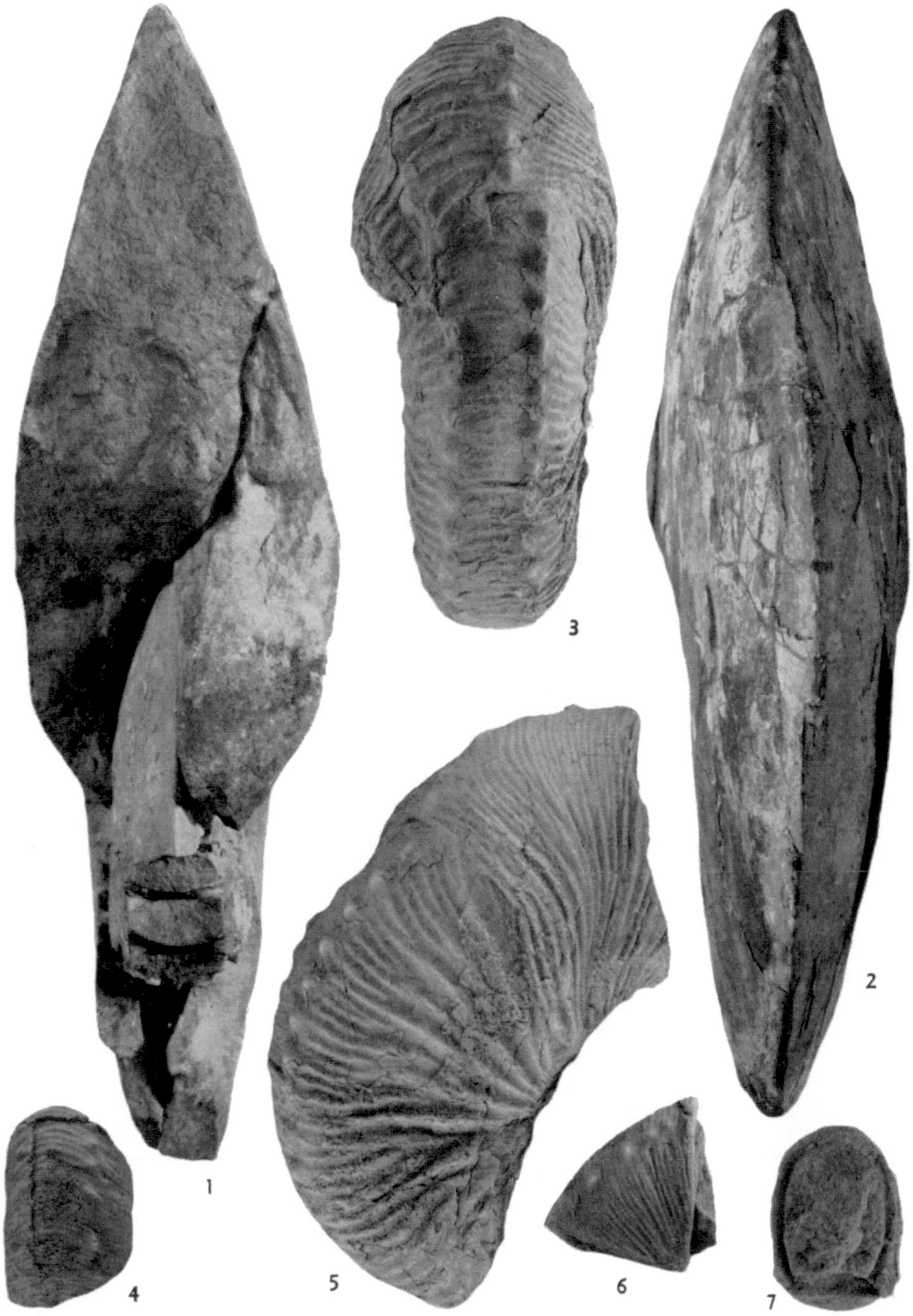

Papers of the Michigan Academy of Science, Arts, and Letters
Vol. XLVII, 1962 (1961 Meeting)

ARTIODACTYLS FROM THE SEYMOUR FORMATION OF KNOX COUNTY, TEXAS

CLAUDE W. HIBBARD, *The University of Michigan*

WALTER W. DALQUEST, *Midwestern University*

Recent collections from the Seymour formation in Knox County, Texas, have yielded a large vertebrate fauna (Hibbard and Dalquest, 1960; and Tihen, 1960). The specimens recovered consist of many hundred individual bones, most of which are very fragmentary. The incompleteness of the material has made the study of the fauna quite difficult. For this reason, and because of the lack of comparative material of the larger Pleistocene mammals at The University of Michigan, only the artiodactyls are reported upon at this time. It is hoped that the study of the invertebrate and vertebrate fauna will soon be completed.

Strickland (1961) discussed the deposition and extent of the Seymour formation in Knox and Baylor counties, Texas. Melton, Dalquest, and Hibbard, in their field work along the valley walls of the South Wichita River, have in a number of localities observed caliche or caliche rubble capping exposures of the Seymour formation. The caliche, when observed in exposures containing the Pearlette ash, occurs in sandy silt and silt, from 3 to 6 feet above the volcanic ash. It is assumed that the caliche is in part of Yarmouth age. It is possible that there might be vertebrate remains in the formation that are post-Pearlette ash and pre-caliche in age, representing part of the fauna that lived in that region during the Yarmouth interglacial. No invertebrate or vertebrate remains have yet been taken from deposits in this interval where the ash is present. In areas where the ash is absent, a careful search of the upper part of the Seymour formation has failed to reveal any fossil remains.

The assignment of the invertebrate and vertebrate remains from below the Pearlette ash to a late Kansan age and a correlation with the Cudahy fauna is based upon the stratigraphic distribution of these fossils in the Crooked Creek formation of southwestern Kansas. To the northwest, in Kansas, *Stegomastodon*, *Megalonyx*, *Nannippus phlegon*, and *Plesippus* are members of the Seger local fauna (Hibbard, 1951), which occurs in the Stump Arroyo member of the Crooked Creek formation. *Nannippus* and *Plesippus*, in the High Plains region, are not known from deposits younger than the Stump Arroyo member. In the Atwater member (Hibbard, 1958) of the Crooked Creek formation and below the Pearlette ash are found the earliest remains of *Equus s. s.* and *Mammuthus* (Hibbard, 1953) in the High Plains region. Since the remains of *Nannippus* and *Plesippus* are unknown from the Seymour formation and those of *Equus s. s.* and *Mammuthus* are fairly common in the sand, gravel, and silt below the Pearlette ash, the age of the fauna is considered equivalent to that of the Cudahy fauna (Hibbard, 1944, and Paulson, 1961). In most of the exposures containing the Pearlette ash there is silt and clay containing remains of mollusks and microvertebrates. It has not been possible to wash enough of this material to give a representative sample of the small vertebrates that lived at that time.

Following is a discussion of the members of the artiodactyls that have been recovered.

Family Tayassuidae

Platygonus cf. *cumberlandensis* Gidley

(Fig. 1)

Remains of the peccary are not common in the Seymour formation. Parts of 4 lower jaws and a metatarsal have been recovered from the sand and gravel. Three of the jaws and the metatarsal were found on the Burnett ranch north of US Highway 82 and east of Farm Road 267, along the south valley wall of the south fork of the Wichita River. Dalquest recovered the first specimen, a right jaw, UMMP 39364, with $P_{\overline{3}}$–$M_{\overline{3}}$, April 14, 1956. The alveoli of $P_{\overline{2}}$ are present. The alveolar length of $P_{\overline{2}}$–$M_{\overline{3}}$ is 93.0 mm. Melton and Hibbard, during March, 1958, found parts of 2 other jaws in this area. The better jaw (UMMP 39365, Fig. 1), with $P_{\overline{4}}$–$M_{\overline{3}}$, is from an older individual than the above specimen. The jaw is so broken that it is impossible to take the alveolar length of the cheek tooth series. The greatest length of $P_{\overline{4}}$–$M_{\overline{3}}$ is 67.8 mm.; that of the younger specimen is 71.2 mm. The third specimen is a fragment of a left jaw with $P_{\overline{4}}$, UMMP 39369. Part of a right jaw, UMMP 39367, broken just posterior to $M_{\overline{2}}$, of an old individual was taken from a sand and gravel pit west of Farm

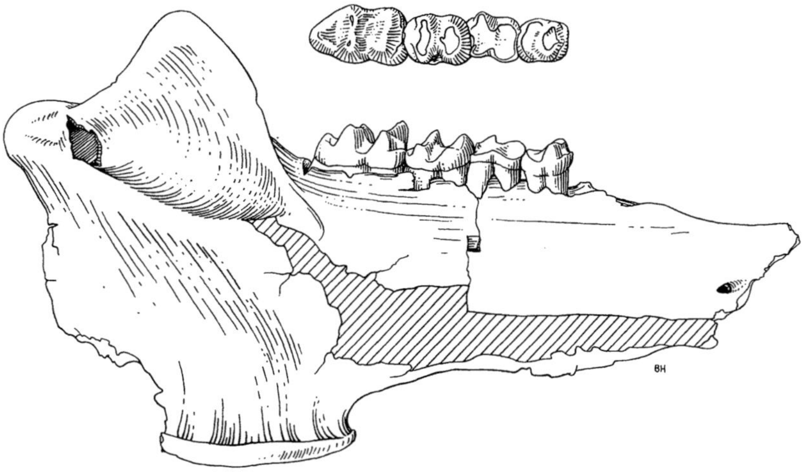

FIG. 1. *Platygonus* cf. *cumberlandensis*, UMMP 39365, right lower jaw, $P_{\overline{4}}$–$M_{\overline{3}}$. Labial and occlusal views. × 0.5.

Road 267 on the Patterson ranch. It contains the alveoli of $P_{\overline{2}}$, and $P_{\overline{3}}$–$M_{\overline{2}}$. The teeth are badly worn. The anterior part of the jaw is missing. The depth of the jaw just posterior to the symphysis is 39.5 mm.

Discussion.—The specimens have been compared with those of *Platygonus bicalcaratus* Cope from the Blanco local fauna. The Blanco specimens differ in having larger teeth and much larger and deeper jaws (see Meade, 1945; Hibbard and Riggs, 1949). The Seymour specimens are distinctly larger than *P. compressus* Le Conte, with a more pronounced and concave angle on the lower jaws. Simpson (1949) gives a good discussion of the variation observed in *P. compressus.*

The specimens have been compared with a topotype of *Platygonus cumberlandensis* Gidley, USNM 8912, from the Cumberland Cave local fauna of Maryland (see Gidley and Gazin, 1938). The lower jaws and dentitions agree in most characters. In the Seymour specimens $M_{\overline{1}}$ is somewhat shorter and more crowded. Although the Cumberland Cave and Seymour peccaries cannot be separated specifically on the basis of the lower jaws, greater differences might be found if complete skulls or skeletons were available.

The Seymour specimens compare in size with *Platygonus pearcei*

Gazin (1938) from the Hagerman local fauna but differ in that they have a greater concave angle to the lower jaw and a deeper excavation on the labial side of the jaw at the base of the coronoid process.

Family Camelidae

Camels are among the common animals in the Gilliland local faunule. Their remains are outnumbered only by those of the horses, glyptodons, and turtles. The fragmentary bones indicate that three major groups of the camels lived in the region. There was a giant form, probably *Gigantocamelus*, as well as *Camelops* and *Tanupolama*.

Genus *Gigantocamelus*? Barbour and Schultz

In the collection is the distal end of a radius, UMMP 33527, with a maximum width of 118.0 mm.; the greatest articular width is 113.0 mm. The distal ends of a number of very large metapodials were observed. These are represented in the collection by UMMP 37832 and 37833. An astragalus, UMMP 37827, has a length of 113.0 mm. and a width of 68.0 mm. Three phalanges, UMMP 37834, 37836, and 37838, may belong to this genus.

The above specimens were found both in the sand and gravel and in the overlying red sandy silt.

Genus *Camelops* Leidy

(Fig. 2*A*)

The distal ends of two radia-ulnae, UMMP 33528 and 37826, were recovered from the sand and gravel. The greatest width of the articular surface at the distal end is 64.0 mm. and 62.3 mm. respectively. The distal end of a tibia-fibula, UMMP 37826, which has a width of 67.0 mm. is approximately the size of the individuals represented by the above radia-ulnae. Also belonging to *Camelops* are two astragali, UMMP 37829 and 37830, and two phalanges UMMP 37835 and 37837.

Dalquest recovered a right maxillary, UMMP 33533, of a large *Camelops* the size of *C. huerfanensis* (Cragin) from the top of the sand and gravel 3¼ miles south of Gilliland. The maxillary contains P_3–M_3 (Fig. 2*A*). The occlusal length of the tooth series is 166.0 mm.

Other bones recovered were parts of vertebrae and scapulae.

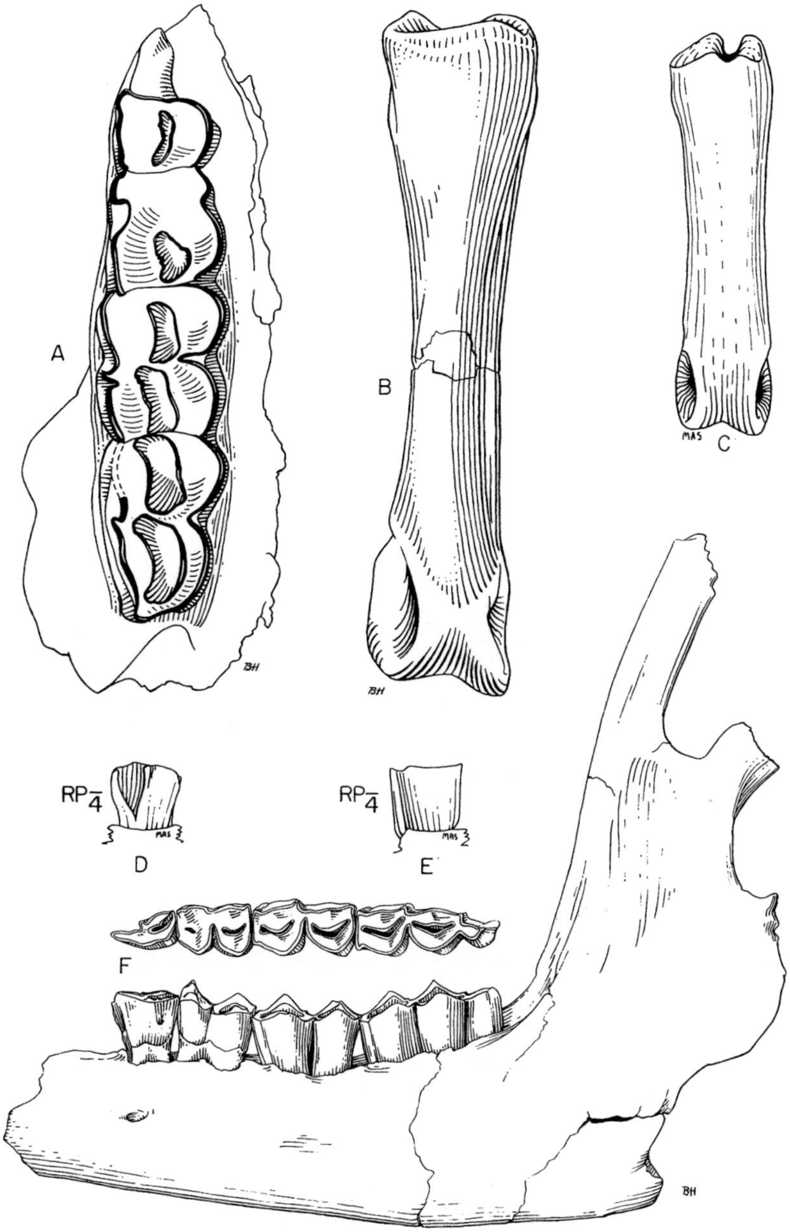

FIG. 2. Pleistocene camels and antilocaprid. *A*, *Camelops* sp., a right maxillary with P^3–M^3, UMMP 33533. Occlusal view. × 0.5. *B*, *Tanupolama* sp., phalange, UMMP 33540. Anterior view. × 1. *C*, *Tetrameryx?*, phalange, UMMP 33539. Anterior view. × 1. *D* and *F*, *Tanupolama* cf. *blancoensis*, UMMP 37824; *D*, lingual view of right P_4 ; *F*, left jaw, P_4–M_3 . Labial and occlusal views. Both × 0.5. *E*, *Tanupolama seymourensis* sp. nov., paratype UMMP 37818, right P_4 . Lingual view. × 0.5.

Genus *Tanupolama* Stock

The small llama-like camels in the Seymour formation are represented by two species. Since one of the species is considered to belong to an undescribed form from the Pleistocene, a brief summary is given of the earlier described species.

Tanupolama macrocephala (Cope)

Holomeniscus macrocephalus Cope, 1893, Geol. Surv. Texas, 4th Ann. Rep. (1892): 85–86, pl. 23, figs. 5, 5a.

Camelus americanus Wortman, 1898, Bull. Amer. Mus. Nat. Hist., 10: 133–134, fig. 21.

Lama stevensi Merriam and Stock, 1925, Carnegie Inst. Wash., Publ. No. 347: 39–42, figs. 1–4.

Tanupolama stevensi (Merriam and Stock), Stock, 1928, Carnegie Inst. Wash., Publ. No. 393: 29–37.

When Cope described *Holomeniscus macrocephalus* from a nearly complete lower jaw (No. 18621, University of Texas, Vertebrate Paleontological Collections), he noted the antero-external and the antero-internal styles on the anterior face of M_2 and M_3. Cope remarked that "*Holomeniscus macrocephalus* approaches more closely the existing lama than any other extinct species of North America."

Wortman (1898) assigned *H. macrocephalus* to *Camelops vitakerianus* Cope at the time he described *Camelus americanus*; the two specimens, therefore, were not compared. Wortman based his decision upon the figures published by Cope, which did not show the antero-external buttress (style) noted by Cope in his text.

Merriam and Stock, at the time they described *Lama stevensi*, compared it with *Camelops hesternus* (Leidy) and the recent llama. When Stock (1928) proposed the genus *Tanupolama*, he recognized the close similarity between *Tanupolama stevensi* and *T. americana* but considered them distinct, since the anterior border of the symphysis of *T. stevensi* does not extend as abruptly upward from the lower border of the ramus as it does in *T. americana*.

We have observed no characters that will separate the three species, and consider the differences to be individual or sexual variation.

The holotype of *Tanupolama macrocephala* lacks the lower canine, a variable character of specimens previously assigned to *T. stevensi*. P_3 may be present or absent.

TABLE I

MEASUREMENTS (IN MM.) OF LOWER DENTITIONS OF HOLOTYPES

	*T. stevensi**	*T. macrocephala*	*T. americana*
$I_{\bar{1}}$, transverse width	9.5	10.0	10.5
$I_{\bar{2}}$, transverse width	8.7	11.4	10.5
$I_{\bar{3}}$, transverse width	8.5	8.7	9.7
$P_{\bar{4}}$, greatest anteroposterior length	14.7	14.5	15.8
$P_{\bar{4}}$, greatest transverse width	9.2	11.0	8.7
$M_{\bar{1}}$, greatest anteroposterior length	18.8	20.0	18.0
$M_{\bar{1}}$, greatest transverse width	13.1	15.5	14.0
$M_{\bar{2}}$, greatest anteroposterior length	26.2	25.0	23.0
$M_{\bar{2}}$, greatest transverse width	17.0	18.5	17.0
$M_{\bar{3}}$, greatest anteroposterior length	36.6	34.0	34.4
$M_{\bar{3}}$, greatest transverse width	16.1	16.5	14.6
Length of symphysis	63.0	71.0	73.5
Greatest depth below anterior border of $P_{\bar{4}}$	27.0	32.0	40.5
Greatest depth below posterior border of $M_{\bar{3}}$	51.0	52.0	—
Distance from posterior side of $P_{\bar{1}}$ to anterior side of $P_{\bar{4}}$	57.0	61.0	61.5
Length from anterior side of $P_{\bar{4}}$ to posterior side of $M_{\bar{3}}$	95.3	92.4	91.0
Length from anterior side of $P_{\bar{1}}$ to posterior side of $M_{\bar{3}}$	161.0	161.5	160.0

* After Merriam and Stock, 1925.

Tanupolama mirifica Simpson

Tanupolama mirifica Simpson, 1929, Bull. Amer. Mus. Nat. Hist., 56: 593–596, 3 figs.

The specimens studied by Simpson are mostly isolated teeth. The holotype is an associated left $M_{\bar{1}}$–$M_{\bar{3}}$. Measurements made from a figure (17A) published by Simpson show a slightly greater anteroposterior length for $M_{\bar{1}}$–$M_{\bar{3}}$ (82.0 mm.) than that of *Tanupolama macrocephala.* The lower jaw (UF 1093) figured by Bader (1957, p. 60) has an anteroposterior length of 77.0 mm. for $M_{\bar{1}}$–$M_{\bar{3}}$. Bader, after studying material collected in Florida since 1929, concluded that most of the teeth agreed well with those of *T. mirifica.*

The few specimens of *Tanupolama mirifica* that we have seen possess $P_{\bar{3}}$'s with two roots, but as previously stated, $P_{\bar{3}}$ may be absent or present. We have been permitted to study a lower jaw (UF

2833) in the University of Florida collection recovered since Bader's study. It contains $RI_{\bar{1}}$ and both the right and left $P_{\bar{3}}$–$M_{\bar{3}}$. The jaw is that of an old adult taken from the Reddick locality, Marion County, Florida. We have failed to find characters in the $P_{\bar{3}}$–$M_{\bar{3}}$ series that would readily distinguish *T. mirifica* from *T. macrocephala*, but if the Reddick specimen is typical of *T. mirifica*, that species is readily distinguished from *T. macrocephala* by the much longer diastemal region. The specimen (UF 2833) measures 99.0 mm. from the posterior border of the alveolus of $P_{\bar{1}}$ to the anterior border of $P_{\bar{4}}$. In the holotype of *T. macrocephala* this distance is 61.0 mm. The distance from the tip of the jaw (between the first incisors) to the anterior border of $P_{\bar{4}}$ is approximately 149.0 mm. In *T. macrocephala* it is 112.0 mm., and in the holotype of *T. americana*, 115.0 mm. *T. mirifica* appears to be a much longer jawed llama. Other isolated teeth recovered in Florida, questionably referred by Simpson (1929) and Bader (1957) to *T. americana*, indicate that a much larger *Tanupolama* also occurs in the fauna. It is too large to be assigned to *T. macrocephala.*

Comments on other described species of Tanupolama.—*Tanupolama parva* Olson (1940) is known only from an $M^{\underline{2}}$ from a late Pleistocene fissure filling near Herculaneum, Missouri. Olson states it is "About one-fourth smaller than previously described species of the genus."

Gazin (1942, p. 515) assigned *Procamelus longurio* described by Hay (1921) from a Pleistocene fissure deposit at Anita, Arizona, to the genus *Tanupolama. T. longurio* (Hay) is represented by a cervical vertebra and foot bones. It is impossible to assign jaw or skull elements with certainty to this species. Gazin (1942) questionably referred a lower jaw from the Curtis Ranch local fauna to *T. longurio.* Savage (1951, p. 267) thought that Hay's cannon bone offers no traits worthy of a specific diagnosis, and that *T. longurio* should be considered indeterminate.

Hay and Cook (1930) described *Lama? hollomani* from the Holloman gravel pit near Frederick, Oklahoma. The holotype is parts of a right cannon bone. Meade (1953) gives a good description of the specimen recovered from the Holloman gravel pit, and assigns the *Tanupolama* material to *T. hollomani* (Hay and Cook). The anterior part of the right jaw (Texas Memorial Museum 934–47) shows that it belonged to a short-jawed form the size of *T. macrocephala.* The unworn $M_{\bar{3}}$ in the fragment of jaw has an anteroposterior length, at the alveolar border, of 33.0 mm. The tooth compares in size with those assigned to *T. macrocephala*. It is smaller than the specimens from the Seymour formation of Knox County, Texas, and the specimen described by Gazin from the Curtis Ranch local fauna.

Tanupolama seymourensis, sp. nov.

(Figs. 3, 2*E*)

Holotype.—UMMP 39379, part of a right lower jaw with $P_{\overline{3}}$–$M_{\overline{3}}$. Collected by Walter W. Dalquest (Field No. 786) March 30, 1957. Paratype: UMMP No. 37818, part of left jaw with $P_{\overline{4}}$–$M_{\overline{3}}$.

Horizon and type locality.—From the red sandy silt just above the sand and gravel, Seymour formation, Gilliland local faunule of late Kansan (2nd glacial) age. Taken on the east side of Farm Road 267 and north of US Highway 82, along the south valley wall of the South Wichita River, Bruce Burnett ranch (approximately 6 miles southeast of Gilliland, Knox County, Texas).

Diagnosis.—Larger than *Tanupolama macrocephala* but approximately the size of *T. blancoensis* Meade, with higher crowned teeth. $P_{\overline{4}}$ large and slightly rounded on the labial side. There is no groove on the posterolabial part of the tooth. There is no ridge present on the posterolingual border of the tooth (Figs. 2*E* and 3).

Description of holotype.—The jaw is that of a young adult and lacks the anterior and posterior parts. The $P_{\overline{3}}$ is two-rooted and larger than in *T. macrocephala*. The greatest anteroposterior length is 12.5 mm. and the greatest width is 6.0 mm. $P_{\overline{4}}$ is large and slightly rounded on the labial side, with the anterior edge deflected lingually. A long enamel lake is present on the posterior half of the crown. The posterior face of the tooth is wedged into the anterior face of $M_{\overline{1}}$. The first molar has a poorly developed anteroexternal style. There is no evidence of an anterointernal style. The $M_{\overline{2}}$ and $M_{\overline{3}}$ have well-developed anterointernal and anteroexternal styles (Fig. 3). The lingual part of the second lobe extends posterior to the junction of the third lobe. This condition would disappear with wear. The lobes overlap on the lingual side for a distance of 4.0 mm. With occlusal wear the third lobe would increase in size but still remain on the same plane as the second lobe, for it is not separated by a groove, or shifted labially as in *Tanupolama blancoensis* or *T. macrocephala*.

The paratype, UMMP 37818, was taken just west of Farm Road

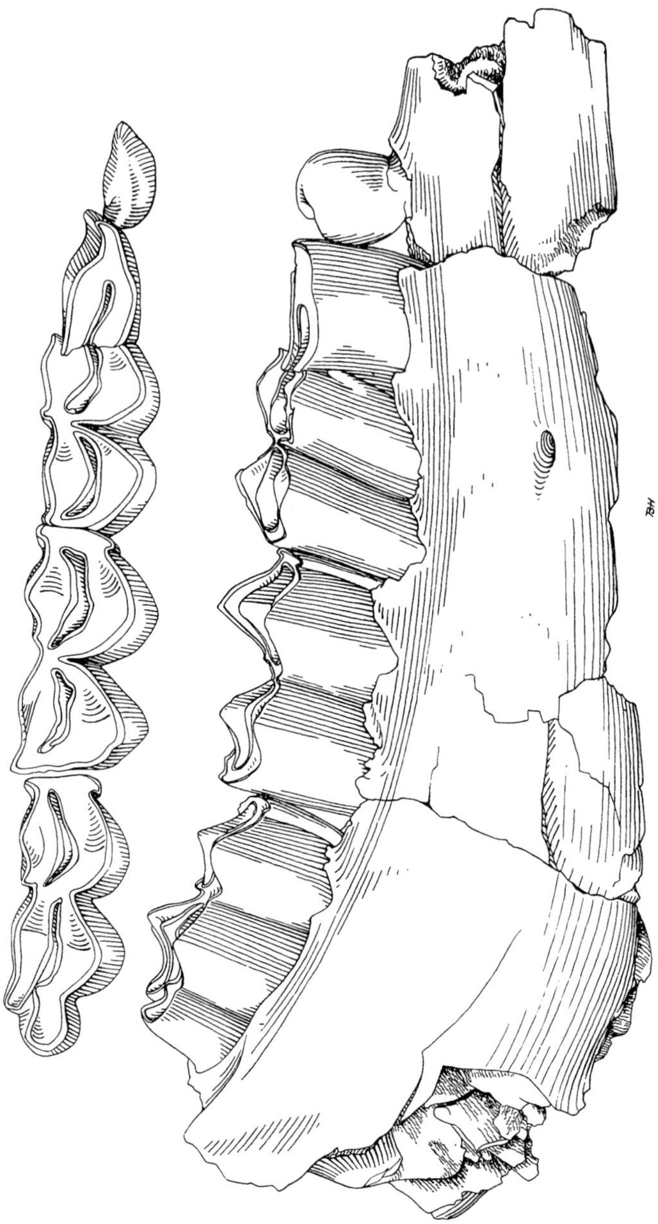

FIG. 3. *Tanupolama seymourensis*, sp. nov. holotype, UMMP 39379, part of left lower jaw, $P_{\overline{3}}$–$M_{\overline{3}}$. Labial and occlusal views. × 1.

TABLE II

MEASUREMENTS (IN MM.) OF *Tanupolama seymourensis* SP. NOV.

	Holotype No. 37379	Paratype No. 37818
$P_{\bar{4}}$ greatest anteroposterior length.......	17.8	17.9
$P_{\bar{4}}$ greatest width.......................	9.8	9.5
$M_{\bar{1}}$ greatest anteroposterior length.......	24.7	25.2
$M_{\bar{1}}$ greatest width.......................	16.0	16.4
$M_{\bar{2}}$ greatest anteroposterior length.......	32.3	31.4
$M_{\bar{2}}$ greatest width.......................	16.8	—
$M_{\bar{3}}$ greatest anteroposterior length.......	39.0	39.5
$M_{\bar{3}}$ greatest width.......................	15.2	15.0
$P_{\bar{4}}$–$M_{\bar{3}}$ occlusal length....................	106.7	107.5
$P_{\bar{4}}$–$M_{\bar{3}}$ alveolar length....................	112.6	114.0

267 on the O. L. Patterson ranch by Dalquest (Field No. 295), April 2, 1956, from the red sandy silts just above the sand and gravel. It is part of a left jaw lacking both the anterior and posterior parts. Alveoli of $P_{\bar{3}}$ are present and contain the two roots of that tooth. The dentition is that of a young adult llama-like camel and agrees in all details with that of the holotype, except that the posterolingual border of $M_{\bar{3}}$ is broken and it is not possible to note whether the occlusal surface of the second lobe extended posteriorly past the anterior part of the third lobe.

Tanupolama cf. *blancoensis* Meade

(Figs. 2*D* and *F*)

Two associated lower jaws, UMMP 37824, with the symphysial region missing were taken by Melton and Hibbard, March 28, 1958, from the sand and gravel near the base of the Seymour formation on the Burnett ranch just east of the type locality of *Tanupolama seymourensis*. This llama-like camel has teeth with shorter crowns and molars with more distinct lobes than those of *T. seymourensis* and *T. macrocephala*. The teeth are similar in structure to those of *T. blancoensis*. The $P_{\bar{3}}$ is lacking and there is no evidence that it was ever present. The $P_{\bar{4}}$ is bilobed, with a groove along the labial side just opposite the anterior end of the enamel lake (Figs. 2*D* and *F*). The labial sulcus is triangular in shape with the apex at the base of the crown. The deep, broad sulcus makes the anterior part of $P_{\bar{4}}$

blade-like in a dorsal view. The greatest anteroposterior length of $P_{\overline{4}}$ is 17.0 mm., the greatest width is 9.6 mm. The anteroposterior length of $P_{\overline{4}}$–$M_{\overline{3}}$ is 104.0 mm. There is no indication of an anteroexternal style on $M_{\overline{1}}$, although there is a slight anterointernal style. The anterior lobes of $M_{\overline{2}}$ and $M_{\overline{3}}$ have well-developed posterointernal ridges, which give the anterior lobes of these teeth a more lingual position than the posterior lobes. This is a character of *T. blancoensis*. The third lobe is more labially placed than the third lobe of *T. seymourensis* and *T. macrocephala*.

The specimen is certainly from a younger deposit than *T. blancoensis*, but it is not as advanced in tooth structure as *T. seymourensis*. Until more material is available for study it seems best to refer it to the earlier species.

Discussion.—Dalquest recovered part of the symphysial region, UMMP 37819, of a *Tanupolama* from the sand and gravel. If it belongs to the same species as the above, it was a short-jawed llama-like camel of the *T. macrocephala* type. The specimen is broken at the posterior edge of the alveolus of $P_{\overline{1}}$ and anterior to the large foramen, which is slightly posterior and ventral to $P_{\overline{1}}$. The right part of the ramus extends to the posterior alveoli of $P_{\overline{3}}$ or $P_{\overline{4}}$. The tooth was two-rooted. The distance from the anterior alveoli of $P_{\overline{3}}$ or $P_{\overline{4}}$ to the posterior border of the foramen ventral to $P_{\overline{1}}$ is 61.0 mm. It is possible that this specimen belongs to *T. seymourensis*.

The distal end of a tibia, UMMP 37821, and the proximal end of a metapodial, UMMP 37823, of a *Tanupolama* were recovered from the sand and gravel. A toe bone, UMMP 33540 (Fig. 2*B*), with a length of 92.0 mm. was taken from the red, sandy silts above the sand and gravel. Fragments of other foot and limb bones were observed.

Family Cervidae

A few scattered fragments of a small deer have been found chiefly in the outcrops of the sand and gravel from south of Gilliland eastward to one mile north of Vera, Knox County, Texas.

Odocoileus sp.

(Figs. 4*C* and *F*)

Material.—Parts of three antlers, UMMP 37908, 37909, and 37911; and the distal end of a metapodial, UMMP 39362 (Fig. 4*C*), which was taken one mile north of Vera from the red sandy silt just above the sand and gravel. A distal end of a humerus, UMMP 37912, was found on the Bruce Burnett ranch just east of Farm Road 267, and part of a left lower jaw, UMMP 37910, was taken two miles north of Vera on the Strudth ranch.

Remarks.—The material is from a small species of *Odocoileus* and compares well with the white-tail deer. The jaw (Fig. 4*F*) contains $P_{\overline{4}}$–$M_{\overline{2}}$ and the alveoli for $P_{\overline{2}}$ and $P_{\overline{3}}$. The teeth are the size of those of *O. virginianus osceola* (Bangs) of Florida. The occlusal length of $P_{\overline{4}}$–$M_{\overline{3}}$ is 57.5 mm. The lower jaw differs from that of the Recent deer in having a greater depth. The depth of the jaw taken between $M_{\overline{2}}$ and $M_{\overline{3}}$ on the labial side is 25.5 mm. The same measurement of a jaw, UMMZ 97529, with teeth of a comparable stage of wear is 21.7 mm.

At least eight species of Pleistocene deer have been described from North America. Because of the wide range in variation in the Recent species, a critical study of the fossil material is needed before the relationships of the Recent and fossil deer can be determined.

Family Antilocapridae

The genera *Breameryx*, *Capromeryx*, *Ceratomeryx*, *Hayoceros*, *Stockoceros*, and *Tetrameryx* have been described from the Pleistocene of North America. Remains of the Recent genus *Antilocapra* have also been reported. Generic differences are based upon the form of the horn-cores. Since our knowledge of the rest of the skeleton of most of the proposed genera is inadequate, it is nearly impossible to assign isolated teeth and individual skeletal parts to more than subfamily.

Skinner (1942) gives an excellent summary of some of the genera and a detailed account of the wide range of variation observed in *Stockoceros* and *Antilocapra*.

Capromeryx sp.

(Figs. 4*A* and *D*)

Among the fossils recovered are parts of a small pronghorn the size of specimens of *Capromeryx furcifer* Matthew, reported from the Cragin Quarry local fauna (Hibbard and Taylor, 1960).

A broken right lower jaw, UMMP 38103, with $M_{\overline{1}}$–$M_{\overline{3}}$ (Fig. 4*A*)

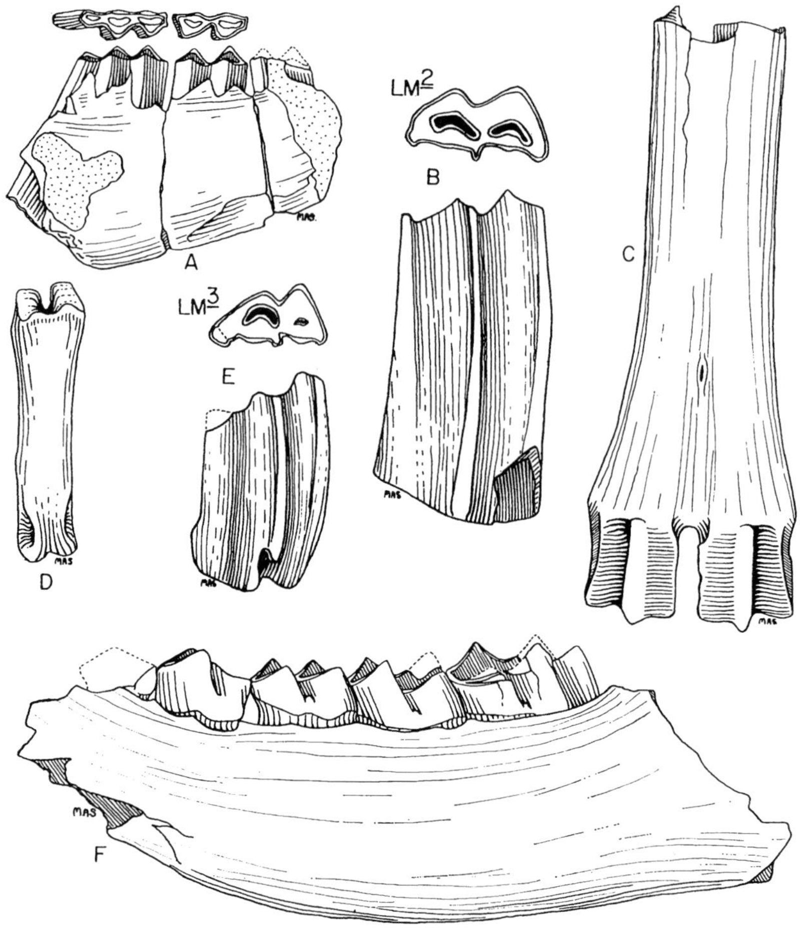

FIG. 4. Pleistocene antilocaprids and deer. *A* and *D*, *Capromeryx* sp. *A*, part of the right lower jaw, UMMP 38130, $M_{\bar{1}}$–$M_{\bar{3}}$. Labial and occlusal views. *D*, phalange, UMMP 38105. Anterior view. *B* and *E*, *Tetrameryx?* sp. *B*, left M^2, UMMP 38102; *E*, left M^3, UMMP 39360. Labial and occlusal views. *C* and *F*, *Odocoileus* sp. *C*, distal end of metapodial, UMMP 39362. Anterior view. *F*, part of left lower jaw with $P_{\bar{3}}$–$M_{\bar{3}}$, UMMP 37910. Labial view. All × 1.

was taken by Dalquest three miles northwest of Red Springs, Baylor County, Texas, from the red sandy silt of the Seymour formation along the south valley wall of the South Wichita River. The occlusal length of M_3 is 13.0 mm. and the greatest width is 4.8 mm.

Other specimens were recovered in Knox County, chiefly on the ranch of Bruce Burnett, east of Farm Road 267 and just north of US Highway 82. These are a first phalange, UMMP 38105, the distal end of a metapodial and two phalanges, UMMP 39361, and two isolated molars, UMMP 38104 and 39363. The first phalange (Fig. 4*D*) has a length of 37.0 mm. The greatest width of the proximal end is 9.0 mm.

Tetrameryx? *knoxensis* Hibbard and Dalquest

This large antilocaprid is known only from the horn-cores of the type specimen, which was recovered from the sand and gravel of the Seymour formation 3½ miles south by east of Gilliland from an exposure along the north valley wall of the South Wichita River (Hibbard and Dalquest, 1960).

In the collection are two isolated upper molars (Figs. 4B and E) taken on the Bruce Burnett ranch. These teeth are larger than those of *Capromeryx* and may belong to this larger pronghorn. The left $M\frac{2}{}$, UMMP 38102 (Fig. 4*B*), has a length of 43.5 mm. The greatest occlusal width is 10.0 mm. The other tooth, a left $M\frac{3}{}$ (Fig. 4*E*) is from an older individual. A first phalange, UMMP 33539 (Fig. 2*C*), is much larger than the one referred to *Capromeryx*. Its length is 52.5 mm. The greatest width of the proximal end is 13.5 mm.

Acknowledgments

We wish to express our gratitude to William H. Burt and Emmet T. Hooper of the Division of Mammals of the University of Michigan Museum of Zoology for permitting us free access to the collection of Recent mammal skeletons.

We also wish to express our appreciation for permission to study specimens under the care of Malcolm C. McKenna, American Museum of Natural History; Clayton E. Ray, Department of Biology, University of Florida; John A. Wilson, Department of Geology, University of Texas; and C. Lewis Gazin, U. S. National Museum.

The field parties are grateful to the landowners who gave us permission to work on their ranches, and especially to Bruce Burnett and O. L. Patterson for allowing us to remove fossil-bearing matrix for washing.

The financial support accorded the senior author by the National Science Foundation, Project G-5635, has made this study possible.

The artists were Bonnie Hall (B.H.) and Margaret Anne Skeels (M.A.S.).

LITERATURE CITED

BADER, ROBERT S. 1957. Two Pleistocene Mammalian Faunas from Alachua County, Florida. Bull. Florida State Mus., 2 (5): 53–75, 6 figs.

COPE, E. D. 1893. A Preliminary Report on the Vertebrate Paleontology of the Llano Estacado. Geol. Surv. Texas, 4th Ann. Rept. (1892): 1–87, 23 pls., 3 figs.

GAZIN, C. LEWIS. 1938. Fossil Peccary Remains from the Upper Pliocene of Idaho. Jour. Wash. Acad. Sci., 28 (2): 41–49, 3 figs.

——. 1942. The Late Cenozoic Vertebrate Faunas from the San Pedro Valley, Arizona. Proc. U. S. Nat. Mus., 92 (No. 3155): 475–518, 2 pls., 9 figs.

GIDLEY, JAMES W., AND C. LEWIS GAZIN. 1938. The Pleistocene Vertebrate Fauna from Cumberland Cave, Maryland. Bull. U. S. Nat. Mus., 171: 1–99, 10 pls., 50 figs.

HAY, O. P. 1921. Description of Species of Pleistocene Vertebrata, Types or Specimens of Most of Which Are Preserved in the United States National Museum. Proc. U. S. Nat. Mus., 59: 599–642, 9 pls.

——, AND HAROLD J. COOK. 1930. Fossil Vertebrates Collected Near, or in Association with Human Artifacts at Localities near Colorado, Texas; Frederick, Oklahoma; and Folsom, New Mexico. Proc. Colo. Mus. Nat. Hist., 9(2): 1–37, 14 pls.

HIBBARD, CLAUDE W. 1944. Stratigraphy and Vertebrate Paleontology of Pleistocene Deposits of Southwestern Kansas. Bull. Geol. Soc. Am., 55 (6): 707–754, 20 figs.

——. 1951. Vertebrate Fossils from the Pleistocene Stump Arroyo Member, Meade County, Kansas. Univ. Mich. Mus. Pal., Contrib., 9 (7): 227–245, 6 pls., 1 fig.

——. 1953. *Equus (Asinus) calobatus* Troxell and Associated Vertebrates from the Pleistocene of Kansas. Trans. Kans. Acad. Sci., 56 (1): 111–126, 3 figs.

——. 1958. New Stratigraphic Names for Early Pleistocene Deposits in Southwestern Kansas. Am. Jour. Sci., 256: 54–59, 1 fig.

——, AND WALTER W. DALQUEST. 1960. A New Antilocaprid from the Pleistocene of Knox County, Texas. Jour. Mammal., 41 (1): 20–23, 1 fig.

——, AND ELMER S. RIGGS. 1949. Upper Pliocene Vertebrates from Keefe Canyon, Meade County, Kansas. Bull. Geol. Soc. Am., 60 (5): 829–860, 5 pls., 11 figs.

——, AND DWIGHT W. TAYLOR. 1960. Two Late Pleistocene Faunas from Southwestern Kansas. Univ. Mich. Mus. Pal., Contrib., 16 (1): 1–223, 16 pls., 18 figs.

MEADE, GRAYSON E. 1945. The Blanco Fauna. Univ. Texas Publ. 4401: 509–556, 8 pls., 4 figs.

——. 1953. An Early Pleistocene Vertebrate Fauna from Frederick, Oklahoma. Jour. Geol., 61 (5): 452–460, 1 pl.

Merriam, J. C., and Chester Stock. 1925. A Llama from the Pleistocene of McKittrick, California. Carnegie Inst. Wash. Publ. No. 347: 37–42, 4 figs.

Olson, Everett C. 1940. A Late Pleistocene Fauna from Herculaneum. Missouri Jour. Geol., 48 (1): 32–56, 1 pl., 10 figs.

Paulson, Gerald R. 1961. The Mammals of the Cudahy Fauna. PMASAL, 46: 127–153, 8 figs.

Savage, Donald E. 1951. Late Cenozoic Vertebrates of the San Francisco Bay Region. Univ. Calif. Publ., Bull. Geol. Sci., 28 (10): 215–314, 51 figs.

Simpson, George G. 1929. Pleistocene Mammalian Fauna of the Seminole Field, Pinellas County, Florida. Bull. Am. Mus. Nat. Hist., 56 (Art. 8): 561–599, 22 figs.

——. 1949. A Fossil Deposit in a Cave in St. Louis. Am. Mus. Nat. Hist., Novitates, No. 1408: 1–46, 10 figs.

Skinner, Morris F. 1942. The Fauna of Papago Springs Cave, Arizona, and a Study of *Stockoceros*; with Three New Antilocaprines from Nebraska and Arizona. Bull. Am. Mus. Nat. Hist., 80 (Art. 6): 143–220, 19 figs.

Stock, Chester. 1928. Tanupolama, a New Genus of Llama from the Pleistocene of California. Carnegie Inst. Wash. Publ., No. 393: 29–37, 6 pls.

Stricklin, Jr., Fred L. 1961. Degradational Stream Deposits of the Brazos River, Central Texas. Bull. Geol. Soc. Am., 72 (1): 19–35, 9 figs.

Tihen, J. A. 1960. Notes on Late Cenozoic Hylid and Leptodactylid Frogs from Kansas, Oklahoma, and Texas. Southwestern Naturalist, 5 (2): 66–70, 6 figs.

Wortman, J. L. 1898. The Extinct Camelidae of North America and Some Associated Forms. Bull. Am. Mus. Nat. Hist., 10 (7): 93–142, 1 pl., 23 figs.

Papers of the Michigan Academy of Science, Arts, and Letters
Vol. XLVII, 1962 (1961 Meeting)

THE MASTODONS AND MAMMOTHS OF MICHIGAN

MARGARET ANNE SKEELS

The University of Michigan

Introduction

Fossil mammals are known only from the Late Pleistocene of Michigan. If earlier Cenozoic deposits ever were present in Michigan they were destroyed by glacial ice that four times overrode the state. The American mastodon and Jefferson mammoth were the largest Pleistocene animals to walk on Michigan soil. Most of the information concerning these large mammals in the state must be sought in newspaper reports of individual finds. However, O. P. Hay (1923) reported upon all Michigan Pleistocene mastodon and mammoth remains known to him. Case and others (1935) reported the occurrences of the Jefferson mammoth from Glacial Lake Mogodore in Cass County. Later in the same year, Case and Stanley (1935) reported on the Bloomfield Hills mastodon. Archie MacAlpin (1940) summarized all known finds of the American mastodon in Michigan but did not publish on the mammoths. The purpose of this paper is to place on record all known occurrences of both the American mastodon and Jefferson mammoth in the state, including those discovered since 1940, and to make available to the citizens of Michigan an account of these interesting mammals.

Most of the fossil remains of these animals found in the state have occurred in swamps and bogs. Such areas owe their origin to the manner in which the last glacier melted; the majority of the lake depressions were formed at the time of the retreat of the last ice sheet. As the ice front retreated large blocks of ice were left behind, buried under the glacial outwash sediments. For a time such ice blocks were protected from the warmth of the sun by the gravel, sand, and silt that covered them. Eventually, however, these ice remnants melted, leaving depressions in their place. Although most of these were of small size, others were quite large. Geologists refer to the smaller ones as kettles or kettle holes. If silt or clay washed into these depressions it sealed them, and water would then accumulate to form a pond or lake. In most cases such depressions contained water only in the spring, but if they went below the water table a permanent pond or lake resulted. Over the years such ponds and lakes continued to receive sand, silt, and clay from flood waters, and additional material from vegetation growing around their edges. In time at least some of these bodies of water, particularly the shallower ones, became filled to levels where aquatic and semiaquatic vegetation could encroach upon the open water. Thus a bog was formed where an open pond or lake had once existed. Wherever water was shallow, plant material extended down through the water to the bottom, but where the water deepened the vegetation often formed a mat that floated over part of the water surface. Partially or completely filled depressions of this sort exist in Michigan today and are called "quaking bogs." The process of filling, however, is one that still continues and depressions in all stages can be found in the state. Floating mats of the nature described are unsafe; a heavy animal venturing too far out on one often breaks through. Many times such an animal is unable to climb back onto the mat, becomes mired, and eventually drowns. These bogs apparently were places of retreat for mastodons and some mammoths during the winter months. They would have been protected from strong winds, and the surrounding forest would have offered abundant food. But the deep muck and peat and the floating mats of vegetation, though partly frozen, were potential traps.

Most mastodon remains in Southern Michigan are associated with muck, peat, and marl. Chances for preservation are good in bogs because there is immediate burial. The body of an animal that dies on high ground is seldom preserved because predators and scavengers scatter the bones. Such exposed bone usually becomes decayed or badly weathered before it can be carried downslope to a lake or stream to be preserved in the sediments deposited there. Bone that comes from a bog is usually very dark colored, dark brown or black, due to the acids produced by the decaying organic matter.

Because of their larger size, mastodons are the mammals most commonly found fossil in Pleistocene deposits of Michigan. Smaller animals such as elk, deer, wolves, and beavers, are probably much more abundant in bog deposits, but because of their smaller size their remains may be overlooked or considered those of recent forms. Moreover, these smaller bones often look very much like the pieces of branches or roots that are sometimes present in the muck or peat. Recovery of the smaller Pleistocene mammals of Michigan would do much to increase our understanding of these Ice Age forms.

The American Mastodon

The ancestors of the American mastodon reached North America from Eurasia via the Bering land bridge during Miocene time, approximately 18 million years ago. These early proboscideans (animals with a trunk) were subtropical forms, not the cold-loving animals that some of their descendants became. During the Ice Age (Pleistocene) the American mastodon gradually became adjusted to cooler conditions, so that by the time of the last of the four glaciations they were well adjusted to life in cold northern (boreal) forests.

Fossil remains indicate that only two species of the Proboscidea lived in Michigan during the Late Pleistocene. These were the American mastodon (*Mammut americanum*) and the Jefferson mammoth (*Mammuthus jeffersoni*). These two forms are usually known, collectively, as "mastodons" or "elephants" to the uninformed person. In reality they represent two very different types of animals. The similarities between them are large size, columnar (graviportal or pillar-like) limbs, a proboscis (trunk), and large tusks (specialized incisor teeth). Important differences occur in skull shape and tooth structure, as explained below. Moreover, they lived in quite different habitats.

It is rather unfortunate that the two generic names, *Mammut* and *Mammuthus*, are so similar. Both words stem from the barbaric word *mammut*, which means "earth-burrower" (Lucas, 1935, p. 157). The term is believed to have originated during the Middle Ages when the Eastern European farmers assumed that these beasts were large burrowing monsters because their bones were commonly found buried in the soil. The peasants had never seen a living elephant.

Cuvier's term *Mastodon* has frequently been used as a generic name for the American mastodon, and "*Mastodon americanus*" is still used in a loose sense today because it is so prevalent in the literature (Jepsen, 1960). However, Blumenbach's (1799) name *Mammut* is the earlier and accepted generic name.

Following is the classification of the Late Pleistocene mastodon of Michigan.

ORDER PROBOSCIDEA

FAMILY MAMMUTHIDAE

Genus *Mammut* Blumenbach, 1799

Mammut americanum (Kerr). American mastodon

Mastodon giganteus Cuvier. Winchell, 1861, Geol. Surv. Mich., First Biennial Rept. (1860), p. 132.

Mammut americanum (Kerr). Hay, 1930, Carnegie Inst. Wash. Publ. 390, 2: 624.

Mastodon americanus (Kerr). MacAlpin, 1940, PMASAL, 25 (1939): 483.

The foregoing are synonymous names, names that have been used in past years for the American mastodon in Michigan.

Characters of Mammut americanum.—The tusks are without enamel. The upper tusks are large and round with slightly indicated annular growth rings (Osborn, 1936). The lower tusks, when present, are small, straight, and circular in cross section (Pl. II, Fig. 3), but they usually were lost by mature individuals. One specimen (UMMP 37811) in the University of Michigan Museum of Paleontology has the two lower tusks present one in each of the lower jaws. The maximum length of the left lower tusk is 241.0 mm.; the maximum transverse diameter, at the base of the tooth, is 47.0 mm. The vertical diameter, at the base of the tooth, is 40.05 mm. The diameter at the tip, which is blunt and polished, is 27.0 mm. The depth of the alveolus (socket) is 170.5 mm. The tusks in mastodons are the upper and lower second incisors; the first and third incisors in each jaw failed to develop.

Tusks of the mastodon and the mammoth consist of ivory, which is a complex type of dentine. Dentine is not so hard as enamel and therefore does not lend itself as well to fossilization. Well-preserved tusks of mastodons or mammoths have not been found in Michigan. Tusks when found are usually very friable so that the layers of ivory split and break off in small pieces.

One of the differences between the American mastodon and the mammoth is the way the tusks leave the skull. In the mastodon they project more or less horizontally from the skull, curving at first outward and later inward (Pl. I, Fig. 2) in the course of growth, whereas in the mammoth the tusks leave the skull more nearly vertically, curve downward and outward, then begin the curve that brings their tips inward toward each other (Pl. I, Fig. 1).

The size of the tusks in both mastodons and mammoths differs between the sexes. The male individuals of both species possessed heavier, larger, and longer tusks.

Mastodons apparently used their tusks to pry off and possibly to break into small pieces the branches upon which they fed, in much the same way that the living African elephant will rip a limb from a tree, stand on it with both feet, and break it in half with one of its tusks. Only one tusk is used in this operation, never both. Elephants and mastodons were thus "right" or "left-handed." When both of the tusks are recovered with the skull of a mastodon, one of them is usually shorter than the other and shows breakage and polish at the tip (Pl. I, Fig. 2). This has often been overlooked in mounted specimens; tusks have frequently been restored equal in length.

The cheek teeth of the American mastodon consist of six on each side of each jaw. The first three teeth of the six are known as milk or deciduous premolars and the last or hinder three as permanent molars. The individual teeth, exclusive of the tusks (incisors), are designated in the upper jaw as $Dp^{\underline{2}}$, $Dp^{\underline{3}}$, $Dp^{\underline{4}}$, $M^{\underline{1}}$, $M^{\underline{2}}$, and $M^{\underline{3}}$. The symbol "Dp" stands for deciduous premolar and "M" for permanent molar. The lower individual teeth are designated as $Dp_{\overline{2}}$, $Dp_{\overline{3}}$, $Dp_{\overline{4}}$, $M_{\overline{1}}$, $M_{\overline{2}}$, and $M_{\overline{3}}$. In a reference to both the upper and lower last molar, for example, the symbol would be written as $M\frac{3}{3}$. In the American mastodon $Dp\frac{4}{4}$, $M\frac{1}{1}$, and $M\frac{2}{2}$ each consist of three transverse ridge crests, or lophs, which are formed by two cusps. $M\frac{3}{3}$ has four lophs and a vestigial heel behind the last ridge crest (Fig. 1B; Pl. II, Figs. 1 and 2). In some individuals, there are five lophs across the third molar. Cuvier's term, mastodon, refers to the nature of these cusps. The word when broken down into its Greek roots means "nipple tooth" (Palmer, 1904, p. 401).

The teeth increase in size from the fourth premolar to the last molar. The size of each tooth varies somewhat in different individuals, as well as in the same individual. See Table I for measurements of American mastodon teeth in The University of Michigan Museum of Paleontology.

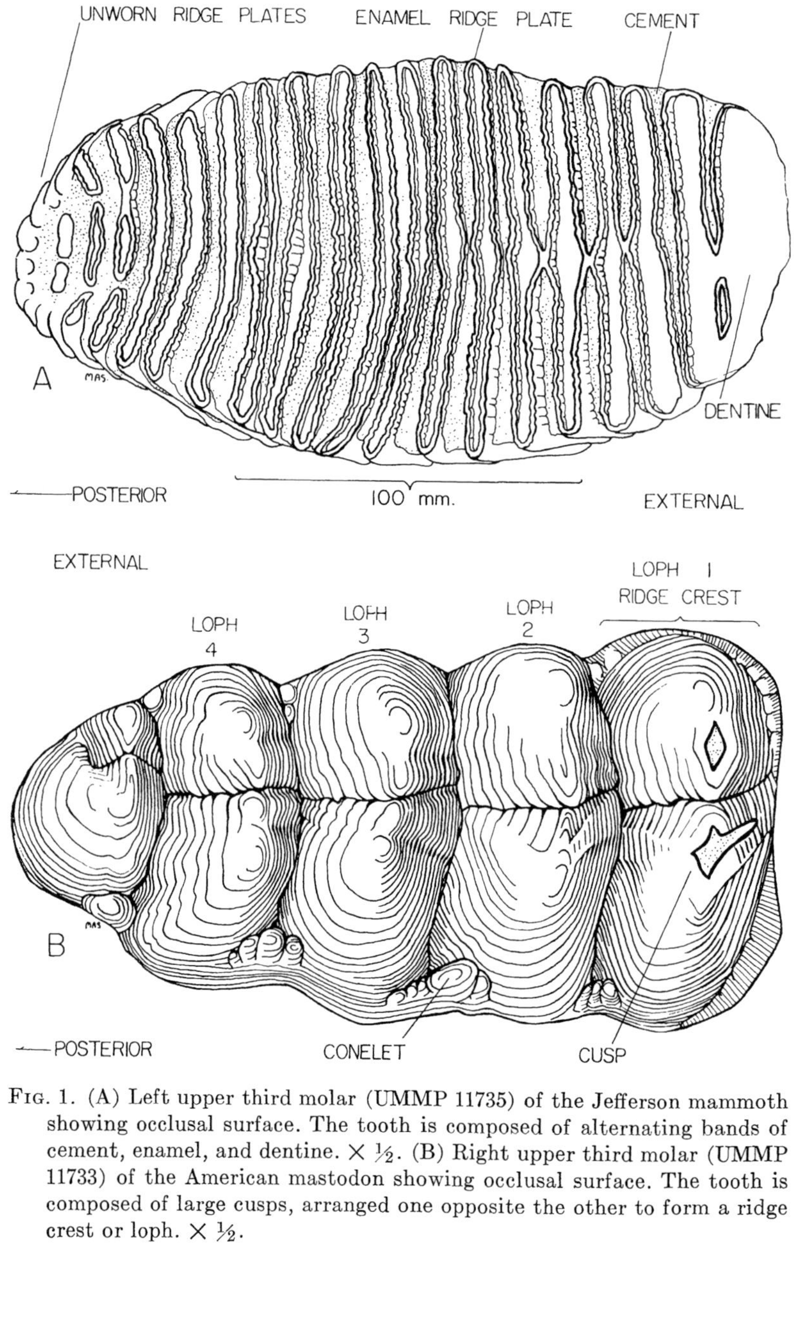

Fig. 1. (A) Left upper third molar (UMMP 11735) of the Jefferson mammoth showing occlusal surface. The tooth is composed of alternating bands of cement, enamel, and dentine. × ½. (B) Right upper third molar (UMMP 11733) of the American mastodon showing occlusal surface. The tooth is composed of large cusps, arranged one opposite the other to form a ridge crest or loph. × ½.

There is no juvenile specimen in The University of Michigan Museum collection. The remains of young mastodons probably are overlooked because of their small size. The recovery of these would contribute greatly to our knowledge of milk teeth and tooth replacement.

A young adult, UMMP 29276, had the fourth upper right premolar in use at the time of death. This tooth is well worn, and the first upper molar shows wear on the anterior pair of cusps. The second molar ($M^{\underline{2}}$) had not moved down to the level of the grinding surface of the first upper molar. The crown of the third upper molar ($M^{\underline{3}}$) is fully developed, though it had not erupted.

In specimen UMMP 11308, the upper dentition of a young male, the fourth upper premolar had been shed and the first molar had moved forward in the jaw and taken its place. This tooth is well worn. The second molar is worn on its anterior cusps and the third had not moved into occlusal position (contact with opposite tooth). Specimen UMMP 24240 has the second upper molar well worn. The third molar is in occlusal position and shows wear. Mastodons ap-

TABLE I

Measurements (in mm.[1]) of Teeth of the American Mastodon

	No. of Teeth	Range in Size	Average
$Dp_{\overline{4}}$, greatest anteroposterior length..	1	73.9	73.9
$Dp_{\overline{4}}$, greatest width..................	1	56.0	56.0
$M_{\overline{1}}$, greatest anteroposterior length...	6	86.0–99.0	91.6
$M_{\overline{1}}$, greatest width..................	6	67.8–76.0	71.5
$M_{\overline{2}}$, greatest anteroposterior length...	11	111.0–129.0	120.1
$M_{\overline{2}}$, greatest width..................	10	84.6–108.1	98.9
$M_{\overline{3}}$, greatest anteroposterior length...	7	184.0–244.3	209.5
$M_{\overline{3}}$, greatest width..................	6	96.4–117.8	104.9
$Dp^{\underline{4}}$, greatest anteroposterior length...	1	70.0	70.0
$Dp^{\underline{4}}$, greatest width..................	1	61.0	61.0
$M^{\underline{1}}$, greatest anteroposterior length....	5	89.3–95.9	92.7
$M^{\underline{1}}$, greatest width..................	5	73.8–82.9	77.8
$M^{\underline{2}}$, greatest anteroposterior length....	13	108.0–139.2	119.9
$M^{\underline{2}}$, greatest width..................	13	87.3–109.9	97.3
$M^{\underline{3}}$, greatest anteroposterior length....	10	174.4–223.2	192.1
$M^{\underline{3}}$, greatest width..................	10	95.5–114.2	105.3

[1] 25.4 millimeters = one inch.

parently had reached maturity by the time the second and third molars were in use. The University of Michigan Museum of Paleontology has no specimen in which only the third molar was present, although this condition did occur in aged animals (Hay, 1914).

Tooth replacement in mastodons continued throughout much of their life. In the majority of mammals milk teeth are pushed out by the upward or downward growth of the permanent teeth that lie at their roots. In the Proboscideans and Sirenians (Manatees and Dugongs) the permanent teeth are formed in the back part of the jaw behind the milk teeth. As the milk teeth are worn down, their replacements have to move forward and upward in the lower jaws and forward and downward in the upper jaws.

Mastodons were neither as large nor as tall as mammoths in overall body size, but the mastodon skull was larger and lacked the high occipital crest so prominent in mammoth skulls (Pl. I, Fig. 1). The mounted female mastodon in The University of Michigan Museum of Paleontology measures approximately 7 feet, 8 inches tall at the shoulders (Pl. I, Fig. 2). Adult males are taller.

Habitat and food.—Mammut americanum was a forest animal, browsing on the forest vegetation along bogs and streams (Hibbard, 1951). In the early days before scientific investigation, however, people who found mastodon teeth thought them to have belonged to some large flesh-eating monster (Jepsen, 1960). Benjamin Franklin, who was one of the first to study such teeth systematically, concluded that it would have been impossible for the American mastodon to have eaten meat.

The following quotations are taken from Warren (1852), who mentions three accounts of mastodon finds associated with materials which were probably stomach contents.

"Near Newburgh, in the State of New York, an American mastodon was discovered in 1845. During the excavation of this animal, a mass of material resembling crushed branches of trees was discovered."

According to Warren, a medical doctor, this material received little attention at the time. Another man, Dr. Prime, was present at the time of excavation and described the mass as follows:

"In the midst of the ribs, imbedded in the marl and unmixed with shells or carbonate of lime, was a mass of matter, composed principally of twigs of trees broken into pieces of about two inches in length, and varying in size from very small twigs to half an inch in diameter. There was mixed with these a large quantity of finer vegetable substance, like finely divided leaves; the whole amounting to from four to six bushels. From the appearance of this, and its situation (position in the skeleton), it was supposed to be the contents of the stomach; and this opinion was confirmed on removing the pelvis, underneath which, in the direction of the last of the intestines, was a train of the same material, about three feet in length and four inches in diameter. Microscopic examination of this matter revealed that the twigs were terminal branches of coniferous trees."

Another mastodon from Chester, near Goshen, Orange County, New York, had been discovered with similar vegetable material. Professor Mitchell described it thus:

"Beneath the bones, and immediately around them, was a stratum of coarse vegetable stems and films resembling chopped straw, or rather drift stuff of the sea; for it seemed to be mixed with broken fibers of conferva, like those of the Atlantic shore."

The third occurrence reported was that of a mastodon from Wythe County, Virginia, that was found associated with "half-masticated reeds, twigs, and grass or leaves," according to Bishop Madison in a letter to Dr. V. S. Barton.

The preservation of masticated vegetable matter from the stomach should not be regarded as improbable because the finds are associated with peat. Peat is an accumulation of vegetable material that has been subjected to compaction for a very long period of time in bogs. It is often well preserved, even to the extent that pollen grains occurring within it can be examined and the genera of plants determined. Peat deposits consist mostly of fine organic materials. Branches of trees are not uncommon in such deposits, but these would never be "broken into pieces of about two inches in length."

Location of finds of mastodon remains in Michigan.—MacAlpin (1940) listed 114 specimens of the American mastodon that had been found in Michigan. He concluded that these represented considerably fewer than half of those collected in the state since 1839. Forty-nine new discoveries have since been reported. All of the recorded occurrences within the state are shown in Fig. 2. Following is a list of the new records since 1939, obtained from newspaper articles, letters, and museum specimens.

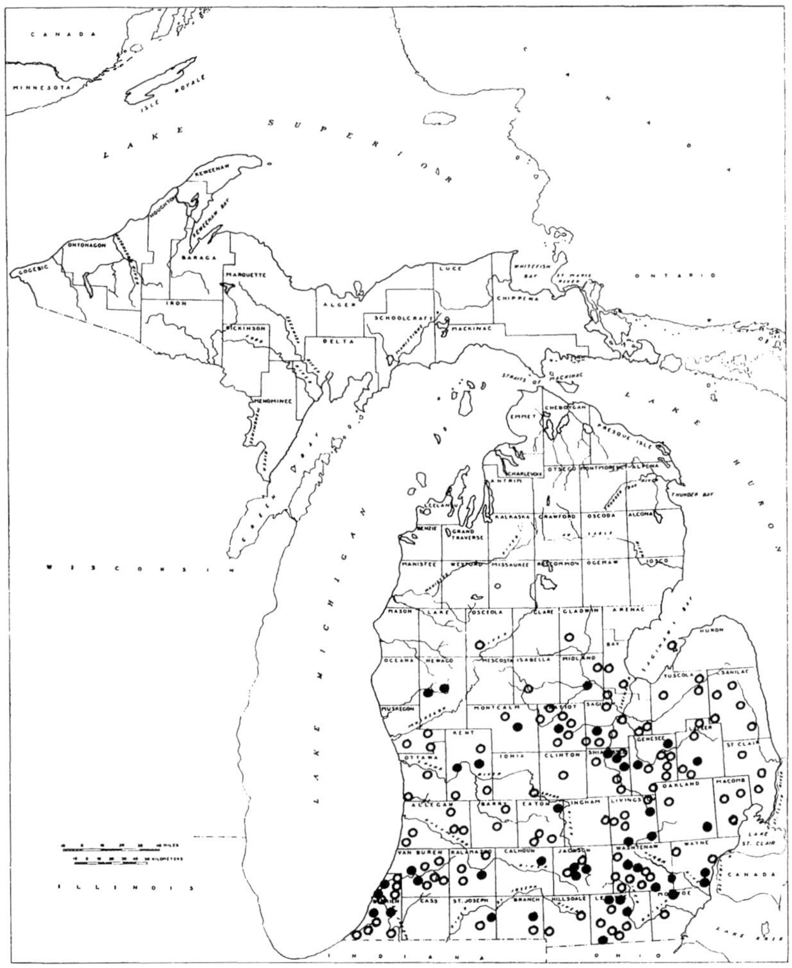

FIG. 2. Distribution of finds of the American mastodon in Michigan. The solid circles represent specimens in museum collections.

Allegan County

1. Plainwell, 4½ miles northeast. Gunplains Township, Sec. 10, T.1 N., R.11 W. Most of a skeleton, lower jaw with teeth, many vertebrae, no skull, fragments of ribs, sacrum, foot-bones, etc. found at a depth of 5½ feet in marl.

2. South Haven, just north. Caseo Township, Sec. 32, T.1 N., R.16 W. Lower jaws, at a depth of 6 feet in marl.

Berrien County

1. Watervliet, NW¼ NE¼, Sec. 1, T.3 S., R.17 W. One skull, UMMP 24240. Donated to The University of Michigan by Mrs. Beulah King and Mrs. Elizebeth Acton.
2. Hager Township, SE¼ NE¼, Sec. 22, T.3 S., R.18 W. Skull fragment with four teeth. UMMP 26865. Gift of Mr. and Mrs. William Walsh McNeill.
3. On the farm of Rudy Adams. Third molar, in marl around 1912.

Calhoun County

1. Along Rice Creek. One tooth, found around 1834 or 1835. Specimen at the Grand Rapids Public Museum, No. 12359.

Clinton County

1. St. Johns, on Avery Road. One tooth.

Genesee County

1. Flushing, W½ W½ of NE¼, Sec. 24, T.8 N., R.5 E. One tooth and possibly other bones, UMMP 23606. Gift of J. R. Roberts.
2. Flint, exact locality not known, but somewhere near Flint. Tooth fragment, in gravel.
3. Davison, exact locality not known. One molar.
4. Richfield Township, no other data. Scapula.

Gratiot County

1. St. Louis, about 3 miles southeast. T.2 W., R.11 N. One tooth, in muck on gravel.
2. Alma, Sec. 17 T.11 N., R.2 E. Skull parts, tooth fragments, axis, three dorsal vertebrae, ribs, and part of pelvis. Specimen at Alma College.
3. Ithaca, north of. E½ SW¼, Sec. 5, T.11 N., R.2 W. Part of a skeleton.

Hillsdale County

1. Near Hemlock Lake, Sec. 5, T.7 S., R.4 W. Vertebrae, UMMP 38622, found in marl while digging a boat canal to Hemlock Lake. Donated by Mr. and Mrs. Herschel H. Hall.

Ingham County

1. Fowlerville, west of. Exact locality not known. Bone scraps, found in sand 23 feet below the surface.

Ionia County

1. Mullikin, 2 miles west and 1 north. Lower jaw and most of the skeleton. Specimen at the Michigan State University Museum, No. 1130.

Jackson County

1. Jackson, 6 miles northwest. Broken tusk. Michigan State University Museum 7745.

Kalamazoo County

1. Near Battle Creek, 12 miles west, at the west end of Sherman Lake. Sec. 30 or 31, T.1 S., R.9 W. Fragment of jaw with two teeth, in a drainage canal.
2. Fort Custer Military Reservation, Sec. 1 or 2, T.2 S., R.9 W. Palatal part of skull with four cheek teeth, tusk, vertebrae, ribs, etc., in a drainage ditch to Eagle Lake. Skeleton at the Kingman Museum, Battle Creek.

Kent County

1. Bowen Township, northeastern part. Three lower molar teeth, in sand.
2. Exact locality unknown. Parts of a skeleton. Specimen at the Michigan State University Museum.
3. Exact locality unknown, but north of Grand Rapids. Humerus and a few carpals. Specimen at the Grand Rapids Public Museum.

Lapeer County

1. Near Lapeer, about 5 miles southeast, Sec. 14, T.7 N., R.10 E. Lower jaw with six cheek teeth and two small tusks, part of pelvis, sacrum, two scapulae, 17 vertebrae, about 20 ribs, foot bones, parts of radius, ulna, humerus, and femurs (UMMP 37811). A C^{14} date was obtained from this specimen. The date is 5950 $\pm$ 300 years before the present (B.P.).
2. Clifford, 2 miles south. NE¼ SW¼ Sec. 21, T.10 N., R.11 E. Part of a lower jaw with one tooth.

Lenawee County

1. Seneca Township, Sec. 5, T.8 S., R.2 E. Maxillary teeth, fourth premolar to third molar, and one tusk, UMMP 29276. A C^{14} date of the inner part of the tusk (No. M-280) is 7,070 ± 240 B.P. Donated by J. M. Bruggeman.

2. Tecumseh, a few miles northeast. Sec. 18, T.5 S., R.5 E. One tooth, UMMP 26864, on the surface of a peat bog. Gift of Ralph Schafer.

Livingston County

1. Putnam Township, SW¼, Sec. 36, T.1 N., R.4 E. About three or four rods north of the Washtenaw-Livingston County line. Neural spine, UMMP 23499.

2. Near Fowlerville. Dorsal vertebrae, UMMP 25780.

Midland County

1. Near the Midland County Club, Sec. 9, T.14 N., R.2 E. Part of the palate with second and third molars, UMMP 27730. Gift of Dan Smick.

Montcalm County

1. Near Stanton, exact locality unknown. Broken humerus, UMMP 31054. Gift of Clifford Wright.

Newaygo County

1. Near Fremont, 6 miles northwest. Broken premolar, UMMP 24836, in gravel. Gift of Don Vander Werp.

Oakland County

1. Near Holly, Sec. 27 or 32, T.5 N., R.7 E. Four teeth and a jaw fragment.

Ottawa County

1. Holland, at Big Creek, a few miles east of Holland. Jaw, teeth, ribs, long bones, and vertebrae.

Saginaw County

1. Near Freeland. W½, NW¼ Sec. 29, T.13 N., R.3 E. Broken jaw, two teeth, and part of pelvis, in marl.

2. Frankenmuth, 8 or 9 miles south. SW corner of Sec. 32, T.10 N., R.6 E. One tooth, in sandy loam.

3. Near St. Charles, locality not known. One molar tooth, UMMP 22271.

Sanilac County

1. Argyle, 2 miles north and 1 east. One rib, a long bone, and several footbones, in a drainage ditch in marl.

2. Croswell, exact locality not known. One tooth, in a swamp.

Shiawassee County

1. Henderson, Rush Township, Sec. 16, T.6 N., R.2 E. Part of skull and most of the skeleton, UMMP 23498, at a depth of 6 or 7 feet in marl.

2. Henderson, Rush Township, Sec. 16, T.6 N., R.2 E. Left ulna, 200 yards from No. 23498.

3. Corunna, New Haven Township. Sec. 24, T.8 S., R.3 E. One-half tooth, in gravel.

4. Bryon, exact locality not known. Part of a tusk, in a drainage ditch.

Van Buren County

1. Near Hartford, 3 miles east and 1 mile south. Two teeth, Michigan State University Museum 5296, in a peat bog.

2. Almena Township, Sec. 33, T.2 S., R.13 W. Part of a tusk, in marl.

Washtenaw County

1. Milan, Augusta Township, Sec. 29, T.4 S., R.7 E. Part of right ramus of lower jaw with six teeth, UMMP 34127, in a drainage ditch. Gift of Joseph S. Zabo.

2. Lodi Township, SW¼ SW¼, Sec. 17, T.3 S., R.5 E. One jaw, atlas, axis, and palate, UMMP 24241. Gift of Fred Grob, 1947.

3. Freedom Township, NE¼ Sec. 4, T.3 S., R.4 E. Fragment of tusk (M-67, C^{14} dated), in Cary outwash gravels.

Wayne County

1. Monguegon Township, NE¼ NE½ Sec. 12. Base of skull, lower jaw, several teeth, one tusk in poor condition, scapula, humerus,

ulna, foot bones, ribs, and vertebrae, UMMP 3483, in a county drainage ditch.

The Jefferson Mammoth

The mammoth arrived in North America from Eurasia much later than the mastodon. The earliest records to date are from the early Middle Pleistocene (Kansan deposits). This ancestral stock probably gave rise to the Imperial mammoth and the Columbian mammoth (both southern forms) and the Jefferson mammoth (a more northern species). The well-known Woolly mammoth, a tundra dweller, arrived in North America much later in the Pleistocene.

Following is the classification of the Jefferson mammoth.

ORDER PROBOSCIDEA

Family Elephantidae

Genus *Mammuthus* Burnett, 1830

Mammuthus jeffersoni (Osborn). Jefferson mammoth

Elephas primigenius Blumenbach. Hay, 1923, Carnegie Inst. Publ. 322, p. 137 (not of Blumenbach).

E. columbi Falconer. Hay, 1923, Carnegie Inst. Publ. 322, p. 151 (not of Falconer).

E. boreus Hay. Sherzer, 1927, Science, 65, No. 1695:616 (not of Hay).

E. primigenius americanus (DeKay). Case and others, 1935, PMASAL, 20, p. 449 (not of DeKay).

Parelephas jeffersonii (Osborn). Osborn, 1936, Proboscidea, V. 2, p. 1083.

A census of the remains of *Mammuthus jeffersoni* has not previously been reported for Michigan. Individual remains of this mammoth have in the past been identified as the "Woolly" mammoth (*M. primigenius*) or the Columbian mammoth (*M. columbi*). Hibbard (1951) only briefly mentioned *M. jeffersoni* as occurring in the state.

Characters of Mammuthus jeffersoni.—The teeth of mammoths, as well as those of the living elephants to which the mammoths are related, consist of alternating cement, dentine, and enamel ridge plates (Fig. 1A; Pl. III, Figs. 1 and 2). These plates are transverse to the long axis of the tooth. The number of ridge plates present in the tooth per 100 mm., the thickness of the enamel, and the distance of the enamel plates from one another are used to distinguish the different species. The premolars are smaller than the molars and hence possess fewer ridge plates. The third molar is the largest tooth and has the most plates; for the Jefferson mammoth, there are generally 25 in the upper and 24 in the lower ($M3\frac{25}{24}$). The symbols denoting the position of the teeth are the same as those used to describe mastodon teeth. For mammoths and elephants, however, after the symbol designating the tooth appears a notation of the maximum number of ridge plates known for the upper and lower teeth. The number of ridge plates observed in the upper tooth is written above the line and the number in the lower tooth below the line.

Osborn (1936) gives the ridge plate formulae for *Mammuthus jeffersoni* as: $Dp3\frac{8+}{?}$, $Dp4\frac{12}{13}$, $M1\frac{11}{?}$, $M2\frac{18-19}{19}$, and $M3\frac{25}{24}$. Seven to ten of these plates occur in the distance of 100 mm. in the third molar (see Fig. 1A).

Very few specimens of the Jefferson mammoth are in museum collections in Michigan. The few remains that have been collected in the state consist mostly of jaws with teeth or of isolated teeth. There are remains of 3 individuals in the Grand Rapids Public Museum, 7 in the Michigan State University Museum, and 9 in The University of Michigan Museum of Paleontology. The most complete specimen yet found in the state is represented by a partial skeleton (No. 8238) at the Michigan State University Museum. This mammoth was found approximately 10 miles southeast of Eaton Rapids on the Losey farm and consists of both lower jaws, one tusk, many vertebrae, most of the ribs, the shoulder blades, and a few leg bones. No skull or pelvis was found.

A lower jaw with two teeth, UMMP 22798, was found one mile north of Assyria, in Barry County. This jaw was donated to the University of Michigan Museum of Paleontology by Robert Hughes in 1942. It is from an old individual; the anterior ridge plates of the last molars ($M_{\overline{3}}$) are worn down to their roots. The greatest anteroposterior length of the left third molar ($LM_{\overline{3}}$) is 265.0 mm. and the $RM_{\overline{3}}$ is 255.0 mm. Both of these teeth have 18 ridge plates forming the occlusal surface. $LM_{\overline{3}}$ has 7½ ridge plates in 100 mm. and $RM_{\overline{3}}$

has 7. The maximum width of $RM_{\overline{3}}$ is 109 mm., as taken across the eighth ridge plate, and the maximum width of $LM_{\overline{3}}$ is 118.5 mm., as taken across the 10th ridge plate. The distance between the grinding surface of the two teeth anteriorly is 35.8 mm. The maximum width of the groove above the rostrum is 70 mm. and the rostrum projects 80.4 mm. beyond the symphysis. The rostral groove between the jaws is constricted dorsally. The greatest width of the jaws is 648.0 mm. and the length is 542.0 mm. The depth of the jaw, taken from the occlusal surface at the 13th ridge plate to the base, is 223.0 mm. The mandibular condyles are 437.0 mm. above the base of the jaw.

Four associated teeth, UMMP 11735, a gift of George Wagner, were found near the Chapel Hill School in Cass County. These are the upper and lower third molars. $LM^{\underline{3}}$ has 26 ridge plates, of which 18 make up the grinding surface. In 100 mm. there are 9 plates. $RM^{\underline{3}}$ consists of 26+ plates, with 18 making up the occlusal surface and 10 occurring within 100 mm. $LM_{\overline{3}}$ consists of 25 plates plus one which has broken off. The occlusal surface consists of 16 plates and there are 7½ plates in 100 mm. $RM_{\overline{3}}$ has 26 ridge plates, 17 of which form the grinding surface. No accurate count could be made of the ridge plates occurring in 100 mm. because the tooth is slightly pathological.

As the above data show, there is considerable variation in ridge-plate number, even in the same individual. The thickness of the enamel forming the ridge plates, the number of these plates per tooth, and the number occurring in 100 mm. are the chief means by which the various species of *Mammuthus* are distinguished. The specific distinctions between the Woolly mammoth, the Columbian mammoth, and the Jefferson mammoth, as set forth by Osborn (1936) may be summarized as follows:

The Woolly mammoth *Mammuthus primigenius* is the animal that today is found frozen in the Arctic regions of North America and Asia. It roamed the tundra in great numbers late in the Pleistocene. It was somewhat smaller than the Jefferson mammoth. The upper and lower third molars have from 24–27 ridge plates, and from 9–13 plates per 100 mm., with an average of 10 in most individuals. In comparison with the third molars of the Jefferson mammoth, those of the Woolly mammoth generally had more plates and were more compressed (which produced the higher plate count per 100 mm.).

Mammuthus columbi, the Columbian mammoth, was restricted to southern North America, where it apparently was very abundant, particularly in Florida. This species was larger than the Jefferson mammoth. The enamel ridge-plate numbers for the third molars are $\frac{18\text{-}19+}{15\text{-}16+}$, and there are usually only 6½ plates per 100 mm. (Osborn, 1936, p. 1071). The enamel forming the ridge plates is thicker than in either the Woolly or Jefferson mammoth.

Habitat and food.—The mammoths were herbivores, as are the living elephants to which all the mammoths are closely related. They grazed on grasses of various kinds. Probably these animals confined themselves to open grasslands and seldom entered deeply forested areas. The Jefferson mammoth was apparently rather rare in Michigan. This scarcity may be accounted for by limited meadow areas, which would reduce the number able to live in the state. The rarity of finds may be due, in part, to the mammoths' preference for an upland habitat, where their remains would seldom be preserved.

Stomach contents are known only from frozen bodies of the Woolly mammoths (Farrand, 1961). Felix (1912) reported a frozen Woolly mammoth from Siberia that had within its stomach various tundra grasses, such as red bent grass and alpine foxtail. The needles of conifers were rare in the material examined. Heintz (1955) mentioned the types of vegetation found in the stomach of a frozen Woolly mammoth as being mosses and grasses. A pollen analysis of the stomach contents of the Beresovka-mammoth (Heintz, 1955) showed that 97.09 per cent of the pollen belonged to grasses and only .17 per cent belonged to trees (fir, alder, birch, willow). The rest of the pollen belonged to various other meadow plants.

There is no reason to believe that the Jefferson mammoth deviated from a herbivorous, grazing mode of life. The species of plants for this mammoth would be different from those forming the diet of the Woolly mammoth, but most were probably grasses. Possibly in the winter months some alder, willow, and birch supplemented this fare. In character the teeth in the Jefferson mammoth are most like those of the Asiatic elephant, which is almost exclusively a grazer.

Location of finds of the Jefferson mammoth in Michigan.—Thirty-two Jefferson mammoth remains have been recorded in Michigan since 1839. As in the case of the mastodon, this number probably represents considerably fewer than the remains actually found in the state. Recorded occurrences are shown in Fig. 3 and listed below.

Arenac County

1. 2½ miles north of Au Gres. One tooth (letter of W. A. Kelly, September 24, 1935). Specimen at Michigan State University.

Barry County

1. Assyria, 1 mile north. NW¼ Sec. 15, T.1 N., R.7 W. Tusk, jaw, scapula, vertebrae, and foot bones (UMMP 22798), in muck. Gift of Robert Hughes in 1942.

Berrien County

1. Galien Township, NE¼ Sec. 6, T.8 S., R.19 W. Lower third molar. Specimen in the Chamberlin Memorial Museum, Three Oaks (Dice, 1920).
2. Bakerstown marsh, south and west of Buchanan. One tooth, found during the digging of a drainage ditch (Hay, 1923, p. 171).
3. Southwest part of Lake Township. Fragment of a tooth, found in the spring of 1943.
4. Three Oaks, 2 miles southeast. Two upper and two lower third molars, found around 1900. Specimens at the Chamberlin Memorial Museum, Three Oaks (Dice, 1920; Hay, 1923).
5. Exact locality unknown. One tooth. Present location unknown (Dice, 1920.)

Cass County

1. Jones, near Chapel Hill schoolhouse. Sec. 32, Porter Township, T.7 S., R.13 W. Two good humeri, broken femur, broken tibia, scapula, pelvis, ribs, some vertebrae, footbones, four teeth, etc. (UMMP 11735). Gift of George Wagner.

Eaton County

1. Eaton Rapids, 2 miles south. Jaws (UMMP 11732) with one tooth each, in the bed of Grand River in 1904. This is No. 8260 reported by Hay (1923, p. 137).
2. Near Grand Ledge. Three teeth and a tusk, a foot below the surface in muck (Lane, 1902; Hay, 1923).
3. Near Eaton Rapids, on Wilcox road. One tooth, in muck.
4. Charlotte, exact locality not known. One left lower third molar (UMMP 13639), in muck of the Thornapple River.

Gladwin County

1. Remains found in Gladwin County, locality not known (Sherzer, 1927).

Gratiot County

1. Ithaca, 7 miles west. Sec. 2, T.10 N., R.4 W. Tooth and parts of the skeleton, 2½ feet below the surface in muck lying on gravel.

Ingham County

1. Leslie, 3 miles west. Scapula, femur, tibia, and ulna, under peat in marl.

Ionia County

1. Hubbardston, bed of Fish Creek near electric power plant. Fragment of tooth (UMMP 13362). Gift of Dr. M. B. Robinson.

Jackson County

1. Near Eaton Rapids, 9½ miles southeast. Both lower jaws; one tusk; cervical, thoracic, and lumbar vertebrae; most of the ribs; both shoulder blades; and two or three leg bones, in marl. Specimen at the Michigan State University Museum, No. 8238.
2. Locality not known, somewhere in the northern part of Jackson County. One molar (UMMP 3163), in peat.

Kent County

1. Grand Rapids, approximately 5 miles south on Plaster Creek. One tooth.
2. Rockford. One tooth (Grand Rapids Public Museum, No. 12356), in a gravel pit.
3. Somewhere on Nine Mile Road. "Remains" (Grand Rapids Public Museum, No. 83226).
4. Grand Rapids, 15 miles northeast. Tooth, femur, and other bones. Specimens at the Grand Rapids Junior College.

Lenawee County

1. Ridgeway, 1½ miles east. One molar tooth (UMMP 3223), in a gravel pit in 1912. Presented to the University of Michigan by K. A. Easlick.

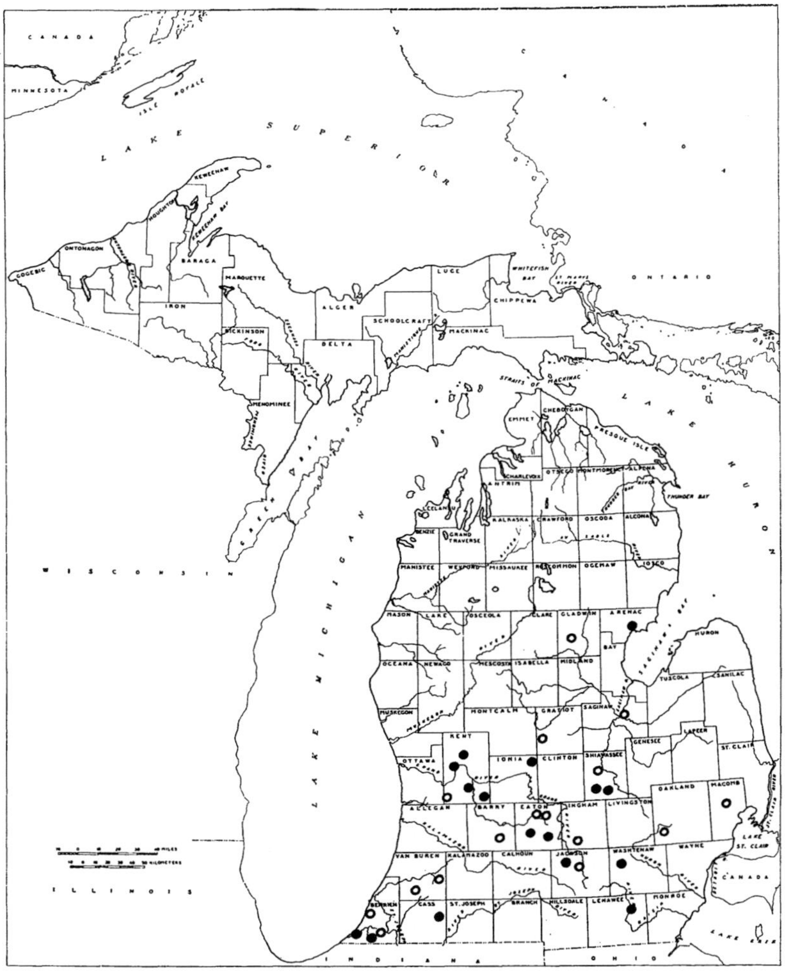

FIG. 3. Distribution of finds of the Jefferson mammoth in Michigan. The solid circles represent specimens in museum collections.

Macomb County

1. "Remains" in Macomb County, locality not known. Tooth, found before 1861 (Lane, 1902, p. 252; Hay, 1923, p. 171).

Oakland County

1. New Hudson, 1½ miles north. Part of a tooth, in a gravel pit. The gravel pit is part of a kame ridge.

Saginaw County

1. East Saginaw, near Père Marquette shaft No. 2. Tooth, found around 1901, three feet below the surface (Lane, 1902; Hay, 1923).

Shiawassee County

1. Near Perry, one mile away. Tooth, on dirt from a ditch dredging. Specimen at the Michigan State University Museum, No. 8115.
2. Morrice, 3 miles south. T.5 N., R.2 E. One tooth (UMMP 23607), on the dump of an old drainage ditch. Gift of Roland Schaadt and Alphons Schlicklin.
3. Owosso, in the city limits. Three teeth, found in 1895.

Van Buren County

1. Decatur, a few miles west. One tooth, plowed up in the spring of 1928. Present location is unknown.
2. Mentha, 1½ miles south, in Sec. 1, T.2 S., R.13 W. Two upper teeth and tusk fragments.

Washtenaw County

1. Chelsea, Sylvan Township, Sec. 13, T.2 S., R.3 E. One half tooth (UMMP 23247). Gift of Virginia Lesser.

Radiocarbon Dating of Fossil Remains

Very few mastodons and mammoths have been collected in the state in recent years, chiefly because of the use of heavy machinery instead of hand labor to dig out muck. When drainage ditches and the like were dug by hand or with oxen or horses, the chances were greater for a bone to be noticed and recovered. Few mastodons and mammoths have become available for radiocarbon (C^{14}) dating since the establishment of the University of Michigan radiocarbon dating laboratory in 1950. The specimens collected before this date were treated with alvar or shellac and are thus worthless for the radiocarbon method (Hibbard and Hinds, 1960). It has been found that ivory, because of its compactness, is far better for use in C^{14} dating than bone. Bone often becomes contaminated by the intrusion of small roots and rootlets of recent plants and hence will give an erroneous date.

Part of a mastodon tusk (M-67, Crane, 1956) was submitted to the University of Michigan Memorial-Phoenix Project Laboratory for radiocarbon dating. It was partially mineralized. Its source was a gravel pit consisting of Cary outwash in the NE¼ Sec. 4, T.3 S., R.4 E., Freedom Township, Washtenaw County, Michigan. The sample from the tusk dated 6100 ± 400 B.P. from carbon black and 6300 ± 500 B.P. from a gas sample. It should be noted that acid soluble carbonates from the tusk were used to obtain the samples. The gravels in the exposure are chiefly limestone, and considerable calcium carbonate cementing had occurred in the area in which the tusk was found. The radiocarbon date, therefore, is in error. Dating of the Cary stage (the age of the glacial outwash gravels), is between 13,000 and 15,000 B.P. (Flint, 1957; Frye and Willman, 1960), which may give a more reliable estimate of the age of the tusk.

The following C^{14} dates for American mastodons were obtained from samples of tusks with one exception, when part of a rib was used. The samples used in dating are given numbers preceded by the letter M, which stands for the University of Michigan radiocarbon dating laboratory. Unfortunately, no dates are available for Jefferson mammoth bones or tusks.

M-347	Tusk (UMMP 37811), taken from muck in Sec. 14, T.7 N., R.10 E., Lapeer County, Michigan (Crane and Griffin, 1959)	5,950 ± 300
M-280	Inside portion of tusk (UMMP 29276), from Sec. 5, T.8 S., R.2 E., Lenawee County, Michigan (Crane and Griffin, 1958)	7,070 ± 240
M-281	Outside portion of the above tusk	7,820 ± 450
M-490	Rib, from muck near Sheridan, Chautauqua County, New York (Crane and Griffin, 1959)	9,200 ± 500
M-694	Tusk (UMMP 24240), from muck in Elkhart County, Indiana	9,320 ± 400
M-39	Tusk fragment, from Nobel County, Indiana (Crane and Griffin, 1958)	12,630 ± 1,000

Extinction

Hester (1960) states that most of the large Pleistocene herbivores in North America became extinct around 8,000 years ago, as indicated by radiocarbon dates. According to his data, the mammoths apparently died out earlier than the American mastodon. The latter probably survived a little later in isolated areas. According to the above radiocarbon dates, the American mastodon became extinct somewhere after 6,000 years ago. This is approximately the time of the beginning of recorded history.

In most cases the causes of extinction are unknown. It is usually impossible to know exactly why a species of animals could not maintain its numbers. Seldom can a single factor be considered the sole cause. The extinction of the American mastodon and the Jefferson mammoth probably was the result of a number of unfavorable conditions that acted at the same time.

It is known that the Late Pleistocene American Indians of the southwest hunted the now-extinct Columbian mammoth (Wormington, 1957), a close relative of the Jefferson mammoth. There is as yet no evidence that early man hunted either the mammoth or mastodon in Michigan.

It is possible that extinction of the American mastodon was partly brought about by the barrier formed by the Great Lakes. Because of this barrier the mastodons might have been unable to follow the northward retreat of the boreal forest environment as the forests followed the retreat of the last glacier. Other barriers were formed by the grasslands and Glacial Lake Agassiz to the northwest. The north outlet to the northeast would offer similar restrictions, since it served as a vast floodway for glacial meltwater.

Recently much work has been done on fossil pollen contained in peat from bogs. The genera of plants represented by the fossil pollen can usually be determined, and radiocarbon dates can be obtained from wood associated with it. From the fossil pollen and its chronology, the type of forest that was present around the bog can be determined for a given time. Two studies from southern Michigan, one from near South Haven, and the other from a bog near Hartford, revealed the following forest changes for these areas (Zumberge and Potzger, 1956). A forest consisting chiefly of spruce and some fir was present in southern Michigan 11,000 years ago. This would point to a cool to cold, moist climate. The end of the spruce-fir forest occurred around 8,000 years ago, followed by an increase of pine. The maximum development of pine in southern Michigan was around 6,000 years ago, indicating a warming and drying climate. Associated with the pine was oak and hickory. Approximately 5,000 years ago pine had declined in importance and there was an increase of oak, chestnut, and hickory. The climate at this time was warmer and dryer than at any earlier time in the Late Pleistocene. Since 5,000 years ago, the forest in southern Michigan has consisted mostly of oak, chestnut, hickory, and other broad-leaved species.

Mastodons in Michigan were adapted to life in a mixed coniferous and broad-leaved forest. A mature evergreen forest, composed of large trees with their lowest branches well off the ground, would provide poor browse for these animals and thus cause a decline in their numbers. A mature coniferous forest would have been barren of young seedlings because of the deep shade the large trees produced. Radiocarbon dating indicates that broad-leaved trees became abundant around 5,000 years ago.

Acknowledgments

I wish to thank the following persons for granting permission to examine the specimens under their care and for access to the records pertaining to these specimens: Edward Brigham, Kingman Museum, Battle Creek; Frank L. Du Mond, Grand Rapids Public Museum; Rollin H. Baker, Michigan State University Museum; Lester E. Eyer, Alma College; and Lewis B. Kellum, University of Michigan Museum of Paleontology. I especially wish to express my gratitude to Claude W. Hibbard for his advice and criticism during the preparation of this paper, and to John A. Dorr, William S. Benninghoff, and Katharine A. Fellows for reading the manuscript and giving helpful suggestions.

I am also grateful to Edwin H. Colbert, of the American Museum of Natural History, for providing the picture of the Jefferson mammoth skeleton for publication. I also wish to thank Herbert W. Wienert for taking most of the other photographs.

LITERATURE CITED

Blumenbach, Johann Friedrich. 1799. Handbuch der Naturgeschichte, ed. 6, 708 pp., 2 pls.

Case, E. C., and others. 1935. Discovery of *Elephas primigenius americanus* in the Bed of Glacial Lake Mogodore, in Cass County, Michigan. PMASAL, 20: 449–454, 2 figs.

Case, E. C., and George M. Stanley. 1935. The Bloomfield Hills Mastodon. Cranbrook Inst. Sci., Bull. 4: 1–8, 3 pls., 2 figs.

Crane, H. R. 1956. University of Michigan Radiocarbon Dates I. Science, 124: 664–672.

——, and J. B. Griffin. 1958. University of Michigan Radiocarbon Dates II. Science, 127: 1098–1105.

——, and ——. 1959. University of Michigan Radiocarbon Dates IV. Am. Jour. Sci. Radiocarbon Supplement 1: 173–198.

DICE, L. R. 1920. The Mammals of Warren Woods, Berrien County, Michigan. Univ. Mich. Mus. Zool., Occ. Pap. (86): 1–20, 2 pls., 1 fig.

FARRAND, W. R. 1961. Frozen Mammoths and Modern Geology. Science, 133 (3455): 729–735, 3 figs.

FELIX, JOHANNES. 1912. Das Mammuth von Borna. Veroff. Stadt. Mus. Volkerkunde, 4: 1–52, 8 tab., 9 figs.

FLINT, R. F. 1956. New Radiocarbon Dates and Late Pleistocene Stratigraphy. Am. Jour. Sci., 254: 265–287, 1 pl., 1 fig.

FRYE, JOHN C., AND H. B. WILLMAN. 1960. Classification of the Wisconsinan Stage in the Lake Michigan Glacial Lobe. Ill. State Geol. Surv. Circ. 285: 1–16, 1 fig.

HAY, O. P. 1914. The Pleistocene Mammals of Iowa. Iowa Geol. Surv., 23 (Ann. Rep., 1912): 1–662, 75 pls., 142 figs.

——. 1923. The Pleistocene of North America and Its Vertebrated Animals from the States East of the Mississippi River and from the Canadian Provinces East of Longitude 95°. Carnegie Inst. Wash. Publ., No. 322: 1–499, 41 pls., 25 figs.

HEINTZ, A. 1955. The Mammoth Teeth from Norway. Norsk. Geogr. Tidsskr. b. 34, h. 2–4: 73–104, 2 pls., 7 figs.

HESTER, J. J. 1960. Pleistocene Extinction and Radiocarbon Dating. Am. Antiquity, 26 (1): 58–77, 1 fig.

HIBBARD, C. W. 1951. Animal Life in Michigan During the Ice Age. Mich. Alumnus Quart. Rev., 57 (18): 200–208, 4 figs.

——, AND F. J. HINDS. 1960. A Radiocarbon Date for a Woodland Musk Ox in Michigan. PMASAL, 65: 103–108, 2 pls., 1 fig.

JEPSEN, G. L. 1960. A New Jersey Mastodon. New Jersey State Mus., Bull. No. 6: 1–20, illus.

LANE, A. C. 1902. Report of the State Board of Geological Survey of Michigan for the Year 1901, 252–253.

LUCAS, F. A. 1939. Animals of the Past. Am. Mus. Nat. Hist. Handbook Series No. 4, 7th ed.: 1–221, 24 pls., 18 figs.

MACALPIN, A. 1940. A Census of Mastodon Remains in Michigan. PMASAL, 25: 481–490, 1 fig.

OSBORN, H. F. 1936. Proboscidea. Am. Mus. Nat. Hist., 1: 1–802, 12 pls., 680 figs.

——. 1936. Proboscidea. Am. Mus. Nat. Hist., 2: 805–1675, 18 pls., 563 figs.

PALMER, T. S. 1904. North American Fauna. U. S. Dept. Agr., Wash., (23): 1–984.

SHERZER, W. H. 1927. A New Find of the Woolly Elephant in Michigan. Science, 65 (1695): 616.

WARREN, JOHN C. 1852. The Mastodon Giganteus of North America. Boston: 1–219, 26 pls.

WORMINGTON, H. M. 1957. Ancient Man in North America. Denver Mus. Nat. Hist. Pop. Ser. No. 4: 1–322, 72 figs.

ZUMBERGE, J. H., AND J. E. POTZGER. 1956. Late Wisconsin Chronology of the Lake Michigan Basin Correlated with Pollen Studies. Bull. Geol. Soc. Am., 67 (5): 271–288, 1 pl., 6 figs.

EXPLANATION OF PLATE I

FIG. 1. Skeleton of the Jefferson mammoth (Am. Mus. Nat. Hist. No. 9950) from the Late Pleistocene of Indiana. Reprinted by permission of the American Museum of Natural History.

FIG. 2. Skeleton of a female American mastodon (UMMP 23498) from the Late Pleistocene of Michigan, University of Michigan Museum of Paleontology.

PLATE I

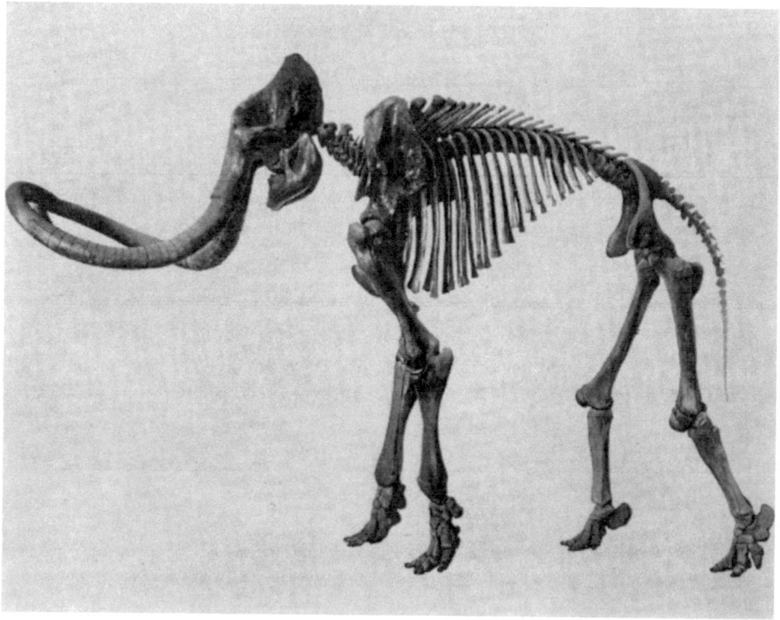

Fig. 1.

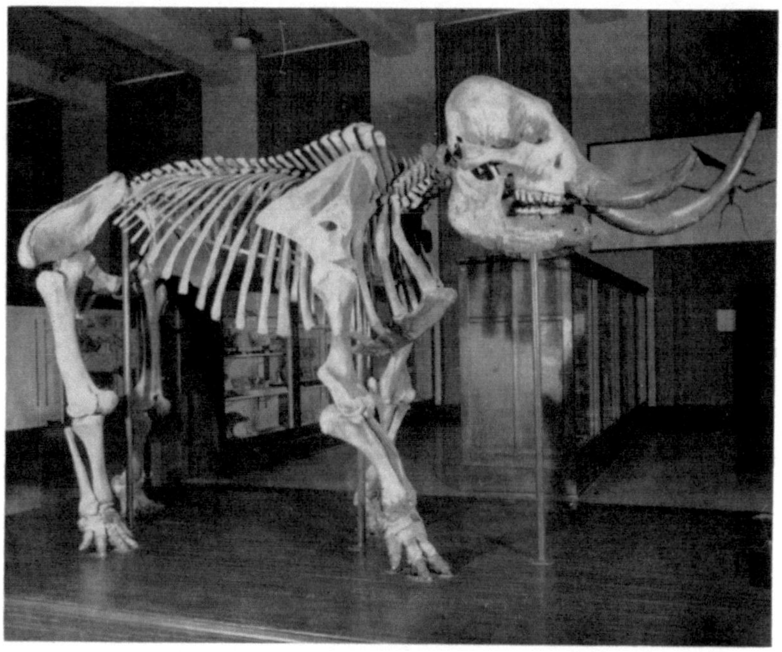

Fig. 2.

EXPLANATION OF PLATE II

FIG. 1. Occlusal view of RM^{3} (right third upper molar, UMMP 11733, occlusal length 220.0 mm.) of an American mastodon. × approx. 1/3.

FIG. 2. Lateral view of the above tooth. × approx. 1/3.

FIG. 3. A lower jaw of a young adult American mastodon (UMMP 37811) showing the lower tusks. The jaw is 35 inches long (889 mm.).

PLATE II

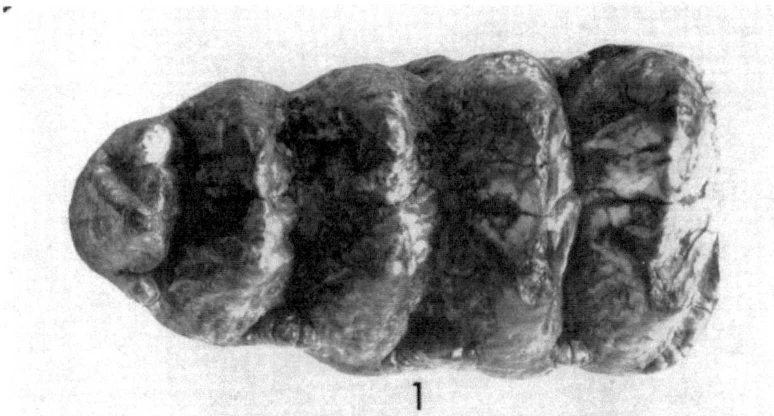

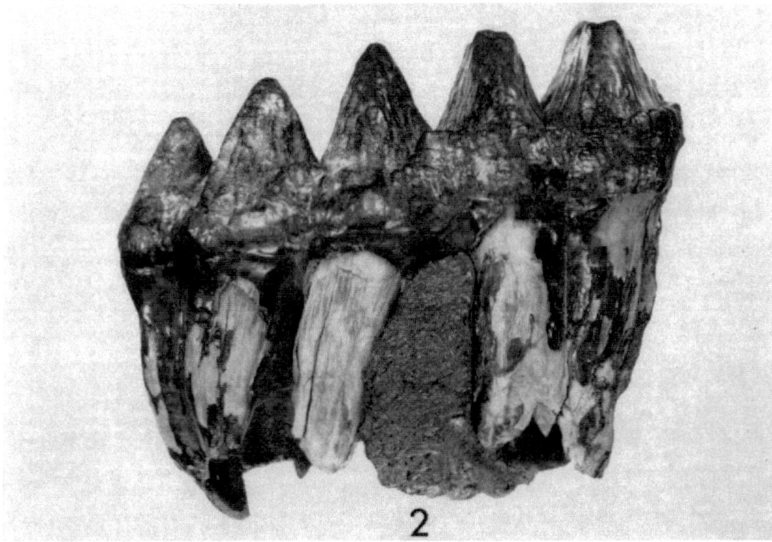

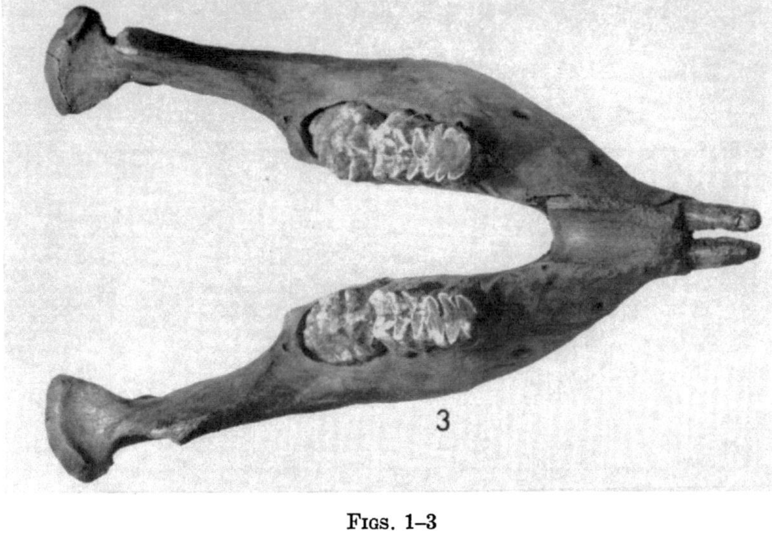

Figs. 1–3

EXPLANATION OF PLATE III

FIG. 1. Third upper right molar (UMMP 11735) of the Jefferson mammoth showing the occlusal surface. The posterior part of the tooth is to the right. × approx. 1/3.

FIG. 2. External view of the above tooth showing the ridge plates and great depth of the tooth. × approx. 1/3.

PLATE III

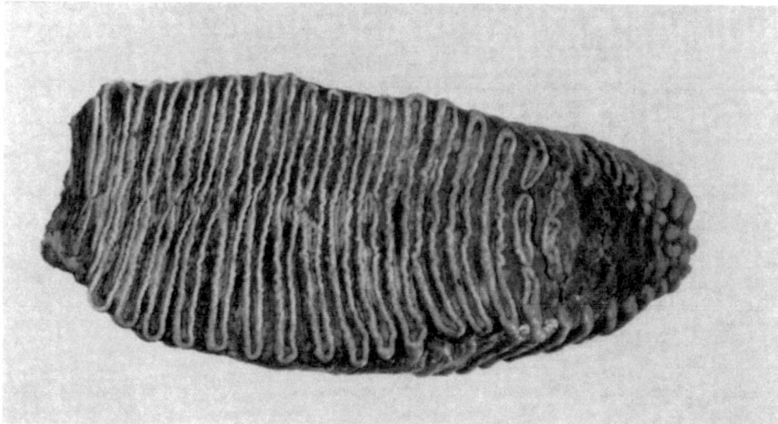

Fig. 1.

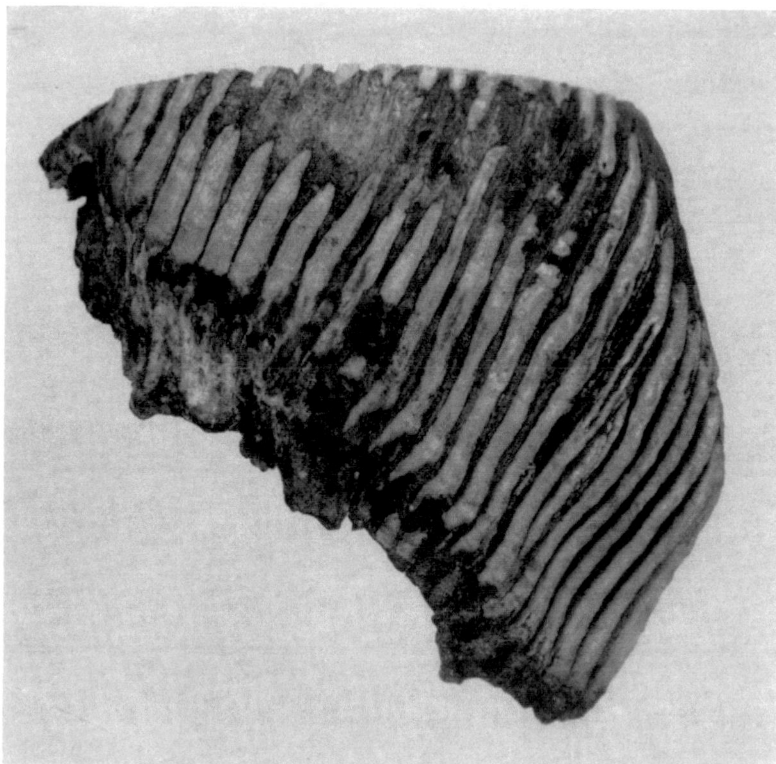

Fig. 2.

Papers of the Michigan Academy of Science, Arts, and Letters
Vol. XLVII, 1962 (1961 Meeting)

A BURIED BIOTIC ASSEMBLAGE FROM AN OLD SALINE RIVER TERRACE AT MILAN, MICHIGAN

RONALD O. KAPP AND WILLIAM A. KNELLER[1]

Alma College and Eastern Michigan University[2]

An organic deposit bearing several types of plant and animal remains was recovered from beneath a terrace of the Saline River, in southeastern Michigan. Careful study of the geologic setting, ecological interpretations of the biotic assemblage, and radiocarbon date of buried wood suggest that burial may correlate with certain postglacial events in this region including changes in the level of Lake Erie.

During the half-century that the Pleistocene history of the Midwest has been under careful scrutiny of geologists, most emphasis has been placed on regional stratigraphic correlation. As detailed local field work is pursued, increasing numbers of postglacial plant and animal remains are recovered, usually from peat deposits or lake and stream sediments. The significance of these isolated local discoveries is often dependent upon the detail with which they are analyzed. The age of local deposits, the ecologic and paleoclimatic implications of a biotic assemblage, and a detailed analysis of the local geologic setting may supplement the regional interpretations in significant ways. Radiocarbon dates combined with pollen analytical studies often provide especially powerful tools for the correlation of geologic and climatic events of postglacial times.

We gratefully acknowledge the assistance of several persons who aided in certain phases of the investigation. Barry B. Miller identified the molluscan remains and made certain ecological inferences based on this group. Dr. Everett E. Ellis of the School of Natural Resources, The University of Michigan, verified the identification of the fossil

[1] Original collections and the detailed study of the geologic setting were made by Kneller; Kapp is responsible for the determinations of wood and fruits and pollen analysis of the silts.

[2] Present address of Kneller: University of Toledo, Toledo, Ohio.

wood. We are grateful to Professors W. S. Benninghoff and Donald Eschman of The University of Michigan for their encouragement and valuable criticism of this investigation. We especially thank Dr. James B. Griffin, director of The University of Michigan Museum of Anthropology, and The University of Michigan Memorial-Phoenix Project's Radiocarbon Laboratory under the direction of Professor H. R. Crane, for making the radiocarbon determination possible.

Location and Geologic Section

During the summer and fall of 1960 the authors visited a section just southeast of the village of Milan, Michigan, exposed along an artificially straightened channel of the Saline River. The locality (Fig. 1) lies in SE¼ of NE¼ Sec. 2, T. 5E., R. 6W., Monroe County, Michigan (Ypsilanti Quadrangle, U. S. Geological Survey). At this location the course of the Saline River is superposed on an ancient delta of glacial Lake Arkona. The exposure under study is the lowest recognizable terrace of the Saline River. The terrace was excavated to straighten the channel for the village sewage disposal system (Fig. 2). Further modification of the terrace occurred when fill was exhumed for the new U. S. 23 expressway.

The pre-excavation surface altitude of the section is 684 feet A.T. (above tide). Fig. 3 is the columnar section indicating the textural variation of the sediments and showing the floral and faunal horizons. The base of the column, 675 feet A.T., consists of very fine to fine gravel and corresponds to faunal zone 1. Faunal zone 2 consists of fine to medium, poorly sorted sands with silt and clay dispersed throughout. In this zone, at an altitude of 676 feet A.T., is situated the log horizon. Leaf and twig litter overlies zone 2 at 677 and 677.5 feet A.T. The plant debris is intermixed with clay and silt and the layers are separated by medium and coarse sand. Faunal zone 3, consisting of medium to fine sand, is above the organic litter. A six-inch layer of coarse sand overlies zone 3. Above this coarse sand is situated faunal zone 4, 681 feet A.T., composed of very coarse sand to fine gravel. The uppermost part of the section grades from medium sand to silt with some clay.

Biota

The only animal remains recovered from these sediments were snails and clams. The number of specimens of the several species of

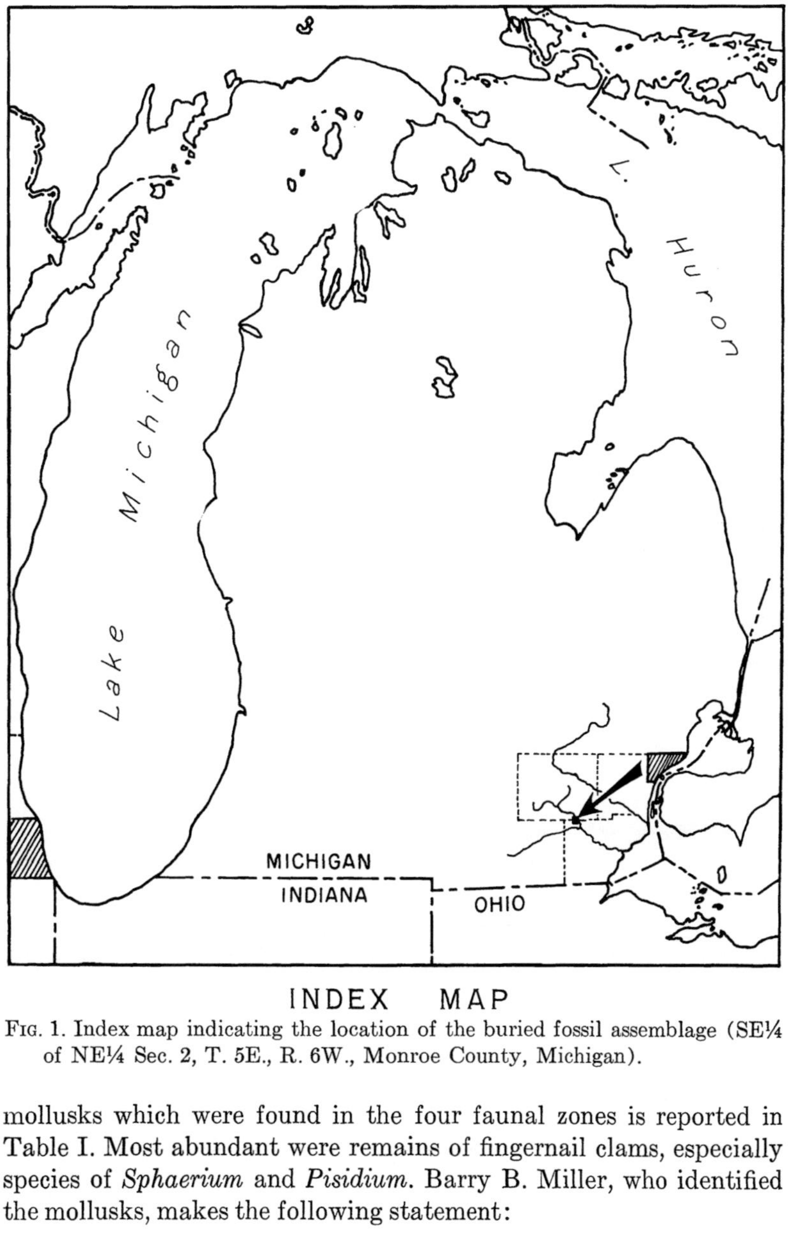

INDEX MAP

FIG. 1. Index map indicating the location of the buried fossil assemblage (SE¼ of NE¼ Sec. 2, T. 5E., R. 6W., Monroe County, Michigan).

mollusks which were found in the four faunal zones is reported in Table I. Most abundant were remains of fingernail clams, especially species of *Sphaerium* and *Pisidium*. Barry B. Miller, who identified the mollusks, makes the following statement:

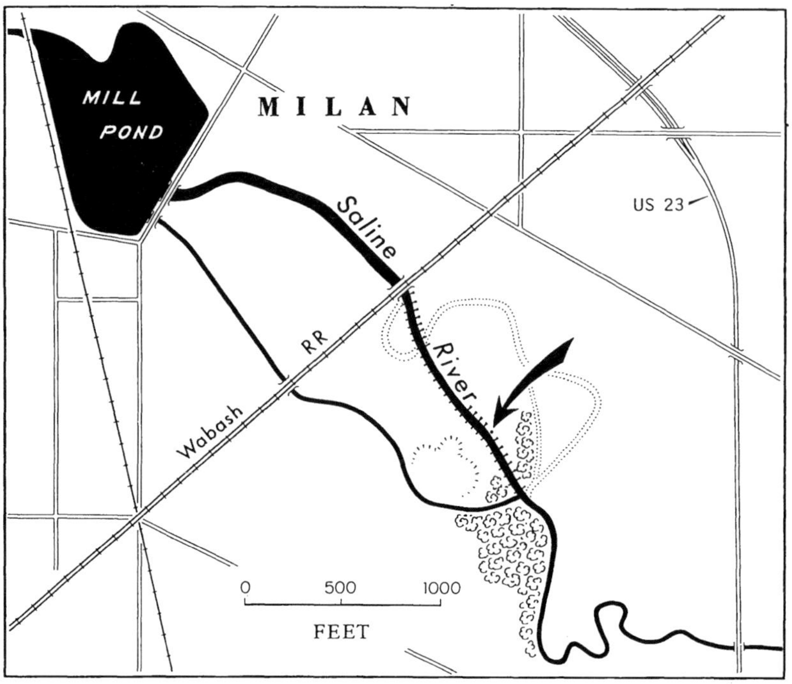

FIG. 2. Map showing the exact location of the buried assemblage in relation to the Saline River, southeast of the village of Milan, Michigan.

"The local environmental conditions can only be reconstructed in a very general way, because of the difficulty of evaluating the presence (or absence) of species, in so small a sample. The predominance of sphaeriids, most of which require permanent water situations, taken together with the relatively large number of *Pisidium dubium* and *P. fallax,* species that are usually found in streams, suggests the presence of a small stream. The few terrestrial species represent mostly woodland forms, which are usually associated with oak-hickory forests."

Lack of abrasion of the shells indicates that the faunal assemblage lived close to the site of deposition. All species in the list in Table I are reported to be present in southeastern Michigan. The absence of certain species which would be expected in the modern stream assemblage in this region is not considered to be of any climatic significance, but simply reflects incomplete sampling from one locality.

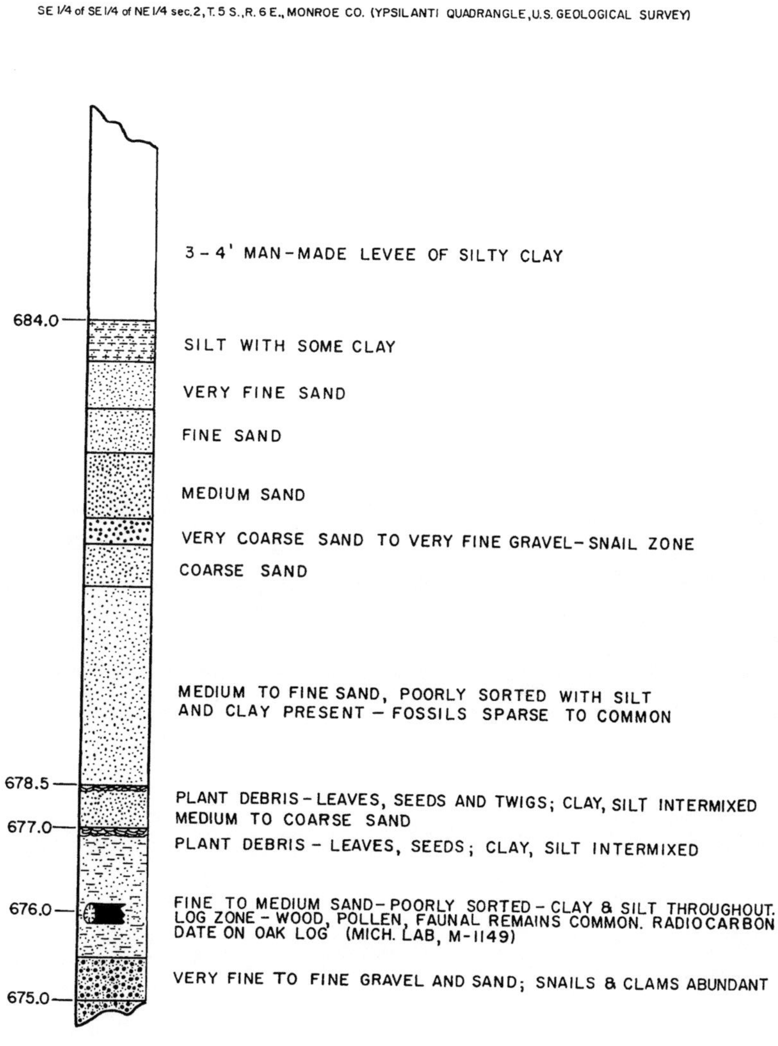

FIG. 3. Columnar section of an excavated Saline River floodplain at Milan, Michigan.

The plant remains which were recovered from the sediments are summarized in Fig. 4. These plant remains include wood, fruits and seeds, and pollen.

The largest log, approximately ten inches in diameter and more

TABLE I
ABUNDANCE OF THE BURIED MOLLUSKS FROM FOUR FAUNAL ZONES

Faunal List	Faunal Zones			
	No. 1	No. 2	No. 3	No. 4
Campeloma decisum (Say)	?1	3	1	—
Amnicola integra Say	12	5	—	—
Pomatiopsis cincinnatiensis (Lea)	7	—	—	—
P. lapidaria (Say)	12	—	—	—
Pyrgulopsis letsoni (Walker)	2	1	—	—
Goniobasis livescens (Menke)	4	?2	—	1
Pleurocera acuta Rafinesque	17	4	—	—
Fossaria sp. ?	1	—	—	—
Ferrissia cf. *tarda* (Say)	1	2	—	—
Physa sp. cf. *P. gyrina* (Say)	3	?1	—	—
Strobilops labyrinthica (Say)	4	?1	—	—
Gastrocopta contracta (Say)	1	2	—	—
Vertigo sp.	1	—	—	—
Anguispira alternata (Say)	3	—	—	—
A. kochi (Pfeiffer)	?3	—	—	1
Discus cronkhitei (Newcomb)	1	—	—	—
Helicodiscus parallelus (Say)	9	4	1	-
Hawaiia minuscula (Binney)	—	1	—	—
Retinella electrina (Gould)	2	?2	—	—
Zonitoides arboreus (Say)	1	—	—	—
Allogona profunda (Say)	—	—	—	1
Stenotrema hirsutum (Say)	4	?1	—	—
cf. *Succinea*	1	—	—	-
Elliptio ?	½	—	—	—
Sphaerium fabale Prime	5/2	2/2	—	?½
S. striatum Lamarck	118/2	13/2	?1	—
Pisidium cf. *casertanum* (Poli)	?½	?9/2	—	—
P. sp. cf. *P. adamsi* Prime	?2/2	—	—	—
P. dubium (Say)	25/2	3/2	—	—
P. fallax Sterki	21/2	4/2	—	—
P. compressum Prime	13/2	19/2	—	—

than eight feet long, is a species of red oak (*Quercus* sp.). A portion of this wood (sample M-1149) was submitted to the University of Michigan Phoenix Memorial Laboratory for radiocarbon dating. The date reported is 4080 ± 200 years B.P. Other logs and branches in-

Tree Pollen		%ΣAP	Fruits	Wood
Picea	(spruce)	0.3		
Pinus cf. strobus	(white pine)	0.9		
Pinus cf. resinosa	(red pine)	0.8		
Acer	(maple)	1.0	×	
Betula	(birch)	0.7		
Carpinus-Ostrya	(blue beech-hornbeam)	0.7		
Carya	(hickory)	5.6		×
Celtis	(hackberry)	0.4		
Cornus	(dogwood)	0.2		
Corylus	(hazel)	0.1		
Fagus	(beech)	4.5	×	
Fraxinus	(ash)	4.4		
Juglans cf. cinerea	(butternut)	6.1		
Juglans cf. nigra	(walnut)	11.0		
Liquidambar	(sweet gum)	0.1		
Liriodendron	(tulip poplar)	0.1		
Nyssa	(tupelo)	0.2		
Platanus	(sycamore)	8.4		×
Prunus	(cherry)	—	×	
Quercus	(oak)	28.0		×
Tilia	(basswood)	16.7	×	
Ulmus	(elm, white and red)	10.3		×
Non-Tree Pollen (NAP)				
Ambrosia	(ragweed)	2.9		
Chenopod-Amaranth	(goosefoot-amaranth)	0.1		
Compositae (exc. Ambrosia)	(aster family)	1.4		
Cruciferae	(mustard family)	0.4		
Graminae	(grass family)	1.2		
Nuphar-Nymphaea	(waterlilies)	0.3		
Sagittaria	(arrowhead)	0.1		
Sparganium	(bur-reed)	1.6		
Thalictrum	(meadow-rue)	0.3		
cf. Parthenocissus	(virginia creeper)	1.0		
cf. Urticaceae	(nettle family)	0.2		
Vicia	(vetch)	—	×	
Spores				
Osmunda cf. cinnamomea	(cinnamon fern)	0.3		
Polypodiaceae	(polypody fern fam.)	0.1		
Undetermined trilete spores (fern?)		0.8		
Fungal remains (spores, sporangia)		4.4		
Unidentified	(mostly NAP)	2.4	4 types	

Pollen: 0 — 10 — 20 — 30 percent of total tree pollen

Total tree pollen (AP)	= 1021
Total nontree pollen (NAP)	= 100
Fern spores	= 11
Unidentified	= 24
Total pollen and spores counted	= 1156

FIG. 4. Plant remains buried by a Saline River terrace at Milan, Michigan.

clude sycamore (*Platanus occidentalis*), hickory (*Carya* sp.) and two species of elm (*Ulmus americana* and *U. rubra*).

Large numbers of matted leaves were found above the log zone, but, due to the advanced state of decay, none was identified. Several maple (*Acer* sp.) fruits were found, in addition to three basswood capsules (*Tilia americana*), two chokecherry seeds (*Prunus* sp.), one legume seed, apparently vetch (*Vicia* sp.), and four types of unidentified fruits and seeds.

A small amount of the sandy silt enclosing the large red oak log was subjected to pollen analysis. Samples were prepared and treated chemically according to the methods described by Erdtman (1943). More than 1100 pollen and spore types were identified and counted; of these, slightly more than 1000 were pollen grains of trees (arboreal pollen). The pollen spectrum derived from the silt is presented in Fig. 4. The frequency of each pollen and spore category was calculated as a percentage of the total number of tree pollen grains (%AP). Of the tree pollen, only 2% was from coniferous trees; the remaining arboreal pollen was derived from nineteen types of deciduous trees. In order of decreasing abundance the hardwood tree pollens found in frequencies greater than 4% were: oak (*Quercus*), basswood (*Tilia*), walnut (*Juglans* cf. *nigra*), elm (*Ulmus*), sycamore (*Platanus*), butternut (*Juglans* cf. *cinerea*), hickory (*Carya*), beech (*Fagus*), and ash (*Fraxinus*).

Nontree pollen (NAP) was present in low frequencies. Ragweed (*Ambrosia* sp.), other Compositae, and grass pollen were found in small but significant numbers. Aquatic plants were represented by pollen of bur-reed (*Sparganium* sp.) and very low frequencies of waterlily (*Nuphar* or *Nymphaea*) and arrowhead (*Sagittaria* sp.) pollen. Certain nontree elements in the pollen spectrum were undoubtedly members of the lake plain forest association of the period. Especially interesting in this respect was the presence of significant numbers of Virginia creeper (*Parthenocissus* cf. *inserta*) pollen. Pollen of this genus is seldom identified in pollen analyses. Meadow-rue (*Thalictrum* sp.), nettles (*Urticaceae*), polypody ferns (*Polypodiaceae*), and the cinnamon fern (*Osmunda cinnamomea*) were other species associated with the hardwood forest.

Discussion

Interpretation of pollen statistics in an attempt to reconstruct the former plant community in an accurate manner is a vexing problem. Pollen frequencies cannot be accepted as representative of the frequency of plants in the community. We know that certain pollen types are indicative of the regional vegetation because they are carried great distances by wind. Other pollen types are disseminated only over short distances and represent plants which are very local in occurrence (Benninghoff, 1960). Over- and underrepresentation also stems from the capacity of certain plants to produce pollen much more copiously than others. Oak and pine are notoriously overrepresented in pollen spectra. A study in Vermont (Davis and Goodlett, 1960) suggests that beech, ash, elm, blue beech-hornbeam (*Carpinus-Ostrya*), and spruce (*Picea*) in pollen spectra are about proportional to their per cent basal area in the community. Others, such as maple (*Acer*) and basswood are seriously underrepresented in pollen spectra. While these data from studies in New England cannot be applied directly to Michigan pollen statistics, certain applications seem valid. Without doubt oak is seriously overrepresented in the pollen spectrum. Certainly maple, and probably basswood, in spite of its high pollen frequency, were more abundant in the vicinity of the depositional area than the pollen statistics would indicate.

The presence of high percentages of walnut, butternut, and sycamore pollen seems rather significant. The distributional range of each of these species extends northward only to central lower Michigan. Apparently minimum annual temperature and length of the frost-free period are critical determiners of the northern limits of these species (U.S.D.A., Forest Service 1957, 1958). High pollen frequencies for these trees in southeastern Michigan are therefore rather surprising. It seems probable that the climate under which this assemblage developed was somewhat warmer than the present. The radiocarbon date on wood from the deposits (M-1149, 4080 ± 200 years B.P.) falls within the postglacial hypsithermal interval (Deevy and Flint, 1957). The pollen spectrum described above is consistent with the prevalent concepts of the type of vegetation in southern Michigan during the hypsithermal interval.

Examination of the columnar section from which the plants and animals were recovered (Fig. 3) reveals that some mechanism in the stream's regimen caused burial of the log zone by eight feet of detrital material. This burial can be interpreted as reflecting a change in the water plane of the Lake Erie basin into which the ancestral Saline River flowed. A rise in the water level of the lake basin apparently caused lateral planation and frequent flooding upstream due to a decrease in stream gradient.

The Lake Stanley-Early Lake Erie stage of Great Lakes history occurred 5,000–6,000 years B.P. (Hough, 1958, p. 283). The streams draining into the Great Lakes at this low water stage must have degraded their valleys. The Nipissing Great Lakes phase was initiated as a result of uplift of the North Bay outlet in Ontario, causing the water level to rise to 605 feet A.T. (Hough, 1953). During the Nipissing phase the upper lakes began to drain through the Port Huron channel into Lake Erie. The rise in the level of ancestral Lake Erie at this time may have caused the streams to aggrade their valleys. The log zone with the associated biotic remains represents a forested floodplain which was destroyed and buried during this aggradation. As Lake Erie later approached its present elevation of 580 feet A.T., the Nipissing terrace was partly excavated by downcutting of the Saline River.

The assignment of this terrace and fossil assemblage to early Nipissing time is substantiated by the radiocarbon date of 4080 ± 200 years (M-1149). This date correlates with the Xerothermic-Nipissing stage, which is accepted by most students of Great Lakes history (Zumberge and Potzger, 1956, p. 277).

Summary

We have described an assemblage of plant and animal remains which was exposed upon excavation of a terrace of the Saline River at Milan, Michigan. The mollusks, mostly of the fingernail clams, *Sphaerium* spp. and *Pisidium* spp., and some woodland snails, suggest a small stream habitat with adjacent oak-hickory forests. The most obvious plant remains are logs and branches of five kinds of deciduous trees. Wood from the largest log, red oak, has been dated at 4080 ± 200 years B.P. (University of Michigan Memorial-Phoenix Project, M-1149). Although some fruits and seeds were recovered, the most complete evidence for the vegetation and climate of the period is revealed by pollen analysis of the silt enclosing the dated log. Oak, basswood, walnut, elm, sycamore, butternut, beech, and ash are predominant members of the pollen flora. High pollen frequencies of sycamore, butternut, and walnut suggest that the climate may have been warmer than at present in southeastern Michigan. The radiocarbon date agrees well with the period of maximum development of vegetation as a response to the climatic trends of the hypsithermal interval which has previously been dated in pollen profiles. It is likely that burial of the assemblage occurred because of aggradation of the stream after a rise in the level of ancestral Lake Erie in Nipissing times.

LITERATURE CITED

BENNINGHOFF, W. S. 1960. Pollen Spectra from Bryophytic Polsters, Inverness Mud Lake Bog, Cheboygan County, Michigan. PMASAL, 45 (1959): 41–60.

DAVIS, MARGARET BRYAN, AND JOHN C. GOODLETT. 1960. Comparison of the Present Vegetation with Pollen-Spectra in Surface Samples from Brownington Pond, Vermont. Ecol., 41 (2):346–357.

DEEVEY, EDWARD S., AND RICHARD F. FLINT. 1957. Postglacial Hypsithermal Interval. Science, 125 (3240): 182–184.

ERDTMAN, G. 1943. An Introduction to Pollen Analysis. Waltham, Mass.: Chron. Bot.

HOUGH, J. L. 1953. Final Report on the Project, Pleistocene Chronology of the Great Lakes Region: Office of Naval Research contract No. N 6 ori-07133, Project NR-018-122. Univ. Ill., Urbana (mimeographed report).

——. 1958. Geology of the Great Lakes. Urbana, Ill.: Univ. Ill. Press.

U. S. DEPARTMENT OF AGRICULTURE, Forest Service. 1957. Miscellaneous Release 22. Central States Forest Experiment Station, Columbus 15, Ohio.

——. 1958. Miscellaneous Release 26 and 28. Central States Forest Experiment Station, Columbus 15, Ohio.

ZUMBERGE, JAMES H., AND JOHN E. POTZGER. 1956. Late Wisconsin Chronology of the Lake Michigan Basin Correlated with Pollen Studies. Bull. Geol. Soc. Am., 67: 271–288.

Papers of the Michigan Academy of Science, Arts, and Letters
Vol. XLVII, 1962 (1961 Meeting)

BOTANY

THE TAXONOMIC STATUS AND ARCHAEOLOGICAL SIGNIFICANCE OF A GIANT RAGWEED FROM PREHISTORIC BLUFF SHELTERS IN THE OZARK PLATEAU REGION[1]

WILLARD W. PAYNE AND VOLNEY H. JONES

The University of Michigan

Archaeologists have excavated a number of bluff shelters of prehistoric Indian occupation in a restricted area in the northwestern corner of Arkansas and adjacent Missouri. M. R. Harrington made extensive excavations in 1922 and 1923 for the Museum of the American Indian, and field parties of the University of Arkansas, under the direction of S. C. Dellinger, carried out a program of excavation during the decade preceding 1938. The primary cultural pattern represented has been designated "The Ozark Bluff-Dweller Culture" (Harrington, 1924, 1960). The major broad archaeological period with which the bluff-shelter manifestation seems to affiliate is that known as Middle Woodland, although there was, perhaps, sporadic occupation earlier, and some persistence into the succeeding Mississippi Period. Griffin (1952, pp. 358–364) outlines these periods. The time period of occupation of the shelters can be bracketed chiefly between the beginning of the Christian era and about 1200 A.D. The bluff shelters are not caves in a strict sense, but comprise areas weathered from beneath the limestone escarpment, which offer protection from the weather and are suitable for domiciliary purposes and storing of food and other goods. Many of the shelters have remained quite dry and have provided an extensive and well-preserved inventory of normally perishable materials.

In some of the shelters fruits of the giant ragweed (*Ambrosia trifida* L.) were found either stored in cache pits or loose in the occupational debris on the floors of the shelters. It has been generally

[1] Publication No. 26 on Atmospheric Pollution by Aeroallergens. Research carried out, in part, under research grant No. E-1379 from the National Institute of Allergy and Infectious Diseases, Public Health Service.

assumed that the fruits contributed to the food supply. In some instances, the circumstances of storage have been interpreted as suggesting that small amounts of the fruits were preserved as "seed" for planting the following spring. Gilmore (1931) also considered that the fruits exhibited color differences and large size, contrasting consistently in these respects with fruits of wild plants, which he took to be further evidence of domestication. Rousseau (1944) reported that giant ragweed fruits from a bluff shelter were so similar to those of an artificial tetraploid which he had created, that they could be assumed to have been produced by a polyploid population. He described a new variety, *A. trifida* var. *polyploidea* Rousseau (1944, p. 215), to accommodate his artificial tetraploids and the presumed tetraploids from the sites. Our purpose here is: (1) to review the archaeological history of the ragweed fruits; (2) to compare the bluff-shelter fruits with those from wild plants to ascertain whether or not they exhibit consistent differences; (3) to examine the taxonomic status of Rousseau's new variety; and (4) to assess the probable economic purpose of the bluff-shelter fruits.

In the shelters which he excavated, Harrington found a complement of conventional Indian crop plants including maize (*Zea mays*), beans (*Phaseolus vulgaris*), pumpkins and squashes (*Cucurbita* spp.), gourds (*Lagenaria siceraria*), and sunflower (*Helianthus annuus*). All of these, except possibly the last, appear to have reached the southeastern United States as domesticated plants from tropical America. Harrington related that quantities of these products were found in cache pits and that smaller amounts also had been placed in bags. From these circumstances, he interpreted the smaller lots as likely seed stock put away for spring planting (Harrington, 1924, pp. 6–7). He also reported that seeds and fruits of several other plants unfamiliar to him were given similar treatment, indicative of their use as seed stock, and presumed that they too were cultivated plants. In a study of Harrington's bluff-dweller plants, Gilmore (1931, p. 86) provided identifications of these additional putative domesticates, listing such genera as *Chenopodium, Amaranthus, Ambrosia, Iva,* and *Phalaris.* All of these genera are constituents of the modern wild flora of the Ozark Plateau, so that Gilmore presumed that at least some of them were local additions to the introduced crop complex. In support of their probable domestication, he repeated the argument of special handling very much as stated by Harrington, but also added the argument that most of these genera are notorious for their commensalism with man.

In discussing the marsh-elder (*Iva*) and the ragweed (*Ambrosia trifida*), Gilmore (1931, pp. 86–87 and 101) also stated that the fruits were of "select appearance" and much larger than those of normal wild representatives. He described the fruits of the giant ragweed as being four or five times the size of any which could be found growing wild. He also referred to their uniformly light color as contrasted to that of fruits of wild plants, which latter he described as variegated and sometimes dark to almost black. He attributed these modifications to cultivation, seed selection, and "improvement" of the species by domestication. Enlarged size was not claimed explicitly for *Chenopodium, Amaranthus,* and *Phalaris*, but seemed to be implied.

In discussing the giant ragweed fruits, Gilmore referred to their "abundance" and to "considerable quantities" of them from the sites, but unfortunately neither he nor Harrington (1924, 1960) gave precise quantitative data, and neither provided the names of the shelters from which they came or the provenience of the fruits within the sites. When Gilmore came to The University of Michigan,[2] he brought with him one lot of 36 fruits of giant ragweed, bearing the Museum of the American Indian catalogue no. 11/7596. It is annotated as having been collected by Harrington, and as coming from pits Nos. 7 and 8, Alum Cave Bluff Shelter, Benton County, Arkansas. A letter (dated March 10, 1961) from Frederick J. Dockstader, director of that museum, to Jones confirmed these data from their catalogue record. Dockstader added that this specimen, from which Gilmore's sample was taken, consisted at that time of about a heaping tablespoonful of fruits, that they were not then in a bag, and that there was nothing in their records to indicate that they were found in a bag. Harrington (1960, pp. 85–90) describes the Alum Cave Bluff Shelter, giving its location as on the White River some 6.5 miles northeast of the town of Rogers. His plate 19 presents a photographic

[2] Melvin R. Gilmore (1868–1940) was associated with Harrington on the staff of the Museum of the American Indian from 1923 to 1928. He joined the staff of the Museum of Anthropology of the University of Michigan in 1929, where Jones worked closely with him from 1931 to 1940. Dr. Gilmore's collections, notes, and correspondence were left to the University and have been available to us.

view of the shelter, and pits Nos. 7 and 8 are discussed (pp. 88–89), but Harrington does not refer to ragweed fruits from them.

Gilmore submitted many of Harrington's bluff-dweller plant materials to the Bureau of Plant Industry, U. S. Department of Agriculture, and received extensive reports from W. E. Safford and others of that organization (*vide* Gilmore, 1931, p. 92). The documents concerning this transaction remain in Gilmore's files. In the inventory and reports, the single ragweed item is the Alum Cave lot, under its proper number. Whatever the actual frequency and amounts of ragweed fruits found in the shelters, we are left with positive knowledge of only this one lot from Harrington's excavations, and concerning it we have no certain affirmation that it was found under circumstances suggesting that it was seed stock.

Safford identified the ragweed fruits from Alum Cave as being *Ambrosia trifida* but was quite perfunctory about them as compared to his comments on some of the other materials. He remarked only on their characteristic form and made no note of anything at all unusual in their size or color. By way of contrast, he was exceedingly puzzled by the *Iva* fruits, which he stated resembled *Iva* but were "much larger than any species we know." It is thus quite clear that Gilmore's reaction to the size and color of the ragweed fruits was his own, and not taken from Safford's report.

The materials which accrued from the University of Arkansas bluff-shelter excavations under the direction of S. C. Dellinger are housed at the museum of that institution. Gilmore visited Fayetteville for several weeks in 1935 and 1936 to study the plant materials, but his extensive notes have not been published. It is clear from these notes that the assemblage of cultivated plants represented is essentially the same as in Harrington's materials, including the same inventory of allegedly locally domesticated plants. Gilmore's notes refer to seven lots of ragweed fruits originating from five different shelters. The shelters and Arkansas catalogue numbers of the lots follow: Butler Bluff Shelter (B-439), Eden Bluff Shelter (E-1643 and E-1706), Green Bluff Shelter (GB-25 and GB-415), Whitney Bluff Shelter (WT-33), and Marble Bluff Shelter (MM-327a). All of these shelters are in Benton County, Arkansas, except for Marble Bluff Shelter, which is in Searcy County. The quantities of fruits and their provenience in the shelters are not stated in the notes, and there are no detailed observations except that the GB-415 is recorded as having "extra large seeds." In general, Gilmore's remarks on the Dellinger collections repeat those published on the Harrington materials, including the claim of large size, light color, and probability of cultivation.

In the fall of 1936, Jones examined the bluff-dweller ragweeds at the University of Arkansas very briefly, selecting samples of three lots to be transferred to The University of Michigan for further study.[3] The lots selected will be listed later in this paper, in which tabulation it will be observed that these samples were found in cache pits or in the occupational debris on the shelter floors, and that none was found in bags or under very special circumstances.

In company with Professor Dellinger (October, 1936), Jones also collected fruits of *Ambrosia trifida* from the wild for comparative purposes. These were obtained from robust plants averaging about 12 feet tall, growing in low ground near the White River, Washington County, Arkansas. It was felt that, as these fruits were obtained from plants growing quite near the bluff-shelter area, they would represent a good sample of modern fruits for comparison. This collection is included in Table I as a part of the Arkansas entry.

The Newt Kash Hollow Shelter, Menifee County, Kentucky, which was excavated by W. S. Webb and W. D. Funkhouser (1936), is similar to the Ozark Plateau shelters in its nature and content. The crop plant inventory as reported by Jones (1936) is essentially the same, but somewhat more restricted in quantity and variety. Only a single fruit of *Ambrosia trifida* was found, and this under circumstances not clearly indicating that it was an economic plant. *Chenopodium, Phalaris,* and *Iva* were common in the debris, and these were also present in quantity in desiccated human feces from the site, along with sunflower seeds. It is interesting to note that the seeds and fruits of these plants were eaten entire, without hulling or grinding, and that the seed coats comprise the bulk of the feces. Indications of the ingestion of ragweed fruits were looked for carefully, but no evidence of this was found. The Newt Kash site appears to be classifiable to the Early Woodland archaeological period (Griffin, 1952, p. 357), and radiocarbon dates of 2600 and 2650 $\pm$ 300 years before the present have been derived for materials from it (Crane, 1956, p. 665, date

[3] We wish to express our gratitude to the University of Arkansas, and to Professor Dellinger personally, for making these specimens available to us and for permitting the use of data concerning them.

M-31). It is thus a somewhat earlier manifestation than the Ozark Plateau shelters.

The lots of ragweed fruits from bluff shelters which were available to us for study are listed below:

Alum Cave Bluff Shelter, Benton County, Arkansas

A lot of 36 fruits (sample from larger quantity) from pits Nos. 7 and 8, collected in 1922 by M. R. Harrington for the Museum of the American Indian, New York.

Museum of the American Indian, catalogue no. 11/7596

Museum of Anthropology, University of Michigan, catalogue no. 13007

Green Bluff Shelter, Benton County, Arkansas

A lot of 22 fruits (sample from larger quantity) from a cache pit about 18 inches deep, lined with grass. The fruits were lying on the grass and covered with mussel shells. Collected by University of Arkansas field party under direction of S. C. Dellinger, *circa* 1935.

University of Arkansas, catalogue no. GB-25

Museum of Anthropology, University of Michigan, catalogue no. 15480

Butler Bluff Shelter, Benton County, Arkansas

A lot of 8 fruits (entire collection) found loose in the occupational debris at a depth of 15″. Collected by University of Arkansas field party under direction of S. C. Dellinger, *circa* 1935.

University of Arkansas, catalogue no. B-439

Museum of Anthropology, University of Michigan, catalogue no. 25138

Eden Bluff Shelter, Benton County, Arkansas

A lot of 2 fruits (entire collection) isolated from a miscellaneous assortment of seeds and fruits found in the fill of the shelter (depth not recorded). Collected by University of Arkansas field party under the direction of S. C. Dellinger, *circa* 1935.

University of Arkansas, catalogue no. E-1706

Museum of Anthropology, University of Michigan, catalogue no. 15478

In 1936, Jones submitted to S. F. Blake, of the Bureau of Plant

Industry, several lots of fruits of *Ambrosia trifida* and *Iva ciliata* from bluff shelters in order to obtain his opinion of their botanical nature and status. In his resulting publication (1939) Blake does not cite the precise origin of the samples, but lumps them together for discussion. The giant ragweed specimens which were sent to him were: Alum Cave (MAI 11/7596), Green Bluff (GB-439), and the single fruit from Newt Kash Hollow, Kentucky. Along with these was forwarded the collection of modern fruits from along the White River as a basis for comparison, and several lots of marsh-elder fruits from both Ozark shelters and Newt Kash Hollow. On preliminary examination of these he remarked (letter to Jones, July 16, 1936):

"In each case [*Iva* and *Ambrosia*], the achenes are about twice as big as the largest wild materials I have been able to find in the National Herbarium. It seems extremely probable that your theory that these achenes come from plants selected and improved by the Indians is correct. They are so huge, as compared with their nearest living relatives, that it seems desirable to describe them as new varieties, and thus draw attention to them and provide them with a handle for reference." He eventually did this for the marsh-elder, creating the new variety *Iva ciliata* var. *macrocarpa* Blake (1939, p. 84), based entirely upon the bluff-dweller specimens, which he interpreted as "an ancient cultivated strain obtained by selection, and now extinct."

Blake (1939, pp. 83–84) remarked on his initial inclination to give the *Ambrosia* fruits varietal distinction, and explained his subsequent decision that this would not be justified. In explaining his change of mind on this point, he commented that Gilmore, in estimating the ragweed fruits as four or five times as large as normal, must have been comparing them with rather small modern ones. Blake describes the range in length of the bluff-shelter ragweed fruits as from 7 to 13 mm., stating that individuals over 10 or 11 mm. long were decidedly exceptional. He was able to collect fresh specimens in the vicinity of Washington, D. C., which ranged from 7.5 to 10.5 mm., and which he considered matched well with the bluff-shelter fruits, except for the few larger ones in the latter. He made no comments on any color difference of the bluff-shelter fruits. His ultimate conclusion was that varietal separation would not be justified, and this, in effect, would seem to commit him to the opinion that no effects attributable to domestication were evident.

In 1940, Jacques Rousseau addressed a letter to Gilmore (dated September 12, 1940) including the information that by the use of colchicine he had produced artificially what he thought (then unverified) to be a tetraploid variety of *Ambrosia trifida* with fruits "twice the diameter" of those of normal (diploid) plants. Rousseau requested from Gilmore fruits of the bluff-shelter ragweeds for comparison. As the letter was received after Gilmore's death, these were provided by Jones. The fruits which were sent to Rousseau were taken from the Alum Cave lot (MAI 11/7596). On receipt of these he wrote (letter, 9/27/40) that the Alum Cave fruits were "of the same size" as those of his presumed tetraploid. In 1944, Rousseau published a report in which he stated that cytological study had verified the polyploid nature of his plants, and recorded his conviction that the bluff-shelter fruits represented chance occurrences of polyploidy which had been selected and maintained by the Indians because of their higher yield. To accommodate the artificially produced tetraploids and the bluff-shelter fruits, he established a new variety of giant ragweed, *Ambrosia trifida* var. *polyploidea* Rousseau (1944, p. 215).

Although Rousseau did not state the source of the diploid plants used in his research, presumably they were collected near Montreal, Quebec, where his laboratories were situated. We were unable to locate specimens of the polyploid giant ragweeds produced by Rousseau or of the parent population from which they were derived. Measurements for these fruits (Tab. I) are extrapolated from a slightly enlarged photograph in his publication (1944, p. 212).

Fruiting specimens of *Ambrosia trifida* were acquired from various herbaria[4] for this study. Added to those collected by Payne, recent collections were available from 29 states, plus Washington, D. C., Alberta, Ontario, Saskatchewan, and Chihuahua. Average measurements of fruits from each of these areas are shown in Table I. Tabulated with these, for purposes of comparison, are measurements from the lots of archaeological fruits, and from Rousseau's diploid and tetraploid populations. In Fig. 1, typical fruits are shown super-

[4] We wish to express our appreciation to the herbaria of the following institutions for providing some of the specimens used: The New York Botanical Garden; Southern Methodist University; University of Arizona; University of Arkansas; University of California, Berkeley; University of Cincinnati; University of Illinois; University of Kentucky; University of Michigan; University of North Carolina; University of Tennessee; Virginia Polytechnic Institute; West Virginia University.

TABLE I

AVERAGE MEASUREMENTS OF FRUITS USED IN THIS STUDY*

Fruits from:	Total Length in mm.	Body Length in mm.	Body Width in mm.	Number of Specimens	Number of Fruits
Massachusetts	15.2	11.5	6.4	3	12
Michigan	12.0	9.5	6.8	2	8
Rousseau's 4n	11.7	9.6	6.6		7
West Virginia	11.5	8.7	6.4	5	19
South Carolina	10.5	8.3	6.1	1	4
Green Bluff (D)†	10.2	8.1	4.9		12
North Carolina	10.2	7.7	5.5	5	20
Illinois	9.9	7.7	5.4	3	12
Tennessee	9.5	7.3	5.6	2	6
Eden Bluff (C)†	9.2	7.4	4.9		2
Connecticut	9.1	7.3	5.0	1	4
New York	9.1	6.9	5.0	4	15
Kentucky	9.0	6.3	4.6	1	3
North Dakota	8.8	7.1	4.0	1	4
Pennsylvania	8.7	7.0	4.3	1	2
New Jersey	8.7	6.5	4.8	3	10
Alum Cave (A)†	8.5	6.6	4.4		36
Indiana	8.4	6.7	4.7	2	8
Iowa	8.1	6.8	4.6	3	12
Butler Shelter (B)†	8.1	6.7	4.6		8
Ohio	8.0	6.0	4.6	2	8
Alberta	7.8	6.2	4.4	1	4
Rousseau's 2n	7.7	6.3	4.4		7
Ontario	7.5	6.4	4.3	1	4
Washington, D. C.	7.2	5.4	4.5	1	4
Missouri	7.0	5.5	3.8	1	4
Wyoming	6.9	5.6	3.3	1	4
Kansas	6.8	5.7	4.2	4	16
Colorado	6.8	5.3	3.4	2	8
Saskatchewan	6.7	5.1	4.0	1	4
Arkansas	6.5	5.1	4.0	7	27
Minnesota	6.0	4.8	3.3	1	4
Nebraska	5.6	4.6	3.1	1	4
New Mexico	5.4	5.0	3.1	1	4
South Dakota	5.4	5.0	3.1	1	2
Arizona	5.3	4.5	3.2	3	12
Florida	5.2	4.2	2.6	1	4
Texas	5.1	4.5	3.3	6	24
Louisiana	4.9	4.3	2.8	2	8
Oklahoma	4.8	4.1	3.1	2	8
Chihuahua	4.5	3.8	3.1	2	8

* Measurements upon which these averages are based were taken from the fruits as follows: total length, length of the entire fruit; body length, length of the fruit minus the beak; body width, widest diameter of the fruit body, excluding the variable spines.

† Fruits from the prehistoric bluff shelters.

imposed upon a map of the United States in the areas from which they were collected. Also shown are fruits selected from the archaeological material. Due to the relatively small number of fruiting specimens available, the figures shown in the table and the photograph represent only approximations of the real average sizes for the localities given.

No fruits from known polyploid parents were available for study, as polyploid individuals have not been reported from naturally occurring populations of giant ragweed. Cytological examinations of several populations of giant ragweed by Payne (unpublished) and other workers have shown only the diploid constitution, $2n = 24$ (Cooper and Mahony, 1935; Jones, 1933, 1943; Mulligan, 1957; Stebbins, 1945). The size range of the presumably diploid plants is such that the fruits from the bluff shelters can be included well within its limits. It is assumed that statements made by the workers mentioned above to the effect that the prehistoric fruits are much larger than any known today stemmed from a lack of familiarity with the fruits of the species as it exists throughout its range.

It can be seen in Fig. 1 that there are two basic forms of the fruits of modern giant ragweeds. The first form is centered in the deciduous forest zone of the eastern United States and southeastern Canada. It is typified by its large size and by an obdeltoid or triangular shape, the latter character being correlated with the possession of variously developed but usually prominent spines. The second fruit type is found in the western and southwestern United States and Mexico, in areas commonly associated with short grass prairie. This form is smaller, ovate to nearly elliptical in shape, with short, appressed or rudimentary spines. Where the two fruit types meet, as they do in the region of the Ozarks, many intermediates are found. It is probably safe to assume that this condition is comparable to that which prevailed at the time of the Bluff-Dweller culture, although civilized man has undoubtedly brought about an even greater mixing of genotypes in recent times. These observations are significant in the discussions presented below.

Several statistics[5] were derived from the measurements taken from

[5] Thanks are due R. D. Remington and J. B. Keller of the University of Michigan Department of Biostatistics for statistical treatments used in this study. Actual figures are not included in the text because of their bulk, but can be supplied to interested readers.

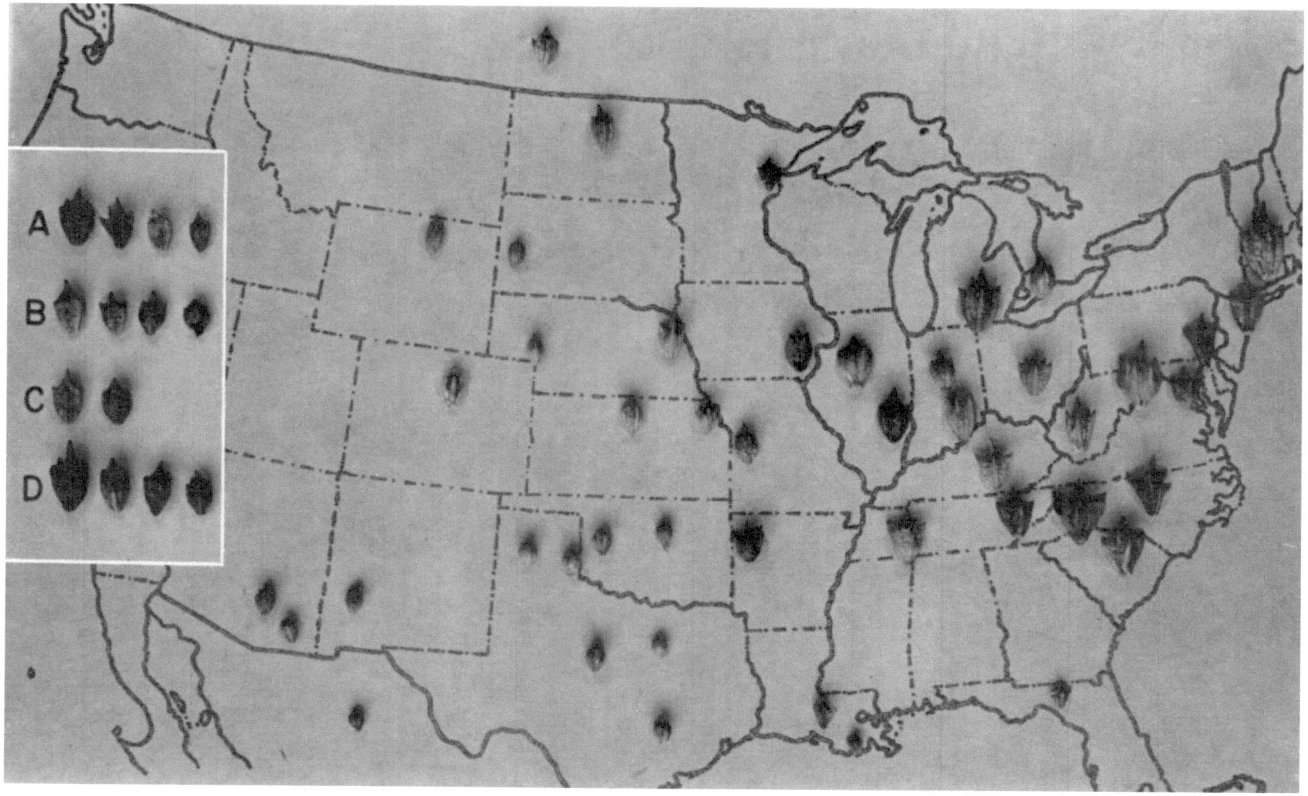

Fig. 1. Fruits of giant ragweed selected from fruiting specimens collected in the area indicated on the map. At the left: *A*. fruits from Alum Cave Bluff Shelter, *B*. fruits from Butler Bluff Shelter, *C*. fruits from Eden Bluff Shelter, *D*. fruits from Green Bluff Shelter. The bluff shelters are all in the northwestern corner of Arkansas in the Ozark plateau.

the fruits used in this investigation. Given below are means and differences for total length (length of the entire fruit), body length (length of the fruit minus the beak), and width (widest diameter of the fruit body, excluding the spines):

	Modern Fruits, Mean	Prehistoric Fruits, Mean	Difference
Total length	7.828 mm.	8.859 mm.	1.031 mm.
Body length	6.205 mm.	6.898 mm.	0.693 mm.
Width	4.421 mm.	4.538 mm.	0.117 mm.

Although the averages are similar, a discriminant function analysis which simultaneously takes into account the differences between all means shows that the means are not the same. The prehistoric fruits are actually somewhat larger, on the average, than the average for the species as a whole. This difference can be satisfactorily explained by assuming that the prehistoric fruits are more closely related to the large-fruited specimens of the east than to the small-fruited specimens of the west. Characters of general morphology and spininess also indicate this to be the case, similar well-developed spine types being found today in both the geologically ancient Appalachian and Ozark regions. The differences in mean sizes need not be interpreted as indicating polyploidy or selection and cultivation by man any more than would larger than average means of fruits from Michigan or West Virginia.

To establish the degree of uniformity in shape among the fruits, correlation between the three measurements was determined by the construction of variance-covariance and correlation matrices. It was determined that there was a high degree of correlation between the measurements of the modern fruits and less for the prehistoric fruits. This result is undoubtedly a reflection of the fact that the prehistoric fruits are from an area located between the two major fruit types. It is to be expected that many of the genotypes from this area would show the result of mixing of the factors governing size expression in the eastern and western forms. These results can also be interpreted as strengthening the arguments against cultivation and selection by the Indians, as such treatment is thought to bring about greater uniformity of form.

One of the best indications of whether or not the Bluff-Dwellers used the cached ragweed fruits as seed sources for spring planting may lie in a consideration of the overwintering requirements of the seeds and the circumstances of the caches. The cache pits from which the fruits were extracted are located well back in the caves where they would be protected from rain and snow, i.e., where their contents would remain dry throughout the winter. In a number of after-ripening experiments made by Payne, it has been observed that the embryos of *Ambrosia trifida* will not germinate without an over-wintering period during which they are kept wet. Similar results have been obtained by Davis (1930). To confirm this observation, tests were made on the germination of ten lots of fruits collected in the fall of 1960 from Wyoming, Colorado, Kansas, Nebraska, Iowa, Louisiana and Michigan. Portions of the collections which had been stored dry over the winter of 1960–61 were tested against portions of the same lots which had been after-ripened with techniques worked out by Payne and Kleinschmidt (1961). Although all of the lots which had been after-ripened wet showed normal percentages of germination, not a single seed germinated from the lots which had been stored dry over the winter. It is probable, therefore, that the giant ragweed fruits from the prehistoric cache pits would have been unsatisfactory sources of seed for spring planting.

As to statements that the fruits from the bluff shelters are uniformly light in color, it can be seen in Fig. 1 that this is not so.

Thus, no good evidence supports either the theory that the fruits from the Ozark shelters were cultivated or that they were polyploid. Rousseau's taxon, *Ambrosia trifida* var. *polyploidea,* has no known natural occurrence in either the past or the present, but is limited, in fact, to the forms produced in his laboratories. As such, it is felt that the establishment of a new variety to include them is not in accordance with currently accepted taxonomic practice as to the designation of varieties.

Whether or not giant ragweed was cultivated by the inhabitants of the Ozark bluff shelters, there still remains the very perplexing problem of the purpose of the fruits found in the sites. It seems clear that the Indians gathered the fruits and brought them into the shelters, and that in some instances the fruits were stored in cache pits. There is very little economic data of any kind on the ragweeds, and the amount on the fruits in particular is almost negligible. Gilmore (1931, pp. 86 and 101) was at a loss to interpret their use. He recalled the claim of an Arikara Indian that the fruits of giant ragweed were used as bait in snaring birds, and that the Arikara and

Omaha Indians used the flower heads to obtain a stain or dye. Gilmore's feeling, however, was that the fruits were probably intended as food, and he refers to them as a "grain crop." He tested the theory by eating a few fruits and reported them to be palatable and rich in food value. Others have also done this and the consensus seems to be that they are edible. However, the thick and spiny, indehiscent coat (involucre) does present a distinct hindrance to their ingestion. This would seem to militate against the economic practicality of using the fruits as anything more than an emergency or famine food.

Gilmore does not support his conjecture with any documentation, and indeed none seems available. None of the dictionaries of food plants or other compendia to which we have had access mentions the use of ragweeds as food. In a compilation of 1,112 species of plants utilized as food by Indians of North America (Yanovsky, 1936), there is no listing for any species of ragweed. The absence of ragweed in the dietary lists of various Indian tribes of the historic period indicates that if the ragweed was formerly used as food, its use dropped out at some earlier time. Ragweed fruits have not been reported from archaeological sites outside of the Ozark Plateau, except for the single fruit from Newt Kash Hollow, Kentucky. In the examination of several thousand lots of plant materials from North American sites, Jones has not observed ragweed fruits other than those from the bluff shelters.

In order to ascertain the food value of giant ragweed fruits, Jones in 1935 sent a quantity of fruits to Elias Yanovsky, Chemist, Carbohydrate Division, Bureau of Chemistry and Soils, U. S. Department of Agriculture. The fruits submitted to him had been collected by Gilmore in eastern Nebraska a year or so previous to that time. The analysis supplied by Yanovsky follows (letter to Jones, 4/19/35):

Moisture	7.7%
Reducing sugars	0.7%
Nonreducing sugars	—
Starch	—
Pentosans and hemicelluloses	—
Crude protein	16.5%
Ash	4.7%

It would appear that the ragweed has little to offer except a fair amount of protein. We have found no other assay of the fruits of ragweed with which to compare this analysis.

It is perhaps suggestive to note that although ragweeds are seldom recorded as Indian food, there is a considerable record of their use in medicine. The giant ragweed is not one of the species so used, and the medical applications involve parts of the plants other than the fruits. A western species, *Ambrosia psilostachya*, was utilized in medicine by several tribes in the Plains, Great Basin, Southwest, and California. Uses included making poultices of the leaves and stems, or brewing a tea or decoction from them. Pueblo Indians at Acoma, New Mexico, drank a tea in cases of difficult labor at childbirth (Swank, 1932, p. 26). The Gosiute of Utah made a poultice of leaves to be applied to sore eyes (Chamberlin, 1911, p. 361). Among the Kiowa of Oklahoma, a decoction is rubbed on sores as a healing agent, and may be used similarly on horses to treat sores which are slow in healing (Vestal and Schultes, 1939, p. 55). The Cheyenne Indians of the central plains drank an infusion of ragweed in instances of bloody stools and in treating colds and cramps of the bowels (Grinnell, 1923, p. 188). Carlson could obtain no data on utilization of *A. psilostachya* by the Comanche Indians of the southern Plains, but the suffix of their term for the plant indicates a medical association (Carlson and Jones, 1940, p. 540). Sparkman (1908, p. 228) says that *A. artemisiaefolia* was used as an emetic by the Luiseño Indians of southern California, but, as this species is not known from that area today, *A. psilostachya* is probably the species involved. There is a possibility that some of these medicinal uses may have been suggested, in line with the "doctrine of signatures," by the "bleeding" or exudation of reddish sap from cut plants, but this association is not apparent for all of the applications.

An Omaha Indian informed Gilmore (1919, p. 132) that their Oto neighbors treated nausea with *Ambrosia elatior*, scarifying the abdomen and applying to it the bruised leaves of the plant.

The only Indian "medicinal" use of *Ambrosia trifida* which we can find is that reported by Smith (1928, p. 210) for the Meskwaki of Wisconsin, who allegedly chewed the root of this plant to drive away fear at night. Vestal and Schultes (1939, p. 55) state that the plant itself was formerly held in fear by the Kiowa.

In view of the rather extensive medicinal use of various species of ragweeds among a variety of Indian groups, and because of the almost total absence of evidence for its use as food, it might be reasonable to suggest that the bluff-dweller fruits were connected in some way with medicine. It is true, however, that there is no direct evidence for the use of the fruits themselves in this way. The hypothesis that the fruits may have been used in medicine argues, again, against the cultivation of giant ragweed, as there is no knowledge of the cultivation of medicinal plants by any Indian groups north of Mexico. Such a practice would seem to be out of line with the status of the bluff-dwellers, and certainly would not have been necessary to ensure a supply of the plant.

LITERATURE CITED

BLAKE, S. F. 1939. A new variety of *Iva ciliata* from Indian Rock Shelters in the South-Central United States. Rhodora, 41: 81–86.

CARLSON, G. G., AND V. H. JONES. 1940. Some Notes on Uses of Plants by the Comanche Indians. PMASAL, 25: 517–542.

CHAMBERLIN, R. V. 1911. The Ethno-Botany of the Gosiute Indians of Utah. Am. Anthrop. Assn., Memoir 11: 329–405.

COOPER, D. C., AND K. L. MAHONY. 1935. Cytological Observations on Certain Compositae. Am. J. Bot., 22: 843–848.

CRANE, H. R. 1956. University of Michigan Radiocarbon Dates I. Science, 124: 664–672.

DAVIS, W. E. 1930. Primary Dormancy, After-Ripening and the Development of Secondary Dormancy in Embryos of *Ambrosia trifida*. Am. J. Bot., 17: 58–76.

GILMORE, M. R. 1919. Uses of Plants by the Indians of the Missouri River Region. Bur. Am. Ethnol. Ann. Rep. 33: 43–154.

——. 1931. Vegetal Remains of the Ozark Bluff-Dweller Culture. PMASAL, 14: 83–102.

GRINNELL, G. B. 1923. The Cheyenne Indians. Vol. 2. New Haven.

GRIFFIN, J. B. 1952. The Archeology of the Eastern United States. Chicago.

HARRINGTON, M. R. 1924. The Ozark Bluff-Dwellers. Am. Anthrop., 26: 1–21.

——. 1960. The Ozark Bluff-Dwellers. Mus. Am. Indian, Heye Foundation, Indian Notes and Monogr. 12.

JONES, K. L. 1933. *Ambrosia bidentata* Michx. X *A. trifida* L. Am. Midland Nat., 14: 720–724.

——. 1943. Studies in Ambrosia III: Pistillate *A. elatior* X *A. trifida* and Its Bearing on Matroclinic Sex Inheritance. Bot. Gazette, 105: 226–232.

JONES, V. H. 1936. The Vegetal Remains of Newt Kash Hollow Shelter. pp. 147–165, *in* Webb and Funkhouser, 1936.

MULLIGAN, G. L. 1957. Chromosome Numbers in Canadian Weeds I. Can. Jour. Bot., 35: 779–789.

PAYNE, W. W., AND W. F. KLEINSCHMIDT. 1961. Maintaining Ragweed Cultures. Jour. Allergy, 32: 241–245.

Rousseau, J. 1944. Reconstitution de l'Ambrosia préhistorique des Ozark. Naturaliste Canadien, 71: 211–216.

Smith, H. H. 1928. Ethnobotany of the Meskwaki Indians. Pub. Mus. City of Milwaukee, Bull., 4: 175–326.

Sparkman, P. S. 1908. The Culture of the Luiseño Indians. Univ. Calif., Publ. Amer. Archaeol. and Ethnol., 8: 187–234.

Stebbins, G. L. 1945. The Cytological Analysis of Species Hybrids II. Bot. Rev., 11: 463–486.

Swank, G. R. 1932. The Ethnobotany of the Acoma and Laguna Indians. Univ. New Mexico Lib. Unpublished master's thesis.

Vestal, P. A., and R. E. Schultes. 1939. The Economic Botany of the Kiowa Indians. Harvard Univ., Bot. Mus., Cambridge.

Webb, W. S., and W. D. Funkhouser. 1936. Rock Shelters in Menifee County, Kentucky. Univ. Kentucky, Rep. Archaeol. and Anthropol., 3: 105–167.

Yanovsky, E. 1936. Food Plants of the North American Indian. U. S. D. A., Miscel. Publ. 237.

Papers of the Michigan Academy of Science, Arts, and Letters
Vol. XLVII, 1962 (1961 Meeting)

NOTES ON WESTERN POLYPORES. II. NEW DISTRIBUTION RECORDS[1]

R. L. GILBERTSON AND J. L. LOWE

State University of New York College of Forestry

As part of a long-term taxonomic study of the Polyporaceae of North America, we have made approximately 5,000 collections of wood-rotting fungi in the western United States and Canada. Most of these collections, the majority of which are polypores, were made in western Montana and northern Idaho, Colorado, the Olympic Peninsula of Washington, and the Chiricahua Mountains of Arizona. Less intensive collecting was done in central and southern Idaho, Wyoming, northern Arizona, Oregon, Alberta, and northern California. In addition to our own collections, we have examined a large number of herbarium specimens from western North America. These are located in herbaria of a number of institutions, including the National Fungus Collections and the Forest Disease Laboratory at Beltsville, Maryland; University of Michigan; University of California; Washington State University; University of Idaho; New York Botanical Garden; Pennsylvania State University; and University of Arizona.

From this field work and study of herbarium specimens we now have records of approximately 225 species of Polyporaceae from western North America. Many of these records extend the known distribution of species previously reported from the West, and some are new reports for the West or for North America. New information on 82 species is presented in this paper.

The most rewarding collecting period in the West was spent in the Chiricahua Mountains of southeastern Arizona. Here, at Portal, the Southwestern Research Station of the American Museum of Natural History provides ideal laboratory and living facilities for biologists. Probably because of the unpromising aspect of the surrounding desert, these mountains have not attracted many collectors of fleshy fungi. The annual precipitation for this area is only about 19 inches.

[1] Contribution from the Department of Forest Botany and Pathology, State University College of Forestry at Syracuse University, Syracuse, New York. The first paper in this series was published in PMASAL, 46: 209. 1961.

However, approximately 9 inches falls in an 8-week period from late July into September. Showers commonly occur almost every day during this period, and conditions become exceptionally favorable for the fruiting of wood-rotting fungi.

At elevations from 4,500–7,000 feet the oak-woodland[2] and pinyon-juniper woodland types, with a great variety of hardwood species along the stream bottoms and in the wet canyons, present an interesting polypore flora. Many species occur here which are typical inhabitants of eastern hardwood forests, but are unknown or extremely rare in the Rocky Mountains to the north and in the Pacific Northwest. These include *Poria overholtsii, P. latitans, P. tarda, Polyporus versatilis* (Berk.) Rom., *P. gilvus* (Schw.) Fries, *Fomes ohiensis,* and *F. robiniae* (Murr.) Sacc. & D. Sacc. In addition, *Trametes colliculosa* (Pers.) Lund & Nannf. is abundant here but unknown elsewhere in North America, and one of only two North American collections of *T. campestris* was made in this zone. More widely distributed species which are conspicuous elements of the flora of this zone are *Poria andersoni* (Ell. & Everh.) Neuman, *Polyporus dryophilus* Berk., *Fomes fraxinophilus,* and *Daedalea juniperina.*

At elevations of 7,000–10,000 feet in the ponderosa pine, Douglas fir, and spruce-fir types a distinctly different polypore flora is encountered. Here one finds many species that are common throughout the Rocky Mountain coniferous forests. Included in this group are *Fomes officinalis* (Vill. ex Fries) Faull, *Polyporus balsameus, P. caesius, P. fragilis, P. tomentosus, Poria aneirina, P. bombycina, P. lenis, P. reticulata, P. subacida,* and *P. xantha* (Fries) Cooke. Some species found here but unknown or rare to the north are *Poria luteoalba, P. subvermispora, Polyporus bresadolae, P. nidulans,* and *P. sector.*

With its great variety of woody substrata and favorable climatic conditions, this limited area in southeastern Arizona provides a rich wood-rotting fungus flora that few areas in the West can equal.

Northward in the Rocky Mountains and Pacific Northwest a number of polypores are found which are unknown or rarely reported elsewhere in North America. These include *Fomes nigrolimitatus, F. nobilissimus* (W. B. Cooke) Lowe, *F. occidentalis* Overh., *Poly-*

[2] Names of vegetation types are those used by Elbert L. Little, Jr., 1950. Southwestern Trees. Agriculture Handbook No. 9. U.S.D.A. Forest Service.

porus alboluteus Ell. & Ev., *P. amarus* Hedgcock, *P. cuneatus* (Murr.) Zeller, *P. leucospongia* Cooke & Harkness, *P. montanus* (Quél.) Ferry, *P. perdelicatus* Murr., *Poria albipellucida, P. albobrunnea, P. carbonica, P. carnicolor, P. similis, P. stenospora, P. weirii* Murr., and *P. zonata.* Virtually all of these species are associated with decays of western conifers.

One of the interesting aspects of the Rocky Mountain polypore flora is the frequency with which unusual substratum relationships are observed. Species that ordinarily occur on angiosperms are found on gymnosperms, or the reverse may be true. Weir (1914, 1917) has pointed out a number of polypores that are found on both types of substrata in the Rocky Mountains. Our attention has been drawn to the frequency with which certain species usually associated with the decay of gymnosperms are found on angiosperms, especially quaking aspen (*Populus tremuloides* Michx.), at higher elevations. These include *Trametes serialis* Fries, *T. odorata* (Wulf. ex Fries) Fries, *Polyporus caesius, P. alboluteus* Ell. & Ev., *Fomes cajanderi* Karst., and *Lenzites saepiaria* (Wulf. ex Fries) Fries. On the other hand, *Polyporus cinnabarinus* Jacq. ex Fries, *P. dichrous* Fries, and *P. semisupinus* Berk. & Curt. are commonly found on angiosperms but also frequently occur on gymnosperms in the Rocky Mountains.

The wood-rotting fungus flora in many areas of the West is inadequately known. Our knowledge of the fungi of New Mexico, Wyoming, Utah, Nevada, Oregon, and California is particularly incomplete. The coastal region of California, with many species of woody plants that are not found elsewhere in the West, is a highly promising area. Vast portions of the western provinces of Canada, the Northwest Territories, Yukon Territory, and Alaska are yet to be explored mycologically.

Most of the field work in the West in 1956, 1957, and 1958 was supported by a grant from the National Science Foundation to Lowe. The use of field facilities generously provided by the U. S. Forest Service and the Southwestern Research Station is gratefully acknowledged. The National Fungus Collections and the Forest Disease Laboratory, U. S. Forest Service, at Beltsville, Maryland, have provided important financial assistance for herbarium work.

Substratum records given in this paper indicate wood of dead fallen or standing trees unless otherwise indicated.

NEW DISTRIBUTION RECORDS

Daedalea juniperina Murr. On living *Juniperus* in Colorado. Common in the Southwest.

Favolus alveolaris (DC. ex Fries) Quél. On *Fraxinus* and *Populus* in Montana.

Fomes connatus (Weinm.) Gill. On living *Acer* in British Columbia.

Fomes everhartii (Ell. & Gall.) von Schrenk and Spauld. On living *Quercus* in Arizona.

Fomes fraxinophilus (Peck) Sacc. A resupinate form of this fungus is common on *Fraxinus* and *Quercus* in the Chiricahua Mountains. It was also found on *Juniperus,* a surprising record as no gymnosperm hosts have been reported previously.

Fomes nigrolimitatus (Rom.) Egeland. On gymnosperms in Utah.

Fomes ohiensis (Berk.) Murr. On *Acer* in Wyoming and *Quercus* in Arizona.

Fomes pini (Thore ex Fries) Overh. On living *Pseudotsuga* in the Pinaleño Mountains of Arizona.

Fomes ribis (Schum. ex Fries) Gill. On living *Lonicera* in Idaho.

Fomes sclerodermeus (Lév.) Cooke. Associated with a white heartrot of living *Quercus* in the Chiricahua Mountains.

Fomes scutellatus (Schw.) Cooke. On dead branches of *Quercus* in Arizona. Very common in the Chiricahua Mountains.

Fomes viticola (Schw.) Lowe. On gymnosperms in Arizona and Wyoming.

Ganoderma applanatum (Pers. ex Wallr.) Pat. On living and dead *Populus* in Arizona. This is one of the decay fungi generally associated with aspen and cottonwood throughout the West.

Lenzites trabea Pers. ex Fries. On gymnosperms in Washington.

Polyporus amorphus Fries. On *Picea* and *Pinus* in Wyoming and Colorado.

Polyporus arcularius Batsch ex Fries. On angiosperms in Arizona.

Polyporus balsameus Peck. On *Pseudotsuga* in Arizona.

Polyporus biennis (Bull. ex Fries) Fries. On *Acer* in Montana.

Polyporus biformis Fries. On *Populus* in Idaho.

Polyporus borealis Fries. On gymnosperms in Arizona.

Polyporus bresadolae (Bourd. & Galz.) Lowe, comb. nov. (*Poria bresadolae* Bourdot & Galzin, 1925). Pileate specimens of this species, previously known only in the resupinate condition and considered to be rare, were collected in Arizona. They were common on decorticated ponderosa pine logs in the Chiricahua and Pinaleño Mountains at an elevation of 8,000–9,000 feet, and associated with a white rot. This species has been placed in synonymy with *Poria purpurea* (Fries) Cooke by some earlier workers (Baxter, 1939, p. 186; Lowe, 1958, p. 106; Pilát, 1941, p. 396). The original authors of the species separated it from *P. purpurea* on the basis of the hymenium not being continuous over the edge of the dissepiments, a character considered by Bondarzew (1953, p. 145) sufficient to validate the species.

The Arizona material includes resupinate, effused-reflexed, and sessile specimens (Pl. I, Fig. 3). The pileus may extend forward up to 4 mm.; it is often fused laterally to form compound pilei extending several centimeters, and it may reach a thickness of 1 mm. The upper surface is grayish-white when fresh, discoloring somewhat pinkish on drying, and more or less villous. At 18 × magnification the edges of the dissepiments appear pale and granulose; under the microscope the hymenial layer does not extend over the edges of the dissepiments.

Context hyphae of the isotype material at Beltsville and of the Arizona material are 3–5 μ in diameter (Pl. II, Fig. 5C), contrasting decidedly with hyphae of a specimen of *Poria purpurea* determined by Fries from Sweden and now at Kew, that are 4–8 μ in diameter. Most American specimens identified as "*P. purpurea*" have the more slender hyphae, and the distribution of the two species in America needs reinvestigation.

The Oregon record of *Poria bresadolae* by Cooke (1942, p. 688) has been referred to *P. luteoalba* Karst. by Gilbertson (1956, p. 81).

Polyporus caesius Schrad. ex Fries. On *Populus* and gymnosperms in Arizona.

Polyporus canadensis Overh. On gymnosperms in Idaho, Montana, Oregon, and Washington. Although described recently from Ontario and not reported elsewhere, this is a widespread and rather common species.

Polyporus cinnabarinus Jacq. ex Fries. On angiosperms in Utah.

Polyporus confluens Alb. & Schw. ex Fries. On the ground under *Tsuga* and *Thuja* in Montana.

Polyporus dryadeus Pers. ex Fries. On *Populus* in Arizona and New Mexico.

Polyporus floriformis Quél. Associated with a brown rot of gymnosperms and angiosperms in Arizona, California, and Colorado. The basidiocarps of this species are frequently quite thick in western collections, although the thin, petal-like form is also found.

Polyporus fragilis Fries. On *Pinus* and other gymnosperms in Arizona. Common in the Chiricahua Mountains at higher elevations.

Polyporus hirtus Quél. On the ground under *Picea* and *Abies* in the Pinaleño Mountains of Arizona.

Polyporus hispidus Bull. ex Fries. On living box elder in the Chiricahua Mountains.

Polyporus immitis Peck. On *Abies, Picea,* and other gymnosperms in Arizona, Montana, and Oregon.

Polyporus minusculoides (Pilát) Lowe. On decayed log of *Tsuga* at Quinalt, Washington. Lowe (1957) recently reported this unusual polypore from North America and this is the first report from the West.

Polyporus mollis Pers. ex Fries. On *Pinus* and other gymnosperms in Arizona and Montana.

Polyporus nidulans Fries. On *Pseudotsuga* in the Chiricahua Mountains. The occurrence of this fungus on Douglas fir is of special interest as it has been rarely found on gymnosperms in other areas.

Polyporus perdelicatus Murr. On gymnosperms in Montana. Common throughout the cedar-hemlock zone of the Northwest.

Polyporus picipes Fries. On gymnosperms in Colorado.

Polyporus planellus (Murr.) Overh. On *Ceanothus* in California.

Polyporus sector Ehrenb. ex Fries. On *Pinus* in the Chiricahua Mountains. This record represents the westernmost extension of this species, which is common in the Gulf States.

Polyporus semisupinus Berk. & Curt. On gymnosperms in Arizona, British Columbia, Colorado, and Washington.

Polyporus subcartilagineus Overh. On gymnosperms in Montana.

Polyporus tephroleucus Fries. On *Alnus* in Arizona. The concept of this species is that of Overholts (1953).

Polyporus tomentosus Fries. On the ground under conifers in Arizona and Utah. Particularly abundant under Engelmann spruce at 9,000 feet in the Chiricahua Mountains.

Polyporus trabeus Rostk. On gymnosperms in Arizona. Lowe and

Lundell (1956) discuss the identity of this species and report it from New York. The Arizona specimens have the typical gloeocystidia.

Polyporus undosus Peck. On gymnosperms in Arizona, Colorado and Oregon. Basidiocarps of this fungus in the West often become much thicker and more extensively effused than are eastern specimens.

Polyporus varius Fries. On angiosperms in Arizona.

Polyporus volvatus Peck. On recently killed gymnosperms in Arizona. Basidiocarps of this species typically appear within one year after the death of a tree due to fire, insects, or other agencies. They frequently appear by the hundreds over the entire length of the trunk and the larger branches.

Poria albipellucida Baxter. On *Platanus* in the Chiricahua Mountains. Baxter (1938) has reported this species from New Mexico, but otherwise it has been known only from the Northwest and California, and has been reported only on gymnosperms. A useful aid in field identification of this species is the bitter taste, a character not mentioned in the original description.

Poria albobrunnea (Rom.) Baxter. On gymnosperms in Washington. Common in the Rockies from Colorado northward but apparently rare in the Southwest.

Poria alutacea Lowe. On gymnosperms in Oregon. Common throughout the Northwest. Specimens of this fungus often consist of a very small basidiocarp associated with an extensive system of branching white rhizomorphs.

Poria ambigua Bres. On angiosperms in California.

Poria aneirina (Sommerf.) Cooke. On *Populus* in Colorado and Arizona. Apparently this fungus is present throughout the West wherever *Populus* is found.

Poria bombycina (Fries) Cooke. On gymnosperms at higher elevations in Colorado and Arizona. This is a common fungus in the spruce-fir zone of the Rocky Mountains, and has basidiocarps that vary in color from lavender to yellow or yellowish brown.

Poria carbonica Overh. On *Pseudotsuga* in Arizona and California.

Poria carnicolor Baxter. On gymnosperms in Colorado. The salmon pink basidiocarps of this species are common in the cedar-hemlock zone of the Northwest.

Poria cinerascens Bres. On gymnosperms in Arizona. Widely distributed and common throughout the Rocky Mountains.

Poria corticola (Fries) Cooke. On angiosperms and gymnosperms in California.

Poria eupora (Karst.) Cooke. On angiosperms in Arizona.

Poria ferox Long & Baxter. On *Arbutus* in British Columbia and California. Previously found only on *Juniperus* in Arkansas and the Southwest.

Poria fimbriatella (Peck) Sacc. Associated with a white rot of *Populus* in the Hoh River Valley on the Olympic Peninsula of Washington. This species, widely distributed in the East, appears to be a rare species in the West, as it has been otherwise reported only from California.

Poria latitans Bourd. & Galz. Recently reported from North America by Lowe and Gilbertson (1961) in Florida, Mississippi, and Arkansas, this species was found on logs of Chihuahua pine in the Chiricahua Mountains. It is associated with a white laminated rot. Macroscopically it resembles the common *P. versipora* (Pers.) Rom., but has narrow allantoid spores about 0.5 μ wide and 3–4 μ long (Pl. II, Fig. 5E).

Poria lenis (Karst.) Sacc. On gymnosperms in Colorado and Arizona, and on *Chamaecyparis* and *Alnus* in California. This is a very common species in the West.

Poria luteoalba (Karst.) Sacc. On *Pinus* in Arizona. A rare fungus in other western areas but common on pine in the Chiricahua Mountains.

Poria mollusca (Pers. ex Fries) Cooke. On *Sequoia* in California.

Poria overholtsii Pilát. On angiosperms, especially *Quercus* and *Platanus* in Arizona (Pl. I, Fig. 4). This species, very common in the Chiricahua Mountains, has not been reported previously from the West.

Poria radiculosa (Peck) Sacc. Associated with a brown cubical rot of *Larix, Pinus, Pseudotsuga,* and other gymnosperms in Idaho, Washington, California, and Arizona. The Idaho and Washington specimens have a very extensive system of stout rhizomorphs.

Poria reticulata (Fries) Cooke. On angiosperms in Arizona and particularly common on *Populus.*

Poria rhodella (Fries) Bres. On *Alnus* and other angiosperms in Arizona. The color of fresh basidiocarps varies from light brown to orange.

Poria salmonicolor (Berk. & Curt.) Cooke. On gymnosperms in

Washington. Previously reported from British Columbia and Idaho as *P. rubens* Overh. & Lowe.

Poria sanguinolenta (Alb. & Schw. ex Fries) Cooke. On gymnosperms in California. This is a very common fungus in the northern Rocky Mountains and Pacific Northwest. The form that does not become discolored on drying (*P. expallescens* (Karst.) Sacc.) is common in Idaho.

Poria šimani Pilát. On *Picea* and *Abies* in the San Francisco Peaks area near Flagstaff, Arizona. This species has not been previously reported from North America. Our specimens have been compared with several from Europe, including isotype material, and found to be identical. The slender, capitately incrusted cystidia are a critical diagnostic feature (Pl. II, Fig. 5A). Pilát (1936–42) illustrates this species as having tramal hyphae with simple septa. His specimens, however, have hyphae with abundant clamp connections. The subiculum hyphae of the species also have abundant clamps (Pl. II, Fig. 5B).

Poria similis Bres. Associated with a white rot of *Alnus* in the Pinaleño Mountains of Arizona. This species is widely distributed in the Northwest, but has not been reported previously south of Idaho and Oregon. Our specimen shows a tendency to become pileate, but has the characteristic hyphae, spores, and cystidia.

Poria sinuosa (Fries) Cooke. On gymnosperms in Colorado and apparently becoming less common southward in the Rockies.

Poria spissa (Schw.) Cooke. On *Quercus* in California and *Pinus* in Arizona. Widely distributed but not often collected in the West.

Poria stellae Pilát. Associated with a white rot of gymnosperms in Arizona, Colorado, Idaho, Montana, and Washington. Our specimens agree with paratype material from Wyoming. The relation of this species to *P. sitchensis* Baxter, stated by Pilát to be very similar, and to *P. odora* Peck and *P. subincarnata* Peck needs further investigation.

Poria subacida (Peck) Sacc. On *Abies* and *Sequoia* in California and *Pseudotsuga* in Arizona. It is very rare in the spruce-fir zone from Colorado southward but present at lower elevations in the ponderosa pine and Douglas fir zones.

Poria subincarnata (Peck) Murr. Associated with a white rot of gymnosperms in the spruce-fir zone of Colorado.

Poria subvermispora Pilát. Very common on logs of Apache pine and ponderosa pine in the Chiricahua Mountains.

Poria tarda (Berk.) Cooke. On angiosperms in Arizona. This is one of the most common fungi on oak and other hardwoods along streams in the Chiricahua Mountains.

Poria taxicola (Pers.) Bres. On gymnosperms in Colorado.

Poria tenuis (Schw.) Cooke. On gymnosperms in Colorado.

Poria terrestris (DC.) Sacc. On *Populus* in Arizona and on angiosperms in Oregon.

Poria unita (Pers.) Karst. On angiosperms in California.

Poria vaillantii (DC. ex Fries) Cooke. On gymnosperms in Arizona and *Sequoia* and other gymnosperms in California.

Poria vaporaria (Pers. ex Fries) Cooke. On *Pinus* in the Chiricahua Mountains.

Poria versipora (Pers.) Rom. On angiosperms in California. An extremely common species along the Pacific Coast.

Poria vulgaris (Fries) Cooke. One of the most common fungi on ponderosa pine logs at higher elevations in the Chiricahua Mountains. This name has had many interpretations in the mycological literature and the concept followed here is that of Lowe (1946).

Poria zonata Bres. On *Tsuga* in Montana and *Tsuga* and *Pseudotsuga* in Washington. This species is known only from western North America.

Trametes campestris Quél. Associated with a white rot of *Quercus* in the Chiricahua Mountains (Pl. I, Figs. 1 and 2). Previously known in North America from a single collection in Idaho.

Trametes serialis Fries. Widely distributed and generally on gymnosperms. In the southern Rockies, however, from Colorado southward into Arizona and New Mexico, this species is commonly found on dead wood of *Populus*.

LITERATURE CITED

Baxter, D. V. 1938. Some Resupinate Polypores from the Region of the Great Lakes. IX. PMASAL, 23: 285–305.

——. 1939. *Ibid.* X. PMASAL, 24: 167–188.

Bondarzew, A. C. 1953. Polyporaceae of the European Part of the USSR and of the Caucasus. Leningrad 1,106 pp. (In Russian)

Bourdot, H., and A. Galzin. 1925. Hyménomycètes de France. (XI, Porés) Soc. Myc. France Bull., 41: 98–251.

Bresadola, J. 1903. Fungi polonici. Ann. Mycol., 1: 65–131.

Cooke, W. B. 1942. Resupinate Pore Fungi in Oregon. Amer. Midland Nat., 27: 677–695.

Gilbertson, R. L. 1956. The Genus *Poria* in the Central Rocky Mountains and Pacific Northwest. Lloydia, 19: 65–85.

Lowe, J. L. 1946. The Polyporaceae of New York State (the Genus *Poria*). New York State Coll. For. Tech. Publ. 65.

——. 1957. *Polyporus minusculoides* in America. PMASAL, 42: 37–39.

——. 1958. The Genus *Poria* in North America. Lloydia, 21: 100–114.

——, and R. L. Gilbertson. 1961. A Synopsis of the Polyporaceae of the Southeastern United States. J. Elisha Mitchell Sci. Soc., 77: 43–61.

——, and Seth Lundell. 1956. The Identity of *Polyporus trabeus* Rostk. PMASAL, 41: 21–25.

Overholts, Lee O. 1953. The Polyporaceae of the United States, Alaska and Canada. Univ. Mich. Press, Ann Arbor, Mich. 466 pp.

Pilát, A. 1936–42. Polyporaceae I–III. *In* Kavina, C. and A. Pilát. Atlas des Champignons de l'Europe. Tome 3.

Weir, James R. 1914. Notes on Wood Destroying Fungi Which Grow on Both Coniferous and Deciduous Trees. I. Phytopathol., 4: 271–276.

——. 1917. *Ibid.* II. Phytopathol., 7: 379–380.

EXPLANATION OF PLATE I

Three polypores from Arizona

FIG. 1. Basidiocarp of *Trametes campestris* (Lowe no. 9195).
FIG. 2. Uniform white rot of oak associated with *T. campestris* (9195).
FIG. 3. Basidiocarps of *Polyporus bresadolae* (9691).
FIG. 4. Basidiocarp and rhizomorphs of *Poria overholtsii* (9004). All × 1.

PLATE I

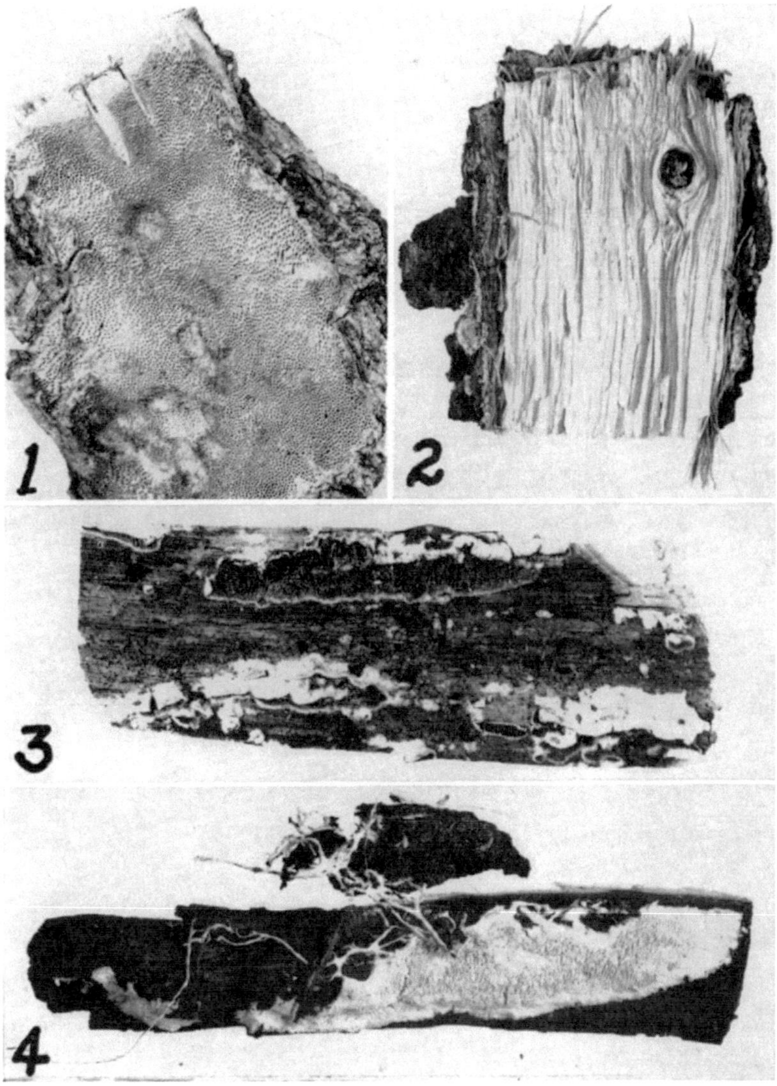

FIGS. 1 to 4

PLATE II

FIG. 5, *A–F*. Microscopic features of three Arizona polypores. *A*. basidia, cystidia, spores, and tramal hyphae of *Poria šimani* (Lowe no. 10147); *B,* subiculum hyphae of *P. šimani* (10147); *C,* subiculum hyphae of *Polyporus bresadolae* (9691); *D,* basidia, spores, and tramal hyphae of *P. bresadolae* (9691); *E,* basidia, spores, cystidium, and tramal hyphae of *Poria latitans* (9253); *F,* subiculum hyphae of *P. latitans* (9253).

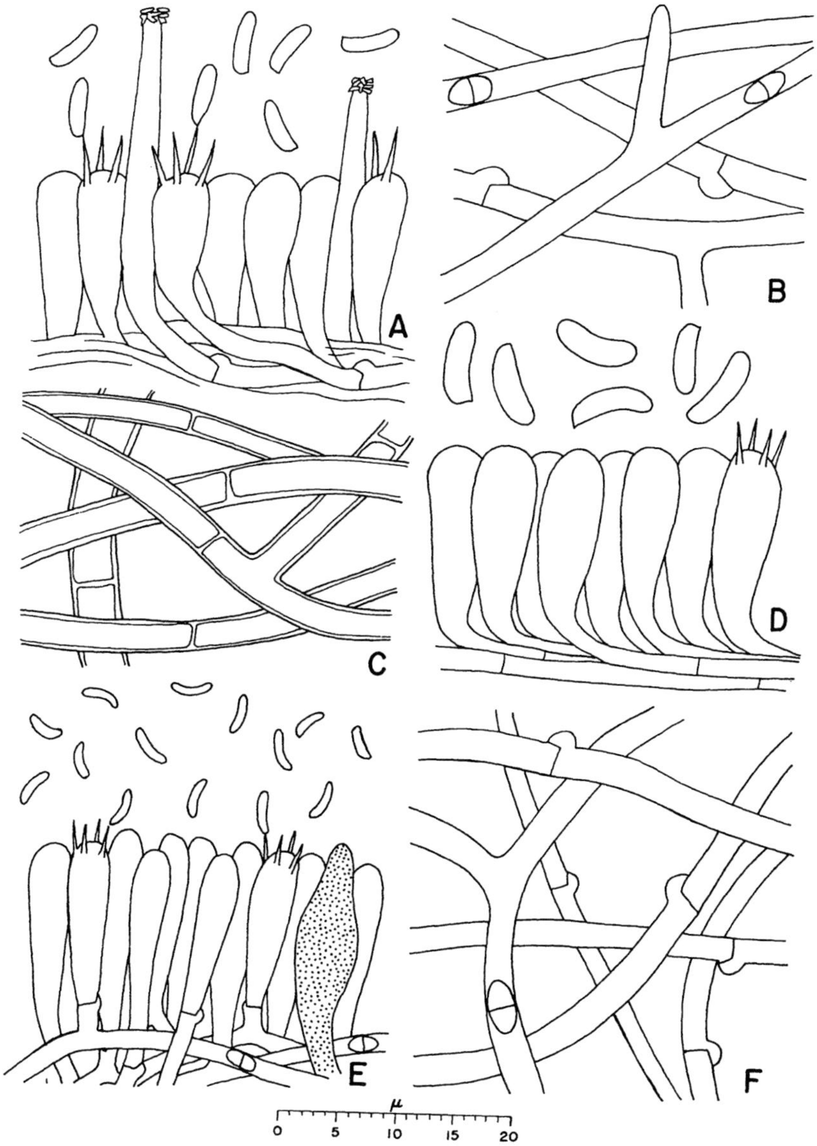

FIG. 5

PAPERS OF THE MICHIGAN ACADEMY OF SCIENCE, ARTS, AND LETTERS
VOL. XLVII, 1962 (1961 MEETING)

STUDIES IN THE GENUS *PORIA*. VI.[1]

JOSIAH L. LOWE

State University of New York College of Forestry

THESE studies, unless otherwise indicated, were made in the summer of 1960 at the Herbarium of the Royal Botanic Gardens at Kew, England. This Herbarium contains the type specimens of many of the species of polypores proposed by Berkeley, Cooke, and other European mycologists, and many additional collections of historical importance. The work was supported by a grant from the National Science Foundation. In this paper 18 species of *Poria* are reduced to synonymy for the first time, and 3 other species are transferred to another genus. The 20 species so involved are indicated by an asterisk(*).

The directors of the following herbaria gave permission to study the specimens under their care: Botaniska Museet, Uppsala Universitet (UPS)[2]; Farlow Library and Herbarium of Cryptogamic Botany, Cambridge, Mass. (FH); National Fungus Collections, Beltsville, Maryland (BPI); National Museum of Prague, Czechoslovakia (PR); Naturhistoriska Riksmuseet, Botaniska Avdelningen, Stockholm, Sweden (S); and the Royal Botanic Gardens, Kew, England (K).

Poria albobrunnea Petch (not *P. albobrunnea* (Rom.) Baxt.). Isotype material from Ceylon has pores 6 per mm.; context pale yellow above and somewhat darker below, the hyphae in KOH solution hyaline above, distinctly yellowish brown below, septate, 3–6 μ in diameter; spores hyaline, smooth, nonamyloid, broadly ellipsoid and

[1] Contribution from the Department of Forest Botany and Pathology, State University College of Forestry at Syracuse University, Syracuse, New York.

The first, third, and fifth papers in this series were published in PMASAL, 32: 99–101, 1948; 33: 3–6, 1949; and 39: 31–35, 1954; the second and fourth appeared in *Lloydia*, 10: 45–59, 1947, and 11: 163–170, 1948.

[2] The institutions are identified by the symbols recommended in *Index Herbariorum, Part I: The Herbaria of the World,* by J. Lanjouw and F. A. Staflu. Utrecht, Netherlands, 1952.

often flattened on one side, 5–5.5 x 3 μ. Similar characteristics are not known on any other species.

Poria albostygia (Berk. & Curt.) Lloyd. The type collection of this fungus, from Cuba, has abundant cystidia which are large and heavily incrusted, and context hyphae 5-7 μ in diameter. This species belongs in the *Polyporus lignosus* Klotzsch complex and is allied to *P. zonalis* Berk., which differs in having wider context hyphae 6–12 μ in diameter.

Poria alutacea Lowe. This species, described from New York in 1946, is widespread in the United States. It occurs in Europe but has not been known under this name. Excellent material of this fungus from Belgium was distributed as *Polyporus vaillantii* DC. by Libert in Pl. Crypt. Arduennae, Fasc. IV, No. 321, in 1837. It is the fungus described by Bourdot and Galzin (1928, p. 674) as *Poria calcea coriacea radicata* Bourd. & Galz.

Poria calcea (Berk. & Br.) Cooke*. The hymenium in the type, from Ceylon, extends across the base of the tubes only, and the dissepiments are entirely sterile. The structure is that of a species of *Porogramme*, not of *Poria*.

Poria cavernulosa (Berk.) Cooke. The type of this species, from Brazil, lacks cystidia and is, unfortunately, sterile. The concept of this as a cystidiate species, as published by Lowe (1958, p. 103), was based on a specimen misdetermined by Berkeley. This, Wright 149 from Cuba in the Curtis Collection at the Farlow Herbarium (FH), is an example of *Polyporus versatilis* (Berk.) Rom.

Poria camaresiana Bourd. & Galz. Isotype material at Beltsville (BPI) forms its pores in the usual manner, by the downward growth of marginal and tramal tissue, and its context hyphae are septate and 2.5–4 μ in diameter. This species is very similar to *P. reticulata* (Fries) Cooke, which has a cupulate habit and hyphae 3–8 μ in diameter. Bourdot and Galzin (1928, p. 664) compare their species with *P. mellita* Bourd., but isotype material of that, in the Lloyd Herbarium at Beltsville (BPI) and at Stockholm (S), has a very much firmer texture and more closely resembles *P. spissa* (Schw.) Cooke externally.

Poria candidissima (Schw.) Cooke. Type material of *Polyporus gordoniensis* Berk. & Br.*, from England, is the same as the earlier described American plant.

Poria carneo-lutea Rodw. & Clel. Isotype material, from Australia, has context hyphae which are apparently nonseptate below the tubes but abundantly clamped in the marginal tissue, and hyaline, smooth, oval spores 4–5 x 3–3.5 μ. The fruiting body does not resemble closely that of any other species of *Poria*.

Poria chrysobapha (Berk. & Curt.) Cooke*. Specimens of the type and isotype (at FH) of this species from Alabama are small and inadequate. Bresadola has annotated the type as follows, "Sans doute *Merulius* sp., ne pas assez distincte de *Merulius pinastri* Fr." The species certainly should be referred to the genus *Merulius*.

Poria collabefacta (Berk. & Br.) Cooke*. The type, from England, appears to be a species of *Merulius*. Pores are 3 per mm.; context hyphae clamped, 2.5–4 μ in diameter; and spores hyaline, smooth, oblong, 4–5 x 2–2.5 μ.

Poria confusa Bres. Type (at S) and isotype (Lloyd Herbarium No. 16615 at BPI) from Hungary is abundantly fertile. The fungus has no distinctive morphological characteristics. The species appears, however, to be better placed as a resupinate condition of a thin, leathery polypore such as is usually placed in the genus *Coriolus*. The spores are cylindrical, 4–5 x 1.5–2 μ, not 5–6 x 3 μ as described by Bresadola.

Poria corticola (Fries) Cooke. Pilát (1936–42, p. 450) considered this common species to be very similar to *P. pearsonii* Pilát, which he segregated because his species grows on gymnosperm wood, becomes finely stratified in age, and has thicker hyphal walls and a firmer consistency. The spore measurements are also described as considerably smaller (3.7–4.2 x 2.6–2.7 μ) than those of *P. corticola*. Spores on material of *P. pearsonii* marked "type" at Kew (K) and Beltsville (BPI) measure 5–6 x 2.5–3.5 μ, the size of those of *P. corticola*. *P. pearsonii** therefore is considered to be *P. corticola* occurring on an unusual substratum but not differing otherwise.

Poria dictyopora Cooke. Most of the context hyphae of the type and isotype (at BPI) specimens of this species, described from Australia, are usually thick-walled and much contorted, septate, and similar to those found in *P. cocos* (Schw.) Wolf, *P. inflata* Overh., and *P. magnahypha* Overh. The species differs from these in having oval spores 6–7 x 4–5 μ. *P. wakefieldiae* Cleland is very similar and may be a synonym; the isotype from Australia is sterile, and the identity of the species is uncertain.

Poria endoxantha Petch. Pores on the isotype from Ceylon average 6 per mm., the context hyphae are septate, 4–7 μ in diameter, and the specimen is sterile. The sporophore is most similar to that of *P. nigrescens* Bres., but differs in having a uniformly pale yellowish brown context which turns to a bright yellow in KOH solution.

Poria flavipora (Berk. & Curt.) Cooke. This is one of the few polypores known to have gloeocystidia. Oval apiculate spores 4–5 x 2.5–4 μ are abundant in the lectotype (No. 102). The paratype, No. 94, is the same species but the material is in fragmentary condition.

The Florida record of *Poria flavipora* (Lowe, 1958, p. 103) should be referred to *P. phlebiaeformis* Berk.

Poria hibernica (Berk. & Br.) Cooke. The type, from England, has pores 3 per mm., context hyphae with clamp connections and 2–4 μ in diameter, and spores cylindrical, slightly curved, 4–5 x 1.5 μ. The species is very similar if not identical to *P. vaporaria* (Fries) Cooke as defined by Lowe (1946, p. 69; 1958, p. 109).

Poria mellea (Berk. & Br.) Cooke. The type, from Ceylon, has pores 1–2 per mm., context hyphae septate, 3–6 μ in diameter, and hyaline, smooth, oblong to oblong-ellipsoid spores 6–8 x 3–4 μ. *P. chlorina* Massee*, type from Christmas Island north of Australia, appears to be the same plant.

Poria pannocincta (Rom.) Lowe. The cartilaginous line at the base of the tubes in this fungus is unique in the genus *Poria*. *P. bourdotii* Pilát*, described from Hungary, is a synonym of this widely distributed species. Resupinate specimens of *Polyporus dichrous* Fries resemble this, but may be readily distinguished by its highly gelatinized hyphae at the base of the tubes and the thick-walled context hyphae. The hyphae of *P. pannocincta* are thin-walled.

Poria phlebiaeformis Berk. This fungus, described from Cuba, is a striking and apparently widely distributed plant in the tropics. It is known from Florida, Cuba, Panama, French Guinea, Brazil, and Ceylon. The Ceylon specimens were described as *P. rubrochorda** by Petch in 1917. They were so named because of the conspicuous rhizomorphs, which are much better developed than in the western examples.

Poria resinascens (Rom.) Lund. & Nannf. Few collections of this inadequately defined species are in the museums. The original specimen was found on *Populus* in Sweden; it has also been recorded on gymnosperm wood by Pilát (1936–42, p. 424). The type of *Poria pseudogilvescens* Pilát* (at PR), described in 1935 from a collection on *Alnus* in Czechoslovakia, seems to be the same thing but, unfortunately, it is sterile. Another specimen identified by Pilát, from Siberia, has spores 2.5–3 μ wide rather than 3–4 μ wide as described for this species. The spores of the Siberian specimen have the same shape and size as are found in *P. resinascens*. Detailed study is needed to find satisfactory characters to separate *P. resinascens* from others of similar morphology, *e.g., P. aneirina* (Sommerf.) Cooke.

Poria rhodella (Fries) Cooke. This extremely variable species has been described repeatedly. Additional synonyms appear to be *P. chakasskensis* Pilát* from Siberia (at UP)[3], *P. turkestanica* Pilát* from Siberia (at PR), *P. rubescens* Petch* from Ceylon, and *P. vinaceo-rosea* Clel. & Rodw.* from Australia. The separate names which have been applied to the various color forms do not appear to be justified. Lundell has proposed a new name for the pink form of this species, *P. excelsa* Lund. & Nannf.* (1946, No. 1328), because Fries' concept cannot be established. Material named *P. rhodella* in European herbaria is rather consistently the same plant as *P. excelsa*, and a new name does not seem necessary.

Poria salmonicolor (Berk. & Rav.) Cooke. This species, recently recognized as valid (Lowe, 1958, p. 111; Lowe and Gilbertson, 1961, p. 56), appears to be the same as and a prior name for *P. oxydata* (Berk. & Curt.) Cooke*. The type of this (No. 6175 from South Carolina) is sterile, but it certainly is not *P. spissa* (Schw.) Cooke as suggested by Lowe (1958, 106). In an annotation on the type packet of *P. oxydata*, Bresadola has noted that he thought this species was not different from *P. salmonicolor*. The paratype, Wright 390 from Cuba, is another plant, *P. sanguinolenta* (Alb. & Schw.) Cooke.

Poria tarda (Berk.) Cooke. The type, from Australia, agrees in all characteristics with the widely distributed fungus in the United States known as *P. fatiscens* (Berk. & Rav.) Cooke* and as *P. semitincta* (Peck) Cooke. The American plant has not been recognized previously from outside North America. *P. interrupta* (Berk. & Br.) Cooke, from Ceylon, is probably the same species, as it agrees except for the less sharply distinct cell walls of the context hyphae. This may be due to treatment with mercuric chloride, which alters the chemi-

[3] Pilát (1936–42, p. 443) reduces *Poria chakasskensis* to the rank of a form of *P. lenis* (Karst.) Sacc. This is an unlikely disposition of the Uppsala specimen, and that may not be an authentic sample, although it is labeled as part of the type collection.

cal nature of the fungous cellulose and often makes comparison difficult.

Poria versipora (Pers.) Rom. The application of names at the specific and varietal level to this extremely diverse and common fungus of worldwide distribution depends upon the taxonomic philosophy assumed. As the variations merge with each other it seems best to consider *P. versipora* a species in which only one variety deserves recognition. Typical material has a soft-fibrous context of loosely arranged hyphae which continue into the trama without change of texture. The following species appear to belong here: *P. versipora versipora, P. jalapensis* Murr. from Mexico, *P. lignicola* Murr. from Cuba, *P. ochracea* Murr. from Virginia, and *P. phellinoides* Pilát* from Siberia (at PR). In other material the trama differs rather decidedly from the context, being waxy-fragile to cartilaginous. Described species which are of this nature are: *P. adpressa* Murr. from Cuba, *P. consobrina* Bres.* from France (isotype at BPI), *P. eyrei* Bres. from England, *P. hypolateritia* (Berk.) Cooke* from India, and *P. subpudorina* Pilát* from Czechoslovakia (at PR). The earliest name that could serve to designate varietal rank in this latter group is *P. hypolateritia* (described in 1886).

Poria vincta (Berk.) Cooke. Detailed study of the sterile type material from Cuba appears to establish that this fungus is a widely distributed tropical species. It has abundant, heavily incrusted cystidia as has been frequently stated. It has also transitional forms to mammillate cystidia such as were found (Lowe, 1948, p. 100) in the type material of *P. carneopallens* (Berk.) Cooke, which appears to be a younger, less weathered specimen of *P. vincta*. Murrill (1921, p. 84) has placed *P. carneopallens* in synonomy with *P. vincta* on external evidence. Such cystidia were found on the type of *P. hyposclera* Berk.* from Tahiti, and that species is therefore placed in synonymy.

LITERATURE CITED

Bourdot, H., and A. Galzin. 1928. Hyménomycètes de France. 761 pp., 185 figs. Paris, France.

Lowe, J. L. 1946. The Polyporaceae of New York State (The Genus Poria). N.Y.S. Coll. For. Tech. Publ. 65. 91 pp., 20 fig.

——. 1948. Studies in the Genus *Poria*. PMASAL, 32: 99–101.

——. 1958. The Genus *Poria* in North America. Lloydia, 21: 100–114.

——. 1961. Southern and Tropical Polypores. II. PMASAL, 46: 205–207.

——, AND R. L. GILBERTSON. 1961. Synopsis of the Polyporaceae of the Southeastern United States. J. Elisha Mitchell Sci. Soc., 77: 43–61.

LUNDELL, S., AND J. A. NANNFELDT. 1946. Fungi Exsiccati Suecici praesertim Upsalienses, Fasc. XXVII–XXVIII. (Nos. 1301–1400). 49 pp.

MURRILL, W. A. 1921. Light Colored Resupinate Polypores—III. Mycologia, 13: 83–100.

PILÁT, A. 1936–42. Polyporaceae I–III. 624 pp., 374 pl. *In* C. Kavina and A. Pilát, Atlas des Champignons de l'Europe, Tome III.

Papers of the Michigan Academy of Science, Arts, and Letters
Vol. XLVII, 1962 (1961 Meeting)

THE GENUS *TRILLIUM* IN MICHIGAN: SOME PROBLEMS OF DISTRIBUTION AND TAXONOMY

FREDERICK W. CASE, JR., *Saginaw Arthur Hill High School*

GEORGE L. BURROWS, IV, *Bridgeport Junior High School*

Gleason, in the *New Britton and Brown Illustrated Flora of the Northeastern United States and Adjacent Canada* (1952), lists ten species of trillium, nine of which occur in Michigan. In this paper, our purpose is threefold; to present a preliminary report on the occurrence of these species of trillium in Michigan, to report on a trillium apparently new to the Michigan flora, and to examine the status of and to suggest an origin for one trillium.

In this paper we shall follow the nomenclature and the species treatment of Gleason (*ibid.*) unless we state otherwise. We do not intend to cover in detail the taxonomic differences between the species; we will omit all but a few useful diagnostic characters. For each county record plotted on our distribution maps, we have seen a herbarium voucher. The herbaria examined were those of the Cranbrook Institute of Science, The University of Michigan, and Michigan State University.[1]

The trilliums of Michigan fall naturally into two groups: those with sessile flowers and those with pedunculate flowers. The sessile-flowered species known from Michigan are

1. *Trillium recurvatum* Beck
2. *T. sessile* L.
3. *T. viride* Beck *var. luteum* (Muhl.) Gl.

The pedunculate trilliums of Michigan are

4. *T. grandiflorum* (Michx.) Salisb.
5. *T. erectum* L.
6. *T. cernuum* L. *var. macranthum* A. J. Eames and Wieg.
7. *T. flexipes* Raf. (syn. *T. gleasoni* Fernald, as used in Gleason)
8. *T. nivale* Riddel
9. *T. undulatum* Willd.

[1] For their help, we are grateful to these institutions, to the directors of their herbaria, and to all others who have given us assistance.

Only *Trillium grandiflorum* and *T. cernuum* appear to occur throughout the state.

1. *Trillium recurvatum* Beck. This species can be separated from the superficially similar *T. sessile* by the strongly clawed base of the petals, by the sharply recurved sepals often adpressed to the stem, and by the definitely petiolate leaves. *T. recurvatum* is native to the general region of the upper Mississippi Valley and adjacent states, extending westward to the prairie borders. In Michigan it occurs in Berrien County (Fig. 1), where it is locally frequent in rich low woods, mainly on alluvial soils. There is a specimen in the Cranbrook Herbarium labeled Wayne County. The collection was made by O. A. Farwell. This last record, a long distance from the normal range of the species in our area, we eye with suspicion. We would be inclined to disregard the record altogether, were there not several other isolated but well-documented records for trillium species in Michigan.

Except for the Farwell record, the occurrence of this species in extreme southwestern lower Michigan offers no special problems of distribution. It seems likely that it will be found to occur in areas adjacent to Berrien County.

2. *Trillium sessile* L. Similar to the preceding species, the sessile trillium has no claw at the petal base, spreading but not recurved sepals, and sessile leaves.

Trillium sessile is a species of the mixed deciduous forest regions west of the Appalachian uplands and north of the coastal plain. It occurs in nearly all forest and brush habitats, especially in rich nonacid soils, occasionally persisting along open roadsides and fence rows throughout its range. Michigan is the northern geographical range limit of the species, and its occurrence here is apparently sporadic. It occurs in Lenawee and Oakland counties (Fig. 1). C. R. Hanes (1947) reports stations in Kalamazoo and St. Joseph counties, but we did not see his specimens. It probably occurs in other counties along the southern border of the state.

3. *Trillium viride* Beck, var. *luteum* (Muhl.) Gl. Several old local Michigan floras list various trilliums under the names *T. viride, T. luteum,* or *T. viridescens.* There are no herbarium specimens referable to any of these entities in the herbaria examined. Yellowish or greenish-flowered plants occur in nearly all large populations of *T. sessile* and *T. recurvatum.* It may be that early botanists collected

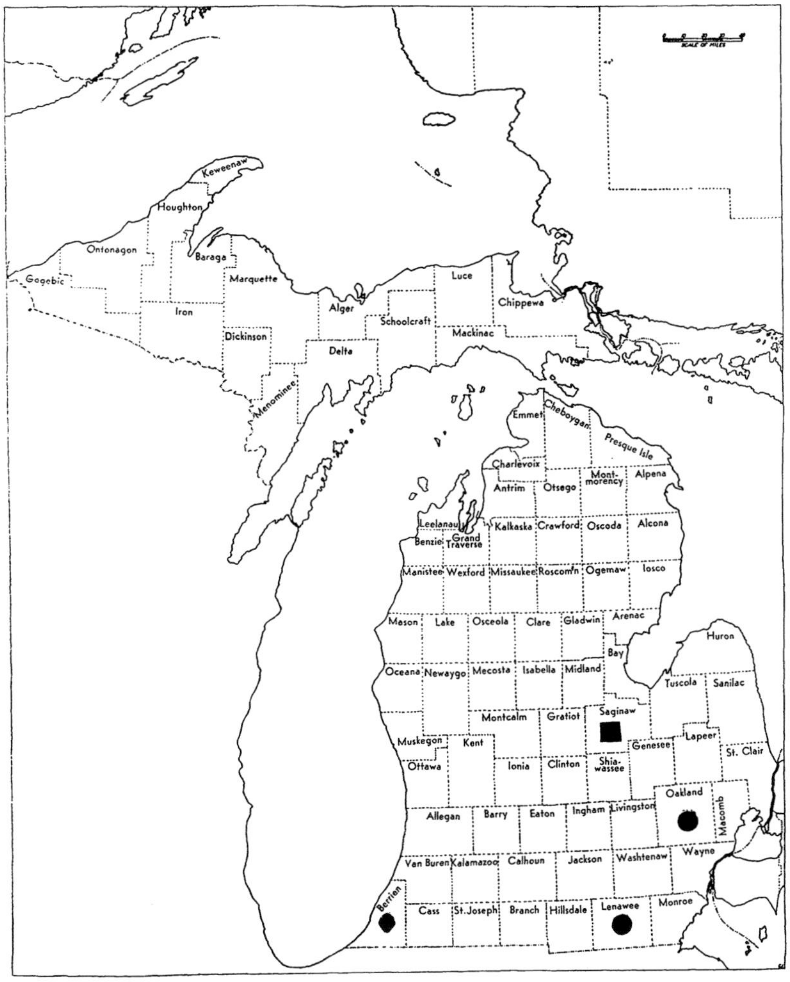

FIG. 1. Distribution of the sessile-flowered trilliums of Michigan. Circles, *Trillium sessile*. Diamond, *T. recurvatum*. Square, *T. viride* var. *luteum*.

these occasional pallid forms from among normal plants of sessile or prairie trilliums and identified them as the species *luteum* or *viride*. Or it may be that these old citations actually represented the species *T. viride* var. *luteum*. At any rate, we have not seen herbarium material of theirs validating the presence of this species in Michigan.

In the spring of 1950, Mr. Burrows discovered a sessile-flowered, yellow-petaled trillium in a woodlot on his family farm in the SW ½ of the SE ¼ of Sec. 6, T. 11 N., R. 5 E. (Bridgeport Twp.) Saginaw County, Michigan. Its size, the length of the petals, the color of the petals, and their lack of a basal claw clearly place this plant with *Trillium viride*, while the smooth upper stem and leaf veins place it with the Appalachian variety *luteum*. We have deposited a specimen together with a Kodachrome slide of this plant with the Herbarium of The University of Michigan. This collection is the first authenticated record for the state and represents an extension of the range some 300 to 400 miles northward. The occurrence of this plant in Michigan is indeed surprising! Is this plant, so far from its general range, a native or is it an introduced species? Without wishing to make an absolute judgment, we present the following information.

The woodlot in which *Trillium viride* var. *luteum* occurs is a mature mixed hardwoods, mainly composed of beech and sugar maple. This woods has been in the Burrows family for over ninety-five years. During that time, to the knowledge of the family, no one has deliberately introduced any exotic species to the woods. The trillium plants there were growing on a hummock formed by the remains of soil thrown up by an overturned tree. There were originally three or four flowering-sized plants and several seedlings. At the time of discovery there was only light shade over the area. Now, eleven years later, the shade is becoming more dense; the plants are declining. If this species is an introduced plant, a railroad line which passes about one city block west of the colony might conceivably have been the agent of introduction. We cannot conceive, however, of trillium as being a typical "railroad weed."

Trillium viride var. *luteum* is an Appalachian-upland and coastal plain species. Many species of plants of these areas have discontinuous ranges northwards. Just enough Michigan records for other species (*Woodwardia areolata, Rhexia virginica*) of the Appalachian or coastal-plain floras exist to suggest that *T. viride* var. *luteum* may represent a native disjunct here, rather than an adventive species.

4. *Trillium grandiflorum* (Michx.) Salisb. The large, white, relatively thin-textured petals, the sharply six-angled ovary, and the sessile leaves (except in forma *lirioides* (Raf.) Vict.) serve to distinguish this species from the other white-flowered Michigan trilliums. Found throughout the state, it occurs in nearly every wooded habitat except the very wet or boggy. It is probably in every Michi-

gan county, although it is less common in parts of the western Upper Peninsula than in the rest of the state. It seems to show a preference for slightly acid, well-drained soils. A large number of teratological forms occur, usually in localized colonies. The cause of this monstrous condition has not been defintely established although some botanists suggest a virus. Many of these freak forms have received taxonomic recognition. We question the value of naming forms which are so obviously abnormal.

No distribution map is included for this species as it is of almost universal occurrence within the state of Michigan.

5. *Trillium erectum* L. This is a rather large, dark red-brown, maroon, or purplish-flowered species with a *straight* peduncle, either erect or angled, and with a small dark purple ovary. The anthers are longer than the filaments, at least in most of the Michigan material we have examined. The flowers have a peculiar, fetid odor not unlike that of a wet dog. The odor increases with the age of the blossom. Throughout its range, *T. erectum* produces occasional plants with yellow, greenish, rose, or white petals. The dark ovary color usually persists and is a strongly reliable "key" character for all color forms. *T. erectum* occurs mostly in rich cool woodlands, often in mixed hardwood-hemlock areas.

Trillium erectum, an Appalachian species, reaches its western limits in Michigan. Here this trillium is sporadic in occurrence (Fig. 3), and its distribution offers some interesting questions to the student of plant geography. The answers probably lie in the character of the soil, or in the postglacial geological history of the area, rather than in climate or soil temperature, for it is absent from certain southern and northern areas, yet present in others. *T. erectum* occurs abundantly in most of the Thumb area of Michigan, from Huron County in the north to the northern edges of Wayne County in the south. Here it prefers deep, subacid humus on the borders of Arbovitae (*Thuja*) swamps. Westward it seems to stop rather abruptly at the Cass River drainage and seems to be absent from the old postglacial lake bed areas of the Saginaw Bay district. It reappears in the Tobacco River drainage in central lower Michigan. This may be a disjunct area. It is also found in the counties bordering the Lake Michigan shore from the Grand Rapids area northward at least to the northern limits of Leelanau County. The records of its distribution in this district are scanty, the full extent of its occurrence unknown.

It seems to be absent from the Upper Peninsula, the northeastern Lower Peninsula and the southwestern part of the state.

Early Michigan reports of *Trillium erectum* confuse it with *T. flexipes* forma *walpolei*. The latter is difficult to distinguish from *T. erectum* in herbarium material or in the older diagnostic keys. Therefore, records for *T. erectum* in the old local floras may be very inaccurate.

6. *Trillium cernuum* L. var. *macranthum* A. J. Eames and Wieg. Typical *T. cernuum* is a species of the coastal plain and the piedmont of the eastern United States and adjacent Canada, and its representative here is the var. *macranthum*. This is barely distinguished from the typical form by the larger over-all size, and by larger anthers, measuring 4–7 mm. long instead of the 3–5 mm. of var. *cernuum*. We have seen a few specimens which approach the var. *cernuum* in stature and anther size, but we suspect these of being depauperate specimens and believe that actually only the var. *macranthum* is present within the state.

The strongly declined or recurved peduncle, the white or barely pinkish petals, the white ovary, and the lavender-gray anthers approximately equal to the filaments distinguish *Trillium cernuum* var. *macranthum* from other Michigan trilliums. This species, a plant of cold soils in mixed woods and in evergreen swamps, is very common northward on the drier edges of cedar-balsam bogs. It is found throughout the Upper Peninsula and most of the Lower Peninsula (Fig. 2), becoming rare and local southward. It rarely forms dense stands such as those so characteristic of *T. grandiflorum*.

7. *Trillium flexipes* Raf. (syn. *T. gleasoni* Fernald, as used in Gleason). Superficially similar to *T. cernuum*, *T. flexipes* usually has a much larger leaf and stem *at anthesis*, although the ultimate size of the two species overlap. In the typical form the petals are white, heavy textured, often much recurved; the anthers are at least twice as long as the filaments, and yellow. The ovary is very strongly six-angled; the peduncle erect, often stiffly so in "pure" stands, or ascending at an angle.

Trillium flexipes occurs mostly in sandy alluvium along small to medium rivers or in rich low woods. While it may occur in peaty soils, it avoids the heavily organic swamp border soils in which *T. cernuum* or *T. erectum* so often occur. *T. flexipes* is widespread in the southwestern part of the Lower Peninsula, reaching the middle of the Lower Peninsula only in the counties along Lake Michigan (Fig. 3).

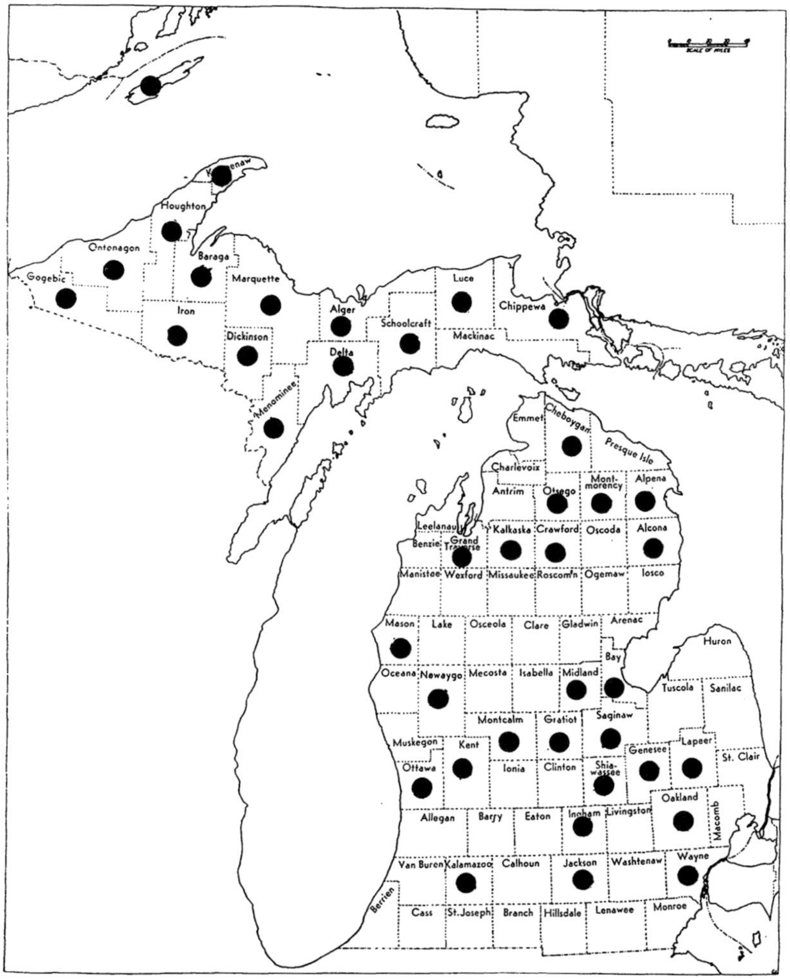

FIG. 2. Distribution of *Trillium cernuum* var. *macranthum* in Michigan.

In type of habitat and in structure, typical *T. flexipes* is a very distinct taxon.

The plant *Trillium flexipes* Raf. forma *walpolei* (Farw.) Fern., though ignored by Gleason, is listed by Fernald and others, and presents an interesting taxonomic problem to the field botanist. The plant, named by Farwell in honor of Bronson Walpole, is separated from typical *T. flexipes* by a declined or recurved peduncle, and by

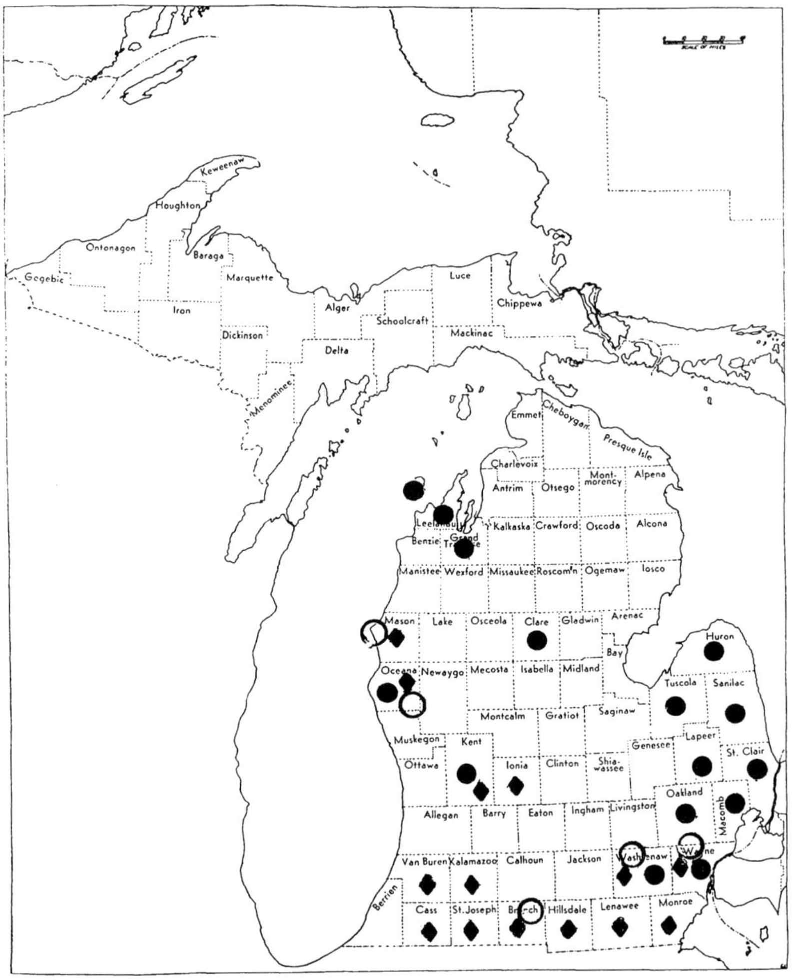

FIG. 3. Solid circles, *Trillium erectum*. Diamonds, *T. flexipes*. Open circles, *T. flexipes* forma *walpolei*. Marks represent only the counties of collections, not the number of times collected.

reddish, maroon, or chocolate petals. On the same day, and in the same locality where he collected the type of his forma *walpolei*, Farwell also collected a form with white petals, reddish at the base. He designated this plant as forma *billingtonii*. We have not seen the latter mentioned in the subsequent literature although the name seems as valid as forma *walpolei.* We have visited areas near the type locality, and a number of colonies of the species in other parts of Washtenaw County. Everywhere it is the same; where one form occurs the other occurs, and many intermediate forms as well. One of the most striking characteristics of these forms is that no two clones are alike. We wish to pose this question: is this great amount of individual variation a forma or variety in the usual botanical sense, or is it rather a typical hybrid swarm hitherto not recognized as such? We believe it is the latter. What then is the ancestry of this hybrid population?

There is only one species of red-flowered pedunculate trillium in Michigan. This is *Trillium erectum.* Genes for maroon or mottled petals almost certainly came from *T. erectum,* as did those for a colored or mottled ovary. It is natural to assume that the very similar *T. flexipes* is the other parent species. Its influence is inferred from the large anthers, much longer than the filaments, the large rhomboid leaves, and the partly white petals of some clones.

In addition to the species characteristics apparently blended in forma *walpolei,* there is evidence for its hybrid nature in the distribution of the supposed parent species and the location of the proposed hybrid swarms. Most collections of the forma *walpolei* occur in the Washtenaw-Wayne County area and in the Oceana-Mason County area. These areas are in the only known zones of contact between the Michigan ranges of *Trillium flexipes* and *T. erectum,* the very zones where one would most expect hybrid populations (Fig. 3). Although *T. erectum* does not seem to occur at present in the immediate vicinity of the hybrid swarms in southeastern Michigan, it is known from stations only a few miles away. At an earlier postglacial date, perhaps, colonies of the two species were sufficiently close for hybridization to occur. There are instances in other plant groups of hybrid swarms persisting in areas where one of the parents no longer occurs (*Aesculus,* Hardin 1957). Information on herbarium labels of specimens from the Oceana-Mason County area indicates that the putative parents and the forma *walpolei* occur together in the same woods there.

In search of further evidence of a hybrid origin for this trillium, we made scrapings of pollen from suitable material of *Trillium flexipes* (including the forma *walpolei*) and a putative parent, *T. erectum.* Pollen samples were stained and examined for irregularities in size and shape. We interpret the presence of smaller, irregular pollen grains as evidence of poor meiosis, usually regarded as being present in hybrids. On counts of *T. erectum* there were approximately 100% uniform large pollen grains. On counts of *T. flexipes*, all from areas of assumed hybridization, samples varied from 100% well-formed regular grains to those with at least 45% small and misshapen grains.

The lower percentage of large, uniformly shaped cells in the *Trillium flexipes* material strengthens our feeling that we are dealing with a hybrid population. Because studies of old, dry herbarium specimens may not be reliable, we shall repeat these studies on living material.

That the origin of *Trillium flexipes* forma *walpolei* may be more complicated than simple hybridization would account for is suggested by the petals, which are frequently longer and narrower than either those of *T. erectum* or *T. flexipes*, and by the very strongly declined or recurved peduncle. Both of these characters are found in *T. cernuum* var. *macranthum*. *T. flexipes* forma *walpolei*, then, may show the influence of a third species. It may be that a form of reticulate evolution is at work, a condition similar in some respects to that demonstrated by Dr. W. H. Wagner, Jr. (1954) for the Appalachian aspleniums.

8. *Trillium nivale* Riddel. This, the smallest of our trilliums, may be separated from other white-flowered species in this area by its much smaller size, its white petals, obtusely three-angled ovary, its oval, strongly petiolate leaves, and its obscurely six-angled stem. The peduncle is erect at anthesis, but becomes recurved in fruit. *Trillium nivale* is a diminutive, early-blooming plant which does not compete with most vegetation and occurs, therefore, mainly on creeping soils or in situations where vegetation is scanty. It frequents limestone ledges, steep, eroding clay banks, or occasionally sandbars along small streams. It is apparently a calciphile and is one of the earliest blooming plants in our entire flora. Variously reported from the Grand River drainage, there are validating herbarium specimens from Hubbardston, Michigan, and the Grand Rapids area (Fig. 4).

Undoubtedly a plant of such a specialized soil condition is very local, even in areas where it occurs; blooming as early as April 1, it may be easily overlooked. It is to be expected in the Grand River drainage, and possibly in other areas of southwestern Michigan.

9. *Trillium undulatum* Willd. The painted trillium may be recognized by its white petals, which have a red blotch and striping at the base. Only *T. flexipes* forma *walpolei* among other Michigan

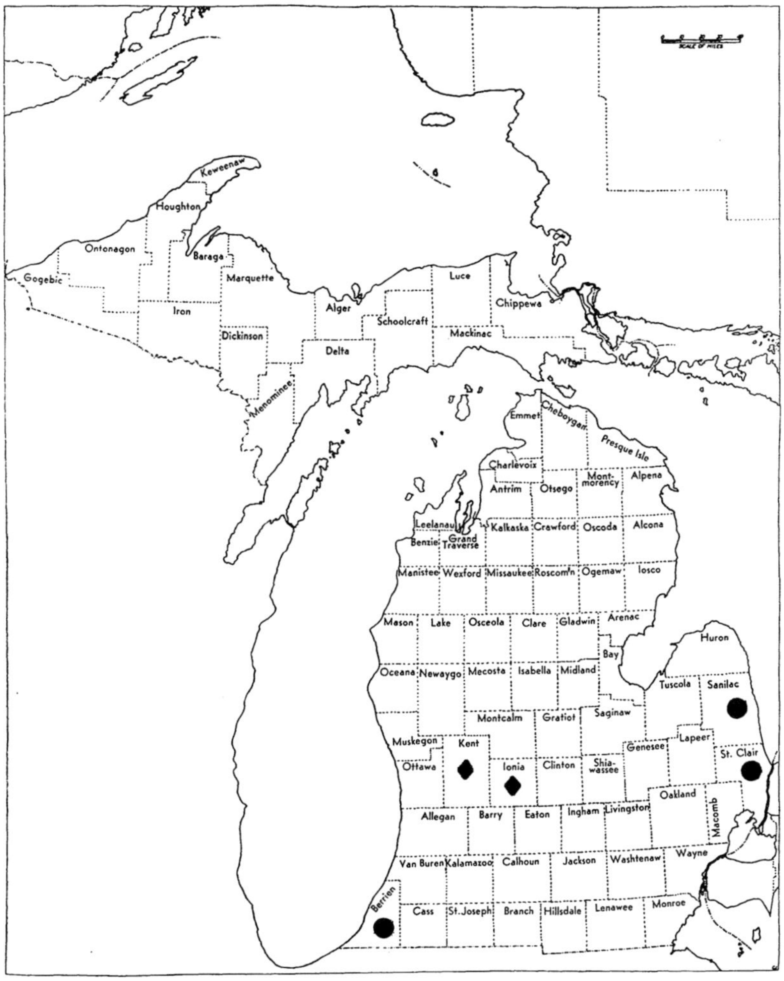

FIG. 4. Diamonds, known distribution of *Trillium nivale*. Circles, *T. undulatum*. The accuracy of the Berrien County record is suspect.

trilliums occasionally has white petals red blotched only at the base. From it, *T. undulatum* is easily separated by the oval, strongly petiolate leaves, and the rounded to obscurely three-angled ovary.

Trillium undulatum was first reported from the vicinity of Port Huron, Michigan, by C. K. Dodge (Beal, 1905). We have seen a specimen from there collected in June 1918, by Cecil Billington. Until collected again by us in May of 1959 and 1960, no one had recorded it since. We have collected it in St. Clair and Sanilac counties (Fig. 4). One sheet in The University of Michigan Herbarium cites this species as from "Niles," Michigan. The plant is this species. *T. undulatum* is an Appalachian-upland plant. Its presence in the Thumb area represents a western outlier for this and some other Appalachian species and is not unexpected there, but reports from westward and southward in the Great Lakes district are unlikely and should be verified. The painted trillium requires intensely acid, organic soils. It grows in Michigan in cold hemlock groves and in mixed stands of red maple, tulip tree, hemlock, and oak. The species blooms late, often not appearing above ground until other trillium species of the same area are fading.

Suitable habitats for *Trillium undulatum* occur in the coastal districts of the Thumb of Michigan and locally northward except in limestone districts. It occurs in Canada westward to the Georgian Bay district; there is a slight possibility that it may occur in other districts of Michigan north of those where it is known.

Summary

Nine species of trilliums are recognized from Michigan. All records of those cited are backed by herbarium specimens. One species, *Trillium viride* var. *luteum*, is apparently new to the Michigan flora. General distribution and some factors related to it are discussed for all the species, and maps showing the known occurrence of all species except the ubiquitous *T. grandiflorum* are presented. The possibility that *T. flexipes* forma *walpolei* is a hybrid is suggested, and evidence is presented for its hybrid origin.

LITERATURE CITED

Beal, W. J. 1904. Michigan Flora. Rep. MASAL., 5: 1–147.

Dodge, C. K. 1921. Miscellaneous Papers on the Botany of Michigan. Mich. Geol. and Biol. Surv. Publ. 31. 234 pp.

Gleason, H. A. 1952. The New Britton and Brown Illustrated Flora of the Northeastern United States and Adjacent Canada. Vol. I. Lancaster, Pa.

Hanes, C. R., and Hanes, F. N. 1947. Flora of Kalamazoo County, Michigan. Vascular Plants. Schoolcraft, Michigan.

Hardin, James W. 1957. Studies in the Hippocastanaceae III. A Hybrid Swarm in the Buckeyes. Rhodora, 59: 45–51.

Wagner, W. H., Jr. 1954. Reticulate Evolution in the Appalachian Aspleniums. Evolution, 8(2): 103–118.

Papers of the Michigan Academy of Science, Arts, and Letters
Vol. XLVII, 1962 (1961 Meeting)

FORESTRY

PHYTOGEOGRAPHY OF A REPRESENTATIVE OUTWASH PLAIN JACK PINE SITE

W. R. BEAUFAIT AND R. T. BROWN

Michigan College of Mining and Technology

Introduction

With increasing frequency, foresters and others concerned with land management are called upon to integrate the multitude of environmental conditions that influence site productivity. Silvicultural treatments and management predictions can be more reliably effected to the extent that these influences and their time sequences are better understood.

Often the comparatively low level of financial return from forest lands, especially those of poor nutrient and moisture availability, has restricted the intensive study required for complete site evaluation. Moreover, in the search for rapid measures of potential productivity, foresters sometimes tend to magnify the importance of relatively few physical characteristics of a site with correspondingly less emphasis on the interactions between environmental elements. Finally, expression of site productivity as a single dependent variable, such as site index, often requires the presence of near-perfect stands of a particular species with similar histories.

On the other hand, multivariate tests of biological productivity on a broad scale are time-consuming and fraught with the pitfalls of mass data-collecting, wherein strange and exotic correlations are sometimes derived. It is the purpose of this paper to summarize progress on a site productivity study of a different sort which attempts to strike a middle ground between the extremes outlined above by making full use of the forest mensurational tools available, as well as by borrowing generously from the procedures used in the companion disciplines of soil science and plant ecology. Thus, instead of searching out widely distributed stands which qualify for site index determination, then sampling edaphic and physiographic influences, we have chosen to study the phytogeographic characteristics of a limited area, isolate the apparent interactions between the soil and plants, and eventually test the derived hypotheses on other outwash sites.

Attention is focused on glacial outwash plains for two major reasons. First, they comprise a respectable proportion of the soils of the glaciated portion of the world. Second, their drouthy habit and nutritional poverty preclude agricultural production and relegate them to forest use, thus reducing the number of critical interactions between soils and treatment history.

PROCEDURES

The study consists of the three corollary phases: soil characterization, tree measurements, and phytosociological observations. Data and collections were taken from 140 equally distributed, mechanically located plots which comprise the Ford Forestry Center's continuous forest inventory locations on two square miles of outwash plain in Baraga County, Michigan. Fig. 1 is a reproduction of aerial photographs of the two sections involved. It shows the variation in forest cover density so characteristic of jack pine sites.

Soils.—Predominant soil types, as recognized by the Soil Conservation Service on the area, are Omega and Vilas sands. The more mesic peripheral types include Ocqueoc, Rousseau, Kinross, and variants of Vilas.

Soils pits were dug at each of the 140 locations, and composite samples were taken from an average of five horizons to a depth of 60 inches. Every profile was described and named by an experienced SCS soil scientist during the sampling procedure. A map showing the soils as described is presented in Fig. 2.

With few exceptions, the soil type lines coincide with changes in physiographic relief. There is a difference of 20 feet in elevation between the top-center very dry site and the wet area in the lower-left portion of the figure. The three islands of Vilas soil types are related to slight depressions in the outwash material.

Soil types have been grouped as follows:

Very Dry—Omega sand, occasionally intergrading toward Vilas sand. Excessively drained.

Dry —Vilas sand. Well-drained.

Moist —Rousseau fine sand, Ocqueoc fine sand, and a moderately well-drained variant of Vilas.

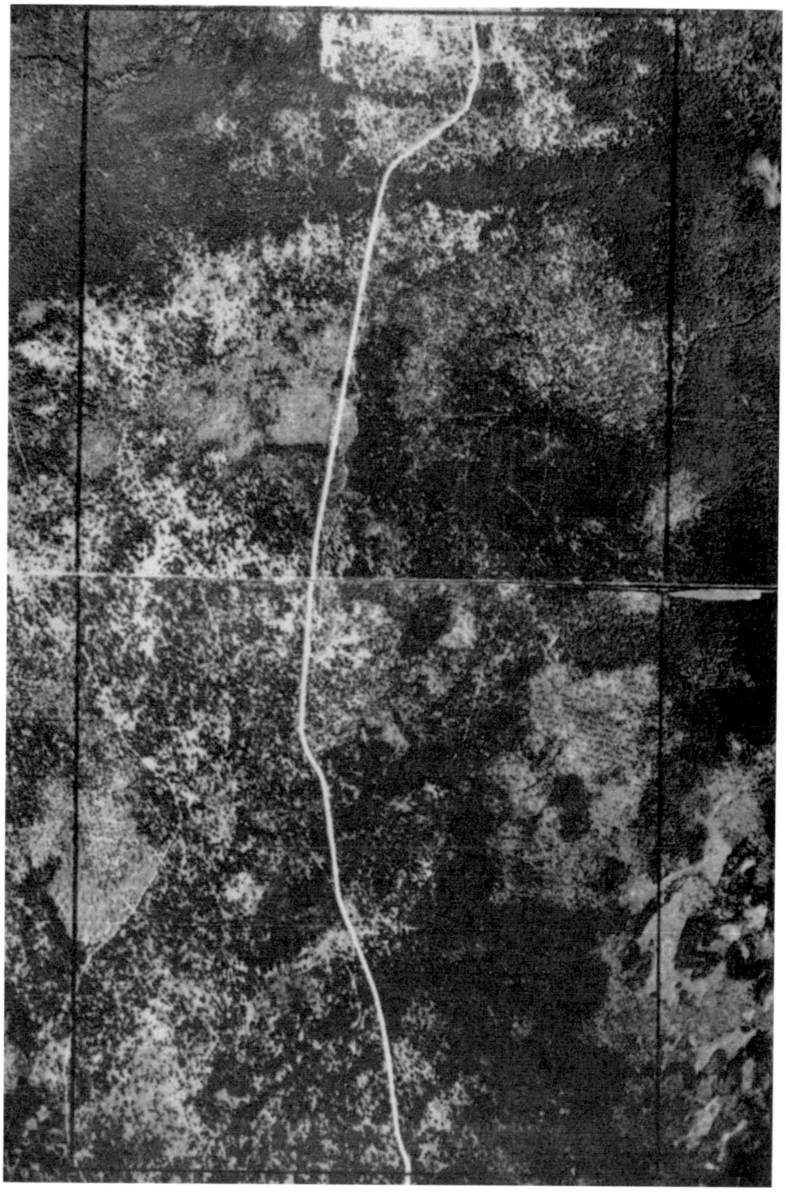

FIG. 1. Aerial photograph of Sections 14 and 15, Township 49N, Range 34W, Baraga County, Michigan.

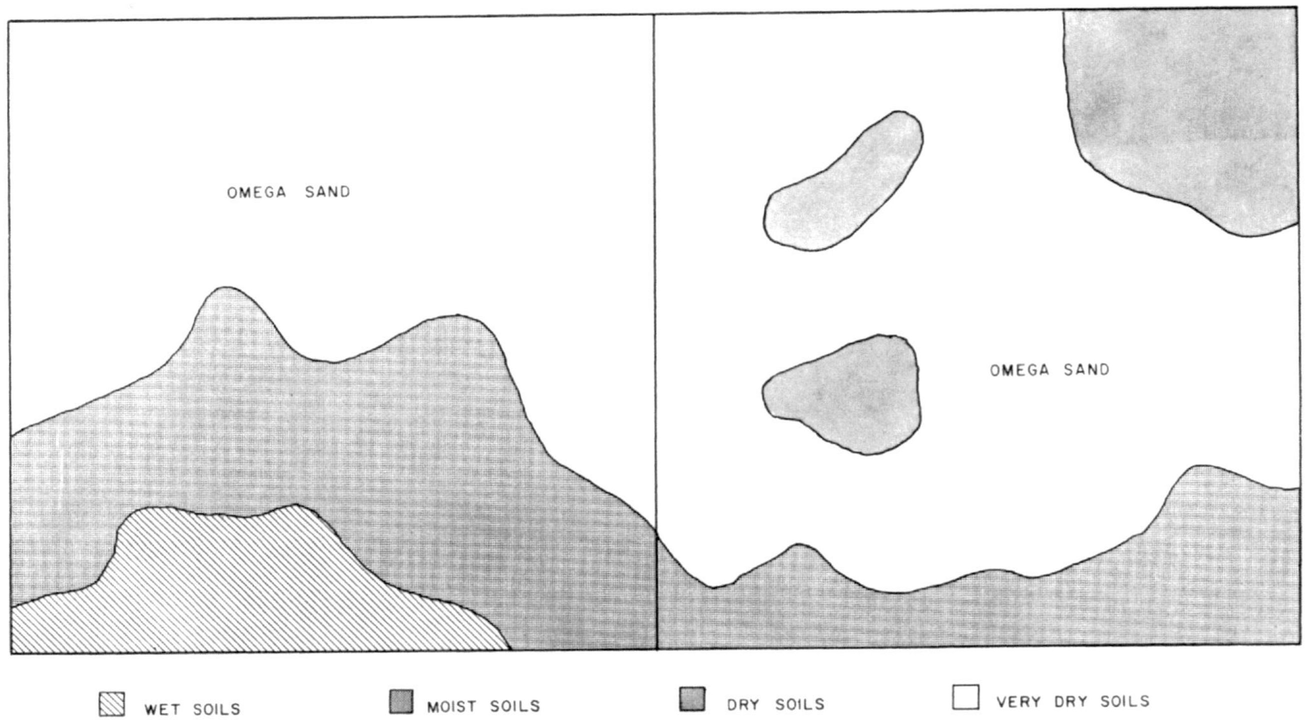

FIG. 2. Soil type map of the two-square-mile study area.

Wet —Imperfectly drained Vilas variant, and Kinross sand.

Tree measurements.—Standard mensurational techniques were employed in the determination of basal areas, size class distribution, and, where applicable, site index. Measurements of topographic position, aspect, and slope were also taken on the 140 one-fifth-acre sampling locations.

The variation in stocking levels seen in Fig. 1 made site index measurements impossible on many plots. Where available, however, they provide for standardization and reinforcement of phytosociological classifications, in which the abundance and sociability of all plants on the area are summarized as expressions of site variation.

Fig. 3 illustrates the distribution of plots where site indices were taken. Considering the similarity between the delineation of moisture regime and the mapped soil types, it might be assumed that site index changes would coincide also. While site index is generally higher in the moister zones, several notable exceptions occur. For example, there is a difference of over 10 feet in the extremes of site index to be found on the xeric (Omega sand) portions of the map. The southeast corner of the area is typified by site indices considerably greater than corresponding vegetation and soil types elsewhere in these two sections.

Phytosociological observations.—At each of the 1/5-acre plots already mentioned, the understory vegetation was described, using the system of Braun-Blanquet. Species-area curves indicated that the minimal plot area should be about 1/10 acre; therefore, 1/5-acre plots are of ample size.

Pinus banksiana is almost the sole tree species, although a few spruces are found on moister sites. Some of the understory plants are ubiquitous; others show a marked preference for a particular moisture regime. The commonest of the former include:

Comptonia peregrina	*Calliergonella schreberi*
Prunus pumila	*Dicranum undulatum*
Vaccinium angustifolium	*Cladonia cristata*
V. myrtilloides	*C. pixidata*
Carex pensylvanica	*C. rangiferina*
Danthonia spicata	*C. uncialis*
Melampyrum lineare	

The total area was divided into three approximately equal parts

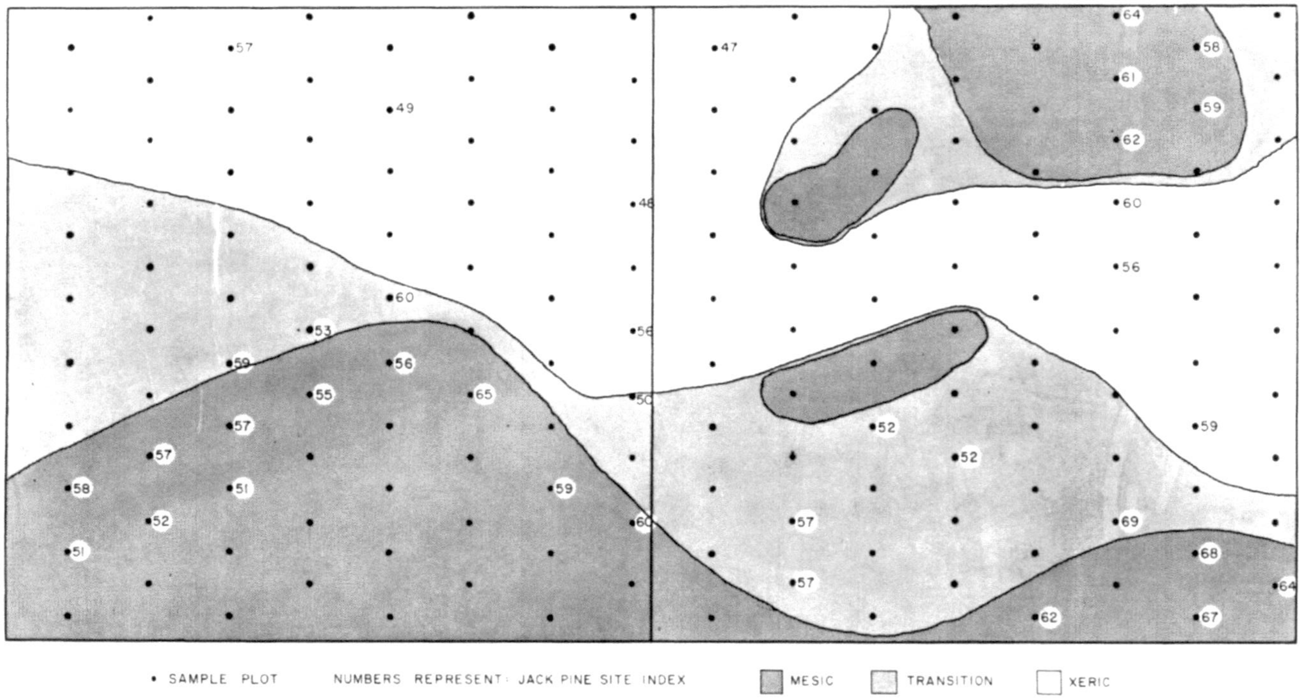

FIG. 3. Map of moisture regime zones as expressed by vegetation.

TABLE I

SUMMARY OF DATA CONCERNING SPECIES TYPICAL OF MOISTURE REGIME ZONES

Legend: r—rare; +—sparsely present, cover small; 1—rather plentiful, cover up to 5%; 2—numerous, cover 5 to 25%; 3—cover 25 to 50%; 4—cover 50 to 75%; 5—cover over 75%.

Species	Habitat					
	Xeric		Transitional		Mexic	
	Abundance Cover	Per cent Presence	Abundance Cover	Per cent Presence	Abundance Cover	Per cent Presence
Pinus banksiana	3	100	3	100	4	100
Andropogon scoparius	2	89	1	52	0	10
Arctostaphylos uva-ursi	1	80	1	71	0	24
Aster leavis	+	73	+	64	0	43
Campanula rotundifolia	+	84	+	67	r	52
Oryzopsis pungens	1	82	+	81	+	57
Apocynum androsaemifolium	0	27	r	52	0	38
Viola adunca	0	45	+	57	0	45
Maianthemum canadense	0	20	r	50	r	50
Pteridium aquilinum	r	52	1	69	2	86
Deschampsia flexuosa	0	48	1	69	1	76
Gaultheria procumbens	0	11	+	62	1	86
Polytrichum juniperinum	0	27	0	48	1	69
Trientalis borealis	0	11	0	43	+	83
Rubus recurvicaulis	0	7	0	21	+	67
Oryzopsis asperifolia	0	5	0	38	+	79
Salix glaucophylloides	0	4	0	19	1	62

on the basis of the moisture regime, as expressed by the understory plants (Fig. 3). Division was accomplished by plotting the geographic distribution of many plants and then drawing lines where vegetation changes seemed sharpest. These changes correspond with moisture conditions observed in the field. We then prepared an association table for each of the three areas; and from the tables we determined mathematically the relationships between plants.

Table I shows the abundance-cover and per cent presence of certain plants found in each habitat. The abundance-cover values given are medians; consequently, any plant present on less than half the plots has a value of zero. Only plants present on at least 50 per cent of the plots in at least one habitat were used in this table. This eliminated many plants which show definite preference for one habitat; but they are not as useful in the determination of site properties.

The xeric habitat, in general the highest topographically, has a ground cover of tufted grasses. Many other plants also tend to be tufted. The five species listed after *Pinus banksiana* are typical of the xeric habitat. Listed in order from most xeric these are: *Andropogon scoparius, Arctostaphylos uva-ursi, Aster leavis, Campanula rotundifolia*, and *Oryzopsis pungens*. Although they are not found exclusively in this habitat, they reach their best development and are most numerous in these open dry forests. The next two species, *Apocynum androsaemifolium* and *Viola adunca*, are more numerous and better developed in the transitional zone than in either of the other types.

The rest of the plants are more typical of the mesic sites. Again in order, these plants are: *Maianthemum canadense, Pteridium aquilinum, Deschampsia flexuosa, Gaultheria procumbens, Polytrichum juniperinum, Trientalis borealis, Rubus recurvicaulis, Oryzopsis asperifolia*, and *Salix glaucophylloides*. In this habitat, *Pteridium aquilinum* grows to a height of three feet and makes an almost complete cover, while in the more xeric places, it is often only six to eight inches tall. *Polytrichum virginianum* often forms a soft green carpet, and the other plants are all more vigorous and flower more readily.

Summary and Conclusions

The plants listed in Table I are indicators of variations in the moisture regime of this outwash plain. The forest type remains the same, but productivity and site index change. Since they are all common and readily identified plants, they should serve as indicators of differences in site.

The variations of site indices observed in otherwise homogeneous soil-plant types suggest the influence of other unidentified variables. This is a major reason for further investigation into the nature and amount of colloidal material within the same and related soil types. It may be that for forestry purposes, soil classification systems will need revision to correspond with observed variation in productivity.

Although we have drawn lines on a map and labeled xeric, transitional, and mesic types, this does not mean that these are discrete plant associations; rather, there is a blending of one into the next, a continuum of vegetation in which each plant has its range of tolerance. In other words, some plants may have a broad range; others a narrow range. The plants with a broad range, we have called ubiquitous. They are of little use in the sharp definition of site properties.

The plants with a narrow range are, however, of much greater importance to foresters and others interested in site classification. Where differences in environment can be related to these or other plants, foresters should have a means for quantifying sites even where no trees now grow.

REFERENCES

BRAUN-BLANQUET, J. 1932. Plant Sociology. New York: McGraw-Hill Book Company, Inc.

FERNALD, M. L. 1950. Gray's Manual of Botany. 8th ed. New York: American Book Co. lxiv + 1632 pp.

GROUT, A. S. 1936–1939. Moss Flora of North America North of Mexico. Published by the author.

PAPERS OF THE MICHIGAN ACADEMY OF SCIENCE, ARTS, AND LETTERS
VOL. XLVII, 1962 (1961 MEETING)

WILDLIFE

MAST CROPS AND FOX SQUIRREL POPULATIONS AT THE ROSE LAKE WILDLIFE EXPERIMENT STATION

WILLIAM R. FOUCH

Rose Lake Wildlife Experiment Station, East Lansing

GAME managers have long recognized the importance of oak and hickory mast—the acorns and nuts—in the diet of the fox squirrel. We know less, however, of the specific effects on squirrel populations and productivity when the mast crop fails. A mast failure can be serious. Allen (1943), for instance, found that after a 1940 mast failure in scrub-oak woodlands in Allegan County, some squirrels died during the winter, and females that survived bore smaller litters than normal the following spring. In this study, encompassing data from 1956 to 1960, I attempt to assess the effect of natural production of mast on fox squirrel productivity and population levels in farm-woodlot range—800 acres of the Michigan Department of Conservation's Rose Lake Wildlife Experiment Station, near Lansing in south-central Michigan. In these 800 acres are seven oak-hickory woodlots totaling 102 acres. Mechanically harvested cornfields occupied a yearly average of 26 acres in the study area.

This paper is a contribution from Federal Aid in Wildlife Restoration Project Michigan W-40-R. I am indebted to the staff of the Rose Lake Wildlife Experiment Station for assistance in various phases of the study and to Dr. Leslie Gysel of Michigan State University for advice on various technical aspects of the problem.

METHODS

Mast measurement.—In determining the amount of mast produced I sampled falling mast by placing beneath selected trees mast traps of Kraft paperboard, 3.3 feet square, in the shape of an inverted pyramid (Easley and Chaiken, 1951). Poultry netting of 1½-inch mesh over the tops prevented entry of squirrels into the traps and reduced the "bounce-out" of nuts and acorns. We collected mast about three times during the fall each year.

Placement of mast traps within the woodlots followed a systematic sampling procedure. First we conducted a 10 per cent cruise in each woodlot to determine the composition of the stand. Then we surveyed a number of strips 20 links (13.2 feet) wide through each woodlot, and within these strips marked all mast trees likely to produce at least a moderate crop. In each woodlot I marked four times as many trees of each species as I intended to sample, selected at random one of the first four trees of each species in the first strip, and placed a mast trap beneath every fourth tree from that point throughout the remainder of the woodlot. I did not sample the mast at the few isolated trees scattered between woodlots. My sample consisted of 34 white oaks (*Quercus alba*), 83 black oaks (mostly *Q. velutina*), and 86 hickories. Because of difficulties of identification I did not differentiate between species of hickories. However, among the more frequent species were pignut (*Carya glabra*) and small-fruited (*C. ovalis*) and shagbark (*C. ovata*) hickories.

After collection I examined all acorns and nuts to determine how many were sound. Unsound mast was classed as insect-damaged or aborted.

Measuring squirrel populations.—I computed the population of squirrels in the seven woodlots by means of the "Lincoln Index." Just before the hunting season each year, we live-trapped and tagged squirrels in each woodlot. The number of squirrels tagged, the total number of squirrels shot, and the number of tags recovered provided the basis for population estimates. The normal bag check at the station gave the necessary data on the number of squirrels killed on the 800 acres, and enabled noting the sex and age of the animals shot, to provide a basis for computing productivity.

Findings

Mast crops.—Total mast collections amounted to 8,131 nuts and acorns (Tab. I). Of these, only 3,983, or 49 per cent were sound, and hence available for squirrel food. The hickories proved to be about 79 per cent sound, black oak 21 per cent, and white oak 66 per cent (Fig. 1). The figure for white oak may not be characteristic, being based almost entirely on one year's collection. Insects were responsible for most of the unsound acorns. Nearly 50 per cent of all black

TABLE I

MAST TRAP COLLECTIONS, 1956–59

	White Oak		Black Oak		Hickory		Total	
	Sound	Unsound	Sound	Unsound	Sound	Unsound	Sound	Unsound
1956	5	40	8	230	85	128	98	398
1957	1,750	607	167	1,009	530	73	2,447	1,689
1958	6	5	458	610	476	104	940	719
1959	91	315	128	967	279	60	498	1,342
	1,852	967	761	2,816	1,370	365	3,983	4,148
Total Number.....	2,819		3,577		1,735		8,131	
Per Cent Sound....	65.7		21.3		78.8		49.0	

oak acorns were so affected. Hollow or undeveloped acorns and nuts I classed as aborted.

Mast crops ranged from an almost complete failure (483 sound acorns and nuts per acre) in 1956 to a bumper crop (12,054 sound acorns and nuts per acre) in 1957, with intermediate production in 1958 and 1959 (Fig. 2; Tab. II). These estimates may be high because of preferential selection of trap sites, but the year-to-year ratios of mast abundance should be reliable. In 1957, white oak acorns made up nearly three-fourths of the total production of sound mast, even though white oaks comprised less than one-fifth of the sample

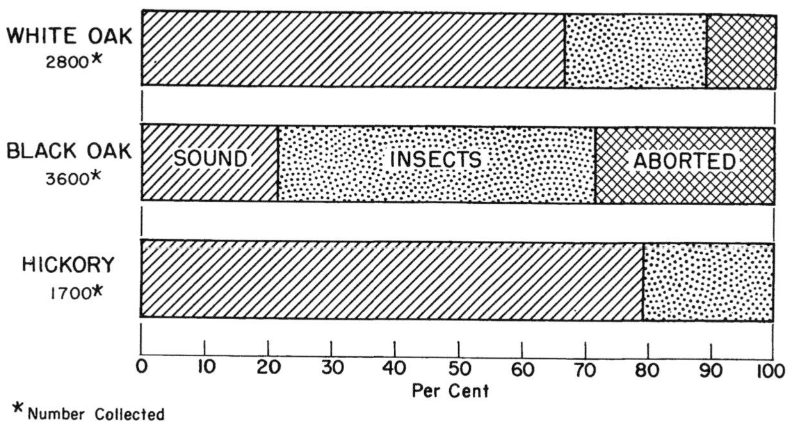

FIG. 1. Soundness of mast collections at Rose Lake, 1956–59.

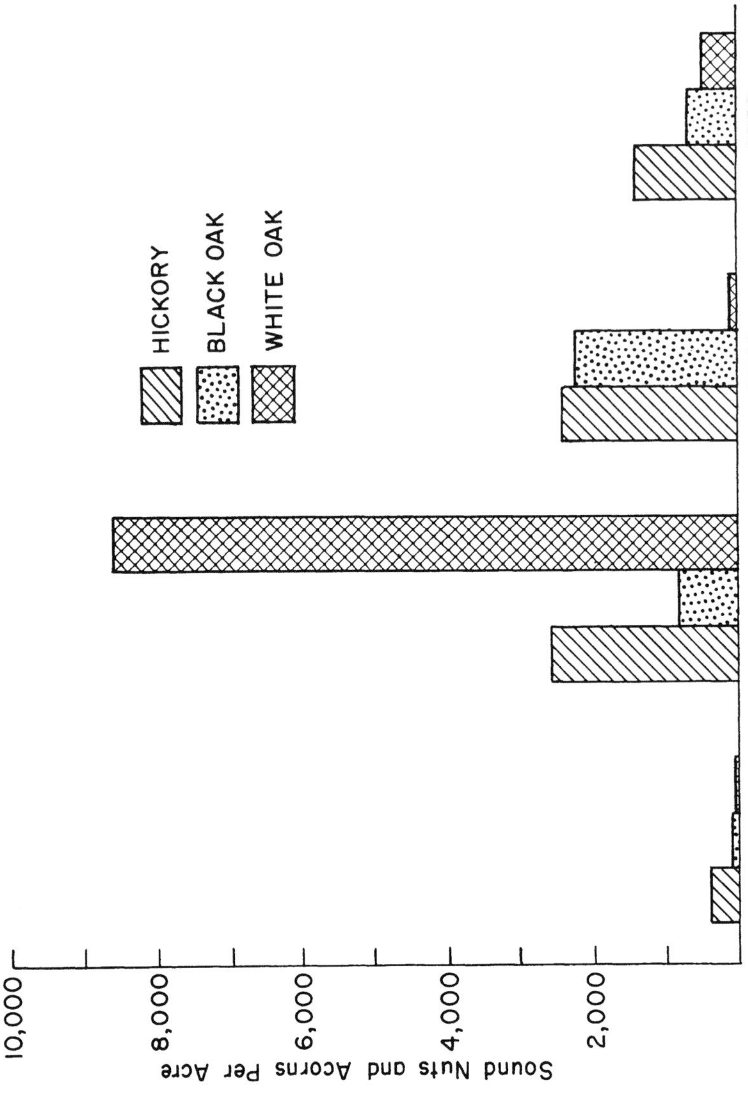

FIG. 2. Sound mast per acre in Rose Lake woodlots.

TABLE II

Sound Nuts and Acorns per Acre, 1956–59

	White Oak	Black Oak	Hickory	Total
1956	25	39	419	483
1957	8,620	823	2,611	12,054
1958	30	2,256	2,345	4,631
1959	455	640	1,395	2,490

TABLE III

Fox Squirrel Populations and Kill, 1956–60

	1956	1957	1958	1959	1960
Population	82	177	185	162	155
Kill					
Total number	41	91	121	96	86
Juv./Ad. Female	1.1	6.2	4.7	5.6	4.0

of mast trees. This is the only year in which white oak made any significant contribution to a mast crop. Hickories during the four years were more reliable mast producers than the oaks.

Squirrel populations.—The squirrel populations as computed by the Lincoln Index ranged from 82 to 185 squirrels in the 102 acres of woodlot, with a very low population in 1956, and generally high populations in the other years. Hunters shot from 41 to 121 squirrels on the 800 acres, the low kill occurring in 1956, and the high, only two years later, in 1958. Productivity as indicated by the number of juveniles per adult female in the kill ranged from 1.1 in 1956 to 6.2 in 1957 (Tab. III).

Interrelationships.—To explore the relationship between mast crops and fox squirrel populations and kill during the study period, I plotted on a graph the production of sound mast in a given year against the population and kill of squirrels the following fall (Fig. 3). Using the squirrel kill as well as the fall population estimates enables measuring population responses to mast fluctuations with two independent indexes of squirrel abundance. It is necessary to correlate mast crops of one year with squirrel population of the following fall, since we assume any effect of abundance or scarcity of mast would be exerted during the critical winter feeding period and would

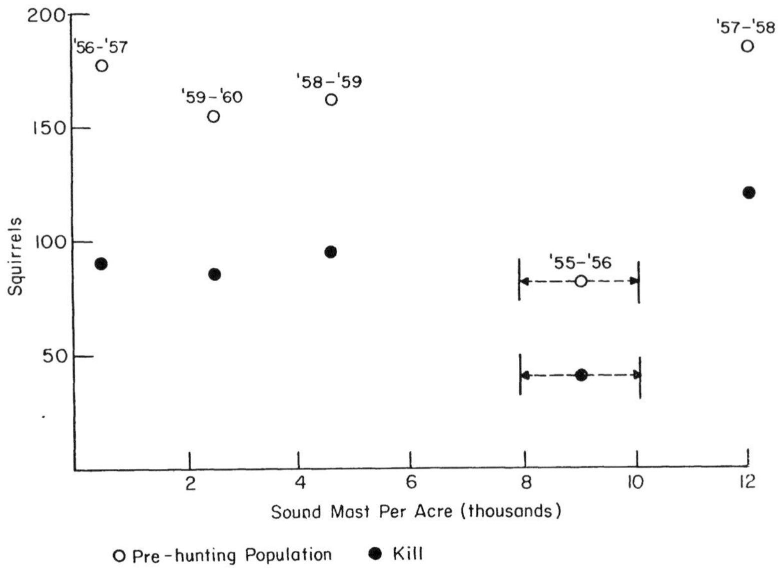

FIG. 3. Relationship of mast crops to fox squirrel population and kill, 1955–60. Each point represents the mast crop for one year and the squirrel population or kill of the following fall.

be reflected in the breeding success the next spring and summer. If there is a positive correlation between mast crops and fox squirrel populations at Rose Lake, one would expect the points on the graph to exhibit such a relationship. Population and kill should drop the fall after a mast failure and increase after a bumper mast crop, i.e., the points would fall in a line from the lower left corner of the graph to the upper right corner. This has not occurred. The coefficient of correlation between mast crops and kill is .025—we are unable to demonstrate a coefficient of linear regression different from zero.

The extremely low kill in the fall of 1956 (Tab. III) aroused our curiosity concerning the 1955 mast crop, the year prior to the commencement of this study. Fortunately, Gysel (1957) in another study had found that black oaks at Rose Lake produced a good crop of acorns in 1955. Also, I had found an abundance of hickory nuts still remaining on the ground in the woodlots in the spring of 1956. I am confident that the mast production would have fallen somewhere within the limits of the dotted lines on the graph. Therefore,

I cannot attribute the shortage of squirrels in 1956 to mast scarcity during the previous winter.

Squirrel productivity rates, based on the hunting season kill sample, show no relationship between breeding success and size of mast crop. The highest productivity (6.2 juveniles per adult female) occurred in 1957, following the poorest of the four mast crops. The squirrel population the preceding year was extremely low, so that a higher productivity would normally be expected; evidently the mast crop failure had no adverse effect on breeding.

Evidently scarcity or abundance of oak and hickory mast do not influence fox squirrel populations at Rose Lake. Supplemental food supplies are evidently available during periods of mast scarcity. At Rose Lake, corn food patches and mechanically picked cornfields provide these foods. The corn patches receive moderate to heavy use by squirrels during the winter. In the farm fields, mechanical pickers leave several bushels of corn per acre on the ground. The squirrel is willing to travel considerable distances if necessary to avail himself of such a food supply. Squirrels will also feed on the bark of the hard maple and the buds of several tree species (Allen, 1943). Acorns and hickory nuts, although certainly a valuable and preferred winter food, were not limiting factors on fox squirrel production in this study.

LITERATURE CITED

Allen, D. L. 1943. Michigan Fox Squirrel Management. Mich. Dept. Cons., Lansing. 404 pp.

Easley, L. T., and L. E. Chaiken. 1951. An Expendable Seed Trap. J. For., 49: 652–653.

Gysel, L. W. 1957. Acorn Production on Good, Medium, and Poor Oak Sites in Southern Michigan. J. For., 55: 570–574.

Papers of the Michigan Academy of Science, Arts, and Letters
Vol. XLVII, 1962 (1961 Meeting)

FISHERIES

SOME ESTIMATES OF PRIMARY PRODUCTION RATES IN MICHIGAN PONDS[1]

ALLEN KNIGHT and ROBERT C. BALL
Michigan State University

FRANK F. HOOPER
Institute for Fisheries Research, Michigan Department of Conservation

Photosynthetic fixation of solar energy in ponds is carried on by the attached algae (periphyton), by the planktonic algae, and by the larger plants growing on the pond bottom (aquatic macrophytes). In the present study we have attempted to: (1) measure independently rates of energy fixation by these three sources; (2) identify environmental factors which favor one source of production over the others; and (3) compare our production estimates with those made in other ecosystems.

It has been demonstrated experimentally that growth of aquatic macrophytes in some way inhibits phytoplankton production (Hasler and Jones, 1949). The mechanism of this inhibition remains obscure. Undoubtedly a variety of factors are involved and the degree of inhibition differs from one ecosystem to another. It has frequently been suggested that aquatic macrophytes compete with plankton algae for nutrient materials (Embody, 1928; Bennett, 1942; Wiebe, 1934). However, shading effects (competition for light) have also been mentioned as mechanisms by which the larger aquatic plants suppress plankton blooms (Hasler and Jones, 1949). The position of periphyton algae in such competitive interaction has not been considered.

The chief methods of evaluating pond production have been measurements of the crops produced, such as bottom fauna and fish (Patriarche and Ball, 1949; Meehean, 1936; Hayne and Ball, 1956). In some instances, an index of the size of the plankton crop has been secured by making plankton counts and transparency measurements. Radioactive carbon (C^{14}) has been used to measure primary pro-

[1] Journal Article No. 2864, Michigan Agricultural Experiment Station.

duction in the ocean and in fresh-water lakes (Strickland, 1960). It affords a convenient and reproducible method of measuring production rates in ponds and should find useful application in pond culture, particularly in situations where one wishes to compare the production of one pond with that of another.

The attached algae have long been recognized as making a major contribution to the primary production of standing water (Young, 1945). However, few attempts have been made to estimate the energy fixed by this segment of the biota. Newcombe (1950) made quantitative measurements of the periphyton that accumulated on glass slides which were suspended in two southern Michigan lakes. His periphyton results tended to confirm his conclusions, based upon plankton and organic matter measurements, that Walnut Lake was more productive than Sodon Lake. Grzenda (1960) and Grzenda and Brehmer (1960) perfected quantitative techniques for studying periphyton. Grzenda showed that the phytopigments extracted from artificial substrates can be used to estimate the production of organic matter. It is appropriate that the C^{14} method and quantitative periphyton techniques be used to attack the general problem of the interrelationship between the production rates of periphyton, phytoplankton, and aquatic macrophytes.

Methods and Materials

The information presented was collected at the Michigan State University experimental farm, Lake City, Michigan between June 15 and September 15, 1960. Routine chemical and physical measurements were made throughout this period. A continuous record of water temperature for the period was obtained from a Taylor thermograph. Hydrogen-ion concentration was determined with a Beckman pH meter (model H). Total alkalinity was measured by titrating samples with 0.02 N H_2SO_4, using methyl-orange as an indicator (Ellis, Westfall, and Ellis, 1948). Dissolved oxygen was determined by the unmodified Winkler method. Total phosphorus was measured throughout the summer, using a modification of the procedure given by Ellis, Westfall, and Ellis (1948). Total iron was determined, using the tripyridyl method (American Public Health Association et al., 1955).

The accrual of periphyton on plexiglass plates yields an estimate of net primary production (Grzenda and Brehmer, 1960). Periphyton was scraped from plates with a glass microscope slide and the phytopigment was extracted with 95 per cent alcohol. The ash-free dry weight of each periphyton sample was determined by evaporating the solvent and igniting the oven-dry samples in a muffle furnace at 520° C. Samples were weighed before and after ignition on an analytical balance. An estimate of ash-free dry weight of organic matter was obtained by subtracting the weight of ash from the dry weight.

The second source of primary production studied was the aquatic macrophytes. Thirty randomly selected samples of vegetation were obtained from each pond. Each sample consisted of all the plants within a one square-foot plot. All parts of the plants including roots and rhizoids were collected. Samples were drained a constant period of time and then weighed. This weight was recorded as wet weight. The last ten samples collected in each pond were taken to the laboratory for determination of dry weight. They were placed in an oven, dried at 55° C., and weighed. This weight was recorded as dry weight.

The third source of primary production studied was that from planktonic algae. The C^{14} technique as described by Steeman Nielsen (1951, 1952) and the oxygen method were both used to measure this source but the latter technique was abandoned when it became clear that results were not reproducible. Water samples were collected in a plastic container. Samples were introduced into glass-stoppered pyrex bottles which were suspended in the ponds at the depth at which the sample was collected. A series of three bottles was suspended at each depth. Two of the bottles were of clear glass and the third bottle was opaque. Bottles were suspended in each pond at depths of 16 and 32 inches. Before the bottles were suspended, two microcuries of C^{14}, in the form of sodium carbonate, were added with a hypodermic syringe to each bottle. Bottles were exposed for a period of 6 hours, which began shortly after 9:00 A.M. At termination of the 6-hour exposure bottles were removed from the pond and a 25 ml. subsample from each bottle was filtered through a type HA, millipore filter pad. The pad was allowed to dry and was then cemented to a planchet. The planchet and attached filter pad were then placed in radiological counting equipment and counts of radioactive carbon were obtained. These counts were converted to total assimilated carbon, using the formulae given by Steeman Nielsen (1952b).

Results were expressed as milligrams of carbon fixed per cubic meter per hour.

Chelated iron (NaFeEEDTA) was introduced into ponds B and C on July 15. The theoretical concentration of chelated iron in ponds B and C, after complete mixing, was one part per million. Diammonium phosphate was added to ponds C and D at a rate of 20 pounds per acre. The theoretical concentration of phosphorus after this treatment was 0.48 mg./l. for pond C and 0.41 mg./l. for pond D. Fertilizer and chelate were applied with a hand-operated garden sprayer. These materials were spread as evenly as possible over the entire surface of the ponds.

Between June 16 and 18, the ponds were drained and were not refilled until June 24. This period, during which the ponds lay fallow, was long enough to desiccate and kill the aquatic macrophytes.

Physical and Chemical

The general trend of the water temperature followed the temperature of the atmosphere very closely. The water warmed slowly until July 12 when we recorded a maximum of 84° F. Thereafter the ponds cooled gradually until September 8, when a sharp decline in water temperature was observed.

During the study period methyl-orange alkalinity varied between 46 and 96 ppm. and pH varied from 8.1 to 9.8. A decreasing trend in alkalinity and increasing trend in pH was observed in all ponds. The mean hydrogen ion concentration for the entire period varied very little from pond to pond. Mean pH values were as follows: pond A—9.03; pond B—9.25; pond C—9.3; pond D—9.1.

The dissolved oxygen concentrations in the ponds ranged from a maximum of 14.97 mg./l. in pond B to a minimum of 8.04 mg./l. in pond D. The dissolved oxygen trend was quite similar in each pond. There was an initial period of low oxygen concentration followed by an increase to a maximum on August 1. Following this peak, oxygen values remained high but fluctuated considerably for the duration of the study.

Influence of Pond Drainage on Chemical Conditions

Draining the ponds prior to the study apparently had a pronounced influence upon the concentration of nutrients in the water.

It is well known that allowing ponds to dry and lie fallow releases nutrients from bottom soils (Mortimer and Hickling, 1954).

The total phosphorus concentration in all experimental ponds reached a peak on precisely the same date (Fig. 1), indicating that some factor was operating simultaneously in all ponds to bring phosphorus into solution. This peak might have been attributed to the addition of chelate and fertilizer in the ponds receiving this treatment were it not for the fact that the control pond not only exhibited this peak but nearly equaled the phosphorus concentration recorded in the ponds receiving artificial enrichment (Fig. 1). This peak in phosphorus concentration is almost certainly an effect produced by drying and disturbing the bottom mud and by release of phosphorus after decomposition of the bottom vegetation. An excess of phosphorus over the actual needs of the plankton algae and bottom vegetation may have been present for a short time at the onset of the study period (Fig. 1), but as the bottom vegetation began to flourish, the concentration of phosphorus in the water steadily diminished, until a plateau level was reached in all ponds.

A situation similar to that observed for phosphorus developed in the case of iron. With fallowing and with death and decay of the aquatic plants, there seems to have been return of iron to the water. Thus we believe that if there were any effects produced by the chelating agent and fertilizer upon iron and phosphorus concentrations, these effects were masked by large releases of iron and phosphorus that occurred upon reflooding.

Periphyton Production

The term periphyton as used in this paper is defined by Odum (1959): "organisms, both plant and animal, attached or clinging to stems and leaves of rooted plants or other surfaces projecting above the bottom." The composition of the periphyton biocoenosis living on the artificial substrates in the ponds was as follows: most abundant forms were representatives of the green algae, such as *Oedogonium, Bulbochaete, Scenedesmus, Tetraedron,* and *Oocystis*; desmids (*Cosmarium, Micrasterias,* and *Staurastrum*) were somewhat less abundant; diatoms, such as *Nitzschia, Fragilaria, Cymbella,* and *Diatoma* were encountered only occasionally.

Net production was measured in all ponds for the period from

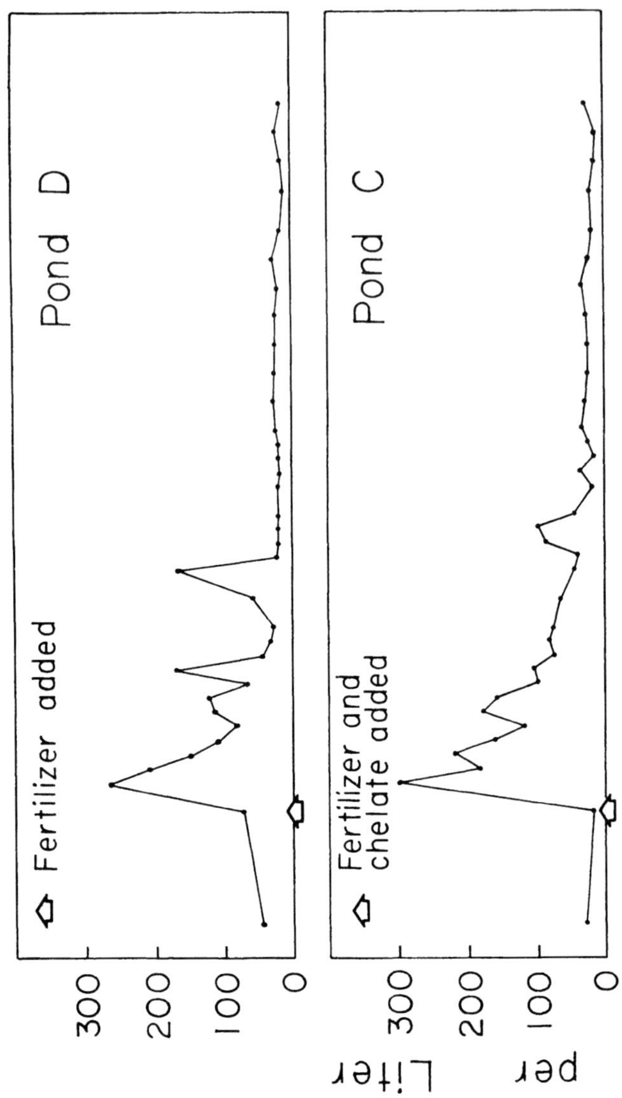

Pond D
Fertilizer added
Pond C
Fertilizer and chelate added
300
200
100
0
per Liter

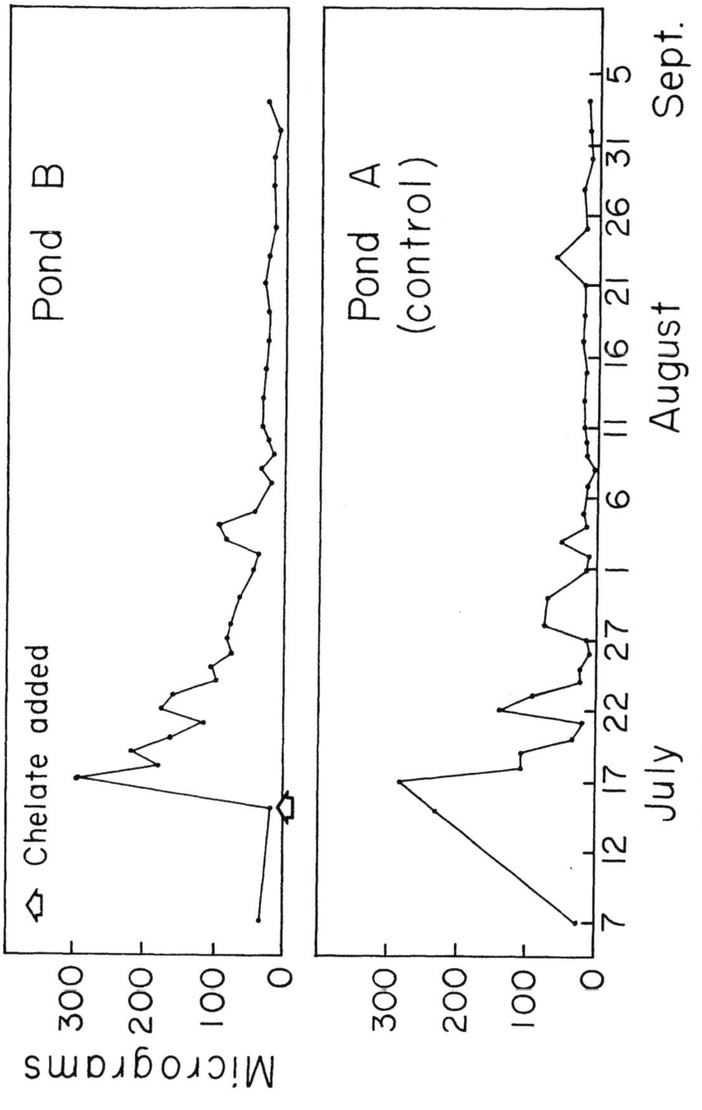

FIG. 1. Total phosphorus concentration in ponds in 1960.

TABLE I

NET PRIMARY PRODUCTION AS DETERMINED BY PERIPHYTON ACCRUAL ON ARTIFICIAL SUBSTRATA (ash-free dry weight)

Collection Date (1960)	Exposure Period (days)	Net Production (grams per sq. meter per day)							
		Mean				Standard Deviation			
		Pond				Pond			
		A	B	C	D	A	B	C	D
7- 5	12	0.212	0.190	0.278	0.301	0.032	0.016	0.043	0.028
7-14	9	0.298	0.259	0.290	0.331	0.000*	0.026	0.068	0.003
7-21	7	0.390	0.332	0.690	0.522	0.039	0.010	0.054	0.017
7-28	7	0.390	0.337	0.528	0.417	0.094	0.025	0.017	0.019
8- 4	7	0.450	0.297	0.586	0.394	0.031	0.026	0.091	0.012
8-11	7	0.428	0.325	0.404	0.437	0.036	0.051	0.018	0.035
8-18	7	0.394	0.372	0.395	0.411	0.031	0.030	0.063	0.032
8-25	7	0.266	0.332	0.394	0.351	0.038	0.034	0.008	0.032
9- 1	7	0.296	0.242	0.265	0.280	0.011	0.034	0.002	0.029
Mean		0.347	0.298	0.426	0.383				

* Single sample.

June 24 to September 1, 1960 (Tab. I). Each pond followed a slightly different week-to-week trend in accrual of organic weight, but a similar general trend was observed in all ponds. Low production in the latter part of June and the first part of July was followed by maximum production during midsummer and a decline in production in late August and early September. Mean net production rates for the summer were as follows: in pond A it was estimated to be 0.347 grams ash-free dry weight per square meter per day; in pond B, 0.298; in pond C, 0.426; and in pond D, 0.383 grams of ash-free dry weight per square meter per day. These values represent production during the summer, hence are probably much higher than a yearly average.

PRODUCTION BY AQUATIC MACROPHYTES

The principal macrophytes in the ponds, in order of their importance, are as follows: *Chara* sp., *Potamogeton pectinatus*, *P. zos-*

TABLE II

NET PRIMARY PRODUCTION OF POND MACROPHYTES DURING THE 1960 GROWING SEASON, AS DETERMINED BY THE HARVEST METHOD

Pond		Weight			
		Mean			Standard Deviation
		Pounds per acre	Grams per sq. meter per day	Grams per sq. meter	Grams per sq. meter
A	Wet	11,771		1,322	107
	Dry	1,047	1.45	118	42
B	Wet	38,533		4,326	330
	Dry	2,360	3.27	265	86
C	Wet	25,274		2,838	530
	Dry	2,042	2.83	229	181
D	Wet	22,272		2,501	331
	Dry	4,329	6.00	486	263

teriformis, P. natans, Carex sp., *Scirpus* sp., *Typha latifolia, Sagittaria* sp., *Elodea canadensis, Equisetum fluviatile, Eleocharis* sp., and *Najas flexilis.*

The mean net productivity value obtained for each pond is given in Table II. Values ranged from a high of 6.00 grams dry weight per square meter per day in pond D to a low of 1.45 grams dry weight per square meter per day in pond A.

PHYTOPLANKTON PRODUCTIVITY

The predominant planktonic forms were rotifers, desmids, and flagellates. Rotifers were represented by *Keratella* sp. and *Notholca* sp. Desmids were represented by *Cosmarium* sp., *Staurastrum* sp., and *Micrasterias* sp. Flagellates were represented by the dinoflagellate (*Ceratium hirundinella,* Stein), and three genera from the order Volvocales (*Eudorina* sp., *Pleodorina* sp., and *Volvox* sp.). Other forms frequently observed were copepods and cladocera. The copepods were chiefly *Macrocyclops* sp., whereas the cladocera were represented by *Bosmina* sp. Diatoms were rare in the pond plankton samples.

The radioactive carbon method used in the present study yields a measure of photosynthesis that is somewhere between the net and gross value, possibly nearer to the former (Strickland, 1960). These data are presented in tabular form in Table III. A maximum carbon assimilation value of 23.7 mg. of carbon per cubic meter per hour was recorded in pond D at the exposure depth of 16 inches on July 24, 1960. The minimum carbon assimilation value was 2.75 mg. of carbon per cubic meter per hour in pond A on July 10, 1960, at a depth of 32 inches.

Comparison of Production Estimates

Production rates as measured by the methods outlined above show variations from pond to pond (Tab. IV). There is, however, a close agreement in the order of various ponds when they are ranked using the C^{14} method and using the dry weight of macrophytes. Both of these estimates suggest that pond A is least productive, followed by pond C. Pond B has a somewhat greater mean production than either pond A or C, but is considerably less productive than pond D. The four ponds cannot be ranked in the same order using the periphyton method of measuring primary production. The periphyton method indicates that the lowest production rate occurred in pond B and the highest in pond C. Ponds A and D were intermediate. An estimate of total accrual of periphyton in the ponds cannot be calculated without information on the average area of substrata suitable for colonization. Thus rates of accrual on artificial substrates are only indices of the intensity of this source of production. If the heavy growth of macrophytes in pond D provided the largest surface area for periphyton growth, then this pond might have had the highest total of periphyton production.

Table IV gives some estimates of primary production in other ecosystems as well as a summary of estimates for the ponds. Values for production in terms of carbon have been converted to organic matter using a factor of 2.0 (Ryther, 1946). Comparing phytoplankton production of the ponds with other ecosystems suggests that production is comparatively low in the ponds. The ponds compare favorably with primary production reported for some parts of the ocean and with values reported for an unproductive lake in Michi-

TABLE III

PRIMARY PRODUCTION MEASUREMENT IN PONDS, USING THE C^{14} METHOD

(Assimilated carbon is expressed as milligrams of carbon per cubic meter per hour.)

Exposure Date	Pond								Weather Conditions during Exposure	Water Temperature during Exposure (degrees F.)
	A		B		C		D			
	Depth (inches)		Depth (inches)		Depth (inches)		Depth (inches)			
	16*	32	16	32	16	32	16	32		
7-20-60	4.35	4.42	6.66	8.90	12.54	12.68	7.29	11.16	Very bright, clear	74
7-27-60	4.86	3.52	6.05	4.83	9.36	6.11	8.41	12.71	Clear, few clouds	78
8- 3-60	5.83	4.32	12.44	10.27	10.46	8.69	14.83	15.60	Partly cloudly	79
8-10-60	3.68	2.75	7.68	7.57	8.66	7.53	14.35	15.15	AM—overcast PM—partly cloudy	73
8-17-60	7.87	6.82	10.17	9.15	8.36	9.59	15.76	22.60	AM—clear PM—overcast	74
8-24-60	4.89	4.36	7.72	7.76	9.30	6.90	23.75	22.70	AM—clear PM—overcast	76
8-31-60	7.44	7.96	11.59	9.15	8.26	6.83	11.37	10.60	Overcast to partly cloudy	78
9- 7-60	7.11	7.14	11.33	13.76	8.90	8.68	18.38	11.99	Slightly overcast	82
Mean value	5.75	5.17	9.21	8.92	9.48	8.38	14.27	15.31		
Mean of 16- and 32-inch depths	5.46		9.07		8.93		14.79			

One dark bottle and two light bottles were exposed for six hours at each depth.

* All exposure depths were measured from the surface of the pond to the sample.

TABLE IV

COMPARISON OF PRIMARY PRODUCTIVITY ESTIMATES FOR VARIOUS ECOSYSTEMS

Location	Method	Production Rate: Grams of organic matter per sq. meter per day — Phytoplankton	Production Rate: Grams of organic matter per sq. meter per day — Periphyton	Production Rate: Macrophytes (Dry weight) — Grams per sq. meter per day	Production Rate: Macrophytes (Dry weight) — Pounds per acre	Source
Pond A		0.30	0.35	1.45	1,047	
Pond B		0.48	0.30	3.27	2,360	
Pond C		0.44	0.44	2.83	2,042	
Pond D		0.64	0.38	6.00	4,328	
Blind Lake, Michigan		1.20				Schelske, 1960
Barents Sea	C^{14}	0.56				Corlett, 1957*
North Sea (annual range)	C^{14}	0.20–3.00				Steel, 1957*
South Atlantic	C^{14}	1.00–8.00				Steeman Nielsen, 1954
Red Cedar River, Michigan	Periphyton accrual		0.56			Grzenda, 1960
Silver Springs, Florida	Organic weight			7.40		H. T. Odum, 1957
Sargasso Sea	Organic weight	0.26				Riley, 1957
Seaweed beds, Nova Scotia	Harvest			1.00		Tamiya, 1957
Wheat (world average)	Harvest			2.30		Woytinsky and Woytinsky, 1953
Green Lake, Wisconsin	Harvest				1,590	Rickett, 1924
Lake Mendota, Wisconsin	Harvest				1,801	Rickett, 1922

* Quoted by Strickland, 1960.

gan (Blind Lake). Periphyton accrual rates are similar to those observed in a warm-water stream in southern Michigan (Tab. IV). Net production of larger plants in the most productive ponds was 2.4 times greater than that in Lake Mendota and 2.7 times greater than that in Green Lake (Tab. IV). Since these Wisconsin lakes are of high productivity it would seem that the level of production by aquatic macrophytes in the ponds is relatively high.

From the foregoing it would appear that production rates of phytoplankton in these ponds are low but that the rates of larger plants are rather high. It is hypothesized that the comparatively low production rates of phytoplankton as compared to macrophytes can be explained in terms of adaptive mechanisms available to larger plants and absent in phytoplankton. The rate of photosynthesis by phytoplankton cells is strongly influenced by light intensity. It increases with increasing intensity until a saturation value is reached. Above the saturation value, the rate remains constant with increasing intensity up to a critical intensity at which it decreases due to light inhibition of the photosynthetic process (Strickland, 1960). Inhibition is brought about by photooxidation of critical enzyme systems (Steeman Nielsen, 1952c). At high intensities, there is a bleaching of the chlorophyll in algal cells.

In shallow, transparent ponds the light intensity at all depths is a large fraction of that at the surface. According to Oosting (1958), in the presence of intense light larger plants exhibit elongated palisade cells. Cutin is thicker and the amount of supporting tissue is greater in intense light. Oosting (*ibid.*) indicates that during intense illumination chloroplasts arrange themselves along the side walls of palisade cells and thus receive a minimum of direct insulation. The possession of mechanisms adapting aquatic macrophytes to intense light may have enabled them to grow in luxuriant beds, while the phytoplankton, without such mechanisms may have been light-inhibited. Flooding ponds in early summer when there is little cloudiness may favor growth of aquatic macrophytes as compared to phytoplankton whereas the lower light intensities of a cloudy spring may favor growth of phytoplankton.

LITERATURE CITED

American Public Health Association, American Water Works Association, and Federation of Sewage and Industrial Water Association. 1955. Standard Methods of the Examination of Water, Sewage and Industrial Wastes, 10th ed. Waverly Press Inc., Baltimore. 522 pp.

Bennett, George W. 1942. Management of Small Artificial Lakes. Bull. Ill. Nat. Hist. Surv., 22: 357–376.

Corlett, J. 1957. Measurement of Primary Production in the Western Barents Sea. Paper presented at a Symposium of the International Council for the Exploration of the Sea. Bergen, 1957. Preprint C/no. 8.

Cushing, D. H. 1957. Some Experiments Using the C^{14} Technique. Paper presented at a Symposium of the International Council for the Exploration of the Sea. Preprint C/no. 11.

Ellis, M. M., B. A. Westfall, and M. D. Ellis. 1948. The Determination of Water Quality. U. S. Dept. Int., Fish Wildl. Ser. Res. Rep., 9: 122 pp.

EMBODY, G. C. 1928. Principles of Pond Fertilization. Trans. Am. Fisheries Soc., 58: 19–22.

——, AND MORRIS L. BREHMER. 1960. A Quantitative Method for the Collection and Measurement of Stream Periphyton Limnol. Oceanog., 5: 191–194.

GRZENDA, A. R., AND MORRIS L. BREHMER. 1960. A Quantitative Method for the Collection and Measurement of Stream Periphyton. Limnol. Oceanog., 5: 191–194.

HASLER, ARTHUR D., AND ELIZABETH JONES. 1949. Demonstration of the Antagonistic Action of Large Aquatic Plants on Algae and Rotifers. Ecol., 30: 359–364.

HAYNE, DON W., AND ROBERT C. BALL. 1956. Benthic Productivity as Influenced by Fish Predation. Limnol. Oceanog., 1: 162–175.

MEEHEAN, O. LLOYD. 1936. Some Factors Controlling Largemouth Bass Production. Prog. Fish.-Cult., 16: 1–6.

MORTIMER, C. H., AND C. F. HICKLING. 1954. Fertilizer in Fishponds. Colonial Office, Fisheries Publ. No. 5, Her Majesty's Stationery Office, London.

NEWCOMBE, CURTIS L. 1949. Attachment Materials in Relation to Water Productivity. Trans. Am. Microscop. Soc., 68: 355–361.

ODUM, E. P. 1959. Fundamentals of Ecology, 2d ed. (In collaboration with H. T. Odum.) W. B. Saunders Co., Philadelphia, 546 pp.

ODUM, H. T. 1957. Trophic Structure and Productivity of Silver Springs, Florida. Ecol. Mono., 27: 55–112.

OOSTING, HENRY J. 1958. The Study of Plant Communities, 2d ed. W. H. Freeman, San Francisco. 440 pp.

RICKETT, H. W. 1922. A Quantitative Study of the Larger Aquatic Plants of Lake Mendota. Trans. Wis. Acad. Sci., 20: 501–527.

——. 1924. A Quantitative Study of the Larger Aquatic Plants of Green Lake, Wisconsin. Trans. Wis. Acad. Sci., 21: 381–414.

RILEY, GORDON A. 1957. Phytoplankton of the North Central Sargasso Sea. Limnol. Oceanog., 2: 252–270.

RYTHER, JOHN H. 1956. The Measurement of Primary Production. Limnol. Oceanog., 1: 72–84.

SCHELSKE, CLAIRE LAVERNE. 1960. The Availability of Iron as a Factor Limiting Primary Productivity in a Marl Lake. Doctoral dissertation, Univ. Mich.

STEEL, J. H. 1957. A Comparison of Plant Production Estimates Using C^{14} and Phosphate Data. J. Mar. Biol. Assoc. United Kingdom, 36: 233.

STEEMAN, NIELSEN E. 1951. Measurement of the Production of Organic Matter in the Sea by Means of Carbon-14. Nature, 167: 684.

——. 1952a. The Use of Radio-active Carbon for Measuring Organic Production in the Sea. J. Cons. Int. Explor. Mer., 18: 117–140.

——. 1952b. Production of Organic Matter in the Sea. Nature, 169: 956.

——. 1952c. On the Detrimental Effect of High Light Intensities on the Photosynthetic Mechanism. Physiol. Plantarum, 5: 334.

——. 1954. On Organic Production in the Oceans. J. Cons. Int. Explor. Mer., 19(3): 309.

STRICKLAND, J. D. H. 1960. Measuring the Production of Marine Phytoplankton. J. Fish. Res. Boards of Canada, Bull. No. 122, 172 pp.

TAMIYA, HIROSHI. 1957. Mass Culture of Algae. Ann. Rev. Plant Physiol., 8: 309–334.

WIEBE, A. H. 1934. Nocturnal Depressions in the Dissolved Oxygen in Fishponds with Special Reference to an Excess of Coarse Vegetation and of Fertilizers (Texas). Trans. Am. Fisheries Soc., 64: 181–188.

WOYTINSKY, W. S., AND E. S. WOYTINSKY. 1953. World Population and Production. The Twentieth Century Fund, New York.

YOUNG, ORSON WHITNEY. 1945. A Limnological Investigation of Periphyton in Douglas Lake, Michigan. Trans. Am. Microscop. Soc., 64: 1–20.

Papers of the Michigan Academy of Science, Arts, and Letters
Vol. XLVII, 1962 (1961 Meeting)

THE POPULATION OF SEA LAMPREY LARVAE IN EAST BAY, ALGER COUNTY, MICHIGAN

WILBERT C. WAGNER AND THOMAS M. STAUFFER
Michigan Department of Conservation

Larvicides have proved their effectiveness by virtually eradicating ammocoetes[1] of the sea lamprey (*Petromyzon marinus*) in streams tributary to Lake Superior.[2] It has been demonstrated, however, that ammocoetes also occur in the Great Lakes or in tributary inland lakes, where larvicide treatment by present methods would be prohibitively expensive. Hansen and Hayne (in press) estimated that the population of ammocoetes in a small bay in Big Bay de Noc, Delta County, exceeded 30,000; Stauffer and Hansen[3] found larvae at four other areas in Lake Michigan in 1957; Thomas (1960) collected ammocoetes in Lake Superior; and we (unpublished) have collected larvae in Huron Bay (Lake Superior) and in Au Train Lake (a tributary of Lake Superior). It thus appears that ammocoete populations are not uncommon in the Great Lakes, and that their presence may retard the sea lamprey control program, even though these lentic populations are presumably derived from tributary streams (Hansen and Hayne, in press).

The present study was undertaken in 1960 to provide additional information on the magnitude and characteristics of populations of ammocoetes in lentic environments. The population of ammocoetes in East Bay, Alger County (Fig. 1) that was selected for study, was of especial interest because the Sucker River (a tributary of East Bay and the presumed source of the bay population) had been successfully treated with larvicide in 1959.

East Bay and the Sucker River are located about 75 miles west

[1] "Ammocoete" and "larva" refer to the sea lamprey, unless stated otherwise.

[2] Programs and Progress, 1959. Mimeographed report of Great Lakes Fisheries Investigations, Bureau of Commercial Fisheries, Fish and Wildlife Service, U. S. Dept. of Interior.

[3] 1958. A preliminary report on the migration of sea lamprey ammocoetes in Michigan. Mich. Dept. Cons., Inst. Fish. Res. Report No. 1535 (unpublished).

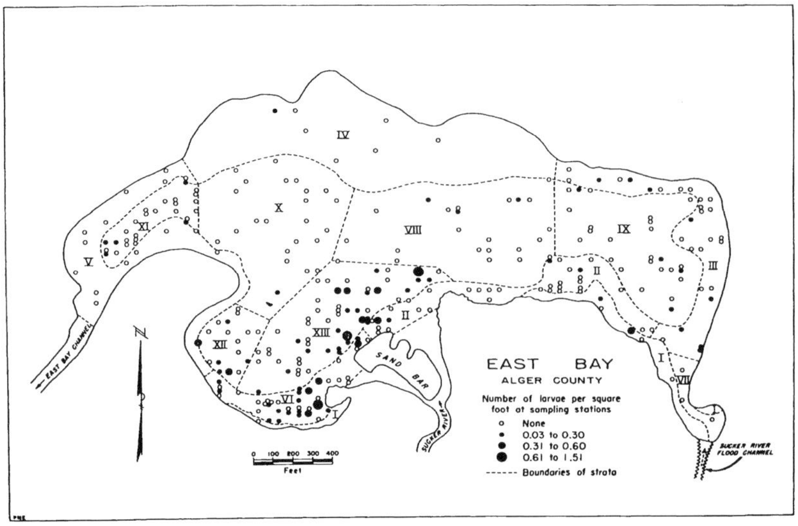

FIG. 1. East Bay, showing strata (I–XIII), and the number of ammocoetes collected per square foot at different sampling stations.

of Sault Ste. Marie, Michigan. The bay is a 78-acre lake, connected by ⅓-mile-long East Bay Channel to West Bay, which in turn opens on Lake Superior. East Bay has a maximum depth of 46 feet and a predominantly sand bottom to depths of 20 feet; at greater depths, the bottom is mostly organic silt. In general, the shoreline is barren and wind-swept, and the shoal area has only small amounts of emergent and floating vegetation. There was apparently no thermal stratification of the lake during the summer of 1960. The predominant fishes in the lake, in addition to lampreys, were yellow perch (*Perca flavescens*), northern pike (*Esox lucius*), rainbow trout (*Salmo gairdneri*), white sucker (*Catostomus commersoni*), and Johnny darter (*Etheostoma nigrum*).

The Sucker River supported a moderate population of brook trout (*Salvelinus fontinalis*) and rainbow trout, and was one of the three largest producers of sea lampreys among Michigan tributaries of Lake Superior (Stauffer and Hansen, 1958). The stream divides into two channels near the mouth and enters East Bay (summer discharge about 30 c.f.s.) on the south side; a third, smaller, intermittent channel enters the bay in the southeast corner during the spring runoff.

In October 1958, the Sucker River was treated with a larvicide by the U. S. Bureau of Commercial Fisheries, but a drop in temperature prevented a complete kill. The river was retreated in August 1959, and no lampreys were found during post-treatment surveys in September 1959. In July 1956, before the river had been treated, 5 hours of electrofishing along the shoreline of East Bay revealed that ammocoetes of the sea lamprey and the American brook lamprey (*Lampetra lamottei*) were relatively abundant near the main mouths of the river and along the south shore toward East Bay Channel, but were comparatively rare along the southeast, east, and north shores. In May 1960, collections were again made with a direct-current shocker along most of the shoreline of East Bay, to determine whether the ammocoete population was still present after larvicide treatment of the "parent" stream. The catch of ammocoetes per hour in 1960 was similar to the catch in 1956 (suggesting a substantial ammocoete population) so a detailed population study was undertaken from July 22 to September 1, 1960.

Methods

An orange-peel dredge was used to estimate the ammocoete population of East Bay by the area-density method (Rounsefell and Everhart, 1953). A physical survey of the lake was made in June 1960, to facilitate the random sampling required for the population estimate. Depths and bottom types were determined and recorded on a shore outline map drafted from an aerial photograph.

Selection of stations.—For the population estimates, East Bay was divided according to bottom type and depth into two areas. Further subdivisions into a total of 13 strata (Fig. 1) were based on the expected concentration of ammocoetes as estimated by the preliminary sampling, and on the distance from the mouth of the Sucker River (ammocoete concentration was expected to be greater near the stream mouth). The one exception was Stratum I, which arbitrarily included all areas with emergent vegetation, regardless of the location in the bay or the estimated abundance of ammocoetes. (Emergent vegetation was restricted to water depths of less than 3 feet.)

To establish station locations in each stratum, grid lines were drawn at 50-foot intervals on the map of the bay. Within each stratum, each grid-line intersection was numbered and became a possible sampling station, subject to random selection. To assure random sampling in time among strata, the order in which collection stations in the bay were sampled was determined at random.

The number of stations to be sampled in each stratum was adjusted so that the strata with the higher populations were sampled more frequently than strata with lower populations. For the first complete series of samples (25 lifts of an orange-peel dredge at each of 101 stations), the number of stations in each stratum was based on the anticipated abundance of ammocoetes as indicated by the earlier electrofishing in the bay, on the distance of the stratum from the mouth of the Sucker River, and on the area of the stratum. During the second round of sampling (15 lifts at each of 195 stations with an orange-peel dredge and 7 stations with a metal enclosure, as described below), the number of stations in each stratum was determined on the basis of the size of the ammocoete population suggested by the first round of sampling.

To facilitate location of each sampling station in the bay (after random selection of the grid point), markers were placed along the shoreline at intervals of 150 to 250 feet, and 14 buoys were anchored in the bay. All markers and buoys were then plotted on the map. To locate the first 143 stations, distances from known points (shore markers and buoys) to the sampling stations were estimated; this method may have introduced some error in the location of sampling sites within intended strata for stations within 200 feet of shore where depth and bottom type changed rapidly in relation to distance from the shore. At subsequent stations the distances from shore to all stations within 200 feet of the shore were measured with a floating line to remove this possible bias. Distances to stations more than 200 feet from shore were estimated; here depth and bottom type were relatively uniform and it was unlikely that error in location of sites would affect the accuracy of the population estimate. For all stations, the direction of both estimated and measured distances to the station was determined by "lining up" with markers and buoys of known location.

Sampling procedure.—In Strata II-XIII (Fig. 1), 15 or 25 lifts were taken at each station with an orange-peel dredge that was operated with a boom and winch from a 16-foot trapnet boat; care was taken to avoid superimposed lifts. The dredged material was emptied into a 28- by 60-inch screen box, 10 inches deep, of 1/8-inch mesh. The fine sediments were washed out by a water-sprinkler system supplied by a 2,000-gallon-per-hour centrifugal pump. Ammocoetes, debris, and coarse particles were retained by the screen. The screened material was carefully examined before it was discarded. All ammocoetes were preserved in 10-per cent formalin and later identified and measured.

The dredge sampled varying areas of the bottom in each lift, depending upon bottom type. In organic silt (81 stations), where the dredge penetrated deepest, the area sampled was judged to be 0.92 square foot (the maximum capacity of the dredge). In hard sand bottom (90 stations), the dredge was calibrated[4] and the average area sampled by one lift was found to be 0.76 ± 0.004 square foot. In heterogeneous bottom types (125 stations), the estimate of bottom area sampled was derived by interpolation between the above values for silt and sand. Not infrequently, the jaws of the dredge were prevented from closing by sticks or other objects. Lifts in which the jaws did not close and in which little bottom material was retained were not used in the population estimate.

In Stratum I, where emergent vegetation was present, the dredge would not penetrate the heavy root mat. To sample this stratum, a 55-gallon steel barrel from which both ends had been removed was used. Mower blades were attached to the bottom rim of the barrel to cut through the roots, and a 10-foot steel pipe was attached across the top to use as a lever to rotate the barrel and thus facilitate penetration of the lower edge into the bottom. In sampling, the enclosure was placed over a randomly selected sampling site and slowly rotated until the lower edge had penetrated 8 to 12 inches into the substrate. Vegetation, roots, and other debris were first removed and examined for ammocoetes. Then a lamprey larvicide, the sodium salt of 3-trifluormethyl-4-nitrophenol, was introduced at a concentration of 40 to 60 ppm. After a waiting period of 1 to 2 hours, a scap net was passed repeatedly through the water in the enclosure to recover any ammocoetes that were present. The water was then pumped from the enclosure into a screened box and the substrate was examined for ammocoetes to a depth of about 2 inches. (No larvae were found in the substrate in any of the samples, however.) The area

[4] One dredge lift was taken at each of ten areas in sand bottom where ¼-inch ball bearings had been placed in a 2-inch grid pattern covering 9 square feet. The average area sampled was determined from the average number of bearings (each of which represented an area of 4 square inches) picked up.

sampled by the enclosure was 2.64 square feet. Three enclosure samples were taken at each station, representing a total area of 7.92 square feet.

Estimation of population.—The estimated number of sea lamprey ammocoetes in each stratum was the product of the mean number of ammocoetes collected per square foot (mean number per station divided by mean station area), and the total area of the stratum, in square feet. The total estimated population was found by summing the estimates for the different strata. A variance for the ratio estimate was calculated by methods given by Cochran (1953).

Most of the sampling on East Bay was done with the orange-peel dredge. Hansen and Hayne (in press) compared dredge collections in the Ogontz River with collections made there with a metal enclosure (which presumably caught all ammocoetes present), and found that the dredge collected only about one-third of the ammocoetes of the sea lamprey and American brook lamprey present in a given area. Thus, for East Bay, estimates based on orange-peel dredge samples were judged to be far too low.

Reasons why the orange-peel dredge does not capture all lampreys within the area covered by the open dredge include the following: (1) The dredge does not close completely on a bottom containing detritus; this bottom type is a preferred habitat for lampreys (see below). (2) Jaws of the dredge, as they close, slice into the substrate at an angle, thus reducing the area which is sampled effectively. A correction factor for this bias was determined for sand bottom by calibration but was assumed to be nil for silt bottom (possibly an erroneous assumption). (3) There may be some escapement through an opening (used for cleaning) at the top of the dredge. (4) Some lampreys may escape from the area to be dredged during the short interval of time between the placement of the dredge and the closure of the jaws.

Population Estimate

A total of 214 ammocoetes were taken from the 303 randomly selected dredging and enclosure stations. The estimated population of sea lamprey larvae in East Bay was 96,300 ± 20,500 of which 47, 400 ± 8,500 were from strata in water depths of more than 20 feet (Tab. I). The percentage of the population that was metamorphosing was apparently small, since only a single transforming specimen

TABLE I

Areas of Strata, Sampling Stations, Larvae Collected, and Estimated Population of Sea Lamprey Larvae in East Bay, Alger County, 1960

Stratum Number	Areas of Strata (acres)	Number of Stations	Number of Larvae Collected	Population Estimates	
				Number of larvae	Standard error
*Shallow water**					
I	2.50	7	4	7,900	5,900
II	3.69	32	11	4,400	2,900
III	4.88	28	13	6,700	2,500
IV	14.20	7	3	16,500	16,600
V	5.44	13	0	—	—
VI	2.50	40	66	13,400	4,500
VII	0.86	3	0	—	—
Total	34.07	130	97	48,900	18,600
*Deep water**					
VIII	10.76	22	3	3,600	2,000
IX	7.50	25	4	3,000	1,400
X	9.39	24	1	900	900
XI	2.13	28	12	2,600	1,300
XII	2.08	17	13	4,300	2,000
XIII	7.37	57	84	33,000	7,800
Total	39.23	173	107	47,400	8,500
Grand total	73.30	303	214	96,300	20,500

*Shallow water = 0–20 feet; deep water, 20–46 feet.

was taken during the sampling. The relatively small average length (3.6 inches, S. D. = 0.66) of the larvae collected also indicated that few were nearing metamorphosis. Applegate and Brynildson (1952) found that the average length of newly metamorphosed sea lampreys migrating downstream in the Carp Lake River, Michigan, was 5.7 inches (S. D. = 0.42) in 1948–49 and 5.7 inches (S. D. = 0.46) in 1949–50. Few ammocoetes in East Bay were near this size.

A total of 121 American brook lamprey ammocoetes (average length, 4.6 inches; range, 2.0–6.0) were also collected at the dredging stations. The population of American brook lamprey larvae was estimated as 72,500 ± 13,700, of which 30,900 ± 6,200 were in deep water. Although American brook lampreys apparently outnumbered the sea lampreys in the Sucker River by about 3 to 1 during 1955–56 (based on 18 collections in July and August with an electric shocker), they were less numerous than sea lampreys in the bay in 1960.

Density of Ammocoetes

As an average for the entire bay, the estimated density of ammocoetes was 0.030 per square foot; among strata the density varied from nil to 0.123 per square foot. Density was closely associated with the distance from the presumed sources of larvae (the three mouths of the river) and also seemed to be associated with bottom type and/or depth. The last two factors were so closely interrelated, however, that separate effects were difficult to differentiate.

Fig. 1 shows that the density of ammocoetes per square foot was closely associated with distance from the stream mouths. The average density of ammocoetes at stations within 400 feet of stream mouths was 0.108 per square foot as compared to an average density of 0.023 at stations 400 to 1,600 feet away.

The collecting stations were divided into three groups, on the basis of the two main bottom sediments present (silt and sand), as follows (number of stations in parentheses): silt, with less than 20 per cent sand (106); silty sand, with at least 20 per cent sand and 20 per cent silt (49) and sand, with less than 20 per cent silt (148). Larvae occurred most frequently in silty sand and less frequently in silt than in sand; the differences among the three groups were significant (chi-square = 11.26). Judging from many collections in streams, we suspected, however, that detritus (which occurred in varying amounts in all three soil types) may have contributed to the differences. To determine the possible effects of detritus, each of the three main groups (silt, silty sand, sand) were divided into two subgroups—one with 2 per cent or more detritus and one with no detritus (i.e., less than 2 per cent). In silt and in sand, larvae occurred at significantly larger numbers of stations with detritus than at stations with none (chi-square = 7.90 and 12.52, respectively) but no difference was observed between the two subgroups in silty sand. There was no significant difference in occurrence of ammocoetes among the three subgroups of stations with detritus, or among the three subgroups without it; therefore, all stations without detritus

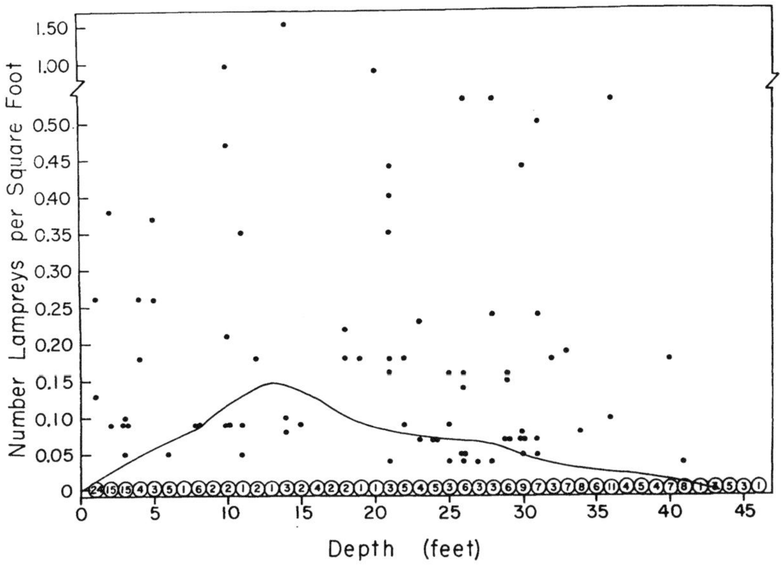

FIG. 2. The estimated density of sea lamprey larvae in relation to depth, East Bay, Michigan, 1960. Dots represent average number collected per square foot at individual stations. Encircled numbers along the bottom of the graph are numbers of stations at which no sea lampreys were collected. The curve represents average concentration in relation to depth of water.

were combined, as were all stations with detritus. Overall, ammocoetes occurred more frequently at 108 stations with detritus than at 195 stations without it (chi-square = 27.40). On the basis of this high chi-square value, and from our general observations while collecting ammocoetes in many Michigan streams, we believe that a substrate containing detritus is a preferred habitat for ammocoetes.

The average density of larvae at the collecting stations in relation to depth of collection is shown in Fig. 2. In 5-foot depth intervals, the average density of larvae per square foot ranged from 0.002 (41–46 feet in depth) to 0.144 (11–15 feet).

LENGTH OF AMMOCOETES

Hansen and Hayne (in press) showed that there was no difference in length between larvae (all species) collected with enclosures and those collected with the orange-peel dredge. Furthermore, in East Bay the average lengths of ammocoetes collected (within the

5-foot contour) with a direct-current shocker and with an orange-peel dredge were not significantly different. Thus, in the present study, we assume that the dredge was not selective for different sizes of ammocoetes.

The average length of the 214 ammocoetes collected by dredging and enclosures was 3.6 ± 0.14 inches. For each stratum, the estimated number of larvae in each 0.1-inch size group was calculated from the lengths of ammocoetes collected and the population estimate. The totals from all strata were then summed to obtain the length-frequency distribution of the population in the entire bay (Fig. 3). The length-frequency distribution of the population approximated a normal curve, with three outstanding variations at 1.8, 2.8, and 4.2 inches. These abnormally high points in the distribution reflect disproportionately light sampling in Stratum IV in relation to the population. The average size of the estimated population (3.4 inches) approximates that of the ammocoetes which were collected.

The length of ammocoetes was related to depth, but apparently not to distance from the mouth or to bottom type. Multiple regression analysis (Snedecor, 1956) was used to test the effect of depth and

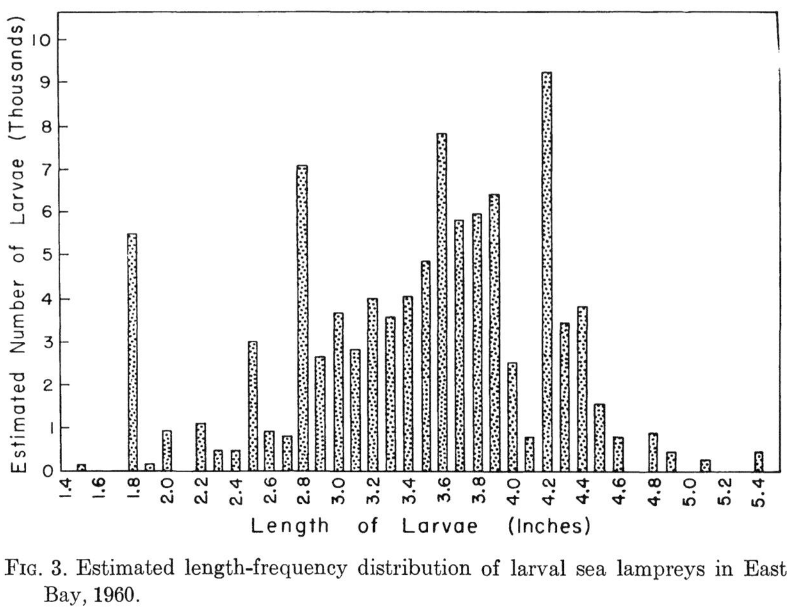

FIG. 3. Estimated length-frequency distribution of larval sea lampreys in East Bay, 1960.

distance from the mouth (independent variables) on the length of larvae (dependent variable). The multiple regression equation ($L = 3.275 + 0.0172\,X - 0.00019\,Y$, where L = length of larvae in inches, X = depth of water in feet, and Y = distance from the mouths in feet) showed that the average length of larvae increased 0.017 inches with each foot of depth and decreased 0.019 inches with each hundred-foot increase in distance from the mouths of the stream. The increase in length associated with depth was significant ($F = 16.35$) but the decrease in length associated with distance from the sources was not significant. An analysis of variance indicated that there was no difference among average sizes of ammocoetes collected in the six subgroups of bottom type mentioned in the preceding section (silt, silty sand, or sand; each considered with or without detritus).

LITERATURE CITED

Applegate, V. C., and C. L. Brynildson. 1952. Downstream Movement of Recently Transformed Sea Lampreys, *Petromyzon marinus*, in the Carp Lake River, Michigan. Trans. Am. Fish. Soc., 81(1951): 275–290.

Cochran, W. G. 1953. Sampling Techniques. New York: John Wiley and Sons. xiv + 330 pp.

Hansen, M. J., and D. W. Hayne. In press. Populations of Sea Lamprey Larvae in Ogontz Bay and Ogontz River, Delta County, Michigan. Trans. Am. Fish. Soc.

Rounsefell, G. A., and W. H. Everhart. 1953. Fishery Science, Its Methods and Applications. New York: John Wiley and Sons. xii + 444 pp.

Snedecor, G. W. 1956. Statistical Methods, 5th ed. Ames: Iowa State College Press. 534 pp.

Stauffer, T. M., and M. J. Hansen. 1958. Distribution of Sea Lamprey Ammocoetes in Michigan Tributaries of Lake Superior 1955–1957. Mich. Dept. Cons., Inst. Fish. Res., Misc. Publ. 11, 25 pp.

Thomas, M. L. H. 1960. A Modified Anchor Dredge for Collecting Burrowing Animals. J. Fish. Res. Bd. Can., 17(4):591–594.

Papers of the Michigan Academy of Science, Arts, and Letters
Vol. XLVII, 1962 (1961 Meeting)

ZOOLOGY

THE AMERICAN SPECIES OF THE GENERA *MELIERIA* AND *PSEUDOTEPHRITIS* (DIPTERA: OTITIDAE)

GEORGE C. STEYSKAL

Grosse Ile, Michigan

Paucity of revisionary work has made the flies of the family Otitidae difficult to determine. Until now no key to the Nearctic species of *Melieria* has appeared and additional data has made the revision of *Pseudotephritis* by Malloch (1931) seriously out-of-date. A key to the American species of *Melieria* is here presented; *M. obscuricornis* (Loew) is newly synonymized under *M. cana* (Loew); a new species, *M. sabuleti,* is described from Michigan; and the male postabdomen is described. A new key to the known species of *Pseudotephritis* is presented; *P. cribrum* (Loew) is newly synonymized under *P. cribellum* (Loew) and *P. conjuncta* Johnson under *P. vau* (Say); two forms of *P. inaequalis* Malloch and four forms of *P. vau* (Say) are distinguished by nonnomenclatural names in the uncertainty that they are actually subspecies; the genus is recognized for the first time from the Palaearctic region in the species *P. corticalis* (Loew) = *Myennis trypetoptera* Hennig, syn. nov., and *P. millepunctata* (Hennig), comb. nov. from *Myennis*; and the male postabdomen is described and used to distinguish species groups.

Genus *Melieria* Robineau-Desvoidy

1830, Essai Myod.: 715, type, *Musca crassipennis* Fabr. (as *gangrenosa* R.-D.), by designation of Rondani, 1856; Coquillett, 1900, N. Y. Ent. Soc., Jour., 8: 22; Aldrich, 1905, Smithsonian Misc. Coll., 46: 592; Becker, 1905, Cat. Pal. Dipt., 4: 39; Séguy, 1934, Faune de France, 28: 62; Hennig, 1939, *in* Lindner, Die Fliegen der Pal. Reg. 5, Lfg. 126/128: 44.

Melieria is a genus of the Otitinae as defined by Steyskal (Ent. Soc. Amer., Ann., in press) and as shown by the male postabdomen (Fig. 3), which is very similar in all of our species. The aedeagus

1

2

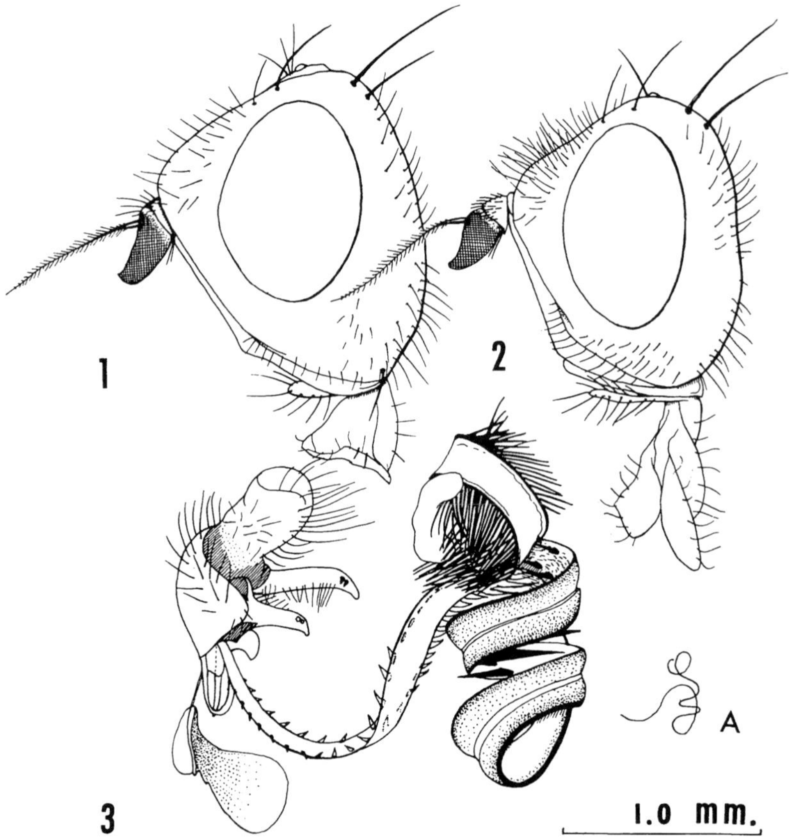

3

Details of *Melieria* Species

FIG. 1. *Melieria sabuleti* sp. nov., female head, allotype, Oceana Co., Mich.
FIG. 2. *M. cana* (Loew), female head, Martin, S. Dak.
FIG. 3. *M. occidentalis* Coq., male andrium; A, diagram of convolutions of aedeagus.

is heavily armed with a double row of teeth in the basal section, bare in the next section, then with a section furnished with long blades on one side only, and finally with long spinose bristles on both sides of the apical section, excepting the extreme tip or glans, which is a membranous sac. The surstyli, more or less fused with the epandrium, bear two small but heavy conical teeth apico-mesally and a mesal row of hairs. The aedeagal apodeme is wedge-shaped, divided into two arms supporting but not extending beyond the basal ring of the aedeagus. The ejaculatory apodeme is broad and flat, with a thick neck and a hyaline, membranous sac.

Nothing is known about the immature stages.

The American species may be distinguished as in the following key:

KEY TO THE AMERICAN SPECIES OF *MELIERIA*

1 (4). Wings with basal brown crossband extending into cell C; cheeks less than 1/3 eye-height; legs and antennae yellow *crassipennis* group

2 (3). Infuscation at tip of wing vein R_5 extending proximally but slightly farther than level of tip of R_3; abdominal tergites with dark brown apical margins *M. similis* (Loew)

3 (2). Infuscation at tip of R_5 extending proximally as far as level of *im* crossvein; abdominal tergites usually lacking dark apical margins . . . . *M. occidentalis* Coq.

4 (1). Wings with cell C entirely hyaline (except in some specimens of *M. picta*), or wing pattern faint or lacking.

5 (10). Wings with distinct pattern of brown spots; cheeks less than 1/3 eye-height . . . . *picta* group

6 (9). Abdominal tergites with dark apical margins; third segment of antennae usually yellow or orange; body pruinosity usually yellowish.

7 (8). Femora yellowish . . . . *M. ochricornis* (Loew)

8 (7). Femora at least medially blackish . . . . *M. picta* (Meigen)

9 (6). Abdominal tergites lacking dark apical margins, wholly gray pruinose, as is body generally; femora, especially fore ones, usually blackish; third segment of antennae usually blackish . . . . *M. cana* (Loew)

10 (5). Wings without distinct pattern of dark spots, usually merely with dark sections on veins; cheeks half as wide as eye-height; abdomen and legs wholly yellowish (*acuticornis* group) . . . . . . *M. sabuleti* Steyskal, sp. nov.

It is quite likely that some of our forms may be conspecific with Palaearctic ones. The entire genus has considerable morphological uniformity and both plastic and chromatic characters show much variability. It is also possible that both in the Palaearctic and Nearctic regions too many species have been recognized, but a solution of the problem must await knowledge of the eastern Palaearctic fauna, at present very little known. Hennig's table may be recast on the pattern of the foregoing one to show relationships of the species of the two faunas.

PALAEARCTIC SPECIES OF *MELIERIA*

1 (2). Wings with basal brown crossband extending into cell C; abdominal tergites with dark apical margins (*crassipennis* group) .... *M. crassipennis* (Fabr.)

2 (1). Wings with cell C entirely hyaline or wings without distinct pattern of spots.

3 (12). Wings with distinct brown spots, at least with spot at fork of *Rs* .... *picta* group

4 (5). Abdominal tergites with dark apical margins; legs largely blackish .... *M. picta* (Meigen)

5 (4). Abdominal tergites without dark apical margins.

6 (9). At least femora blackish medianly.

7 (8). Crossvein *rm* beyond tip of R_1 .... *M. occulta* Becker

8 (7). Crossvein *rm* at or before tip of R_1 .... *M. cana* (Loew)

9 (6). Femora wholly yellow.

10 (11). Antennae with nearly bare arista; fore tibiae and tarsi blackish apically; wing spots weak *M. nigritarsis* Becker

11 (10). Arista distinct hairy; antennae bright orange; wing spots strong or weak .... *M. nana* (Lw.); *M. omissa* (Mg.); *M. unicolor* (Lw.); *M. latigenis* Hendel; *M. obscuripes* (Lw.)

12 (3). Wings with spots very weak or absent; abdominal tergites wholly pale pruinose .... *acuticornis* group

13 (14). Antennae with third segment yellow; long slender species; ovipositor sheath 1½ times as long as wide *M. immaculata* Becker

14 (13). Antennae blackish; more robust species; ovipositor sheath not longer than wide .... *M. limpidipennis* Becker; *M. acuticornis* (Lw.)

A. *crassipennis* Group

M. similis (Loew)

Ceroxys similis Loew, 1873, Smithsonian Misc. Coll., 256: 127.
Melieria similis (Loew) Aldrich, 1905, Smithsonian Misc. Coll., 46: 592; Cresson, 1924, Am. Ent. Soc., Trans., 50: 232; Sturtevant, 1925, N. Y. Ent. Soc., Jour., 33: 213 (spermathecae).

Loew's types were from Connecticut and Quebec. I have seen specimens from B. C., Alta., Man., Ont., Que., N. B., N. S., Minn., Wis., Ind., Mich., Ohio, N. Y., N. H., and Mass. As Loew noted in the original description, the only difference between this species and *M. crassipennis* seems to be in the dark coloration of the femora of the European form. It is likely that both *M. similis* and *M. occidentalis* are but races of *M. crassipennis*; no report of the latter species, however, is available from the eastern Palaearctic region.

M. occidentalis Coquillett

(Fig. 3)

1904, Invertebrata Pacifica, 1: 29; Cresson, 1908, Ent. News, 19: 98, pl. VI, figs. 4 and 5 (as *M. obscuricornis*); Cresson, 1924, Am. Ent. Soc., Trans., 50: 231; Cole and Lovett, 1921, Calif. Acad. Sci., Proc., (4) 11 (15): 328; Cole, 1927, Calif. Acad. Sci., Proc., (4) 16 (14): 493, fig. 247.

I have seen the type from Ormsby County, Nevada, as well as other material from B. C., Alta., Idaho, Wyom., S. Dak., Calif., Nev., Utah, Ariz., N. Mex., and Texas. Cresson reports localities in Colo., Kans., and Minn., and Cole and Lovett list Oregon. Cole figured the basal and apical parts of the aedeagus. The entire andrium is shown in Fig. 3.

B. *picta* Group

M. cana (Loew)

(Fig. 2)

Ortalis cana Loew, 1858, Berlin. Ent. Zeits., 2: 374.
Ceroxys canus (Loew), 1873, Smithsonian Misc. Coll., 256: 128.
Melieria canus (Loew) Coquillett, 1900, Wash. Acad. Sci., Proc., 2: 459.
M. cana (Loew) Aldrich, 1905, Smithsonian Misc. Coll., 46: 592; Cresson, 1924,

Am. Ent. Soc., Trans., 50: 231; Séguy. 1934, Faune de France, 28: 64; Hennig, 1939, *in* Lindner, Die Fliegen der Pal. Reg. 5, Lfg., 126: 45.

Ceroxys obscuricornis Loew, 1873, Smithsonian Misc. Coll., 256: 126. *New synonym.*

Melieria obscuricornis (Loew) Aldrich, 1905, Smithsonian Misc. Coll., 46: 592; Cresson, 1924, Am. Ent. Soc., Trans., 50: 230 (*pro partim*); not Cresson 1908, v. *occidentalis*).

Hennig gives the distribution of this species as a broad area extending from central and southern Europe to northern China. I have seen material from Alaska, B. C., Alta., Sask., Man., Idaho, Mont., Wyom., N. Dak., S. Dak., Minn., Nev., Utah, Colo., Iowa, Ind., and Ariz. A figure of the head is shown in Fig. 2 as typical of the species with narrow cheeks.

M. ochricornis (Loew)

Ceroxys ochricornis Loew, 1873, Smithsonian Misc. Coll., 256: 126.

Melieria ochricornis (Loew) Aldrich, 1905, Smithsonian Misc. Coll., 46: 592; Cresson, 1924, Am. Ent. Soc., Trans., 50: 230 (*pro partim*).

Cresson's synonymy of this species with *M. obscuricornis* is certainly erroneous. The paler color of the antennae and legs, as well as the distinct, although often narrow, dark apical margins of the abdominal tergites, distinguishes this form from *M. obscuricornis*, which as indicated above is indistinguishable from *M. cana.* While occupying a wide range, it occurs farther eastward than does *M. cana.* Loew's type was from Wisconsin. I have seen material from Alaska, B. C., Alta., Sask., Man., Ont., Que., Calif. (Humboldt Co.), Idaho, Mont., Colo., N. Dak., S. Dak., Minn., Wis., Iowa, and Mich. It has been recorded by others from N. B. and northern N. Y.

M. picta (Meigen)

Ortalis picta Meigen, 1830, Syst. Beschr., 5: 276.

Melieria picta (Mg.) Séguy, 1934, Faune de France, 28: 65; Hennig, 1939, *in* Lindner, Die Fliegen der Pal. Reg. 5, Lfg. 128: 50 (résumé of literature).

A series of 38 specimens taken by Clifford O. Berg at Anchorage, Alaska, July 19 and August 2, 1952 (Cornell University and my collections) shows wide variation in wing markings, but all agree in having dark apical margins on the abdominal tergites and all femora and tibiae black except at the knees, a combination found in no hitherto recorded American form. The specimens agree so well with both Séguy's and Hennig's keys and descriptions that there can be little doubt of their conspecificity with the European form. The wing markings are well developed, probably more pronounced than in any other American material that I have examined; in one specimen the basal band extends even into the costal cell. The marks at the apex of the wing show the most variation, in most specimens consisting of separate marks at the tips of veins R_3, R_5, and M_1, with a tendency for the mark at the tip of R_3 to be narrowly isolated from the costa, while in a few specimens these marks are all fused to form a broad reniform blotch.

Hennig gives a résumé of the European data and cites records from most of Europe and from Kurdistan.

C. *acuticornis* Group

Several Palaearctic forms of *Melieria* are characterized by strongly reduced or absent wing markings. These may be conveniently grouped under the name of the oldest described species *M. acuticornis* (Loew). *M. unicolor* is also of this appearance but there is a distinct spot at the fork of *Rs,* and it is therefore placed in the *picta* group. The following new species, apparently related to *M. immaculata* Becker, from Tibet, constitutes the first record of the group from the Nearctic region.

M. sabuleti Steyskal, sp. nov.

(Fig. 1)

Male. Length of wing, 24.5 mm. General color clayey yellow, only the third antennal segment blackish. The larger part of the thorax and abdomen seems to be dark in color under the heavy pruinosity. Wings whitish, with traces of brown markings at the fork of *Rs,* apex of pterostigma, both crossveins, and the ends of veins R_3, R_5, and M_1, where the veins are dark brown. All bristles are black.

Head as in Fig. 1, the eyes shorter (twice as high as width of cheeks) and the cheeks and parafacials broader than in any other American form. The third antennal segment of the holotype male is rather shorter and more pointed than in the female figured. Arista with short hairs.

Postabdomen essentially as figured for *M. occidentalis* (Fig. 3), surstyli slightly longer than the dorsoventral diameter of the epandrium.

Female. Similar to male but somewhat larger; length of wings, 27.0–29.5 mm. The seventh abdominal segment (ovipositor sheath) is 1.3–1.35 wide at base by 0.8–0.9 mm. long. Head as in Fig. 1.

Holotype (male) and allotype, Silver Lake sand dune, Oceana County, Michigan, June 19, 1942; paratypes, 4 females, Sleeping Bear sand dune, Leelanau County, Michigan, July 12, 1936. All specimens were collected by Curtis W. Sabrosky, to whom I am indebted for the privilege of describing them and to whom the specimens will be returned for deposit in the U. S. National Museum collections.

Genus *Pseudotephritis* Johnson

1902, Ent. News, 13: 144 (change of name for *Stictocephala* Loew, 1873, Smithsonian Misc. Coll., 256: 134, type, *Ortalis vau* Say, by original designation; not *S. Stål*, 1869); Hendel, 1909, Deutsch. Ent. Zeits., 1909: 40; Malloch, 1931, U. S. Natl. Mus. Proc., 79 (art. 34): 1–6 (revision).

Pseudotephritina Malloch, 1931, U. S. Natl. Mus. Proc., 79 (art. 34): 5. Subgenus; type, *P. cribellum* (Lw.), by original designation.

Approximately 630 specimens have been examined in the course of my studies on this genus, including much material from the Western States. The male postabdomen has proven valuable in defining species and species groups. The additional material gathered since the Malloch revision has enabled me to delimit the species more adequately and to recognize categories that may ultimately prove to be subspecies.

The genus belongs to the Otitinae and to the group that has been sometimes given family rank as the Pterocallidae, although my studies lead me to believe that the group is not sufficiently distinct from the Otitinae to deserve even tribal rank.

The male postabdomen (Figs. 4–8) shows the characteristic Otitinae aedeagus armed with bristles and spines except at the extreme base and at the tip, or glans, which is a pilose sac. The aedeagal apodeme is a simple Y-shaped structure, the arms of which do not project beyond the base of the aedeagus. The shape and armature of the surstyli vary from species to species and provide characters used in the key to species. The epiphallus is well developed as a

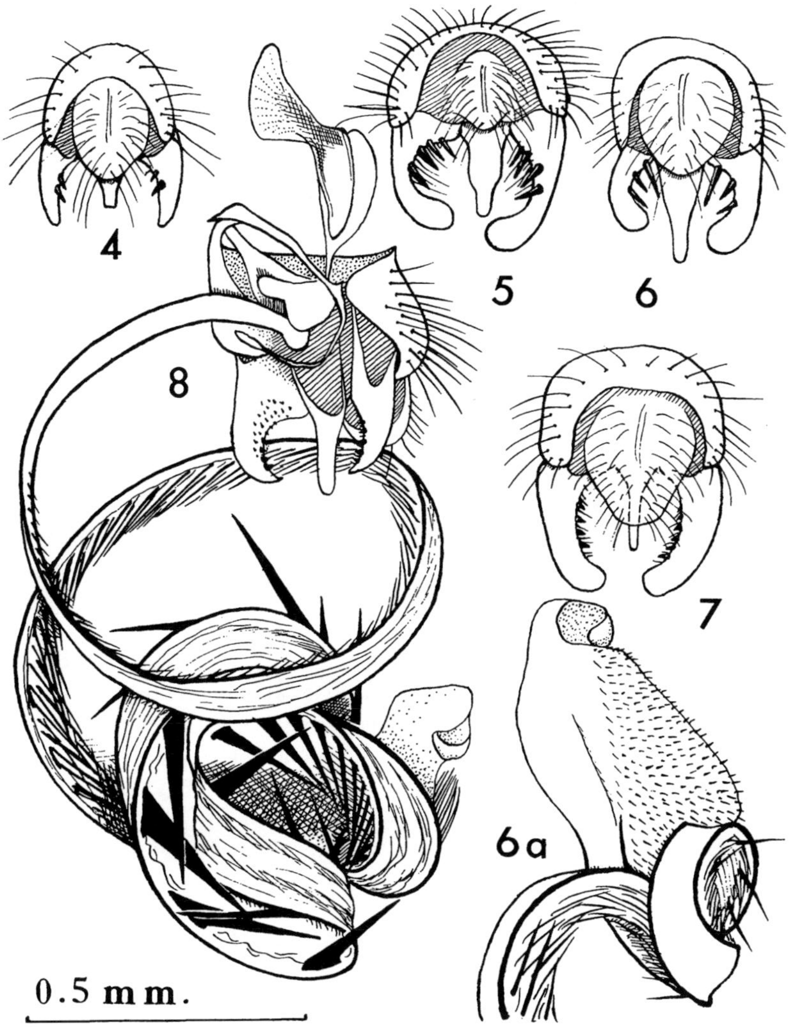

Details of Andria of *Pseudotephritis* Species

FIG. 4. *Pseudotephritis corticalis* (Loew), Cabin John, Md.
FIG. 5. *P. cribellum* (Loew), Detroit, Mich.
FIG. 6. *P. inaequalis* Mall., Twin Falls, Idaho (form *davisensis*); 6a, Castle Creek Hot Springs, Ariz. (form *inaequalis*), tip of aedeagus.
FIG. 7. *P. vau* (Say), Oxford, Miss. (form *vau*).
FIG. 8. *P. approximata* Banks, Lafayette, Ind.

"median peg" between the surstyli; it also exhibits specific characters. The presence of this type of epiphallus may provide the basis for the recognition of a group in the Otitinae that will have to be founded upon the genus *Pterocalla,* since it is found in several species of that genus, in *Callopistromyia annulipes,* and in *Dyscrasis hendeli,* but the many and various other genera that have been referred to the Pterocallidae remain unexamined.

Body and wing color patterns, wing venation, and chaetotaxy are all subject to a considerable amount of intraspecific variation. Some notes on variation in one population of *Pseudotephritis cribellum* are given below.

In all of the species examined a peculiar sexual dimorphism is found in the posterior postalar bristles. In the males, the bristle of each side is erect and surrounded by only inconspicuous short hairs; in the females, the bristle is replaced by a group of 10–15 somewhat weaker bristles lying prone against the thorax.

The slit and specialized hairs on the hind tibiae mentioned by Malloch (1931, p. 4) are also found in *Callopistromyia annulipes, Pterocalla strigula,* at least a few Neotropical species of *Pterocalla,* and *Dyscrasis hendeli.*

The larvae live under the bark of dead or moribund trees. Adults may be found on the bark of the trees, the females ovipositing in cracks in the bark. The kinds of trees involved seem to be various deciduous species, notably species of poplar (*Populus*). Available records of the trees are cited below.

The Palaearctic genus *Myennis* and the Neotropical genus *Neomyennis* are closely related to *Pseudotephritis,* but exact differences cannot be cited at present.

KEY TO THE KNOWN SPECIES OF *PSEUDOTEPHRITIS*

1 (8). Scutellum wholly pruinose; 2 *Sc*; front lacking well developed bare areas at bases of hairs; no spur on posterior side of vein M_1 extending into discal cell. . . . . Subgenus *Pseudotephritis* s. s.

2 (3). Crossvein *rm* in hyaline area, surrounded by several dark spots; cell *C* with small round spot between humeral crossvein and large dark spot in cell; apex of wing hyaline; male postabdomen with surstyli short, tapering to

subacute tips and furnished with only two distinct denticles (Fig. 4) .... *P. corticalis* (Loew)

3 (2). Crossvein *rm* in dark brown or yellowish area, without small dark spots in vicinity; apex of wing with well developed brown spot; male postabdomen with surstyli long, curved mesad, with rounded tips and furnished with 3–4 small denticles near middle of mesal edge; median peg slender, from conical base (male of *P. millepunctata* not known).

4 (5). Extension of anal cell and area immediately anterad to it hyaline; apical wing spot extending into cell R_3 ; crossveins *rm* and *im* in common brown marking (Central Asia) .... *P. millepunctata* (Hennig)

5 (4). Extension of anal cell and spot anterad to it brown.

6 (7). Costal cell with small round spot between large spot and humeral crossvein; apical spot extending into cell R_3 ; dark mark behind *acr* bristles a narrow, parallel-sided stripe; large form of eastern No. Amer. (wing 5.3–6.9 mm. long); male postabdomen as in Fig. 8

P. approximata Banks

7 (6). No small round spot in costal cell; apical spot not extending into cell R_3 ; male postabdomen as in Fig. 7

P. vau (Say)

8 (1). Scutellum with two large shining black marks; 1 *Sc*; front gray-pruinose with well developed bare spots at bases of hairs; vein M_1 with short spur extending into discal cell midway between crossvein *rm* and base of cell; no dark marks behind *acr*; male postabdomen with surstyli well curved mesad, with rounded tips and 3–4 large denticles in basal part; median peg stout (Figs. 5, 6)

Subgenus *Pseudotephritina* Malloch

9 (10). Propleural hairs black; second *dc* large; male postabdomen with denticles of approximately equal size; median peg shorter than surstyli (Fig. 5) ...... *P. cribellum* (Loew)

10 (9). Propleural hairs yellow; second *dc* small; male postabdomen with basal denticle distinctly the largest; median peg as long as or longer than surstyli (Fig. 6)

P. inaequalis Malloch

Subgenus *Pseudotephritis* Johnson

P. corticalis (Loew)

(Fig. 4)

Stictocephala corticalis Loew, 1873, Smithsonian Misc. Coll., 256: 136.
Pseudotephritis corticalis (Loew) Hendel, 1909, Deutsch. Ent. Zeits., 1909: 40; Greene, 1917, Ent. Soc. Wash., Proc., 19: 155; Malloch, 1931, U. S. Natl. Mus., Proc., 79 (art. 34): 4.
Myennis trypetoptera Hennig, 1939, *in* Lindner, Die Fliegen der Pal. Reg. 5, Lfg. 128: 75. *New synonym.*

The Loew type was from New York. Greene described the immature stages from material taken under bark of tulip tree (*Liriodendron tulipifera*) in Virginia. The species has been cited also from N. J., Conn., D. C., Pa., and Alta. I have seen material from all of the previously cited areas as well as Ohio and Md. Collection labels read "under dead poplar, Washington, D.C.; *Amelanchier canadensis* bloom, Lyme, Conn.; *Quercus*, Falls Church, Va." Hennig (personal communication) confirms the above synonymy from comparison of American material sent him; his type was from Luga, Tolmatchevo, Russia (near Leningrad).

P. millepunctata (Hennig), comb. nov.

Myennis millepunctata Hennig, 1939, *in* Lindner, Die Fliegen der Pal. Reg. 5, Lfg. 128: 73.

It is obvious that this species should be referred here and that it is closely related to *P. vau*. The type was from "St. Siza, Ussuri Gebiet," which is in far eastern Siberia. Two specimens in the U. S. National Museum collections are from "Okeanskya," Siberia, August, 1923 (T. D. A. Cockerell).

P. approximata Banks

(Fig. 8)

1914, Ent. Soc. Wash., Proc., 16: 138 (as *appoximata*; corrected to *approximata* in index to volume); Johnson, 1921, Boston Soc. Nat. Hist., Occas. Pap., 5: 15 (as var. of *vau,* with *P. metzi* as syn.); Malloch, 1931, U. S. Natl. Mus., Proc., 79 (art. 34): 4.

The type, which ostensibly should be in the U. S. National Museum, has not been located. It was from Falls Church, Va. Because of confusion with *P. metzi*, subsequent records must be doubted. I have seen material from Pa., Mich., Ind., Miss., and Iowa, with none of which any ecological or biological data is associated.

The species is quite close to *P. vau*, but the stouter and longer median peg of the male postabdomen and the apparent reliability of the color characters lead me to consider the species distinct.

P. vau (Say)

(Fig. 7)

Ortalis vau (Say), 1830, Acad. Sci. Phila., Jour., 6: 184.
Stictocephala vau (Say) Loew, 1873, Smithsonian Misc. Coll., 256: 138.
Pseudotephritis vau (Say) Johnson, 1902, Ent. News, 13: 144; Hendel, 1909, Deutsch. Ent. Zeits., 1909: 40; Weiss and West, 1920, Biol. Soc. Wash., Proc., 33: 11 (occurrence on fungi); Johnson, 1925, Boston Soc. Nat. Hist., Proc., 38: 96 (note on habits); Sturtevant, 1925, N. Y. Ent. Soc., Jour., 33: 215 (seminal receptacles); Frost, 1928, Ent. News, 39: 171; Frost, 1929, Ent. News, 40: 85 (dates attracted to bait); Townsend, 1928, Ent. Soc. Am., Ann., 21: 121 (seasonal and vertical distribution); Malloch, 1931, U. S. Natl. Mus., Proc., 79 (art. 34): 3.
P. vau Say, var. *conjuncta* Johnson, 1921, Boston Soc. Nat. Hist., Occas. Pap., 5: 15. *New synonym.*
P. metzi Johnson, 1915, Psyche, 22: 49; 1921, Boston Soc. Nat. Hist., Occas. Pap., 5: 15. *New synonym* (cf. also below).

Considerable variation has been noted in this wide-ranging species, but no discernible differences have been found in the male postabdomen. Specimens have been seen from Quebec and B. C., south to Va., Miss., Ark., Tex., Idaho, and Calif. Four forms (here designated by names of nonnomenclatural status, although some may eventually be found to deserve subspecific rank), are here distinguished, separable as follows:

1 (2). Crossveins *rm* and *im* in common yellow cloud; yellow cloud anterad of anal cell closely approaching brown spot anterad of extension of anal cell; apical spot not attaining tip of vein R_3 ; brown mark behind *acr* bristles as broad as distance between the bristles . . . . form *vau* s. s.
Distribution: N. H. (White Mts.) ; R. I. (Westerly) ; N. J. (Boonton; Riverton); N. Y. (Ithaca; Long Id., type of

conjuncta); Pa. (Allegheny; Angora; Lawndale; Overbrook; Philadelphia); Md. (College Park; Plummers Id.; Takoma Park); D. C.; Va. (Chain Bridge; Fairfax Co.; Falls Church, resting on *Quercus*; do., reared from *Castanea dentata* Camp Peary); Mich. (Bay; Berrien, in ammonia trap for apple maggot fly; Eaton; Ingham; Leelanau; Livingston; Midland; Muskegon; Roscommon; and Wayne Cos.); Ohio (Athens Co.); Ind. (Lafayette); Miss. (Oxford); Ark. (Danville; Hot Springs); Tex. (Dallas; Waco); Idaho (Gooding); Calif. (Siskiyou Co.); Wash. (Yakima Co.); Que. (Outremont); Alta. (Edmonton); B. C. (Kaslo).

2 (1). Crossveins *rm* and *im* usually in separate yellowish markings; cloud anterad of anal cell well removed from brown spot anterad of anal cell extension.

3 (4). Brown mark behind *acr* quite narrow; apical wing spot usually not attaining vein R_3 ; mark over crossvein *im* not connected with preapical costal mark; second antennal segment usually dark brown . . . . form *idahoana*
Distribution: Idaho (Gooding; Weiser); Wash. (Wapato; Yakima Co.).

4 (3). Brown mark behind *acr* as broad as distance between the bristles; second antennal segment usually distinctly paler than third segment.

5 (6). Apical wing spot usually attaining vein R_3 ; mark over crossvein *im* usually connected with preapical costal spot form *metzi*
Distribution: Maine (Cherryfield); N. H. (White Mts.); Mass. (Springfield); N. Y. (Ithaca; Long Id., type of *P. metzi*); Md. (Plummers Id.); Va. (Falls Church, resting on *Quercus*; do., reared from *Castanea dentata*; Great Falls); Pa. (Lehigh Gap); Mich. (Berrien, in ammonia trap for apple maggot fly; Charlevoix; Kent; and Wayne Cos.; Isle Royale); Minn. (St. Paul; St. Anthony Park); Ont. (Perry Sound; Waubamic).

6 (5). Apical wing spot usually not attaining vein R_3 ; mark over crossvein *im* well disjunct from mark over crossvein *rm* and from subapical costal spot . . . . form *californica*

Distribution: Calif. (Lake; Mendocino; Napa; San Benito; Santa Clara; Santa Cruz; San Joaquin; San Luis Obispo; and Yuba Cos.).

Subgenus *Pseudotephritina* Malloch

P. cribellum (Loew)

(Fig. 5)

Stictocephala cribellum Loew, 1873, Smithsonian Misc. Coll., 256: 134.

Pseudotephritis cribellum (Loew) Hendel, 1909, Deutsch. Ent. Zeits., 1909: 40; Malloch, 1931, U. S. Natl. Mus., Proc., 79 (art. 34): 5.

Stictocephala cribrum Loew, 1873, Smithsonian Misc. Coll., 256: 135. *New synonym.*

Pseudotephritis cribrum (Loew) Hendel, 1909, Deutsch. Ent. Zeits., 1909: 40; Malloch, 1931, U. S. Natl. Mus., Proc., 79 (art. 34): 6.

Information concerning the types of *P. cribellum* and *P. cribrum*, received through the kindness of P. J. Darlington, confirms the suspicion originally expressed by Loew and later by Hendel that the species are synonymous. *P. cribellum* has page priority, and is therefore considered the valid name.

Much confusion has been evident between this species and *P. inaequalis*; therefore only type localities and data from material I have examined are listed: "Middle States" (type of *P. cribrum*); Pa. (Philadelphia); Mich. (Clare; Ingham; Monroe, on poplar trunks; Wayne Cos.); Ind. (Lafayette, on maple trunk); Miss. (Oxford); Nebr. ("Nebr.," type of *P. cribellum*; Lincoln); Tex. (Denton, on *Quamassia*; Victoria); Kans. (Doniphan; Douglas; and Riley Cos.); Colo. (Lamar); N. Mex. (Verlarde, on peach).

Some measurements of wing venation have been compiled on 21 female specimens of a Monroe, Mich., population. Six of the 21 lack the stump on vein M_1 of both wings. The total length of the discal cell, measured on vein M_1, varied from 1.60 to 2.29 mm., with mean at 1.98 mm. The ratio of the distance from crossvein *rm* to *im* to total length of discal cell (*rm-im*/total length) varied from 0.20 to 0.295, with mean at 0.237. The ratio of the distance from *rm* to *im* to length of *im* (*rm-im*/im) varied from .800 to 1.26, with mean at 1.05, and frequently with considerable difference in the ratios of the right and left wings of the same specimen.

P. inaequalis Malloch

(Fig. 6)

1931, U. S. Natl. Mus., Proc., 79 (art. 34) : 5.

The color characters and length of scutellar bristles used by Malloch have been found inadequate for the definition of this species, but the male postabdomen shows distinguishing characters and the color of the propleural hairs seems constant.

This species is also widespread. I have seen specimens from Indiana to California and Texas, north to South Dakota and Washington. Two forms, which may be subspecies in spite of apparently overlapping distribution, can be distinguished by venation and color pattern of the wing, characters which in *P. cribellum* do not seem to be distributed in this fashion. The names used must be considered provisional and not of nomenclatural status.

1 (2). Crossveins *rm* and *im* well separated, the dark fascia covering them disjunct .... form *inaequalis* s. s.
Distribution: Ind. (Lafayette) ; Kans. (Riley Co.) ; Nebr. (Lincoln) ; S. Dak. (Custer Co.) ; Colo. (De Beque, type) ; Tex. (El Paso; Presidio) ; N. Mex. (Albuquerque; Belen; Los Chaves; Las Cruces; Mesilla Park; Sandoval; Tularosa; Vado) ; Ariz. (Bisbee; Castle Creek Hot Springs; Douglas; Florence; Glendale; Lowell; Maricopa Co.; Nogales; Patagonia, dead willow branch; Peoria; Phoenix; Tempe; Tucson) ; Calif. (Meloland, Imperial Co., cottonwood; Riverside Co.; Sanger).

2 (1). Crossveins *rm* and *im* close together in common dark band form *davisensis*
Distribution: Kans. (Onaga Co.) ; Tex. (Presidio, wild sunflowers) ; Idaho (Caldwell; Gooding; Nampa; Twin Falls) ; Nevada (Carson City; Fernley; Reno; Sparks) ; Calif. (Davis, reared from cottonwood *Populus* bark; Melones; Redlands, San Diego Co.) ; Wash. (Granger; Prosser; Sunnyside; Wenatchee; Yakima Co.; Zillah).

Many specimens of both of these forms were captured in traps in peach and quince orchards. The Davis, Calif., and Nevada material, consisting of 87 specimens, is very uniform. Those from Twin Falls, Idaho, 24 specimens, include 6 which approach the typical form. The specimens of the typical form from Imperial and Riverside Cos., Calif., are unusually pale in color.

PART II

SOCIAL SCIENCE

Papers of the Michigan Academy of Science, Arts, and Letters
Vol. XLVII, 1962 (1961 Meeting)

HISTORY

WILLIAM AUSTIN BURT: MICHIGAN'S MASTER SURVEYOR

ALAN S. BROWN

Western Michigan University

When Frederick Jackson Turner and his followers drew attention to the frontier they also focused interest on the men and women who created it—the American pioneers. Thus the pioneer settler, his life, trials, and tribulations are a well-known part of our history. But one pioneer group, the government land surveyors, has not received much notice or credit for its role in settling our frontiers. Before the settlers came, these "strange men with chain and compass . . . across the valley pass'd, through field and woodland . . . turn'd not aside, but kept straight on their way."[1] Their work, in large measure, opened the public lands for settlement.

William Austin Burt was a government land surveyor. His work in the period between 1833 and 1855 marks him as Michigan's finest example of the pioneer land surveyor. Not only did Burt extend the frontier, locate township and section lines for settlers, and discover and report on the mineral wealth of our state; he also invented an instrument, the solar compass, which enabled several generations of surveyors to survey more accurately the public domain of the United States.

The fifth child of Alvin and Wealthy Burt, William Austin was born in Petersham, Massachusetts, on June 13, 1792. His Scotch-English forebears had come to New England in the seventeenth century. After service in the American Revolution Alvin Burt took his growing family to New York state, where a home was established at Broadalbin in Montgomery County. Here young William Austin continued the schooling he had begun in Massachusetts, but education was a luxury on the frontier, and his formal school attendance ended shortly after he reached his fourteenth year.

Learning and study did not stop for Burt, however, and he taught

[1] Quoted in R. Carlyle Buley, *The Old Northwest: Pioneer Period, 1815–1840*. (Indianapolis, 1951), I, 115.

himself the principles of astronomy, mathematics, and navigation in between periods of work for his father on the family farm. By his own account his "mind took a mechanical turn" at an early age. While he diligently pursued his scientific inquiries his spiritual education was not neglected. The religious training he received from his mother was never forgotten. Throughout his life Burt was a faithful Baptist with a reputation for honesty in all his dealings.[2]

By 1812 Burt had taught himself the basic principles of surveying and had practiced them on lands in the Holland purchase area of New York. The next year he took another important step when he married Phoebe Cole, the daughter of an old family friend.[3] For the rest of his life Phoebe was his "Dear Companion," and together they brought five sons into the world. These sons, John, Alvin, Austin, Wells, and William, followed their father in his work on the public surveys of Michigan, Wisconsin, and Iowa.[4]

Shortly after his marriage Burt served in the New York militia for a brief period, and again in 1814 he enlisted for sixty days' service at Buffalo. Public service as justice of the peace, school inspector, and postmaster in Erie County followed this brief military episode.[5]

After the War of 1812 many young men from the east moved westward to the lands of Ohio, Indiana, and Illinois. Burt's interest in this region was aroused after a trip made in 1817. Fertile lands and opportunity beckoned, and a few years later he went out to Auburn, Oakland County, in the "wilderness of Michigan." Here he earned a living as a mill builder, surveyor, and general carpenter.[6] He investigated the possibility of surveying for the Federal Government but failed to secure an appointment. This failure was not because Burt lacked ability but was more likely because the government had too many

[2] Biographical information in this and the preceding paragraphs is found in the following: Silas Farmer, *A History of Detroit and Michigan* (Detroit, 1889), II, 1178–81; William A. Burt, "Autobiography," *Michigan Pioneer and Historical Collections* (40 vols., Lansing, 1877–1929, hereafter cited as M.P.H.C.), XXVIII, 646–647; George H. Cannon, "The Life and Times of William A. Burt of Mt. Vernon, Michigan," *ibid.*, V, 115–123.

[3] William Burt, "Biographical Sketch of William A. Burt," MS in Burt Papers, Marquette County Historical Society, Marquette, Michigan.

[4] Farmer, *History of Detroit*, II, 1179–80.

[5] *Ibid.*, p. 1179.

[6] *Ibid.*

applicants for the amount of surveying then available in Michigan.[7]

In 1824 Burt moved his family to Michigan. Within a short time they were settled on land he purchased in Washington Township, Macomb County.[8] Other pioneers followed, and in a few years the settlement of Mt. Vernon developed. Burt found his new surroundings pleasant and fertile, but, as was true of all early Michigan settlers, he and his family had their share of fever and ague. Since speculation and inflation had not yet developed, pioneers needed ready money for their lands. Burt thought this had "a good effect—that of settling the country with wealthy and industrious citizens. And it settles with a rapidity unparalleled by any other new country in the Union."[9]

In his early years in Michigan Burt engaged in a variety of occupations which included local surveying, farming, and mill building. His neighbors added to this variety when they elected him to the Territorial Council in 1826 and 1827. He was also appointed postmaster for Mt. Vernon, and in April 1833 he acquired the title he was to hold for the rest of his life when he was named associate judge on the Macomb Circuit Court bench.[10]

Despite the press of his various duties Burt found time in this period to produce the first of several inventions associated with his name. This invention, a "typographer," was patented in 1829. Unfortunately for Burt, frontier Michigan and the United States in general had slight use for a typewriter. Although some historians of the writing machine fail to mention Burt's invention, the fact seems clear that his was the first American typewriter to be patented.[11]

The lifework of Judge Burt, however, was to be surveying, and in the 1830's the Federal Government began to press forward its

[7] As late as 1828 there were as many as eighty applicants for the slight amount of survey work available in Michigan Territory. See Edward Tiffin, Sur. Genl., to Lucius Lyon, Feb. 16, 1828, Lyon Papers, Clements Library, Ann Arbor.

[8] Farmer, *History of Detroit,* II, 1179.

[9] Wm. A. Burt to brother and sister, Mar. 14, 1828, Burt Papers.

[10] Farmer, *History of Detroit,* II, 1179.

[11] For details concerning the Burt typewriter see Horace E. Burt, "William Austin Burt, Inventor," *Michigan History Magazine,* VI, (1922), 175-193; Bruce Bliven, Jr., *The Wonderful Writing Machine* (New York, 1954), pp. 25–35. In Richard N. Current, *The Typewriter and the Men Who Made It* (Urbana, Ill., 1954), no mention is made of Burt and his invention.

surveys in Michigan to the point where more men were needed for this important task. Burt was already known to the U. S. Surveyor General as qualified, "honest and intelligent, well acquainted with surveying."[12] On November 25, 1833, he received his appointment as a federal deputy surveyor.[13] For the next twenty years no man was better known or more respected in the public surveys than William Austin Burt.

Land surveying in Burt's time was a task that challenged the skill and fortitude of its practitioners. With compass, chain, and ax, the government surveyor penetrated country that few but Indians had ever seen. Relentlessly, across swamps, through woods and thickets, over brush and plain, the surveyor ran hundreds of lines to establish township and other boundaries so that those who followed could accurately locate their lands. Meticulous field notes were kept as each township and section line was run, but such notes usually fail to record the incredible hardships endured by the men who ran the lines.[14]

Occasional letters written from the field do give some account of the surveyors' problems. And not the least of these were mosquitoes. As the survey parties moved across Michigan they met the "burdensome loads of countless millions of myriads of musquitoes. . . . you can have no idea how troublesome the little, slender, soft-billed, weak-looking, long-legged gentry are; but I tell you that they bite a man through his clothes in a thousand places at a time . . ." In short, mosquitoes were said to "swarm about us in a cloud, not to guide us through the wilderness, but to devour."[15]

The mosquitoes and other difficulties encountered by federal surveyors in Michigan were not always met head on. Rather than con-

[12] John Mullett to Samuel Williams, Chief Clerk, Sur. Genl. Office, Cincinnati, Ohio, April 6, 1827, Williams Papers, Clements Library; Lucius Lyon to Micajah T. Williams, Sur. Genl., Aug. 31, 1833, Burt Papers.

[13] The original certificate of appointment is in the Burt Papers.

[14] For general accounts of surveyors and surveying life see Dwight L. Agnew, "The Government Land Surveyor as a Pioneer," *Mississippi Valley Historical Review,* XXVIII (1941–42), 369–382; C. S. Woodard, "The Public Domain, Its Survey and Surveyors," M.P.H.C., XXVII, 306–323.

[15] Lewis Clason to R. T. Lytle, Sur. Genl., Aug. 13, 1837; J. A. Rousseau to R. T. Lytle, Sept. 1, 1837, U. S. Congress, House Exec. Doc., 25th Cong., 2nd Sess., VII, No. 197, pp. 6–9.

tinue work under such conditions some individuals turned in fraudulent survey reports. When William Austin Burt received his first federal contract he was warned that better work was expected than "that which has heretofore been practised in the north part of Michigan."[16] As far as Burt's work was concerned the fears of the government were groundless. His first contract to survey lands in township 9 N R 13 E (Sanilac County) was returned in February 1834 and met with the entire satisfaction of his superiors. Indeed, the Surveyor General wrote Lucius Lyon that "Your friend, Mr. Burt, proves to be an excellent surveyor. For a first contract he has returned the most satisfactory work I have yet met with."[17] Burt's reputation as an able and accurate surveyor increased with experience, and in later years he was frequently called upon to resurvey less accurate work by some of his colleagues.

In May 1835, Burt was surveying in Wisconsin Territory where local attraction acted strongly on the magnetic compass that was then standard survey equipment. Frustrated, he observed that "the aberrations of the needle are truly perplexing. I have to correct very many of my north and south lines, and it is most annoying, this inability, as yet, to discover a method for doing away with the difficulty or cause thereof."[18] Night after night Burt and his son, Alvin, sat around their campfire after the day's survey discussing the problem of compass variation and the need for a more accurate instrument. That summer at his Mt. Vernon home Burt turned out a working model of the famous solar compass. This instrument, first patented February 25, 1836, was manufactured for many years by Burt's friend, William J. Young, of Philadelphia.[19] It was tested by a committee of the Franklin Institute, "who after full examination of its principles and merits, awarded the inventor a premium of twenty dollars and a Scott's Legacy Medal."[20] Scientifically and in practical use, the solar compass met all tests for many years there-

[16] Micajah T. Williams, Sur. Genl., to William A. Burt, November 26, 1833, Burt Papers.

[17] William A. Burt to Micajah T. Williams, Feb. 17, 1834; Lucius Lyon to Wm. A. Burt, Oct. 25, 1834, *ibid.*

[18] Farmer, *History of Detroit,* II, 1180.

[19] Austin Burt., "Burt's Solar Compass," M.P.H.C., XXXVIII, 114–116.

[20] William Austin Burt, *A Key to the Solar Compass* (Philadelphia, 1858), preface.

after, but financially its inventor received little more than the original twenty dollars.[21]

For purposes of this paper little technical discussion of the solar compass is necessary, Briefly, it was Burt's knowledge of astronomy—gained by arduous self-instruction—that suggested the instrument to him. Armed with a knowledge of the sun's daily motion, the user of the solar compass caught the sun's image on the instrument's silver plates. Then, taking the sun's position as a fixed point, he established his latitude and the sun's declination and thus found the true meridian. The instrument was first used to supplement the magnetic needle when local attraction threw it off, but after Burt perfected it the solar compass was mounted on a tripod with the standard compass and used in the actual running of township lines.[22]

For several years after its first patent the solar compass was used in the public surveys by Burt, his sons, and other deputy surveyors, including the well-known John Mullett. These field trials brought it official notice in December 1840, when the surveyor general at Cincinnati reported that several of his deputies were using it and recommended it "to all who feel interest in the advance of science".[23] In April 1841, it was recommended for surveying. "The [solar] compass gives unerringly the true meridian every time it is set in the light of the sun, and can in no wise be affected by any change of variation of the needle, or any degree of local attraction. This constitutes its superiority over the common compass . . . "[24]

In January 1840 William Austin Burt received instructions and a contract to begin surveys that would take him to Michigan's wild and rugged Upper Peninsula. It was in this largely unknown area that Burt performed his greatest public service, and it was also here that his solar compass proved almost indispensable for proper survey of the mineral regions. By June he had carried out his instructions

[21] In 1923 Horace E. Burt, a grandson, stated that Wm. A. Burt never realized more than $80 during the patent period (1836–50) for his compass. Horace E. Burt to Lew Allen Chase, Mar. 15, 1923, Burt Papers.

[22] William A. Burt, *Description of the Solar Compass Together with Directions for Its Adjustment and Use* (Detroit, 1844); George H. Cannon, "The Life and Times of William A. Burt," M.P.H.C., V, 118–120.

[23] Burt, *Description of the Solar Compass,* p. 26, quoting letter of E. S. Haines, Sur. Genl., Dec. 17, 1840.

[24] Lowell O. Stewart, *Public Land Surveys, Instructions and Methods* (Ames, Iowa, 1935), pp. 88–89.

to run the meridian line north of the Straits of Mackinac "in the most accurate manner."[25]

The wild country above the Straits was a trial even for the experienced and able Burt. Although he seldom complained concerning the hardships of his work, he did write that his "Coat & Pantaloons are most gone" and requested his wife to make him a new outfit "of the strongest kind of Bedticking" to wear in the heavy brush.[26] Years later when he published a manual for use with his compass Burt remembered the rugged country of his own surveys and gave instructions for proper clothing and equipment to use in such country. For field work he suggested stout boots, heavy trousers, "flannel for underclothes . . . for all seasons and all kinds of weather," and added that a large silk handkerchief "to tie over the ears and neck, is a good protection from flies and musquitoes"[27]

In the late spring of 1841 Burt resumed his surveys in the Lake Superior country.[28] On this survey, as in other work in the Upper Peninsula, a major problem was maintaining a flow of supplies to the men who ran the section and township lines. At times Burt had as many packmen working as he had men on the lines. Despite supply and terrain difficulties the work proceeded, and by early August Burt had reached the boundary limit set by existing Indian treaties.[29] An important aspect of this 1841 season's work was Burt's experiment, undertaken at Douglas Houghton's request, in combining linear surveying with observing and specimen collecting planned to show the geology of the area.[30] As Burt later observed, the combination of linear and geological surveys had the advantage of securing "a large amount of interesting and useful knowledge . . . at the least possible expense, and which could not be done so well in any other way".[31] The Federal Government was impressed and authorized

[25] George H. Cannon, *op. cit.*, p. 120.

[26] William A. Burt to Phoebe Burt, June 24, 1840, Burt Papers.

[27] William A. Burt, *Key to the Solar Compass*, pp. 78–84.

[28] William A. Burt to E. S. Haines, April 16, 1841, Burt Papers (photostat).

[29] William A. Burt to E. S. Haines, Aug. 9, 1841, *ibid.*

[30] William A. Burt to E. S. Haines, Oct. 28, 1841, *ibid.* See also Douglas Houghton's report, dated Jan. 25, 1842, in which he credits Burt's assistance, George N. Fuller (ed.), *Geological Reports of Douglas Houghton: First State Geologist of Michigan, 1837–1845* (Lansing, 1928), p. 668.

[31] William A. Burt, draft of letter to Genl. Land Office, Feb. 22, 1844, Burt Papers.

Houghton, Michigan's first state geologist, to put such a plan into effect in 1844.[32]

Burt spent most of the working months of 1842–43 in resurveying fraudulent or inaccurate work done by other deputies, and he also engaged in pioneer surveys in Iowa Territory.[33] By now his solar compass was in such demand that he took time to publish a manual describing the instrument and giving instructions for its use. Not only was the compass finding more users, but men of long experience in the public surveys found it "indispensable."[34]

One of the most important developments in the history of Michigan's Upper Peninsula came in the late summer of 1844. This was the discovery of iron ore by Burt's survey party in the Marquette-Negaunee area. This find set off a chain of reactions leading to the economic development of the Upper Peninsula and intimately associated with the rise of the United States as an industrial giant. For William Austin Burt, however, it proved once and for all the great value and utility of his solar compass.

The survey party that made the iron discovery proceeded up the Escanaba River and began work by closing the township lines Burt had run in 1841. By September 15 they reached Lake Superior, and three days later were camped at the east end of Teal Lake (T 47 N). The next morning, September 19, they began running lines south between Ranges 26 and 27. The magnetic needle, used in conjunction with the solar compass, showed repeated fluctuations. As the different variations of the needle were reported Burt became increasingly interested in the performance of his own instrument. "How would they survey this country without my compass? What could be done here without my compass?"

The great moment when iron outcrops were discovered came after the compass man announced "a variation that will beat them all." Then in words frequently quoted, Burt called out, "Boys, look around and see what you can find." Within a few minutes all members of

[32] For details concerning adoption of the Houghton-Burt proposal for combined geological-linear surveys by the Federal Government see Edsel K. Rintala, *Douglas Houghton, Michigan's Pioneer Geologist* (Detroit, 1954), pp. 77–79. Despite some statements to the contrary Michigan was the first state to employ this method; see Fuller, *op. cit.*, p. 16.

[33] E. S. Haines, Sur. Genl. to William A. Burt, April 11, 1842, Burt Papers; *ibid.* for correspondence and Burt's notes on Iowa surveys (1843).

[34] William A. Burt, *Description of the Solar Compass,* p. 26 for letter of John Mullett, Feb. 27, 1844.

the party found numerous specimens of ore which Burt duly catalogued and turned over to state authorities upon completion of the survey.[35]

It is fortunate that several members of the Burt party left descriptions of this historic moment, so full of consequences for Michigan's future. For Burt the main interest was in the accuracy obtained by means of his invention. Thus, his field notes fail to convey any great excitement or comment on the importance attending his discovery. Describing the survey of the east boundary of T 47 N Range 27 W he noted that "two good Solar Compasses were used . . . [and that] Spathic and hematite iron ore abound on this line".[36]

In the next few years survey parties led by Burt worked in a northwesterly direction toward the Keweenaw Peninsula. In official reports Burt described fourteen iron ore beds which he estimated to contain one-seventh of the iron ore in the Upper Peninsula. This region, he wrote, "far excels any other portion of the United States in the abundance and good quality of its iron ores." As a practical man Burt also urged that the Federal Government take steps to unlock the mineral wealth he had discovered. For a start he suggested building macadamized roads or even a railroad from the iron region to Keweenaw Bay. Later, when mining operations on the Marquette range were underway, Burt and his son, John, played important roles in opening up transportation in the Upper Peninsula by their support of the Sault Canal project.[37]

During the surveying seasons of 1845–46 Burt, with sons Austin and Wells, ran hundreds of miles of line under extremely difficult conditions in the Keweenaw and Ontonagon areas. The great rush to the copper lands was on—the first mining rush in the United States—and Burt noted that speculation was rife. Accurately, he predicted losses for many, but added that some fortunes would

[35] For accounts of the iron ore discovery of Sept. 19, 1844, by participants see Harvey Mellen to William Burt, Nov. 1, 1869, Burt Papers; Jacob Houghton to *Mining Journal* (Marquette), ca. Oct. 16, 1869; Houghton's account is also printed in Alvah Bradish, *Memoir of Douglas Houghton, First State Geologist of Michigan* (Detroit, 1889), pp. 105–106.

[36] William A. Burt's field notes are in Lands Division, Michigan Conservation Dept., Lansing, Michigan.

[37] William A. Burt, Survey Report 1846, U. S. Congress, Sen. Doc., 31st Cong., 1st Sess., III, p. 852. For Burt's work concerning the Sault Canal, including a preliminary survey made in Oct. 1852, see Charles Moore (ed.), *The Saint Marys Falls Canal* (Detroit, 1907), pp. 82, 122.

surely be made from Michigan's copper.[38] Despite all his experience Burt found these copper country surveys the most difficult he had ever attempted. Adding to the problems presented by terrain were the aberrations of the needle caused by mineral deposits and to make matters worse, "the thick forest prevents the rays of the sun falling on the Solar Compass in many places and in the early or latter part of the day high hills and knobs sometimes intervene between the instrument and the sun."[39]

Regardless of obstacles Burt turned in a survey so accurate that surveyors half a century later found few discrepancies between their resurveys and Burt's original lines.[40] This work in the copper regions was the first official test of a combined linear and geological survey. It had begun under Douglas Houghton's direction, but upon his tragic death Burt assumed the task of completing and reporting the survey. Burt's meticulous report provided the Federal Government with extensive data concerning the principal types of rocks in the copper country. In addition he described the best harbors in the area. Of considerable interest was his comment respecting the Portage River, near present-day Houghton and Hancock. Here, Burt wrote, "Doubtless at some future day, these two miles will be canaled for the passage of larger craft through this channel into Lake Superior. Near the mouth of the river, on the right bank, is an eligible site for a town, and when the harbor is improved, as it deserves to be, will probably become a place of some importance."[41] Thus did Burt predict the site of the major urban areas in the copper region of Keweenaw land.

As the nineteenth century neared mid-point the federal surveys of Michigan public lands approached completion, and Burt's field trips became less frequent. There remained, however, one important task for the veteran surveyor. This was to complete the survey of the Michigan-Wisconsin boundary. For two reasons this was a necessary and delicate operation. First, in 1846 Wisconsin received an

[38] William A. Burt to Phoebe Burt, Aug. 28, 1845; William A. Burt to Alvin Burt, Oct. 15, 1845, Burt Papers.

[39] William A. Burt to Lucius Lyon, Sur. Genl., Aug. 28, 1846; William A. Burt to Samuel Williams, Chief Clerk, Office of Sur. Genl., Jan. 30, 1845, *ibid.*

[40] Charles D. Lawton, Mich. Commr. of Mineral Statistics, to John Burt, Feb. 16, 1886, *ibid.*

[41] William A. Burt, "Topography and Geology of the Survey of a District of Township Lines South of Lake Superior, 1845," U. S. Congress, Sen. Doc. 357. 29th Cong., 1st Sess., VII, pp. 3–12.

enabling act to form a state government; and in the second place, the boundary line was needed so that survey parties in the western Upper Peninsula could close their township lines. That William Austin Burt received this contract was but further proof of his experience and ability. On May 14, 1847, his party of approximately thirty left Detroit on board the *Sam Ward* bound for the Sault.[42]

The great Sault Canal was still in the future, so Burt's party portaged to Lake Superior where they boarded the propeller-driven *Independence.* On Sunday, May 23, they arrived at L'Anse, on Keweenaw Bay, and Burt attended worship at the Reverend John Pitezel's Methodist mission. The next day he was the first to shoulder his pack for the seventy-five-mile march to Lac Vieux Desert. Here camp was established, and the survey began on June 7.[43]

Marking the boundary line was tedious. There were numerous lakes and streams, foul weather delayed operations, and finally rations were in short supply. Burt urged his men on, but by early July several were too weak to continue. Just in time, on July 5, packers arrived with supplies from L'Anse, and the survey was completed.[44] None of the hardships encountered were mentioned in Burt's field notes submitted to Lucius Lyon, Surveyor General, that November. In matter-of-fact language Burt reported that he had "surveyed, marked, and established so much of the boundary line between Michigan-Wisconsin as lies between the source of the Brule river and the source of the Montreal river . . . "[45]

For this important work Burt received $18 per mile—the highest rate ever paid on the Michigan public surveys. Yet, as a contemporary noted, it was a "small amount of money for a large amount of work."[46] In 1928–29 the state engineer of Wisconsin reported that "the original Burt line is one of the best early surveys." Indeed, there were generally only a few minutes variation between the modern resurvey lines and those by Judge Burt.[47] Such accuracy was a tribute not only to the surveyor but to his invention, the solar compass.

[42] Lucius Lyon, Sur. Genl., to William A. Burt, May 8, 1847, Burt Papers; George H. Cannon, "Our Western Boundary," M.P.H.C., III, pp. 244–261; E. C. Martin, "Leaves from an Old Time Journal," *ibid.,* pp. 405–409.

[43] E. C. Martin, *op. cit.*

[44] George H. Cannon, *op. cit.*

[45] William A. Burt, field notes transmitted to Lucius Lyon, Nov. 20, 1847.

[46] George H. Cannon, *op. cit.,* p. 261.

[47] Lowell O. Stewart, *Public Land Surveys,* pp. 57–58.

In December 1849, Burt went to Washington, D. C., to consult with officials of the General Land Office "in regard to the best method of preventing, in future, fraudulent surveys, and the best method of conducting surveys hereafter." At this time he made a decision concerning his solar compass that was to trouble him and his heirs for many years thereafter. Upon discussions with Land Office personnel and certain members of Congress, Burt decided against renewal of patent rights to his invention. In retrospect it was not a wise move; at the time it seemed so. In essence he decided to allow his compass, then officially endorsed and proved indispensable in the public surveys, to become public property. As compensation he would rely on a special act of Congress.[48]

The sequel was not a happy one for Burt and his family. Time after time, from 1850 to 1900, petitions and special bills were brought before Congress. Time after time, despite impressive testimonials from users of the solar compass, and despite favorable reports from the Committee on Public Lands, the Burt claims for compensation failed.[49]

In August 1858, William Austin Burt died at his Detroit home. His famous compass was then required equipment in the public surveys as they moved across the territories of the United States.[50] To Burt, with his deep interest in scientific advance and with his zeal for public service, this may have been some compensation to offset the failure to realize monetary gain from his invention. But to a generation many years removed from the nineteenth century, and able to view with more perspective and appreciation the services of William Austin Burt, it does not seem just that the Federal Government failed to reward the inventor in his lifetime.

[48] For documentation for this paragraph see William A. Burt to Phoebe Burt, Dec. 18, 1849; William A. Burt to John Wilson, Commr. Genl. Land Office, Dec. 20, 1849, Burt Papers; John Burt, *History of the Solar Compass, Invented by William A. Burt* (Detroit, 1878), pp. 9–10.

[49] For details see numerous documents, letters, etc., in the Burt Papers; also Hiram A. Burt, *Facts Upon Which Is Founded the Claim of the Heirs of Wm. A. Burt, Deceased* . . . (Wash., D. C., 1886). This matter of the Burt family attempt to secure compensation for their claims is the subject of an article in preparation by the present writer.

[50] On subsequent use of the solar compass see letter of D. K. Parrott, Actg. Asst. Commr. Genl. Land Office, April 29, 1920, printed in Horace E. Burt, "William Austin Burt, Inventor," *Mich. Hist. Mag.*, VI (1922), pp. 186–188.

Papers of the Michigan Academy of Science, Arts, and Letters
Vol. XLVII, 1962 (1961 Meeting)

THE INTEREST OF THE UNITED STATES IN CENTRAL EUROPE: APPOINTMENT OF THE FIRST AMERICAN CONSUL TO HUNGARY

GEORGE BARANY

University of Denver

During the first half of the nineteenth century, the establishment and expansion of American diplomatic and consular missions in Central Europe were stimulated chiefly by the commercial interests of the United States. With the Austrian Empire, diplomatic relations were formally established in 1819, when the first Austrian diplomatic agent, Count von Stürmer, was sent by the court of Vienna to the United States. In the next year he was succeeded by Baron Lederer, who as consul general was in charge of Austrian affairs until the exchange of the first ministers plenipotentiary between the two countries in 1838.[1]

The first American consul accredited to the Austrian Empire, John George Schwarz, was appointed to his post in Vienna in 1829.[2] In the same year, a Treaty of Commerce and Navigation between the United States of America and His Majesty the Emperor of Austria was signed by Secretary of State Van Buren and Baron Lederer. This treaty included among other things the "most favored nation clause," with regard to Austro-American commercial relations (Arts. V and IX), and also provided for mutual exchange of consuls and other agents (Art. X).[3] In May 1838, a report of an

[1] Albert B. Faust, *Guide to the Materials for American History in Swiss and Austrian Archives* (Washington, D. C.: Carnegie Institution, 1916), pp. 10; 185–216. For additional bibliographical information, see Samuel Flagg Bemis and Grace Gardner Griffin, *Guide to the Diplomatic History of the United States, 1775–1921* (Washington, D. C.: U. S. Govt. Print. Off., 1935), esp. pp. 240–247 and 928–931; R. Adelaide Hasse, *Index to United States Documents Relating to Foreign Affairs, 1828–1861* (Washington, D. C.: Carnegie Institution, 1914–21), 3 vols.

[2] The reports of John George Schwarz were published in *Doc. No. 71,* pp. 1–85, vol. 9 of *House Ex. Docs.,* 31st Cong., 1 sess.

[3] Hunter Miller, ed., *Treaties and Other International Acts of the United States of America,* III (Washington, D. C.: U. S. Govt. Print. Off., 1933), 507–521. — Concerning the first Austro-American trade negotiations under Joseph II,

American tobacco agent advised the State Department to appoint "*natives*, who exercise a commanding influence over commerce and the business community, and [have] an important stake in the development of the commercial resources of the country" [italics added] as consular representatives of the United States to varied parts of the Austrian Empire, primarily to Fiume, Milan, Pesth, and Prague.[4]

With regard to the last two cities, the report submitted the names of two "such persons as would agree to fulfil all the conditions requisite." Both persons were financiers of high reputation who knew English. The report implied that these circumstances would facilitate the getting of information in addition to that received through official channels.[5]

As to the character of the information, the author of the report, Nathaniel Niles, was instructed concerning his duties by Secretary of State John Forsyth in a letter dated June 7, 1837:

> Sir.
>
> The President, anxious to procure the abolition or modification of the heavy duties and restrictions by which the trade of this country with the various European nations in the article of tobacco has been hitherto burdened, has appointed you a Special Agent for the Empire of Austria to collect information, and perform such other services as are herein after designated, having a tendency to promote the interests of this trade within the limits of that Empire.[6]

The establishment of an American legation in Vienna was strongly motivated by similar considerations. This is obvious from the instructions sent to Henry A. Muhlenberg, first envoy extraordinary

---

see the documentation and references in Hubert van Houtte and Edmund C. Burnett, "American Commercial Conditions, and Negotiations with Austria, 1783–1786," *American Historical Review*, XVI (April 1911), pp. 567–587.—In the negotiations America was represented by Franklin and Jefferson, whereas the ambassador of Joseph II to Louis XVI, the Count Mercy d'Argentault, was the representative of Austria.

[4] Niles to Forsyth, May 10, 1838. *Doc. No. 246*, pp. 43–44, in vol. 4 of *Sen. Docs.*, 27th Cong., 2d sess., 1841–42. In the extracts published, Pesth is repeatedly misspelled as Perth, probably due to a misreading of the original handwritten report.

[5] For Pest, the report suggested the appointment of Albert Wodianer; for Prague, the name of Moriz Zdekauer was submitted. For biographical data, see Constant von Wurzbach, *Biographisches Lexikon des Kaiserthums Oesterreich* (Wien: Universitäts-Buchdruckerei, 1856–91), vol. LVII, 201–203, and LIX, 235, resp.

[6] *Natl. Archives, Records of the Dept. of State, Austria,* 1: 1 (Microfilm No. 77, Roll 13).—Hereafter cited as *Instructions.*

and minister plenipotentiary of the United States to Austria. Besides stressing the new minister's "duty generally to take care of the interests of the United States, and their citizens," the instructions emphasized:

> The amicable and commercial relations that have long subsisted between the United States and the Austrian Empire appeared to call for the establishment of a closer connexion by means of Diplomatic Representatives, and it is hoped that the mutual alacrity with which this new relation has been formed, will prove a happy augury and lead to permanent good results by enlarging the commercial intercourse between the two countries, and perpetuating those feelings of amity and good will which it will be the desire and aim of the United States to cultivate by all means within their power.
>
> The prominent object of your mission, that to which the attention of the Government and citizens of the U. States (a large and respectable class of whom is deeply interested) has been of late so strongly drawn, is, the present condition of our tobacco trade, and the best methods by which the commerce in this staple article may be relieved from the burdens which have been laid upon its introduction into the Austrian Empire.[7]

Forsyth also referred to a "Report of a Select Committee of the House of Representatives at the 2d session of the 24th Congress," which appropriated the necessary sums for the expansion and strengthening of American diplomatic activities,

> with a view, as was understood, to the appointment of Special Agents, to negotiate with the Governments of those countries in which American tobacco is subjected to duties and restrictions, and in which the U. States have no accredited Representatives, and provision was also made for missions to Austria and Prussia, it is believed, with a special view to the extension of this particular trade.[8]

Congress established the yearly salaries of tobacco agents, who

[7] Forsyth to Muhlenberg, Apr. 20, 1838. *Instructions*, p. 13. Cf. Daniel Jenifer's report on the resolution of the "select committee on the subject of the memorial of certain tobacco planters of Maryland and Virginia" and the resolution accepted by the House of Representatives, *Cong. Globe*, IV, 24th Cong., 2d sess., 1836–37 (Feb. 28, 1837), p. 213; also, discussion of the Civil and Diplomatic Appropriation Bill for the year 1837, esp. Jenifer's remarks, *Register of Debates in Congress*, XIII, Pt. II, 24th Cong., 2d sess., 1836–37 (Feb. 28, 1837), pp. 2059–61.—For biographical data on Henry Augustus Philip Muhlenberg, see *DAB*, XIII, 309–310.

[8] *Instructions*, p. 14. Cf. President Van Buren's message to Congress on Dec. 3, 1838. James D. Richardson, ed., *A Compilation of the Messages and Papers of the Presidents*, III (Washington, D. C.: Bureau of National Literature, 1912), 1706.

had no diplomatic status, as $3,000 plus an allowance of $500–$1,000 for contingent expenses. Chargés d'affaires received $4,500 and $500, ministers plenipotentiary $9,000 and $800, as well as an outfit equal to one year's salary, plus a quarter's salary for their return home.[9] These were rather generous appropriations, since the yearly salary of a court councilor in Vienna was around 3,000 florins at the time, whereas an average official in the administration or a secondary schoolteacher made only 300–400.[10] The envoy's salary, ca. 20,000 florins, was about equal to the yearly net income of Count Stephen Széchenyi, who, with his offer of 60,000 florins, became the founder of the Hungarian Academy in 1825.

It is beyond the scope of the present study to describe the efforts of American representatives aimed at changing rigid Austrian trade restrictions and, in Forsyth's words, "oppressive sanitary regulations."[11] One example will suffice to indicate that obsolete Austrian quarantine regulations were a real nuisance, preventing the development of foreign trade with the United States.

> My Dear Sir: Since I last had the pleasure to write your uncle, I have had occasion to speak with Prince Metternich, Count Kollowrat, and the Baron Eichoff, on the subject of the quarantine regulations, which have so long served as an insuperable barrier in the way of commerce between this empire and the new world; and I am happy to be able to inform you that all these distinguished and enlightened gentlemen are as surprised at, as they were ignorant of, the existence of such absurd and fatal regulations. They all see the importance of their being immediately revised, and are anxious to have the subject come up from Trieste for consideration.[12]

As evidenced by the correspondence, however, it took the Austrian authorities about two years to put into practice the promised "immediate" changes in the quarantine regulations, the reasonableness

[9] Forsyth to Niles, June 7, 1837; Forsyth to Muhlenberg, Apr. 20, 1838; Webster to Jenifer, Aug. 30, 1841; Webster to Dodge, May 9, 1842; Buchanan to Stiles, Apr. 30, 1845, *Instructions*, pp. 3; 11; 21–23, and 39.—Reports from the Viennese legation repeatedly complained of the high cost of living in the Austrian capital.

[10] According to a report of Schwarz, dated July 8, 1837, one florin was equivalent to 44 cents. *Doc. No. 71*, p. 35, vol. 9, *House Ex. Docs.*, 31st Cong., 1 sess.

[11] To Muhlenberg, Apr. 20, 1838, *Instructions*, p. 16.

[12] Niles to Murray, Feb. 14, 1838, *Doc. No. 246*, p. 21, vol. 4, *Sen. Docs.*, 27th Cong., 2d sess., 1841–42.

of which was apparently recognized by all leading statesmen of the Empire.[13]

The reports sent to the State Department contain many details suggesting that American representatives did their best to convince the Austrian officials of the necessity of modernizing the system of imperial customs and quarantine regulations. These efforts seem to have brought but meager results, perhaps in part because the Austrians, too, had some grievances against the commercial policies of the United States Government.[14] In any case, Congress decided to "reduce" the legation of the United States at Vienna in 1845, and the Department of State recalled Daniel Jenifer, who had succeeded Muhlenberg as minister plenipotentiary in 1841. The instructions relative to this matter read:

> Sir: You are aware that Congress, at its last Session, reduced the Austrian mission, and that this country will, hereafter, be represented in Austria by a Chargé d'Affaires. In giving official notice of this change, I am directed to appraise you that the President has appointed Mr. William H. Stiles, of Georgia, to succeed you in the capacity just named. I presume that the chief reason which influenced Congress to reduce the mission to one of inferior rank was, that the Imperial Government had not deemed it proper to accredit another Minister Plenipotentiary to the United States since the departure of Baron Mareshal [!]; but had entrusted its interests here to a Chargé d'Affaires.
>
> ...Upon taking leave of the Austrian Government, you will embrace the occasion to convey to the Emperor the assurance of the desire felt by the President to cultivate and extend the most friendly relations with the Government and people of Austria;—and his sincere hope that the reduction in grade

---

[13] See the reports of Dec. 16, 1837 (Moore to Niles); March 10 (Bruck to Niles); March 16 (Niles to Metternich); Apr. 14 (Niles to Forsyth); July 29, 1838 (Niles to Forsyth); May 14 (Muhlenberg to Forsyth); Apr. 20, 1839 (Muhlenberg to Metternich); Feb. 6 (Muhlenberg to Forsyth), and Jan. 10, 1840 (Metternich to Muhlenberg). *Doc. No. 246,* pp. 10–11; 31–37; 42; 44–45; 51; 62–64; 68–69, resp. *Sen. Docs.,* 27th Cong., 2d sess., 1841–42.

[14] E.g., Austria repeatedly insisted on reducing American customs duties on Austrian (i.e., Hungarian) wines on the basis of the favorable treatment granted to the French wines imported to the United States. Cf. Muhlenberg to Forsyth, No. 15 and 31 [*sic*], 1838, pp. 5 and 1 of reports, resp.; Feb. 15, 1839, p. 7 and June 16, 1839, p. 3 of reports. Also: Clay to Forsyth, Feb. 15, 1841, pp. 2–3 of report. *Natl. Archives, Records of the Dept. of State, Despatches from U. S. Ministers to Austria* (Microfilm No. T-157, Roll 1), vol. I.—Hereafter cited as *Despatches.* Cf. Muhlenberg to Forsyth, June 16, 1839, *Doc. No. 246,* pp. 65–66, vol. 4, *Sen. Docs.,* 27th Cong., 2d sess., 1841–42; Schwarz to Forsyth, March 10, 1835, *Doc. No. 71,* p. 29, vol. 9, *House Ex. Docs.,* 31st Cong., 1 sess.

of the Diplomatic Agents, representing the two Countries, respectively, may not have the least influence in defeating or postponing the mutual advantages confidently *anticipated, at the time those relations were established.* [Italics added.][15]

It would be misleading to attribute all the difficulties in the initial phase of Austro-American diplomatic relations to a lack of mutual understanding concerning the principles of commercial policies or to the clumsiness of Austrian diplomacy. From the very be-

[15] Buchanan to Jenifer, Apr. 18, 1845, *Instructions,* pp. 36–37. For the desire of seeing Austria represented by a minister plenipotentiary in the United States, see Legaré to Jenifer, May 10, 1843, *Instructions*, p. 31 and Jenifer to Buchanan, July 19, 1845, *Despatches,* pp. 2–3 of report.—Baron Mareschal, first Austrian minister plenipotentiary and envoy extraordinary to the United States left his post for that of Lisbon in 1841. He was succeeded by the Chevalier de Hülsemann, former secretary of the legation, who was appointed chargé d'affaires after Mareschal's departure. Faust, *op. cit.*, pp. 10–11.—Incidentally, the usefulness of American diplomatic missions both to Austria and Prussia was questioned in Congress as early as 1841. C. J. Ingersoll, representative from Pennsylvania, was particularly anxious to cut the expenses of foreign missions by reducing the number of American ministers abroad. Saying that "a consul would be sufficient at Vienna and Berlin, where we now had full ministers, and at Naples and Sardinia, where we had unrequited chargés," Ingersoll moved in April 1842 to reduce "the sums of appropriation for the missions to Austria and Prussia one-half" and to reduce "the whole system ... to Republican standard." True, the annual salary of the nine American ministers to foreign countries amounted to little more than $80,000 at the time, but in Ingersoll's opinion America "suffered vastly more" by "imported foreign ministers" than by those whom she paid because foreign ministers to the United States "had revolutionized this country. The gold-spoon influence which came from that quarter infected every body here; no man could escape it, neither Presidents, Secretaries, nor all the rest. These foreign missions were therefore injurious, not only because they cost us so much money, and add so much to the Presidential patronage in the unnecessary numbers sent abroad, but infinitely more in those they bring back on us ... the representatives of foreign courts here, whose access to our own Executive was perfectly familiar, exercised a great influence." Ingersoll's puritanical views and his efforts to restrict America's foreign relations were rejected by an overwhelming majority in the House of Representatives. In an earlier debate held in August 1841, the representative from Massachusetts, Cushing, recommended that "we should draw our relations close with the secondary powers, as well in the Mediterranean as the Baltic," for this could serve the interests of the United States in the case of a war between the European great powers, England, France, and Russia. *Cong. Globe*, X, 27th Cong., 1 sess. (Aug. 27, 1841), pp. 393–395, and *Cong. Globe,* XI, 27th Cong., 2d sess. (Apr. 14, 1842), pp. 422–425.—For biographical data on Daniel Jenifer, see *NCAB,* XIII, 324.

ginning of official relations, certain political considerations seem to have interfered even with the economic connections of the two countries. As indicated by the appointment of the first American diplomatic and consular representatives and the instructions given to them, the United States was interested at the time in Austria and in Central Europe generally primarily from the point of view of developing her foreign trade. Nonetheless, there was a certain "image" of America of which ruling circles in the absolutistic monarchies were unable to rid themselves. Republican and revolutionary, this image seemed to haunt members and servants of the old dynasties, who often tried hard indeed to indicate their sympathy to the representatives of the experiment in the New World. The first two American ministers plenipotentiary, Muhlenberg and Jenifer, "had every reason to be satisfied with the polite and kind attention"[16] with which they were received by the Emperor, in court, and by high-ranking officials. A rather characteristic passage of Muhlenberg's letter dated November 31 [*sic*], 1838, reported:

> ...I have been received by all the different members of the Imperial family with marked kindness and respect. All the Archdukes expressed a desire that our commercial intercourse might be increased and a belief that it would prove mutually beneficial. The Archduke Francis, [!] added to me complimentary remarks concerning the integrity and enterprise of our merchants and the beauty, order and cleanliness of our vessels, having had an opportunity of seeing some of them!!

But in his next report, dated February 15, 1839, the American minister, talking about his interviews with Metternich "in relation to American affairs generally and, more particularly, in relation to the promotion of a mutually beneficial commerce between the United States and this country," already noted:

> Our Institutions—our strength and rising importance—our productions—our wants—our commercial regulations and their causes—the extent of our trade and the mode in which it is conducted—I find, are not as well understood as might have been expected. Upon all these points and others connected with them, I have endeavored to communicate correct information and to make an impression as favorable as possible not only upon the Prince Metternich himself, but also upon all others more or less intimately connected with the Court

[16] *Despatches.* Muhlenberg to Forsyth, Aug. 20, 1838, p. 5 of report, Nov. 13 and 31 [*sic*], 1838, pp. 4–5 and 1 of reports; Jenifer to Webster, May 22, 1842, pp. 2–3 of report.

and likely to have an effect upon public opinion which, I find, is here, as well as with us, a matter of great importance.

The prejudices evidently existing against us as Republicans, I will not say with the Imperial Family or the Prince Metternich, but with the diplomatic corps generally and the great mass of those connected with and dependant upon the Court and whose influence is overpowering, I have labored to counteract by dilating upon the well established truth that tho we believed our Institutions best calculated to promote the welfare and happiness of our own country and should therefore strenuously endeavor to uphold them there, yet that we had laid it down as a fixed rule, from which we never deviated, not to interfere in the internal affairs of other nations, believing every one best calculated to judge of what was proper and most beneficial to itself; that we were far from being propagandists as had been proved by our whole course during the wars arising out of the French Revolution—during the troubles in South America and latterly in the neighbouring English Colonies. The Prince was pleased to say that he had no reason whatever to doubt my assertions, that as far as he knew our conduct had always been correct and even laudable and did not doubt that the most amicable relations might subsist between us and Governments constituted upon entirely different principles, that as far as he and Austria were concerned there should certainly be no cause for complaint as no Government could be more friendly disposed towards us than this.[17]

Muhlenberg's apology for American institutions and official policies was not at all out of place. Despite Metternich's polished attitude and reassuring words concerning the feasibility of "peaceful coexistence," officials including Metternich considered Americans as dangerous "propagandists" of the democratic disease and refused, whenever possible, to grant passports for America to citizens of the

[17] Muhlenberg to Forsyth, Feb. 15, 1839, *Despatches,* pp. 1–3 of report. Besides discussing the possibilities of the export of American cotton, tobacco, rice, and colonial wares to the Austrian Empire, and besides explaining the Austrian protective tariff, Muhlenberg also transmitted in this report an "unofficial" request by the Swiss chargé d'affaires at Vienna asking "if the United States could find it convenient to honor them with a Chargé d'Affaires, or, at least, a Consul General, with Treaty making powers" (pp. 8–9 of report). It is also worth mentioning that in order to increase Austrian interest in American agricultural products, the United States was constantly described by its representatives during the eighteen thirties and forties as a "most important outlet" for Austrian industrial products in the next twenty-five years. Cf. Muhlenberg's interesting letter of Feb. 28, 1839, addressed to Metternich and attached to his report to Forsyth, dated March 16, 1839, in *Despatches.* Cf. Jenifer to Webster, May 22, 1842, pp. 4–5 of report, *ibid.*—Muhlenberg's reports of Feb. 15 and March 16 were published in *Doc. No. 246,* pp. 52–62, vol. 4 of *Sen. Ex. Docs.,* 27th Cong., 2d sess., 1841–42.—My citations follow the orthography of the original handwritten reports.

Austrian Empire. In 1821, for example, when Count Stephen Széchenyi[18] intended to visit "the land of freedom" (his words) with his friend the Baron Nicholas Wesselényi, Austrian authorities were shocked. Count Bellegarde, president of the War Council, protested: "ah Mr le Comte c'est une fantasie que je desavoue—à moins que vous ne reviez Constitution—car sans cela je l'avoue je ne vois qu'une caprice." In Metternich's opinion, Széchenyi's American travel plan was "bizarre" and impossible to put into practice. Whereupon Széchenyi exclaimed in his diary: "Beym Himmel es gibt Menschen, die nicht begreifen, dass man gerne ein freies Land Besucht!"[19]

The appointment of the first American consul to Hungary well illustrates the imperial officials' traditional fear of close connections with "the land of freedom." In the summer of 1839, Muhlenberg "paid a visit to Hungary descending the Danube in a Steamboat one hundred and thirty English miles to Pesth and Ofen, on opposite sides of the river and returning by land." Whereas Muhlenberg's valuable report on his Bohemian journey of 1840 was printed in the *Senate Documents*[20] the description of his trip to Hungary has never been published, thus escaping even the scrutiny of Hungarian historians:

> To Presburgh [!] the present capital, where the Diet is now in Session, distant from Vienna fifty English miles, the country is rather poor and its agri-

[18] Concerning Count Stephen Széchenyi, see the centennial Széchenyi number of the *Journal of Central European Affairs*, XX, no. 3 (October, 1960), and George Barany, "The Emergence of Széchenyi and Hungarian Reform until 1841" (unpublished Ph.D. dissertation, University of Colorado, 1960, Boulder, Colorado).

[19] Viszota, Gyula, ed. *Gr. Széchenyi István naplói* (Count Stephen Széchenyi's Diaries), II (Budapest: Magyar Történelmi Társulat, 1926), 211–215; 217 and 219 (entries of Oct. 10; 17; 20; 27, and Nov. 8, 1821). Vol. XI of *Gróf Széchenyi István Összes Munkái* (Count Stephen Széchenyi's Complete Works), published by the Hungarian Historical Association in the series *Fontes Historiae Hungaricae Aevi Recentioris*. Referred to henceforth as *Diaries*. In 1824, Baron Lederer supported his request for furlough by pointing out that he intended to find an adequate educational institute for his two sons who were "to be protected in their gentle and critical age of 8 and 10 years from the principles prevailing here." Lederer's German request of Aug. 30, 1824, was printed in Faust, *op. cit.*, p. 191.

[20] To Forsyth, July 14, 1840, *Doc. No. 246*, pp. 69–72, vol. 4 of *Sen. Docs.*, 27th Cong., 2d sess., 1841–42.

cultural produce of no great amount. Below Presburgh and down to Pesth the soil is very productive yielding immense quantities of the finest wheat, rye, barley, oats and Indian corn with some Tobacco. There are also very extensive and flourishing vineyards producing richly excellent red and white wines with more spirit than those of France or Germany and therefore, as far as I can judge, well qualified for exportation.

Between the grain and wine districts there are found large tracts of excellent grazing land covered with herds of fine oxen, hardy horses, and a variety of sheep. The wool produce of Prince Paul Esterhazy alone is said to yield him an annual revenue of 250,000 American dollars. Below Pesth the country is said to be still more productive, particularly the so called Bannat [!] (formerly part of Transylvania) now incorporated with Hungary. Here immense quantities, as well of grain as of tobacco and Hemp are grown. Of the richness of Hungarian mines I need not inform you. With reasonably divided estates and an improved agriculture Hungary might be a granary to supply all Europe. The estates are however so large that much land is in a waste state and the balance not cultivated with that care which is bestowed in other portions of the Empire or even in many districts of America. The great want of roads, markets, and manufactures leaves the inhabitants, perhaps, without the necessary stimulus as the present abundant produce of this rich Country is comparatively speaking of little value. It is indeed a fine country and pleases me better than any other part of Europe I have yet seen. Many of its nobles are well informed and quite respectable both in manners and appearance. The same may be said of the merchants and dealers of the larger cities—but not of the great mass of the population. It seems but one degree removed from a savage state and is filthy in person, habits and appearance beyond description.

Pesth is at present the commercial emporium of this Country, which may at no very distant day become an important point of Europe for many reasons which will strike you without detailing them particularly if you cast your eye upon Russia. The city suffered greatly about eighteen months since by an ice freshet but is rapidly rebuilding and in an improved style. It promises to become a large place in some years particularly if navigation of the Danube should be improved which is however connected with considerable difficulty. In America the difficulties would not be considered great, but here they are almost insuperable. Pesth has now very considerable warehouses for wool, tobacco, wine and grain and more are erecting. I may note as something unexpected that I had the pleasure of drinking some beer at this place brewed with American *hops* which had been purchased with some advantage to the establishment.

Permit me respectfully to suggest for consideration the propriety of appointing an American Consul at Pesth. He should be selected from *the native citizens* of information, respectability and influence. The situation would not be one of profit but honor is a great thing in this Country and no doubt many proper persons would be found willing to accept of the situation for the honor. I would mention the name of Frederick Kappel, an intelligent merchant of high standing. Some others equally well suited might be found. Another such an appointment should be made in Prague the centre of the great manufacturing enterprise of that section of the Empire—and probably, also one at Milan where it is said manufactures are progressing with rapid strides. Of Milan I shall be able to say more if I can accomplish my intended trip to that place. These Consuls would be particularly useful in furnishing statistical information and it can scarcely be acquired in any other way in this country where such things are seldom published, where even the Government has it not, or does not see proper to promulgate it. They would also be the proper mediums of operating upon the community and Government when changes in the commercial system or tariff are desired as the Government never makes a first move in such cases, but always waits to be called upon and only acts when calls have been repeated, when all parties likely to be affected have been heard and when something like a general consent has been obtained.[21]

This evaluation of Hungarian conditions was supplemented by data relative to Hungarian tobacco, wool, and wine production and prices. Muhlenberg's report recognized both the Hungarian land problem and the inefficiency of the government in dealing with the problems of trade and industrialization. It also returned to Niles' previous suggestion of May 1838 concerning the appointment of natives as American consuls. Eight months later Muhlenberg again reminded Secretary of State Forsyth:

Permit me again to call your attention to the necessity of appointing a consul at Pesth in Hungary. I have already mentioned the name of Mr. Frederick Kappel as a suitable person to fill that post. He is a gentleman of great worth and respectability, an eminent and wealthy merchant who speaks English with considerable fluency. Trade with Hungary appears to be on the increase. Many goods are sent from Pesth, the principal market, not only to Trieste, but also up the Danube and thence across to the Elbe and Hamburg. There is besides considerable travelling up the River from Constantinople to Vienna and thence to the north and south of Germany and travellers are often in want of advice and assistance at such a central point as Pesth.[22]

Muhlenberg was right in evaluating the growing importance of Pest as a prospective commercial center of Hungary. The economic and social reforms and the writings of Széchenyi initiated a Hungarian national renaissance noticed by many a foreigner visiting the country. One of them, an English writer by the name of Miss Julie Pardoe, could not help warning her compatriots concerning certain

[21] Muhlenberg to Forsyth, Aug. 8, 1839, *Despatches,* pp. 2–6 of report.

[22] Muhlenberg to Forsyth, Apr. 21, 1840, *Despatches,* p. 4 of report.

"American speculations" in Hungary, adding that, according to some rumors, an American "trade consul" was to be appointed to Pest.[23]

Miss Pardoe's visit to Hungary coincided with that of Muhlenberg. Her book on Hungary was published in London in 1840. By that time, Congress and the State Department gave their approval to Muhlenberg's suggestion and appointed Frederick Kappel as American consul to Hungary. The relative instructions, sent to Muhlenberg by Forsyth on July 1, 1840, read:

> Sir.
>
> Your letter of the 21 April, no. 12, has been received and submitted to the President, who in pursuance of the recommendation conveyed in it, has thought proper to appoint Frederick Kappel, Consul of the U. States for Pesth in Hungary. I now transmit his commission on the receipt of which you will apply to the Austrian Government for an exequatur recognising him as Consul, and forward it, when obtained, with the commission, to Mr. Kappel.[24]

Frederick Kappel, well-to-do merchant and financier,[25] greatly contributed to the improvement of Hungarian sheepbreeding and wool production in the early nineteenth century. In order to promote the quality of Hungarian tobacco, he imported the seeds of the best American tobacco plants and distributed them to producers. An early associate of Széchenyi, he participated in and financially supported the enterprises of the "father of Hungarian reform."[26] A leading member of the Casino, Kappel, along with Széchenyi, was one of the three first shareholders in the Bridge Association created by Széchenyi to construct a permanent suspension bridge between Buda and Pest. He also assisted Széchenyi in establishing a modern cylinder mill and a port and dockyard in Pest. Like Széchenyi, Kappel was an elector of the city of Pest. He was consulted by the Hungarian Diet concerning the law of exchange then under consideration and concerning other financial operations. He was also highly esteemed by Viennese officials. Yet they never consented to his appointment as consul of the United States of America.

In early November 1840, Randolph Clay, who became chargé d'affaires ad interim after Muhlenberg had resigned his post in Vienna

[23] Julie Pardoe, *The City of the Magyar or Hungary and Her Institutions in 1839–40,* III (London: G. Virtue, 1840), 305.

[24] Forsyth to Muhlenberg, July 1, 1840, *Instructions,* p. 20.

[25] Cf. Wurzbach, *op. cit.*, X, 451.

[26] In his private diaries Széchenyi mentioned having seen Kappel on a number of occasions between 1827 and 1845. See *Diaries,* III–VI.

in mid-1840, addressed a note to the Chancery of State "endorsing Mr. Kappel's Commission and requesting the Austrian Government to grant him the customary Exequatur." He also forwarded the official documents informing Kappel that he had been appointed consul of the United States "for the port of Pesth in Hungary." The Hungarian financier "sometime afterwards came to Vienna and purchased his Seals of Office."[27]

Already in his report to Webster of April 19, 1841, Clay expressed his doubts concerning the wisdom of appointing an American consul to Hungary and the procedure followed by Muhlenberg. This part of his report is highly characteristic of Austro-Hungarian relations; I therefore print it all:

I am unable to say whether Mr. Muhlenberg informed Prince Metternich of his intention to recommend to the Government of the United States to appoint a Consul at Pesth: or whether he spoke to the Prince upon that subject; but, as no mention is made of such a conversation in any of Mr. Muhlenberg's despatches, I presume he did not. It is, I think, unfortunate that he did not consult the Austrian Minister of Foreign Affairs as he would then have known his opinion in relation thereto and saved him and, probably, this Legation from some embarrassment. Five months have passed since I addressed the note to the Chancery of State to obtain an Exequatur for Mr. Kappel. The lapse of such a length of time without an answer would fully justify an official request that the matter might be attended to and a reply be given on the part of the Austrian Government. Nevertheless, from information that I have received, indirectly, I think it neither politic, nor proper to press the matter without your instructions. The Austrian Government will probably decline granting the Exequatur, and assign some plausible reason for doing so. Whatever may be the comity apparent in its reply the refusal will, in fact, be made for the following reasons.

Hungary, although it forms a part of the Austrian dominions, is, in many respects a separate Kingdom. The Emperor of Austria is *King* of Hungary: there is a diet, composed of two Chambers, which must be consulted before measures, affecting the privileges, or general welfare of the Kingdom, can be adopted; there is a special Chancery, at Vienna, charged with the conduct of Hungarian affairs; the levy of recruits made, almost arbitrarily, in the other states of the Empire must, in Hungary, be first voted by the diet: even that powerful engine for the insurance of submission to the existing form of Government—the police—dares not, openly act there, although, secretly, it doubtless extends its subtle authority throughout the Kingdom. Hungary being therefore, in some respects, independent of the central Government of the Empire and a rather formidable opposition to the Imperial will existing in the sittings

[27] Clay to Forsyth, Dec. 8, 1840; Clay to Webster, Apr. 19, 1841, *Despatches*, pp. 6 and 1–2 of reports, resp.

of the diet, I am inclined to believe the Ministry, always suspicious of innovations, are fearful that the acknowledgment of a Consul of a foreign Government to reside at the Capital of Hungary might serve as a pretext for the discontented members of the diet to attach to the measure some claim for further independence which, though it would not succeed, might give trouble to the Cabinet and increase the dissatisfaction, to the existing order of things, already felt in many parts of the Kingdom. Connected with these facts is a second reason for declining the proposition of the United States. Were Mr. Kappel recognised, the French and British Governments would most probably appoint Consuls at Pesth and the Austrian Ministry could not refuse their Exequaturs after having granted one to an Agent of the United States. The same policy which induced Austria to take exception to the appointment of British and French Consuls at Cracow would, a fortiori, exclude such agents from Pesth.

This Government may refuse the Exequatur on the ground that as Pesth is not, strictly speaking, a *port,* the United States cannot claim the *right* of appointing a Consul to reside there under the tenth article of the treaty between the two Nations. Should it be thought proper, however, on the part of the United States to discuss the question, an answer might be found, to the objection referred to, in the fact that the United States have already a Consul at Vienna which is less a port than Pesth. The interests of the United States would, in my opinion, be better consulted if the matter were allowed quietly to drop in case it shall appear, from the answer returned to my note of the 7th of November last, that there is a repugnance on the part of Austria to recognize foreign Agents within the limits of the Kingdom of Hungary.[28]

Clay's well-informed appraisal of the domestic and international implications of Kappel's appointment was correct. Under circumstances then prevailing in the Austrian Empire, the appearance of an American consul residing at Pest would have been considered as a political act going far beyond the original intentions of the government of the United States to promote by this appointment its commercial relations with the Empire. The Hungarian Diet of 1839–40, referred to by both Muhlenberg and Clay, witnessed an important shift in the political balance of the conservative and liberal forces favoring the latter. The emergence of a radical opposition in the Chamber of Magnates, joining the liberal opposition which had had the majority in the Chamber of Deputies since the Diet of 1832–36, could only stiffen the administration's resistance to any "unnecessary" innovation prone to boost the overconfidence of the opposition to Vienna. Although the Diet itself ended in a reconciliation between

[28] Clay to Webster, Apr. 19, 1841, *Despatches,* pp. 2–6 of report. Enclosed with the report is a copy of the French original and also the English translation of Clay's note addressed to the Austrian Chancery of State, dated Nov. 7, 1840.

throne and nation, the political truce did not last long. By the second half of 1840, it became obvious that the political leadership of the Hungarian reform movement had shifted from the moderate Széchenyi to the representatives of a younger generation, Count Louis Batthyány and Louis Kossuth. The spring of 1841 saw in Hungary the beginning of a bitter public debate between Széchenyi and Kossuth.[29] The polemic which was to dominate Hungarian political life until the outbreak of the revolution in March 1848, was already in the background of the Austrian government's refusal to acknowledge Kappel's appointment as United States consul for Hungary.

Clay's report of June 26, 1841, essentially confirmed the information and suggestions contained in his letter of April 19. Knowing that the Austrian government was "decidedly opposed to what Prince Metternich terms an innovation," he again advised Secretary of State Webster "to let the matter remain as it is—that is to say—as though no proposition had ever been made on the part of the United States . . ." Clay emphasized:

> I do not think the conduct of the Austrian Government indicative of indifference to the interests or the wishes of the United States. Did not the Cabinet of Vienna fear that the residence of Consuls of other nations, within the limits of Hungary, might become dangerous to the existence of its authority in that country, in case such Agents were to use their influence to increase the dislike already felt, by a large class of the population, to the present state of things, I cannot doubt that the nomination of Mr. Kappel would have been accepted.[30]

Clay may have overestimated Vienna's fears of the "dangers" of foreign consuls residing in Pest because of the language of Metternich's answer to the American note of November 7, 1840. In his note, Metternich politely but firmly refused to recognize foreign consuls with a permanent residence in Hungary, claiming that according to the unanimous opinion of the domestic authorities,

> the necessity and utility of such an innovation do not appear in any manner demonstrated. It could, on the contrary, produce inconveniences of a grave nature which it is the duty of every Government to avoid.
>
> It is also for these reasons that the Imperial and Royal Court has already found itself obliged to decline definitively more than one proposition of this kind

[29] For bibliographical references and an English summary of background, events, and aftermath of the Diet of 1839–40, see Barany, *op. cit.*, pp. 533–577.

[30] Clay to Webster, June 26, 1841, *Despatches*, pp. 1–3 of report.

which has come to it from the side of other foreign Governments and it could not, consequently, defer to the wish of the United States of America without justly exciting complaints and pretensions sufficiently embarrassing, on the part of the said Governments.[31]

Along with Metternich's answer Clay also enclosed with his report to Secretary of State Webster copies of two letters written by him to Kappel. In the first of these letters, dated June 17, Clay informed Kappel that the imperial government declined to acknowledge him as consul of the United States at Pest "not because there would be any objection to you personally, but for political reasons."[32]

The next day, however, Clay felt it necessary to send another warning upon the receipt of a letter written by Kappel on June 14.

...As you maintain in your letter to me that you have received the Consular Seal, flag and arms for the use of the office, through Mr. Moore, American Consul at Trieste, and, as I find your letter to me is written with all the forms of an official document and sealed, as well as the packet addressed to the Secretary of State, with the Arms of the United States I deem it necessary to repeat to you the instructions, contained in my letter of yesterday (June 17) that you should not perform any act as Consul of the United States at Pesth; as such act would not be recognized by the Austrian Government and would only tend to embarrass the relations between the two Countries. You will also refrain from affixing the arms of the United States over the door of your office, or from hoisting the flag to indicate your official residence, until you shall have received some further instructions from the Department of State or from this Legation in case the Representative of the United States at this Court be authorised by the Secretary of State to give them upon the subject.[33]

Kappel loyally complied with Clay's instructions.[34] Following Clay's suggestions, the State Department did not force the issue against the will of Austrian authorities. But a great part of the po-

[31] A copy of Metternich's original French note, dated June 16, 1841, and its English translation were enclosed with Clay's report of June 26, 1841.

[32] Clay to Kappel, June 17, 1841, *Despatches.*

[33] Clay to Kappel, June 18, 1841, *Despatches.*

[34] Clay to Webster, July 20, 1841, *Despatches*, p. 1 of report. See also fn. 38.—Incidentally, connections of the American legation in Vienna were not completely severed with Kappel. This is evidenced by an instruction given more than twenty years later to the American chargé d'affaires, Henry R. Jackson. Jackson was directed to pay an account submitted by Kappel to his predecessor, W. M. Curdy. W. L. Marcy to Jackson, Feb. 20, 1854, *Instructions*, p. 96.—Concerning the account in question or the character of Kappel's services to the legation in the eighteen-fifties, I have thus far been unable to find any information.

litically articulate Hungarian public, including many leaders of the liberal opposition desirous of national independence and reluctant to submit to policies dictated from Vienna, upheld the fiction of an American consul at Pest. Two years *after* Metternich had refused to recognize Kappel, A. J. Blackwell, a British agent in Hungary who himself wanted to become English consul at Pest, reported to Sir Robert Gordon, British ambassador to Vienna:

> I was very much surprised to hear of an American Consul at Pesth. I heard it from the celebrated Bezeredj [!], one of the most gentlemanly members of the Lower House, to whom I was introduced yesterday. It is I trust needless for me to observe, that I have not spoken to anybody here, respecting the contemplated establishment of a British Consulate at Pesth. I was therefore agreably surprised when Bezeredy, in the conversation I had with him on commercial affairs, said somewhat abruptly "Why does not your Government establish a Consulate at Pesth? We have all wondered that one was not established there long ago. The Americans have had a Consul there for some time, a Mr. Kappel, a resident merchant." I asked him whether Mr. Kappel had obtained an exequatur from the Austrian Government. He said "No, but the Americans don't mind that." "That may be," I replied, "but unless he has an exequatur he cannot be called a Consul: and even supposing, for the sake of argument that the English Government wished to establish a Consulate at Pesth, they would not think of appointing a Consul before they had ascertained whether the Austrian Government would grant him an exequatur," to which Bezeredy significantly replied "What your Government wishes, it wills, a few words would suffice to obtain half a dozen Consular exequaturs."[35]

Blackwell never obtained the exequatur he was longing for. Although he had a great sympathy for Hungarians, his principles were, as he put it, conservative

> ...in the true signification of the term and I am sure that Prince Metternich himself cannot be more averse to democratic agitation than I am ... But ... the fact is that every Englishman is tainted in their eyes. To have been born and bred under a constitutional Government is enough to conjure up ideas of

[35] Blackwell to Sir Robert Gordon, July 1, 1843, "Anglo-Hungarian Documents, 1841–1850. Reports of Blackwell on the Hungarian Parliament of 1843–1844," *South Eastern Affairs,* I (1931), p. 141. Joseph Andrew Blackwell attended the Hungarian Diet of 1843–44 as an agent of Sir Robert Gordon, British ambassador to Vienna. He again visited Hungary in 1845–46, and also attended the Diet of 1847–48. A friend of Hungary, he worked for the strengthening of Anglo-Hungarian commercial relations, wishing to be appointed British consul to Pest. His reports are in the MS collection of the Hungarian Academy. Cf. Jenö Horváth, "Anglia és a magyar szabadságharc" (England and the Hungarian Freedom Fight), *Századok,* LX (1926), 599–608.

agitation and what not in their somewhat obscure vision . . . the policy of the Austrian Cabinet in regard to Hungary is truly pitiable. . .[36]

To be sure, the Hungarian desire to have foreign consuls residing at Pest was certainly strongly motivated by the wish to strengthen Hungary's economic ties with non-Austrian territories. The imperial tariff policy was a long-standing grievance of the Hungarian estates who, up to 1840, were in favor of the abolition of the internal customs line between the Austrian and Hungarian halves of the Empire. This attitude, reflecting the interests of agrarian circles, was changed under the influence of Frederick List's doctrines. By 1843 Kossuth and the majority of liberals were in favor of a protective Hungarian tariff line, linking the issue of Hungary's industrialization with the ideas of the country's economic *and* political independence from Austria.[37]

The maintenance of the fiction of an American consul at Pest[38] and the idea of establishing a British consulate were both aimed at stressing Hungary's right and effort to strengthen her ties with the West independently from and, if need be, in spite of her imperial obligations. Obviously, such a trend had political overtones, and the imperial authorities did not fail to recognize them. Frederick Kappel, consul designate of the United States, was an Austrian citizen who

[36] Blackwell to Gordon, July 9, 1843, *ibid.*, p. 144.

[37] See e.g., Blackwell's reports of June 23; 29; July 1; 4; 9; Aug. 5; 24, and Sept. 13, 1843, *ibid.*, pp. 137–49.

[38] Blackwell repeatedly mentioned "Mr. Kappell, the soidisant American Consul." In addition to his report of July 1, see also his reports of July 9 and Aug. 5, 1843, *ibid.*, pp. 144 and 147. In the latter, he wrote to Sir Robert Gordon: "Several persons connected with the mercantile classes, whom I have spoken with here regard the establishment of a British Consulate in this city as the necessary preliminary step for opening a trade between England and Hungary. Mr. Cappel [*sic*], the American Consul, is out of town. He is one of the most respectable merchants of the place but is said to be a very man. A friend of mine, who is well acquainted with him, tells me that he received his credentials, in due form, upwards of a year ago and is in daily expectation of receiving his exequatur, which, he assures his friends, the American Chargé d'Affaires at Vienna will not fail to obtain from the Austrian Government.—Meanwhile, a large Consular seal attached to a screw press is displayed in his Counting House as well as an American flag which he intends to hoist when he receives his exequatur!!!"—Incidentally, the necrology published upon the death of Kappel in the *Pesth-Ofner Zeitung* during the "absolutistic" Bach regime in 1855 mentioned the fact that "the Government of Washington had appointed him Consul to Pesth and Hungary." Wurzbach, *op. cit.*, X, 451.

had never visited America and whose loyalty to the Empire had never been questioned by Viennese officials. Thus, as Clay remarked, there were no objections to him personally. But his appointment involved an ideological and political issue which, far beyond the original intentions of American representatives in Austria or the State Department, could not but arouse the suspicion of Metternich and other imperial authorities.

If one examines the Austro-Hungarian documents relative to the Kappel affair, one cannot find the slightest reference to the personality of the American Consul designate.[39] On the other hand, the foreign and domestic political implications of the case were stated rather bluntly in the correspondence of the imperial authorities. This correspondence involved the Chancery of State (*Geheime Haus-Hof- und Staatskanzlei*), the Court Chamber (*Allgemeine Hofkammer*), the Hungarian Court Chancery (*Ungarische Hofkanzlei*), the Council of Lieutenancy (*Consilium Locumtenentiale*), and the Police Agency (*Polizei- und Bücher- Censur Hofstelle*). All of these Austrian and Hungarian authorities agreed that the American chargé's request for an exequatur for Kappel should be turned down.

Regarding Clay's note of November 7, 1840, the memorandum of the Chancery of State of November 22 referred to the Chancery's previous position expressing "manifold doubts" (*vielfache Bedenklichkeiten*) in connection with a proposal of the Porte to establish a Turkish Consulate at Semlin (*Zimony*). According to this position, the granting of the Turkish demand might serve as a basis for similar requests from other powers such as Russia, France, and Great Britain. For the same reasons, the memorandum suggested that "the proposal of the North American Government be rejected in a decent way" (*auf schickliche Art*).[40]

Baron Von Kübeck, president of the Court Chamber, sent the memorandum of the Chancery of State, with his approval,[41] to the

[39] The Austrian and Hungarian documents relative to the appointment of the first American consul to Pest are in the *Haus-, Hof- u. Staatsarchiv* in Vienna. I should like to express my gratitude to State Archivist Dr. Richard Blaas and his co-workers for having kindly provided me with microcopies of these documents and of Prince Metternich's two secret instructions referred to below.

[40] Nov. 22, 1840, ad 6801/P.P. *Abschrift Einer Note der geh. Haus-, Hof- u. Staatskanzlei an das Presidium der allgemeinen Hofkammer.*

[41] Dec. 8, 1840. Kübeck to Mailáth.

Hungarian chancellor, Count Anthony Mailáth. Mailáth asked for the opinion of the Hungarian domestic authorities,[42] submitting the answer of Archduke-Palatine Joseph, along with the rest of the documents, to Count Sedlnitzky, president of the Police Agency, on February 25, 1841.[43]

In his Latin answer[44] to Mailáth's German request, the Archduke-Palatine Joseph, president of the Hungarian Council of Lieutenancy expressed the view that the answer to the question of admitting an American consul to Pest must depend on "whether it is expedient to admit such consuls of foreign powers to non-maritime trading places?"[45] Agreeing with the imperial authorities as to the dangers of creating a precedent, the Archduke-Palatine concluded:

> ...the commercial relations between the United States of America and the City of Pest are hardly of such quality or quantity as would require the establishment of special diplomatic agents.[46]

On May 30, 1841, the chief of the Police Agency, Count Sedlnitzky, advised Metternich to turn down the American demand

> ...and to see to it also in the future that no consulates be established by any foreign power either in Hungary, Transylvania, or in the imperial-royal military frontier [district].[47]

Besides strongly supporting the suggestions of other authorities concerning the necessity of refusing the request of the U. S. Government, Sedlnitzky also advanced some interesting arguments reflecting upon the domestic political situation within the Austrian Empire. If it were not feasible, he reasoned, to eliminate the existing foreign

[42] Dec. 26, 1840. Mailáth to Archduke-Palatine Joseph.

[43] Feb. 25, 1841. Mailáth to the *Polizei- und Bücher- Censur- Hofstelle.* (2440/180)

[44] Feb. 11, 1841. No. 257. 2440/180 841. Archduke-Palatine Joseph to Mailáth.

[45] "...num in genere Consules exterarum [*sic*] Potentiarum ad emporia non maritima admittere expediat?"

[46] "...me quoque opinioni Principis Metternich et Praesidis Camerae Caesareo [*sic*] Regiae Unitae Aulicae in eo depromtae, ut ab admissione idmodi Consulum penitus supersedeatur, tanto magis adstipulari, quod praescindendo etiam ab usu hactenus obtinente, praebendaque per similem admissionem aliis quoque Potentiis formandarum aequalium praetensionum opportunitate, commerciales correlationes inter Status Americae foederatos et Civitatem Pestiensem intercedentes haud ejus indolis aut extensionis sint, quae constitutionem pecularium diplomaticorum Agentum exposcerent."

[47] May 30, 1841. Sedlnitzky to Metternich, p. 3.

consulates from Austrian seaports, then it would be highly desirable at least to bar them from the interior provinces. According to Sedlnitzky the consular agents, admitted to promote commercial relations, "almost always" behaved as political agents, arrogantly intriguing and interfering with affairs beyond their competency, and trying to claim diplomatic privileges and immunities for themselves. Therefore, the chief of the Austrian police stressed in particular the "disadvantages" of tolerating similar conditions in Hungary,

> where there are so many elements susceptible to agitation and temptation, and where, like in the Grandduchy of Transylvania and in the imperial-royal military frontier [district], there is no adequate police authority to control the machinations (*Antriebe*) of foreign agents, and where the necessary intervention can be exerted only through indirect means.[48]

Metternich was far from disagreeing with his police chief. The secret instructions which he sent to the Baron von Neumann, Austrian chargé d'affaires in London on July 22, 1842 (one year after the rejection of the American demand), are most revealing in this respect.[49]

[48] *Ibid.*, p. 2.—Cf. Clay's report of April 19, 1841.

[49] July 22, 1842. Metternich to Neumann, nos. 9 and 10.—The two instructions, especially no. 9, parts of which had been crossed out, rewritten, and repeatedly rephrased by Metternich, show that in the Chancellor's judgment the issue of admitting foreign consulates to Hungary was an extremely delicate and politically sensitive problem. The second of the instructions, marked "Secret," reads (in part) as follows:

> M[r] le Baron
>
> Je juge nécessaire d'ajouter à la dépêche précédente, pour votre connoissance personnelle et exclusive, que j'ai lieu de croire que [crossed out: le sens dans lequel] Sir Rob. Gordon écrit à son Cabinet sur l'affaire de la nomination d'un Consul Brit. à Pesth, dans un sens qui me prouve qu'il ne tient pas assez compte de la nature des circonstances dans lesquelles nous nous trouvons placés vis à vis de la hongrie [!]. Cet ambassadeur écoute les paroles d'une faction qui le pousse à fin d'obtenir [two lines crossed out] l'admission des consuls étrangers, dans la vûe de s'en faire [crossed out: un point d'appui pour] un moyen pour créer des embarras au Gouvernement . . .

The situation was the more embarrassing from the point of view of Metternich because Ambassador Sir Robert Gordon happened to be the younger brother of Lord Aberdeen, foreign secretary of Her Majesty's cabinet.—See July 14, 1850, Blackwell to Palmerston. "Anglo-Hungarian Docs.", *South Eastern Affairs,* IV, (1934), 214. Blackwell underestimated Metternich's determination to prevent the British from establishing a separate consulate at Pest. In 1850

Mentioning that "for a number of years" British commercial circles had been attracted by the possibility of developing direct trade relations with Hungary, Metternich absolutely refused to agree to the establishment of a British consulate in Hungary. He instructed the Baron Neumann to decline firmly any British initiative in this direction and went into rather elaborate detail concerning the potential embarrassment such an initiative might create. The Chancellor's lengthy arguments emphasizing the precarious conditions prevailing in Hungary need not be cited; the following reference to the Kappel case makes his point.

The granting of the British government's request to create a consulate in Pest, Metternich insisted,

> would immediately result in an analogous demand on behalf of France and other powers, until we would be unable to decline the requests *(démarches)* which, in the same sense, had already been attempted in favor of admitting a Consul of the United States of America. In view of the deplorable [crossed out: political] activity displayed by France [crossed out: French agents] in other countries with the direct aid of alleged commercial agents whose real task, however, is political, it is absolutely impossible for us to admit them to Hungary before things will be prepared there so that their presence should not cause any noteworthy trouble . . .[50]

Metternich's reference to the diplomatic step taken by the United

Blackwell, referring to Sir Robert Gordon's active role in promoting the cause of the consulate at Pest wrote:

> After some delay, Lord Aberdeen was induced—I believe chiefly through the exertions of Mr Macgregor and Lord Canning—to consent to the measure, and as the Board of Trade had recommended the establishment of a Consulate at Pesth, and as Sir Robert flattered himself that he had sufficient influence to induce Prince Metternich to grant an exequatur, I took my wife and son with me and left them at Gratz in the full expectation that ere many months had elapsed, we should be settled in the Hungarian Capital.

This passage deals with events that occurred in the autumn of 1843. But on September 1, 1842, five weeks after he had sent his instructions to Baron Neumann, Metternich wrote to Kübeck: "Die Idee der Anstellung eines englischen Consuls in Pesth habe ich nicht todtzuschlagen gebraucht, denn MacGregor ist vollkommen mit der Wichtigkeit der Sache einverstanden. *Sie Wird Nicht geschehen.*" [Italics added.] Max v. Kübeck, ed., *Metternich und Kübeck, Ein Briefwechsel* (1910), quoted in László Bártfai Szabó, *Adatok gróf Széchenyi István és Kora Történetéhez* (Contributions to the History of Count Stephen Széchenyi and His Era), I (Budapest: 1943), 416.

[50] July 22, 1842. Metternich to Neumann, no. 9.

States but a year and a half before was, of course, justified. The insertion of his remark between two rather outspoken attacks on France, as well as the allegation of political activities by French agents, can also be explained. Back in 1791–92, Viennese radicals, Freemasons, and *illuminati* had had a club called the American-French-German Union, members of which had decided to promote the ideas of the French and American constitutions. They took upon themselves the obligation "to defend with all their strength, in writing, verbally or even with the sword the present conditions in France and America against all despots."[51] As far as the older generation was concerned, French and American revolutionary ideals did not seem very different after the French wars and Napoleon's fall. To arch-conservatives, American political institutions were anathema, whereas young Széchenyi and the liberals looked upon them as the incarnation of political wisdom and a guaranty of true social progress. Thus the American image appealed to the supporters of reform while it frightened "reactionaries." But in either case, it was very real.

For the reformer Széchenyi, disciple of Franklin and admirer of Washington and Jefferson, America was an inspiring experiment.[52] For those imperial officials who prevented him from visiting the United States in the eighteen-twenties, the whole country seemed to be "subversive." Széchenyi divined the rising star of America in his formative years. In 1819 he put forward a "frontier-theory" of his own, comparing the foundation stones of the "Land of the Future" (*das werdende Land*) to those of the Roman Empire. He claimed that just as once civilization had come from Asia to Europe, in the same way culture and perfection were moving successively to America; in the Roman Empire, however, everything was on a small scale, whereas in the United States everything was being started in hitherto unknown proportions. Széchenyi, to be sure, was impressed with the ideas of the Americans as well as with the increasing power of the New World.[53] In 1829, he wrote:

[51] Kálmán Benda, ed., *A magyar jakobinus mozgalom iratai* (Documents of the Hungarian Jacobin Movement), I (Budapest: Akadémiai kiadó, 1957), Doc. 37, pp. 578–99. Concerning the masonic lodge called Franco-American Union, *ibid.*, vol. II, 36–37. Benda's three-volume work was published in the series *Fontes Historiae Hungaricae Aevi Recentioris.*

[52] Barany, *op. cit.*, pp. 182; 275–276; 572.

[53] *Ibid.*, pp. 179–181; 395. Cf. Széchenyi's letter to Pierre Du Ponceau, president of the American Philosophical Society, *Journal of Central European Affairs*, XX (Oct. 1960), 312–313.

> America is the country, where the Rights of Mankind are most equal, where the Constitution is best, — and I consider it as my duty, since I had devoted my life to such a noble task, to visit the fountain where from this essence of Justice floweth...[54]

This passage indicates that the American experiment had a great appeal to early nineteenth century liberal reformers in Central Europe well before the time of de Tocqueville.[55] Often underestimated today, this appeal was duly appreciated by the contemporary opponents of progress who feared it just as much as their American counterparts dreaded the infection of foreign envoys' "gold-spoon influence" in Washington.[56]

Besides the problem of creating a precedent, it was this impact of the American image upon the thinking of imperial officials that stood in the way of establishing an American consulate in Pest in 1841. This image had been there long before the arrival of the first United States minister in Vienna. Muhlenberg and other American officials sent to Central Europe in the *Vormärz* with the purpose of strengthening their country's commercial relations with the area were soon to realize "the prejudices evidently existing against us as Republicans." A decade later, in the wake of the revolutions of 1848 and Kossuth's visit to the United States, Americans were reminded, this time by Abraham Lincoln, that "...the American people cannot remain silent without justifying an inference against our continued devotion to the principles of our free institutions..."[57] But it took another century or so before the public at large in the United States was prepared to face the fact that in foreign relations, too, the "Land of the Future" was supposed to live up to the expectations of the oppressed and those desirous of freedom all over the world.

[54] *Lebens Regeln und das Rath eines treuen Freundes,* 1829, Annex I, 10a, *Diaries,* III, 486.

[55] Cf. Barany, *op. cit.*, pp. 321–322.

[56] Cf. fn. 15 in the present study.

[57] Roy P. Basler, *The Collected Works of Abraham Lincoln,* II (New Brunswick, N. J.: 1953), 115 ("Resolutions in Behalf of Hungarian Freedom," Jan. 9, 1852).

Papers of the Michigan Academy of Science, Arts, and Letters
Vol. XLVII, 1962 (1961 Meeting)

POLITICAL SCIENCE

RECENT EFFORTS TOWARD LEGAL REGULATION OF POLITICAL PARTIES IN WESTERN GERMANY

ROBERT W. MILLER*

Wayne State University

The West German Basic Law of 1949, one of the most important and interesting constitutional documents of the postwar period, has been the object of experts' attention ever since the Federal Republic's entry into the European states-system. The unique feature of this *Grundgesetz*, as far as the evolution of German political life is concerned, is contained in three brief paragraphs of Article 21 whereby political parties are constitutionalized for the first time in Germany's constitutional history.[1] A special place of prominence in the institutional and political make-up of the West German state is thus assigned to political parties by the concise wording of the article:

> (1) The political parties participate in the forming of the political will of the people. They may be freely formed. Their internal organization must conform to democratic principles. They must publicly account for the sources of their funds.
> (2) Parties which, by reason of their aims or the behavior of their adherents, seek to impair or destroy the free democratic basic order or to endanger the existence of the Federal Republic of Germany are unconstitutional. The Federal Constitutional Court decides on the question of constitutionality.
> (3) Details will be regulated by federal legislation.

For the purposes of this paper, the central constitutional issue relating to the public position of German political parties in terms

* For the use of various materials in the preparation of this study the author wishes to express his thanks to Professor James K. Pollock, Department of Political Science, University of Michigan, and to Dr. Bernhard Wegmann, Director of the Bundesrat office in Bonn.

[1] For a broad treatment of the constitutional position of political parties in the Bonn charter, see Carl J. Schneider, "Political Parties and the German Basic Law of 1949," *Western Political Quarterly* 10, No. 3 (September 1957), 527–540.

of their duties and obligations is to be found in the fourth sentence of the first paragraph of the article cited above.

It may be said with a good deal of certainty that the sad lessons of the last days of Weimar Germany stood out in the minds of members of the Parliamentary Council whose task it was to draft a basic law for the new West German state from the fall of 1948 to the spring of 1949. Mindful of the dangerous precedent of the 1930's when secret financial support had been accorded by powerful industrial and financial groups to the National Socialist party (NSDAP)[2] the Bonn parliamentarians insisted upon a constitutional requirement calling for the publication of the sources of party funds.[3] A good deal of significance may be attached to an attempt in an earlier draft of Article 21 to link the requirement of democratic internal order in parties with that of public financial accountability.[4] It is this aspect of the Basic Law that has given rise to a number of political controversies of the first magnitude. By early spring, 1960, the problem of party income could be counted among the most controversial issues to reach the floor of the lower house of parliament (*Bundestag*) in recent years. In its broader aspects, the controversy touches upon issues of fundamental constitutional significance, i.e., of giving legal expression to the letter and spirit of Article 21 in all its ramifications.[5] In more specific terms the controversy involves a struggle between the governmental party (Christian Democratic Union—CDU) and the chief opposition party (Social Democratic Party of Germany—SPD) over the question of party financing from extra-party sources.

[2] The fate of the Weimar Republic's parties is most ably dealt with in a recent comprehensive work, "Das Ende der Parteien 1933," *Aus Politik und Zeitgeschichte* (Suppl. to *Das Parlament*), B 23–24/60 (June 8 and 15, 1960); see also Karl Dietrich Bracher, *Die Aufloesung der Weimarer Republik* (Stuttgart and Duesseldorf: Institut fuer Politische Wissenschaften, 1957).

[3] Principal initiators of the move to incorporate such a provision in the Basic Law were Johannes Brockmann (Center party) and Georg-August Zinn (Social Democratic party). See *Stenographischer Bericht des Parlamentarischen Rates, Hauptausschuss, 57. Sitzung.*

[4] "The internal order of parties must conform to democratic principles and must be assured against undemocratic influences by publication of the sources of their funds."—*Drucksache des Parlamentarischen Rates Nr. 815.*

[5] For a thoughtful legalistic treatment of the constitutional stipulations of Article 21, see Hermann von Mangoldt and Friedrich Klein, *Das Bonner Grundgesetz,* Vol. I (Berlin and Frankfurt: Verlag Franz Vahlen, 1957), pp. 611–633.

THE PRACTICE OF FUND-RAISING IN POSTWAR CAMPAIGNS

A systematic discussion of the events leading to the unsavory situation in the financial field of political party activity in Western Germany would have to be prefaced, it seems, by a brief account of the various methods of raising party funds that have prevailed thus far. Such methods include membership contributions from various sources and party fund-raising activities. In addition, sizable revenues have accrued to the nonsocialist parties through so-called "sponsors" and civic associations, while the Socialist party has derived extensive sums of money from party-owned enterprises. The following paragraphs will elaborate upon these matters.[6]

An examination of the question of party income in the Federal Republic will reveal almost at once two categories of parties: (*a*) those which rely for the major share of their regular revenue on fixed dues and levies collected from party members as an obligation of membership; (*b*) those which show a decidedly heavy reliance on extensive donations of all kinds. Deriving a large portion of their income from regular subscriptions of their own members, the SPD [and the Refugee Party (BHE)] show characteristics of the first category. The nonsocialist parties, CDU, FDP, and DP,[7] on the other hand, depend primarily on other sources of income and are thus typical of the second category.

Insofar as the SPD has been the only major party in the habit of presenting annual reports on its financial structure, the figures published in the SPD Yearbook for 1956–57 (*Jahrbuch der Sozialdemokratischen Partei Deutschlands 1956/57*) can be taken as a fairly reliable index of the party's income. The published figures for 1957, a major campaign year, show a total of SPD receipts in the amount of DM 6,427,000.[8] Of this income, DM 2,258,000 (36% of the total) represents a combined revenue derived from party membership fees

[6] Most of the data concerning the revenues of specific parties are based on the excellent and thorough study of the German party system by U. W. Kitzinger, *German Electoral Politics* (Oxford: Oxford University Press, 1960), Chapter X.

[7] The abbreviation FDP stands for *Freie Demokratische Partei* (Free Democratic Party), DP for *Deutsche Partei* (German Party).

[8] *Jahrbuch der Sozialdemokratischen Partei Deutschlands 1956/1957* (Bonn and Hanover: Neuer Vorwärts, 1958), pp. 253–254 cited in Kitzinger, *op. cit.*, p. 205. (The official rate of exchange was then DM 4.20 = $1.00.)

(1,245,000 DM), dues of Bundestag members (330,000 DM), collections and miscellaneous income (178,000 DM), and so-called "contributions for public relations" (from associated enterprises) (505,000 DM). The bulk of funds which flowed into the party's coffers, however, had been raised from so-called "election contributions" and amounted to the impressive figure of DM 4,170,000 (64% of SPD income).

In view of the size of the two subcategories listed as "election contributions" and "contributions for public relations," it may be well to examine the nature of these sources. The magnitude of the "election contributions" item apparently reflected a massive effort of the SPD toward collecting monies specifically for the 1957 election campaign. This was accomplished by the sale of campaign stamps, by payments into publicly advertised "campaign accounts", and by payments made on blank "campaign checks" (with the party designated as the payee). At the same time, it may be assumed that large and medium-sized amounts came from business firms and legal persons of various types.[9] The other subcategory referred to, "contributions for public relations," may serve as an indicator of the SPD's prominent status as a large-scale entrepreneur. The extensive interests of the SPD in various enterprises and firms are most clearly reflected in the party's newspaper publishing activities. The SPD has retained ownership of most newspapers licensed in the British Zone after 1945 and was instrumental in the establishment of a servicing company called *Konzentration G.m.b.H.*, which maintains its own newsprint importing business and suppliers. The *Konzentration* company combines newspaper companies from all sections of the land, having spread out into areas beyond the former British Zone; with some 15,000 employees among its member firms, it accounted for about 10 per cent of newspaper circulation in the Federal Republic.[10]

Even though the extent of SPD's participation in the field of newspaper publishing and its exaction of stipulated sums from associated enterprises[11] may be subject to criticism for various reasons—*Kon-*

[9] Breweries and cigarette manufacturers, for example, had been known to contribute to the SPD on the theory that a great proportion of their customers were, after all, followers of that party.

[10] Kitzinger, *op. cit.*, pp. 219–220. The author estimates that *Konzentration G.m.b.H.* may have provided the equivalent of $112,000 for SPD purposes.

[11] Under SPD party statutes, such enterprises are obliged to contribute an annual sum fixed by the party's Executive.

zentration G.m.b.H. does perhaps represent a miniature "newspaper empire" controlled by the SPD—there remains the fact that the party has striven to set an example in providing at least a modicum of open policy in its fiscal dealings. Moreover, the specific origin of its funds derived from economic activities is rather well ascertainable through the examination of tax reports which all party-owned or party-controlled establishments must file for taxation purposes.

The provision of the Basic Law calling for an account of the origin (though not of the destination) of party funds has largely remained a dead letter as far as the CDU has been concerned. The financial aspects of its activities—as well as those of the FDP—have remained, for the most part, unclear and hidden from public scrutiny if not secret.[12] According to the best available estimates, the CDU raised close to 3 million DM from membership dues and subscriptions exacted from Bundestag and Landtag members and other officeholders; an additional 1,760,000 DM came from donations and payments of various kinds on a federal or a land level. The size of CDU's 1957 campaign expenditures, estimated at 35 to 38 million DM,[13] gives rise to the impression that a number of other extremely lucrative sources of income must have been at the party's disposal.

Aside from the funds raised by the standard devices indicated above, the CDU has operated a highly profitable enterprise in the form of a weekly economic information broadsheet published under the masthead *Wirtschaftsbild.* This periodical has been supplied to firms at a price greatly exceeding its publication cost or, for that matter, its usefulness to the subscriber. The value of the publication to the customer presumably lay in the confidential nature of its contents, which would yield valuable tips and possible special economic benefits from advance information on the allocation of contracts by public authorities, on governmental purchasing policies, etc. Much to the disappointment of its prosperous subscribers, the *Wirtschaftsbild* failed to live up to its hinted promises. Varying in price from DM 1175 ($280) annually for Edition A to DM 588 ($140) for the regular edition, the publication brought in sizable amounts of money for the

[12] One of the most searching treatments of this problem may be found in Arnold J. Heidenheimer, "German Party Finance: the CDU," *American Political Science Review,* Vol. LI, No. 2 (June 1957), 369–385.

[13] Kitzinger, *op. cit.,* p. 202. (The doubtless exaggerated estimate of the SPD placing the total of CDU expenditures at DM 117 million may be readily dismissed as a propaganda stunt.)

CDU treasury, particularly when annual payments for bulk purchases reached the equivalent of $14,000 or even $22,400.[14]

The most decisive source of the CDU's income (and of the smaller nonsocialist parties' as well) has been the political funds maintained by large West German industrial and commercial concerns. Such political funds fell into two categories: the *Foerderergesellschaften* (Sponsors' Associations), an older, well-established type, and the *Staatsbuergerliche Vereinigungen* (Civic Associations), representing a later variety. Both types of organizations engaged in collecting money from businesses in a manner that secured tax relief for the donor, for after 1952, contributions for political (*staatspolitische*) purposes (including subsidies to political parties) were tax-deductible.[15]

Formally operating as trade associations, the Sponsors' Associations acted as intermediaries between the parties and industrial and commercial firms, collecting their revenue by levies in proportion to payroll (for an industry) or turnover (for a commercial establishment). The Sponsors' Associations have been known to distribute money to several nonsocialist parties simultaneously, although member firms could designate the party to receive such funds. Under the tax structure prior to 1955[16] the Sponsors' Associations were free to allocate only a part (ca. 25% to 33%) of tax-free funds for political purposes, since only a "subordinate proportion" of the funds of a trade or professional association could be passed on to political parties.[17] After covering their own operational costs, they used the remaining income for more general purposes, such as work in public relations in behalf of free enterprise, information services, and political and economic lectures. They represented, in effect, a common front of industry and business, performing the basic political task of defending their sponsors' vital economic interests.

[14] *Ibid.*, pp. 206–7. E.g., the then publicly controlled Volkswagen automobile factory subsidized the CDU in the amount of $14,000 by subscribing to 100 copies of the *Wirtschaftsbild*.

[15] In accordance with an opinion of the Federal Fiscal Court of May 1952, reprinted in *Bundessteuerblatt III*, 5 September, 1952, pp. 228–232.

[16] The Income Tax Law was amended in 1954; the rules governing tax deductions for political contributions were set forth in the Implementing Order of 1955.

[17] The term "subordinate proportion" was used in the Federal Fiscal Court's opinion cited above in note 15.

Under the amended income-tax law of 1954, a new political fund device could be developed, inasmuch as paragraph 49(2) of the Implementing Order to the tax law spoke of tax relief on contributions given to legal persons whose activities were directed exclusively toward political purposes. At the same time, such financial grants to legal persons could be transmitted to a party *in toto*. It was this change in income-tax regulations that brought into existence a central political fund, the *Staatsbuergerliche Vereinigung* (Civic Association) operated from (though not formally by) the headquarters of the Federation of German Industry (*Bundesverband der Deutschen Industrie*) in Cologne. The Civic Association soon came to represent a type of personal union between ordinary industrial and commercial associations on the one hand and the political funds on the other, for the Cologne group and the political funds in the Länder were managed for the most part by the same men who were simultaneously officers of the Federation of German Industry and of Land industrial and commercial associations. With a radical decrease in the activities of the various Sponsors' Associations, which by 1957 had confined their field of operations to the Land level, the Civic Association came to occupy a central position in the disbursement of tax-free political funds to nonsocialist parties. It would be difficult indeed to estimate with reliable accuracy the volume at which this kind of subsidy flowed into the middle-class parties' coffers. While estimates of these party incomes may be relatively low—the CDU, *e.g.*, is believed to have derived some DM 5.17 million or 40 per cent of the cost of its central campaign ($1,232,000) from the Civic Association in Cologne[18]—the figures must command anybody's respect. All in all, the principal four nonsocialist parties must have received over one-half of the combined total of their campaign funds from the Cologne source and the Sponsors' Associations.

The Defeat of the Tax-Exemption Clause

The tax privileges granted on funds disbursed to political parties were, of course, a thorn in the side of the Social Democrats. Moreover, loud rumbling came from small splinter parties which were eager to receive subsidies, but could not qualify under the restriction imposed on them by paragraph 49 (1) of the 1955 Imple-

[18] The FDP received the equivalent of $336,000, the DP $840,000, according to Kitzinger, *op. cit.*, pp. 214–215.

menting Order to the amended income tax law (*Gesetz zur Neuordnung von Steuern, 1954*). The critical subsection (1) provided that only a political party with at least one deputy in the federal or the Land representative assembly was eligible to receive such funds.

The first crack in the wall of tax privileges which had been erected for the benefit of the nonsocialist parties came in February 1957 when the Federal Constitutional Court struck down subsection (1) of paragraph 49 of the Implementing Order. Agreeing with the contention of the plaintiff Dr. Gustav Heinemann, suing in the name of his small All-German People's Party (GVP), the high Court held that the rule in question violated the principle of equality before the law and equality of chance implied in Articles 3 and 21 of the Basic Law.[19] In one passage of its opinion the Court also gave expression to its doubts as to the constitutionality of the tax-exemption principle as a whole: "... constitutional grounds of objection might perhaps be argued against the application of the tax relief provisions of the income tax law to political parties."[20]

Encouraged by these developments, the SPD (through Minister-President Georg-August Zinn of the SPD-governed Land of Hesse) brought an appeal before the Federal Constitutional Court in June 1957, thus launching a frontal attack on the entire tax-exemption system relative to expenditures made for political party purposes. A year later, in June 1958, the 1954 tax law amendment was declared unconstitutional[21] and Minister-President Zinn could take full credit for his role in "frustrating," as *Der Spiegel* put it, "an attempt to bolster political parties economically through governmental tax policies."[22]

The Proposed Party Law (Parteiengesetz) of 1959

By 1958–59, the need for bringing new order into the confused constitutional question of party-state relationships reached its high point. The problem turned, above all, around the desirability of

[19] *Entscheidungen des Bundesverfassungsgerichts* (Decisions of the Federal Constitutional Court), 1 BvG, 241/56. See also *Der Spiegel,* Vol. 11, No. 30 (24 July, 1957), 12–13.

[20] Decision of the Federal Constitutional Court, 1 BvG, 241/56.

[21] By decision of the Federal Constitutional Court, 2 BvG 1/57.

[22] For the history of this case see *Der Spiegel,* Vol. 12, No. 27 (2 July, 1958), 17–20.

establishing an equitable basis for supervising the flow of political funds to and from party treasuries. By the same token, the obligations and the legal status of powerful financial backers of the various parties needed legislative clarification. It was under these circumstances that the government, in December 1959, submitted its draft law concerning political parties to the Bundestag.[23] It came after repeated urging from the FDP and the SPD[24] and after a protracted phase of preparatory work carried out in the Ministry of Interior. The genealogy of the law itself is an extremely interesting one and merits a few comments here.

The Federal Government's concern with a party law dated far back to the early years of the Bonn Republic. Preparatory work for such legislation had been initiated in 1950 and the first draft of a party law was dealt with by the Cabinet a year later. It was rejected as defective, and a similar fate befell the second draft in 1952. A pause in the drafting process ensued between 1953–55, owing to the enormous complexities of this subject and the pressing need for dealing with a reform of the electoral law for the forthcoming 1957 election.[25] At the end of 1955, the Minister of Interior Dr. Gerhard Schröder appointed a commission to study the entire range of problems related to the political party system and to questions of regulation and financing. Composed of a panel of highly competent men (such as Professor Theodor Eschenburg, Dr. Wilhelm Grewe, and Dr. Theodor Maunz), the commission produced by August 1957 a lengthy report under the title *Rechtliche Ordnung des Parteiwesens*[26] (Legal Regulation of the Party System). The report was hailed by politicians and scholars alike as a great contribution toward the clarification of that specific sector of German political life. The Schröder report served as a basis for the new bill, which was approved by the Cabinet in May 1959.

[23] *Deutscher Bundestag, 3. Wahlperiode, Drucksache 1509.*

[24] See *Das Parlament,* 6 August, 1958, p. 3 and 10 June, 1959, p. 7.

[25] For an authoritative treatment of the 1956 electoral law in German, see Karl-Heinz Seifert, *Das Bundeswahlgesetz* (Berlin and Frankfurt: F. Vahlen Verlag, 1957); the most outstanding analysis of the West German electoral law prior to 1956 may be found in James K. Pollock, ed., *German Democracy at Work* (Ann Arbor, Mich.: University of Michigan Press, 1955), and J. K. Pollock, "The West German Electoral Law of 1953," *American Political Science Review,* Vol. XLIX, No. 1 (March 1955), 107–130.

[26] Frankfurt and Berlin: Alfred Metzner Verlag, 1958 (2d ed.).

The draft of the *Parteiengesetz,* having received a few inconsequential alterations in the Bundesrat,[27] finally reached the floor of the lower house on February 19, 1960. In his introductory remarks opening the Bundestag debate, Interior Minister Schröder referred to the new party law as a "constitutional necessity" deriving from the Basic Law's twenty-first article and stressed the Government's firm belief that the proposed law constituted a "bold step into a new unexplored territory, unequaled anywhere in the world."[28]

When the first speaker of the SPD rose to deliver his party's denunciations of the bill, however, it became evident at once that the Government's draft would encounter a stiff opposition. The SPD spokesman happened to be G. Heinemann (who had switched from the GVP to the SPD); he denounced the proposed law as "a deliberate falsification of the intent of Article 21" and pointed to the central principle of this constitutional provision which had been designed "to prevent the infiltration of political parties by anonymous financing."[29]

Unfortunately, the complaints of the SPD were not altogether unjustified. The most objectionable features of the bill, as far as a neutral observer may be able to judge, lie in two fundamental elements of the draft: (1) the questionable attempt to draw a distinction between a party sister organization and an "extra-party" group; (2) the division of party contributions into only two categories, members' contributions and outside contributions, without requiring specific identification of the source and the amount.

In their formulation of paragraphs 5 and 6 of the bill, the Interior Ministry's drafters resorted to a subtle semantic device by distinguishing two varieties of party-affiliated organizations: *Sonderorganisationen* (special organizations) and *Nebenorganisationen* ("collateral" organizations). The former type, which under the law would include "associations of persons bound by decisions of regular party organs" and "independently operated enterprises and facilities which are organizationally, financially and economically integral

[27] See *Drucksache 1509* (fully cited above, n. 23), *Anlage 2, Stellungnahme des Bundesrats,* pp. 39–42. The Government refused to accede, however, to most of the points raised in the Bundesrat's written opinion; see *ibid., Anlage 3,* pp. 43–44.

[28] *Das Parlament,* 9 March, 1960, p. 6.

[29] *Ibid.,* p. 10.

components of a party," would be subject to the requirement of presenting public accounts of the sources of their revenues (par. 5). The "collateral" organizations, on the other hand, were not to be under any legal obligation to list the origin of their revenues, since they were not considered to be component parts of a party (par. 6). "Collateral" organizations of a party included "associations of persons, business enterprises and other establishments" whose objective is to "promote the political principles and aims of a party, to sponsor membership drives and to lend the party their support" without constituting organic parts of the party itself. Quite clearly, the SPD's sister organizations as well as enterprises and firms would fall under the heading of *Sonderorganisationen.* The opposition's anger in the Bundestag was aroused perhaps much more by the stipulations of paragraph 6, whereby the aforementioned Sponsors' Association and particularly the Civic Association in Cologne would not be subject to public scrutiny. It was this issue that provided the best ammunition for the opposition party to cause embarrassment to the CDU.

The second objection to the bill concerned its failure to go beyond the simple division of contributions into outside donations and members' subscriptions. It would have seemed desirable to have made legislative provision for compulsory listing of contributors' names and of the appropriate amounts, so that contributions to parties could become the object of public scrutiny, especially in cases where large amounts might be involved. This the government was unwilling to do and thus exposed itself to serious criticism.[30]

A further point to be raised in this context—it was not brought forward by any of the parties in the Bundestag debate for obvious reasons—is the *Parteiengesetz's* conspicuous omission of any provision requiring a public accounting of the *destination* of political funds. It may be safely assumed, however, that on this score all the parties would close ranks in opposing any type of legislation.

Conclusion

There can be little doubt that the confused picture of German party finance is in dire need of regulation. The problem affects a whole range of intricate issues and interests, it involves a complex web of relationships between the State, parties, interest groups, and the individual citizen. In light of the forthcoming parliamentary

[30] See *Der Spiegel,* Vol. 13, No. 24 (10 June, 1959), pp. 28–31.

election in September of this year, however, there is very little likelihood that the bill will soon emerge from oblivion; and least of all will the Government be inclined to press for a change in the status quo before the election returns are in. In the meantime the juggling of huge sums will go on as before.[31]

[31] At the time of this writing, the bill was still back in Bundestag committee No. 6 (Interior Affairs); see *Bundesanzeiger* for 6 February, 1961, p. 5.

Papers of the Michigan Academy of Science, Arts, and Letters
Vol. XLVII, 1962 (1961 Meeting)

POLITICS IN MICHIGAN: THE CURSE OF PARTY RESPONSIBILITY

NORMAN C. THOMAS
The University of Michigan

The American party system possesses certain identifiable characteristics which distinguish it from party systems in most other democracies.[1] The most obvious of these characteristics is, of course, the overwhelming predominance of the two major parties. A second major aspect of our party system is the absence of any ideological or programmatic orientation by either of the major parties. This situation is an outgrowth, at least partially, of the two-party system and of American federalism. American parties must make a wide appeal to differing social, economic, and regional interests if they are to be victorious in national elections. Consequently each party represents a coalition of divergent interests. On the national scene, the Republicans include within their ranks conservatives such as Senator Goldwater, moderates such as Richard Nixon, and liberals such as Governor Rockefeller. The Democrats, on their part, present an equally broad spectrum, running from Senator Byrd on the right through moderates such as Senator Mansfield to liberals such as Senator Humphrey. The consequences of this situation are well known. There exists a broad area of overlapping interests between the two major parties. There is a consensus regarding the goals toward which our society and its government should move.

A third major attribute of American parties involves a decentralization of authority in party organizations. Our party organizations are generally strongest at the grass roots. They tend to weaken and lose effectiveness as the national level is approached. Most state party organizations are battlegrounds for struggles between competing party factions, while the national party organizations are merely loose confederations of state and local organizations.

Closely related to the lack of centralized authority in party organizations is the absence of party discipline within American leg-

[1] See Clinton Rossiter, *Parties and Politics in America,* Ithaca, Cornell University Press, 1960, Ch. 1.

islatures. Most members of our legislative bodies vote with their respective parties on matters of organization and procedure and then proceed independently with indifferent regard to the official position of the party. This is especially true in the U. S. Congress. In performing their representative functions, American legislators are more responsive to the demands of special interests than to the program of the party leadership. Still another characteristic of American parties is the high degree to which interest groups have tended to align themselves with one or the other of the major parties.

The American party system performs two major functions in the political process. First, and perhaps most importantly, it serves as a vehicle of compromise. Prior to the conduct of election campaigns, compromise is effected within the framework of each major party. Strong attachment to principles and programs is reduced or ignored in the effort to piece together a majority coalition that will carry the day for the party at the polls. Between elections the two major parties attempt, within the governmental framework, to work out compromises and to reach an accommodation on the important issues of public policy. Political parties provide, then, a major means of reducing conflict and settling social and economic differences in a peaceful manner. Rigid, doctrinaire policy stances are generally shunted aside in favor of a willingness to seek out policies which can serve as the basis of accommodation. Neither party expects to win a total victory in the political process. Nor would the American people welcome or countenance such a definitive result. Indeed, one analyst of American politics suggests that moderation, compromise, and an ideological stalemate are in fact what the American people earnestly desire.[2] The conciliatory or compromising function of American political parties constitutes a major contribution to the effective operation of democratic processes in our complex federal system.[3]

In addition to serving as a vehicle of compromise, the American party system also provides a means by which the people, through collective action, can secure control of the government and influence

[2] Samuel Lubell, *The Future of American Politics,* New York, Harper and Brothers, 1952.

[3] See Herbert Agar, *The Price of Union,* Boston, Houghton Mifflin Co., 1950, which stresses the historical role of American parties as effective promoters of compromise; and E. Pendleton Herring, *The Politics of Democracy,* New York, Rinehart & Co., 1940.

the course of public policy. The two major parties each represent coalitions of interests that attempt to assume responsibility for the direction of the affairs of state. Their electoral activities are directed toward fulfillment of this function. Professor V. O. Key asserts that effective competition between the two parties is an important if not essential means of protecting liberty and of assuring responsiveness in government.[4]

These two major functions of American political parties can serve as a basis for a discussion and evaluation of their performance, both at the national level and in the states. Several pertinent questions arise. To what degree does the party system under examination serve as a vehicle of compromise? To what extent is party competition present? Do the parties provide the voters with a clear and meaningful choice between alternative candidates and policies? It might also prove useful to ask whether or not these criteria provide irreconcilable standards. If the standards are contradictory, which function ought to take precedence in a scheme of democratic values? The party system in the state of Michigan will be examined on the basis of these criteria.

Most academic criticism and evaluation of American party systems, at the national and state levels, center on party competition and the issue of party responsibility. American party politics, it is charged, elevates compromise and independence to the detriment of accountability and positive governmental action. American parties do not offer the voters meaningful choices. The committed voter has no way to express himself at the polls. Once elected, American parties do not carry out their platforms. Instead they spread their time compromising and procrastinating. This line of criticism culminated in the 1950 report of the Committee on Political Parties of the American Political Science Association. The committee charged that American parties were incapable of organizing their "members in the legislative and executive branches into a government held together and guided by the party program."[5] Under these conditions, said the committee, party responsibility vanishes. The committee recom-

[4] V. O. Key, Jr., *American State Politics: An Introduction,* New York, Knopf, 1956, pp. 11–12.

[5] Report of the Committee on Political Parties of the American Political Science Association, *Toward a More Responsible Two-Party System,* New York, Rinehart & Co., 1950. See also, E. E. Schattschneider, *Party Government,* Rinehart & Co., 1942.

mended that the party system be overhauled in order to provide a greater degree of centralization and efficiency in party machinery, more meaningful policy alternatives, and more vigorous and responsible execution of party programs. The committee particularly sought to increase the internal cohesion of the parties in Congress.

Although the committee's report has been the subject of much debate,[6] neither Congress nor the parties have moved very far toward implementing its proposals. Party discipline and responsibility, as visualized in the report, are as far removed from national politics in 1961 as they were ten years ago. The question of whether more "responsible" political parties, in the committee's sense of the term, would be as desirable in practice as in theory remains unanswered. With regard to our national parties the question is largely academic, for it is highly doubtful that any action to change the party system will be taken in the foreseeable future.

The party system in the State of Michigan, however, possesses certain qualities that meet the requirements of the committee report. Michigan's parties have developed efficient organizations. In the years since 1948 the Michigan Democratic party, under the leadership of former Governor G. Mennen Williams and former Chairman Neil Staebler of the State Central Committee, built an organization that is the envy of state chairmen throughout the nation. The Michigan Republican organization, having been torn by factional struggles in the mid-1950's, is not quite as smoothly running a machine as its Democratic counterpart. However, Lawrence Lindemer, former chairman of the Republican State Central Committee, succeeded in revamping the party so that in 1960 it narrowly missed recapturing the governorship which the Democrats had held since 1948. The Republican organization is also strengthened by the money-raising prowess of the Republican State Finance Committee.[7] Party organi-

[6] See, for example, Julius Turner, "Responsible Parties: A Dissent from the Floor," *American Political Science Review,* vol. 45 (1951), pp. 143–152; Austin Ranney, "Toward a More Responsible Two-Party System," *American Political Science Review,* vol. 45 (1951), pp. 488–499; and, J. Roland Pennock, "Responsiveness, Responsibility, and Majority Rule," *American Political Science Review,* vol. 46 (1952), pp. 790–807.

[7] For an excellent analysis of Michigan campaign financing in the 1956 election see John P. White and John R. Owens, *Parties, Group Interests and Campaign Finance: Michigan '56,* Citizens' Research Foundation, Princeton, 1960.

zation in Michigan is vigorous and active. It stands in marked contrast to the atrophied organizations found in so many states.[8]

In addition to possessing strong, well-disciplined organizations, Michigan's parties offer the voters of the state clearly distinct alternatives, in terms of candidates and policies. In practically no other state is the ideological distance between the two major parties so well marked. The differences between the two parties are popularly viewed as the reflection of a titanic struggle between the automobile manufacturers on the one hand and organized labor in the auto industry on the other. This popular conception of Michigan politics is demonstrated in Allen Drury's best-selling political novel, *Advise and Consent.* Included among Drury's characters are two fictional senators from Michigan who are of opposite parties. Each of these senators is keenly aware of the forces that dominate political life in Michigan:

> The trouble with the President of General Motors, in the opinion of Roy B. Mulholland, was that he thought he owned the Senators from Michigan, or at least the junior Senator from Michigan, namely Roy B. Mulholland.[9]

And:

> The trouble with the president of the United Auto Workers, in the opinion of Bob Munson, was that he thought he owned the Senators from Michigan, or at least the senior Senator from Michigan, namely Bob Munson.[10]

Even though sophisticated students of Michigan politics are well aware that the Republican party is more than a political subsidiary of the General Motors Corporation and that the Democratic party is not merely an adjunct of the United Automobile Workers, considerable significance must be attached to this popular image. It indicates clearly that Michigan's political parties are a great distance apart.

Generally the Democratic party favors increasing the extent and scope of services provided by the state government. The Republican party advocates programs calling for a much lower level of state services. The parties differ sharply regarding taxes. The Democrats

[8] See Key, *op. cit., passim.*

[9] Allen Drury, *Advise and Consent,* Garden City, Doubleday & Co., 1959, p. 46. Professor John P. White of The University of Michigan cited these passages in his study of Michigan campaign financing, fn. 8 above.

[10] *Ibid.,* p. 54.

have long championed personal and corporate income taxes as the primary source of new revenue for the state. The Republicans have favored increased sales taxation and so-called 'nuisance" taxes on alcohol, tobacco, and certain business transactions. The philosophy of the Michigan Democratic party is dominated by New and Fair Deal liberalism. The Michigan Republicans present a mixture of cautious conservatism and nineteenth century laissez-faire economics. The ideological polarization of Michigan's political parties is a well-known fact and is not surprising. At the conclusion of their study on campaign finance in Michigan, Professors White and Owens state that:

> In the highly charged ideological climate of Michigan politics, the differences between the major parties are more pronounced than in most states. Economic interests divide fairly clearly along political lines, and this is reflected strongly in the financial base of both the Republican and Democratic parties.[11]

Certainly no recent political writer in Michigan has been moved to charge that the state's two major parties offer the voters little more than a choice between Tweedle-dum and Tweedle-dee. Perhaps the ideological distance between Michigan's highly programmatic parties is an inevitable aspect of the state's political life. The major interests represented by the two parties—organized labor and major manufacturing—are sharply divergent. The parties themselves are not loose coalitions of a wide variety of economic and social interests. Rather, they are closely knit, well-organized units whose component interests need not compromise among themselves before entering the legislative struggle.

The parties are well organized in the Michigan Legislature. The party caucus dominates roll call votes in the State Senate to a degree that is almost unparalleled in American politics. The situation in the House of Representatives is almost as rigid. There, however, individual members do upon occasion break away from the party caucus and vote independently. Although neither party requires that its legislative members adhere to the vote of the party caucus, party regularity is the rule rather than the exception. Unlike the situation which exists in Congress, party position is of vital importance in the Michigan Legislature. Roll call votes on which the party caucuses have reached prior decisions are party-line votes with a minimum

[11] White and Owens, *op. cit.*, p. 34.

amount of deviation. Most legislators are willing to follow their party leaders, even though they may personally disagree with the policy course set by the leadership. The action of several Republican members of the State Senate during the tax crisis of 1959 dramatically illustrates the high degree of internal cohesion among legislative parties in Michigan. Certain Republican senators spoke out publicly in favor of a compromise involving a flat rate income tax or a payroll tax. These same legislators, however, voted against such taxes after the Republican caucus in the Senate declared itself in opposition to any compromise. In the Michigan Legislature the parties are highly disciplined and well integrated. Their leadership is effective. They support sharply contrasting policies.

It appears, then, that Michigan's party system possesses many of those characteristics which the American Political Science Association's Committee on Political Parties maintained would produce public officials and party leaders who are highly responsive to the people and a governmental operation which is the epitome of efficiency. Michigan, by all odds, should have the most "responsible" two-party system in the nation. The stormy record of Michigan's political warfare over the past few years indicates that something has gone wrong. Instead of displaying a smoothly running state government, Michigan has presented the nation with a picture of bitter party strife, financial chaos, and governmental stalemate.

The state's political climate during the past several years has been charged with acrimony. The Democrats, holders of the governorship since 1948 and winners in all major statewide contests since 1954, have been frustrated in their efforts to enact their liberal programs into statute by Republican controlled legislatures. They feel that they have been kept from their hard-won birthright as the majority party of the 1950's by a constitutional gerrymander. The Republicans have watched while their once omnipotent position as Michigan's only political force was being eroded away by an onrushing Democratic flood, seemingly composed of hordes of union members. Like the Federalists of old who retreated to the stronghold of the judiciary, the Republicans have entrenched themselves in the Legislature. In such an atmosphere, compromise and accommodation have been almost impossible to secure. Extremists have prevailed in the councils of each party and "rule or ruin" has been the watchword.

The animosity between the parties came to a boil during the

financial crisis of 1959. The recession of 1958 caused the state's sales tax revenues to fall far short of the estimated amount. Additional revenues were urgently needed if the state was to meet its obligations and if the current level of state services was to be maintained. The Democrats, under the leadership of Governor Williams, proposed the adoption of a personal income tax. Republican leaders countered by advocating an increase in the sales tax from 3 to 4 per cent. Neither side would yield. The Democrats refused to cooperate with Republicans in placing on the ballot in the spring 1959 general election a referendum proposal increasing the constitutional limitation on the sales tax from 3 to 4 per cent. This delayed any further referendum action until November 1960. The Republicans subsequently refused to permit any form of income tax legislation to pass through the Legislature. The Republican contingent in the State Senate seemed more intent on doing battle with Governor Williams than on solving the state's financial problems. Meanwhile the state received unfavorable nation-wide publicity. State employees suffered a "payless payday," the Administrative Board held biweekly sessions at which the General Fund was drained, and the fight raged on.

After months of struggle, a temporary compromise was effected in December 1959. Assorted "nuisance" taxes amounting to approximately $65 million, an amount far short of the state's financial needs, were adopted. A referendum proposal to increase the sales tax was placed on the November 1960 ballot. But the state had acquired a monumental deficit, essential services had been neglected, the capital expansion program had come to a halt, and Michigan's national reputation had suffered grievous damage that will take years to repair.

It is difficult to ascertain the extent to which the recent political situation in Michigan is due to the party system. Certainly the parties, with their efficient organizations, their marked ideological polarity, their strong ties with and responsiveness to powerful economic interests, and their highly disciplined legislative contingents, could not be expected to produce a politics of compromise and mutual accommodation. Perhaps it is foolish to have expected them to have behaved other than as they did.

It is equally certain that there are other factors which contributed significantly to the development of the recent political stalemate.

The governmental machinery established in the state constitution played a vital role. The Constitution provides for a system of legislation apportionment that overemphasizes the rural areas. Since Democratic strength is concentrated in the metropolitan Detroit area, the system of legislative apportionment has favored the Republicans and increased Democratic frustrations. The inequities in apportionment are especially severe in the State Senate. Representation in the State Senate is based primarily upon geographical areas. The thirty-four senatorial districts can be changed only by constitutional amendment. The districts are arranged so that the Republicans comfortably control the Senate. For example, in the 1958 election the twelve Democratic senators who were elected received 714,628 votes. The twenty-two victorious Republicans received 668,986 votes.[12] It was the Republican majority in the Senate, rather than in the House, that fought Democratic Governor Williams to a standstill. It was a Republican state Senator who boasted that he "torpedoed the income tax." The Constitution also gives relatively little power to the chief executive, thus making him vulnerable to attack by hostile legislators.

A second factor contributing to the development of political stalemate can be found in the economic geography of Michigan. The industrial concentration in Southeastern Michigan has meant that Michigan politics often turns on the issue of "outstate" vs. Wayne County (Detroit). The fact that Wayne County, with 35 per cent of the state's population, is heavily Democratic has tended to increase the already intense party competition in the state. The party system alone, then, is not to blame for Michigan's stormy politics.

Michigan's party system has failed in one of its functions and succeeded all too well in the other. It obviously has not served as a vehicle of compromise. On the other hand party competition has been extremely intense. The political struggle has become so fiercely competitive that defeat has become intolerable for either side. The consequence has been political stalemate and governmental paralysis.

The Michigan experience suggests several thoughts concerning the operation of political party systems in American democracy. In the first place, it seems that there is a limit to the degree that parties should compete with each other in terms of clear and mean-

[12] "Official Canvass of Votes, Primary and General Elections, August 4, 1958-November 4, 1958," compiled by James M. Hare, Secretary of State.

ingful policy alternatives and in terms of well-disciplined legislative party groups committed to a definite program. Parties have an equally vital role to play when they serve as brokers between conflicting points of view. A reasonable amount of party competition is necessary and healthy but competition should not develop to the point where compromise is no longer a possible solution to a political issue. Compromise and competition are not irreconcilable in the American political scene. Both must be present in the requisite amounts. Without a minimum of competition politics loses its meaning. Without a similar amount of compromise politics is impossible.

A second conclusion suggested by recent Michigan politics is that a party system involving disciplined parties and party responsibility similar to the British model, so obviously admired by the American Political Science Association's Committee on Political Parties, is incompatible with the presidential system of government based on the separation of powers. Without the fusion of legislative and executive power, strong, disciplined parties provide a basis for inaction and frustration rather than for the efficient implementation of party programs according to the will of the voters. In the past several years, Michigan has witnessed one of the most serious weaknesses of the presidential system—divided government. With a vigorous and dynamic governor confronting a united and hostile legislative majority the situation was bound to develop as it did.

The undesirable aspects of divided government in Michigan are illustrated even more clearly when they are contrasted with the situation which has existed at the national level from 1954 to 1961. There the parties are not so polarized ideologically nor so well disciplined as those in the Michigan Legislature. Because each party represents at the national level a coalition of many varied interests, because each congressman acts more or less independently, compromise and accommodation have been possible. President Eisenhower cooperated with a Democratic Congress. This happy state of affairs, in contrast to that in Michigan, existed not because the stakes were any lower in Washington than in Lansing, but because the national party system had not made political defeat an unacceptable result. If the policy objectives of our national parties were as distinct ideologically as those of Michigan's two parties have been since 1948, and if party regularity were as strong in Congress as it has been in the Michigan

Legislature, then the frustrations so common in Lansing in recent years would have been commonplace in Washington.

In spite of its numerous disadvantages, Michigan's programmatic party system is not without its blessings. The two-party system is now firmly established, albeit with an overdose of party competition. The parties compete on a relatively equal basis. This has not always been so. For most of Michigan's history, the Republican party was invincible.[13] Michigan politics took place within the limits of the Republican party. The Democrats were able to win elections only when the Republicans were split by factional struggles. Until 1932 the Democrats served little purpose other than to administer political patronage when the party was victorious nationally. One-party politics is unhealthy in a democracy. A cursory glance at the states south of the Mason-Dixon Line provides ample evidence of the effects of a total absence of party competition on the economic and political life of a state.[14] Michigan's party system does furnish a basis for the development of healthily competitive two-party politics in which compromise plays an essential role.

What, then, are the prospects for the future of Michigan politics? In many respects they are not at all promising. The new Democratic Governor, John B. Swainson, is deeply indebted to organized labor for his victories in the 1960 Democratic primary and in the hard-fought general election. Labor is not likely to pass up this advantage in pressing its views on the Governor. Present indications are that labor will have an important voice in the Swainson administration. Michigan's acquisition of an additional seat in the U. S. House of Representatives, following the 1960 census, may also provide the basis for a full-scale political donnybrook. A leading Republican newspaper recently commented editorially that it does not seem likely that the Republican party will be willing to compromise over the reapportionment of Michigan's congressional districts.[15] "Recent history," stated the paper, "indicates that conservative GOP

[13] For an illuminating account of the development of two-party politics in Michigan see Stephen B. and Vera H. Sarasohn, *Political Party Patterns in Michigan,* Detroit, Wayne State University Press, 1957.

[14] For a definitive analysis of one-party politics in the South see V. O. Key, Jr., *Southern Politics,* New York, Alfred A. Knopf, Inc., 1949.

[15] "Compromise Not Likely on Extra Congressman," editorial. *The Ann Arbor News,* November 17, 1960, p. 4.

legislators in control of the legislature are not very adept at making mutually profitable compromises with the Democrats. Chances are they will miss this opportunity." Nor is the Republican party likely to yield any of its current advantage in the apportionment of the Michigan Legislature. If for no other reason, it is difficult to take an optimistic view of Michigan's political future simply because too many powerful interests have too great a stake in the status quo.

The February 1961 conventions of the Michigan Democratic and Republican parties did not provide much encouragement to persons such as American Motors President George Romney and his Citizens for Michigan organization, who are hoping and working for a thaw in the state's political "cold war." The Democrats, enthusiastic over having survived the departure of Governor Williams from the Michigan scene without the loss of a single state-wide office, met and announced their determination to remain a highly programmatic and issue-oriented party. At the Republican convention conservatives strengthened their position within the party. George Van Peursem, a former speaker of the state House of Representatives and a spokesman for the legislative wing of the party, succeeded the liberal Larry B. Lindemer as chairman of the State Central Committee. The chairman of the platform committee was former Congressman and defeated senatorial candidate Alvin M. Bentley, a confirmed conservative. The convention adopted a platform which was at best "a call for the status quo."[16]

Perhaps the most significant indication of the conservative tone of the Republican convention was its failure to include in the platform an unqualified endorsement of the referendum proposal to call a constitutional revision convention.[17] Under pressure from liberal

[16] "Michigan Politics," by Will Muller, *The Detroit News*, February 12, 1961, p. 14-B.

[17] The proposal will appear on the ballot in the April 1961 general election. If it is approved, the Convention will be held in the fall of 1961. This is the third referendum proposal concerning a constitutional convention since 1958. In 1958, a proposal lost because it failed to carry an absolute majority of the votes cast in the election. The 1960 proposal, which did pass, changed the requirement for passage of a constitutional convention proposal to a simple majority of those voting on the question. The 1960 proposal also changed the system of apportioning delegates to a constitutional convention. Under the former system, three delegates were to be chosen from each senatorial district. Under the new system, one delegate is to be selected from each senatorial district and one from each representative district.

elements, the party had endorsed constitutional convention (con-con) proposals in 1958 and 1960. Meanwhile the Democrats, who had refused to support con-con proposals in 1958 and 1960, gave their unqualified support to the 1961 con-con proposal. This shift of position with regard to one of the most pressing political issues in the state emphasized again the great ideological distance that separates Michigan's two major parties.

On the other hand, there are hopeful signs. The Republican party has been building a liberal wing with its strength based in the State House of Representatives. Even the situation in the State Senate may improve. The Republican majority there recently elected new leaders who are several degrees more liberal than their immediate predecessors. It is not unreasonable to assume that the rise of liberalism within Republican ranks will make it possible for the parties to reconcile their differences in a mutually acceptable fashion.

The Democratic party is also broadening its appeal. During the past ten years Democratic strength has increased markedly in outstate rural counties. In the 1956 campaign the Democrats received major financial support from attorneys, small businessmen, and officeholders in addition to support from organized labor.[18] Political developments since the 1960 election have been encouraging. The new Democratic Governor has made conciliatory moves toward the Republicans and their supporters. Following the approval of a referendum increasing the constitutional limitation on the sales tax, he joined with Republican leaders in advocating the passage of legislation increasing the sales tax to 4 per cent. He has held talks concerning Michigan's economic climate with leaders of business and labor. He has proposed a sweeping revision of the tax structure designed to provide relief for business and industry and for purchasers of food and drugs.

The current drive for constitutional revision also offers some hope. The original supporters of the movement, the League of Women Voters and the Junior Chambers of Commerce, have been joined by the state AFL-CIO. The obstacles facing any constitutional revision convention will be numerous. Tradition and vested interests which are content with the status quo will have to be overcome. If certain essential reforms can be accomplished, primarily the adoption of a more equitable system of legislative apportionment and a strengthen-

[18] White and Owens, *op. cit.*, p. 33.

ing of the governorship, the state's prospects for a more harmonious political future will be greatly increased.

Whether the situation will change depends ultimately upon the people of Michigan. When the demand for a change in the pattern of Michigan politics and in the instrumentalities of state government is sufficiently articulated, that change will occur. If Michigan's political parties can anticipate that change and work to bring it about, they will do much to atone for the record of the past few years. If they resist such a change, if they are unwilling to put aside some of the partisan considerations that have caused such turmoil in the past, they will ultimately be called to account by the people.

Papers of the Michigan Academy of Science, Arts and Letters
Vol. XLVII, 1962 (1961 Meeting)

ECONOMICS

A RECONSIDERATION OF THE WAGES FUND THEORY

WILLIAM BREIT
Louisiana State University

The interpretation usually given to the wages fund doctrine is that it is one of the most sterile and pernicious theories in economic literature. For example, James Bonar called it "the crowning instance of an untrue abstraction" that "has probably done more injury to the reputation of economic theory than any other generalization ever received into economics textbooks and then expunged from them."[1] Writing forty years after Bonar, Paul Samuelson claims he has been impressed by the "falseness and emptiness of the wage fund doctrine," stating that it "constitutes one of the most sterile chapters in that dreary gap between the classical age and the revolutionary neoclassical discoveries of the last third of the nineteenth century."[2]

Such repudiations of the wages fund doctrine are quite general although not easily explained. It seems incredible that the great thinkers of classical economics should have devoted themselves with vigor to the espousal of such a "false and empty" theory. After having been one of the main building blocks of the classical edifice of Smith, Ricardo, Malthus, Senior, Mill, and others, it was vigorously attacked in succeeding decades by Longe[3] and Thornton,[4] the latter's onslaught finally leading to a much-discussed "recantation" by John

[1] James Bonar, *Disturbing Elements in the Study and Teaching of Political Economy* (Baltimore: Johns Hopkins Press, 1911), p. 75.

[2] Paul Samuelson, "Economic Theory and Wages," *The Impact of the Union* (New York: Kelley and Millman, 1951), ed. by David McCord Wright, p. 320, 316.

[3] Francis D. Longe, *A Refutation of the Wage-Fund Theory of Modern Political Economy as Enunciated by Mr. Mill and Mr. Fawcett* (London: Longmans, Green and Co., 1866). Reprinted by Johns Hopkins Press, 1903, as *A Reprint of Economic Tracts,* edited by Jacob H. Hollander.

[4] William Thomas Thornton, *On Labour: Its Wrongful Claim and Rightful Dues, Its Actual Present and Possible Future* (London: Macmillan and Company, 1869).

Stuart Mill in 1869. T. W. Hutchison claims that this event "was one of the more overt signs of the crumbling of the classical system,"[5] and it is sometimes supposed that from this time onward the theory dropped out of economic literature—it "sank without a trace,"[6] one commentator put it. Nevertheless, any appearance that the theory succumbed that soon is purely illusory. At least two major attempts at reconstruction occurred in the last half of the nineteenth century. In a sense they were largely dying gasps. The arguments raised on behalf of the doctrine by Cairnes[7] and Taussig[8] remained almost unheeded by a tranquilized generation of economists.

The origin of the wages fund theory rests on the classical theory of capital, which in turn was based upon a conception of the production process as being discontinuous and time-consuming. If the production process is time-consuming then it follows that the real wages received for the work performed today must have been produced yesterday; what workers produce today reaches the form of finished goods at a later time. In any period shorter than the length of the overall productive process, the aggregate amount of wages goods is fixed; there is a fixed fund which cannot be increased during the specific period. The number of workers divided into the fund gives the average wage rate.

The rejection of the wages fund doctrine by John Stuart Mill, after the attacks by Longe and Thornton, is hard to account for. The cogency of the criticisms of the doctrine cannot explain his retraction, since they contained little analytical content, and there was certainly no empirical refutation. These critics of the doctrine gloriously confused the monetary and real factors. The strange point is that in his eventual repudiation of the doctrine, Mill fell into the same fallacy after having clearly understood the distinction at an earlier date.[9]

[5] T. W. Hutchison, *A Review of Economic Doctrines, 1870–1929* (Oxford: Oxford University Press, 1953), p. 13.

[6] A. C. Pigou, "Mill and the Wages Fund," *Economic Journal*, Vol. LIX, June 1949, p. 177.

[7] J. E. Cairnes, *Some Leading Principles of Political Economy Newly Expounded* (New York: Harper & Brothers, 1874).

[8] F. W. Taussig, *Wages and Capital* (New York: D. Appleton and Company, 1896).

[9] John Stuart Mill, "Thornton on Labour and Its Claims," *Fortnightly Review,* No. XXIX and XXX (May and June 1869), pp. 505–518 and pp. 680–

It is significant, of course, that the wages fund theory flourished in the period when economists were primarily concerned with the conditions prerequisite to capital formation and economic growth. By the close of the nineteenth century, however, the English capitalist system had matured to the point where the emphasis could be shifted from that of capital accumulation to the role of the market mechanism in distributing the products which buyers have requested. The individual unit without reference to the economy of which he was a part was the phenomenon whose habits were of central interest. The "economy as a whole" dropped out of focus, and the "isolated datum" of economic man opened into view.

As we know, the classical economic theory was developed in the early stages of English capitalism and on an economic scene in which landed interests opposed the new and emerging manufacturing class. Hence the problem of capital accumulation and functional distribution had popular interest and practical import. By the twentieth century, however, the economic scene had changed in certain fundamental respects. The latter part of the nineteenth century was one of great progress in which capitalism had firmly established itself. The economy of the Western world was one in which the forces of progress were powerful and in which investment outlets were numerous. The only unemployment was due to fluctuations in the business cycle.[10] Thus the turn of the century saw a shift from concern over the problem of growth to concern over the problem of economic performance. As Leontief has recently pointed out, economists worry first about growth and then about efficiency.[11]

Even as early as 1870, Jevons was able to state the economic problem as follows:

> Given: a certain population, with various needs and powers of production, in possession of certain lands and other sources of material: required, the mode of employing their labour which will maximize the utility of the produce.[12]

---

700. Reprinted in John Stuart Mill, *Dissertations and Discussions*, 2nd ed. London: Longmans, Green, and Dyer, 1873), pp. 25–85.

[10] Alvin Hansen, "Economic Progress and Declining Population Growth," *The American Economic Review*, Vol. XXIX (March, 1939). Reprinted in American Economic Association, *Readings in Business Cycle Theory* (Philadelphia: The Blakiston Company, 1944), pp. 336–84. See especially pp. 367–70.

[11] Wassily Leontief, "The Decline and Rise of Soviet Economic Science," *Foreign Affairs*, Vol. I (January, 1960), pp. 261–272.

[12] William S. Jevons, *The Theory of Political Economy* (London: Macmillan, 1871), p. 267.

In working out his theory, however, Jevons did not depart from the older classical theory. His ideas on capital were basically those incorporated in the wages fund theory.[13] But with the shift in emphasis it is not surprising that the wages fund theory no longer played a central role in economic analysis. When the problem was one of economic development the wages fund theory had a strategic role to play. For rapid development, the resources of an economy must be channeled into capital investment. To this end, wages must be kept down so as to control consumption, and the wages fund theory provided a matrix for the classical theory of economic growth. But when the problem shifted from economic progress to efficiency, the wages fund dropped out of wage theory. Not that economists were anti-wages fund. It would be more correct to say that they were passively non-wages fund. This point is not usually made. It is generally asserted that the theory was forsaken, the abandonment taking place after Mill's recantation.[14] But as we have seen the controversy lasted until the twentieth century, and the theory was a central issue in the writings of such leading economists as Cairnes, Walker, and Taussig.

It should be clear that the wages fund theory was a theory of the average wage rate, and was not intended to say anything about the wages of individual workers. If the main emphasis of the time was on atomistic economic problems, then of course the wages fund theory had no part to play, *at least not as a theory of individual wages*. The point to remember is that the marginal productivity theory did not replace the wages fund theory, since they are not substitute products. Indeed it was just this fact that made Alfred Marshall reluctant to accept the marginal productivity analysis as a theory of wages. In his *Principles*, after illustrating the use of the marginal productivity theory to explain one of the causes that govern wages, Marshall added,

> This doctrine has sometimes been put forward as a theory of wages. But there are no valid grounds for any such pretension. The doctrine that the earnings of a worker tend to be equal to the net product of his work, has by itself no real meaning; since in order to estimate net product, we have to take for

[13] George Stigler, *Production and Distribution Theories* (New York: Macmillan, 1941), p. 29.

[14] cf. A. C. Pigou, "Mill and the Wages Fund," *Economic Journal*, June, 1959, pp. 171–80. Especially p. 177.

granted all the expenses of production of the commodity on which he works, other than his own wages.[15]

Professor D. H. Robertson, one of the most influential adherents of the marginal productivity theory, has called Marshall's qualification of the importance of marginal productivity "a godsend to critics"[16] and admitted that the marginal productivity theory did not replace the wages fund theory, since marginal analysis attempts to solve a micro rather than a macro problem. Robertson says,

Now so long as we are fixing our eyes on a single business or a single industry the assumption that all other factors of production have clearly defined supply prices is perhaps sufficiently nearly valid to give no great trouble to anyone; but what we are in search of is the principle governing the level of wages *as a whole*. . . .[17]

Robertson then goes on to argue that "grumbles" over the productivity doctrine result from attempts to read into the theory more than it is supposed to contain. All it was intended to explain was specific wages.

T. W. Hutchison has interpreted Marshall's reluctance in accepting the marginal productivity theory as a realization that the theory could not explain the level of wages as a whole:

Nowhere than in his theory of wages does Marshall make a more strenuous effort to link up with the classical theory and treatment, though, in fact, the classical, mainly "macro-economic" distribution theory is concerned with quite different questions of distribution than is . . . a precisely and correctly formulated marginal productivity analysis. The two types of treatment and analysis, though not necessarily incompatible, cannot be said, when combined, to produce a very significant synthesis; they remain separate . . . answers to separate questions.[18]

With the development of marginal productivity doctrine there was one important point in the new orthodoxy which left a flank exposed. It is doubtful that marginal productivity analysis can be considered

[15] Alfred Marshall, *Principles of Economics* (New York: Macmillan, 8th ed., 1920), p. 518.

[16] Dennis H. Robertson, "Wage Grumbles," *Economic Fragments* (London: P. S. King & Son, Ltd., 1931). Reprinted in American Economic Association, *Readings in the Theory of Income Distribution* (Philadelphia: The Blakiston Company, 1951), pp. 221–244. Citation from page 227.

[17] *Ibid.*, p. 227.

[18] T. W. Hutchison, *A Review of Economic Doctrines* (Oxford: Clarendon Press, 1953), p. 84.

significant in the determination of aggregate wages. Demand functions for labor are nonadditive since they are derived on the assumption that other elements in the system remain stable. Economists whose expertise in marginal analysis is unquestioned are none the less reluctant to apply their tool kit to macroeconomic analysis.[19]

Notwithstanding the fact that the doctrine of the wages fund was no longer a part of wage theory, it remained a part of capital theory, and it did so in its most pristine form, since the capital theory of the twentieth century is nothing more than a generalized version of the wages fund. In Böhm-Bawerk, for example, the period of production is the parameter manipulated by the capitalist in order to maximize his rate of return. It is the period of time elapsing between the input of labor, on the one hand, and the sale of the final product on the other. Land is disregarded, and labor and capital are considered to be the only factors of production. But the crucial point is that the lapse of time between input and output is made possible by capital, which must be advanced in the form of wages to workers because they cannot wait until the product is sold. Böhm-Bawerk's concept of capital is thus very similar to Taussig's: capital is the stock of commodities building up during the period of production.

In modern capital theory it is postulated that the entrepreneur needs the services of labor, equipment, and other factors of production. His capital fund is his command over the products of other firms. The owners of factors want command over products in exchange for their services; the entrepreneur needs command over the products of other firms to tide him over the initial period during which he has no products of his own to sell. Hence the firm's capital consists of other firms' products. When the capital of all the firms in the economy is added together, we obtain society's real capital: the sum total of all products in the productive system.[20]

[19] For example, Kenneth Boulding claims that he "can hardly help regarding the development of marginalism as a retrograde step as far as the theory of overall distribution—i.e. of the macroeconomics of distribution—is concerned." Kenneth E. Boulding, "Wages as a Share in the National Income," *The Impact of the Union* (New York: Kelley and Millman, 1959), p. 132.

[20] Compare with Scitovsky's definition: "We can say that society's capital consists of the total quantity of products stored up in the productive system in the form of inventories of consumers' goods, plant and equipment of all kinds, goods in the process of manufacture, inventories of parts and raw materials, and so forth." Tibor Scitovsky, *Welfare and Competition* (Chicago: Richard D. Irwin, Inc., 1951), p. 217.

The wages fund played a key part in classical theory because of the tremendous role given to capital accumulation in the spectacle of economic progress. Although technical change was recognized as a cause of economic growth, the classicals minimized the role of technology and concentrated on capital accumulation instead. One of the significant facts about capital formation, however, is that it takes time before consumer goods result from the use of more "capitalistic" methods. In the interim period somebody has to make advances to the various factors producing this capital. These advances are the wages fund. So long as time is required for production, capital in the form of wage goods is necessary. It is significant that Mrs. Robinson's recent work on capital theory involved her in a return to the problem of classical economics and, in consequence, a return to the wages fund. If the focus of attention in economics is on economic growth and capital formation, a return to the wages fund theory can be expected. As Mrs. Robinson says, "The revival of interest in classical economics brings a revival of the classical theory."

> Most kinds of human production consist of processes which take time to complete. The flow of consumption goods becoming available today is the result of work which was done in the past and with the aid of tools and equipment that may have been constructed very far in the past.... Thus in all human economies there must be property in capital goods—equipment and work-in-progress—as well as in territory.... In a capitalist economy property is owned by a small number of individuals who hire the labour of a large number at agreed wage rates and organise their work. The excess of the product over the wages bill then appears as income from property.[21]

Mrs. Robinson's analysis is therefore a return to classical economics and to the wages fund theory. The relationship between her argument and that of the wages fund theory has been noted by Professor Lerner:

> In throwing out the marginal productivity mechanism she is forced ... to a kind of residual theory which says that the real wage is determined by how much is left over to be divided among the workers as they take their turn in the queue after the requirements for accumulation and any consumption by capitalists, and we are back at a strict wages fund which comes back to life in several different contexts but always in response to minor variations of the same spell.[22]

---

[21] Joan Robinson, *The Accumulation of Capital* (London: Macmillan, 1956), p. 4.

[22] Abba P. Lerner, "Review of Joan Robinson, Accumulation of Capital,"

To put this another way: the whole point of capital theory is that there is a lapse of time between input and output. This is made possible by capital, which must "advance" the wages of labor because wage earners cannot wait until the products of their labor have been sold. Viewed from the input side, capital is a wages fund. From the output side, capital is the stock of unfinished products accumulating during the period of production. When the theory is formulated in this fashion there is no inconsistency between capital theory and the wages fund doctrine. Both describe the roundabout process of production: one from the input side, one from the output side.

The role of the wages fund doctrine as part of capital theory is increasingly recognized by wage theorists. Mr. Rothschild, for example, admits adherence to the wages fund theory and, in summing up his findings, presents a very complete description of the wages fund doctrine, showing that the total wage bill cannot be increased in the short run. In his words:

> To improve wages the national income, out of which they are paid, must be increased. This can be achieved by industrialization, i.e., the accumulation of capital. But it will take some time until the newly-formed capital will bear fruit in the form of consumption goods, and in the interim period total real wages will be fixed within fairly rigid limits. The originally existing capital, capable of turning out consumption goods, will constitute something like a wage-fund.[23]

It is clear, therefore, that the marginal productivity theory has in no way superseded the wages fund theory. The marginal analysis tells us a good deal about what determines the demand for individual workers of a certain skill; it tells us very little about total wages. Instead of replacing the classical analysis of distribution, it has simply created a vacuum in its place.

There is a special sense in which the Keynesian analysis has filled the vacuum, for Keynes' analysis in effect rehabilitated the wages fund theory. In showing that labor can determine its money wage but has no control over its real wage, Keynes' analysis made clear a point which follows directly from the wages fund theory: real wages

---

American Economic Review, Vol. XLVII (September, 1957), p. 699. See also, in this connection, Nicholas Kaldor, "Alternative Theories of Distribution," *Review of Economic Studies,* Vol. 23, Second Series, 1955–1956, pp. 83–100.

[23] K. W. Rothschild, *The Theory of Wages* (Oxford: Basil Blackwell, 1956), p. 10.

are not determined in the labor market but are a function of a quite different set of decisions. Keynes denied that labor has any control over real wages. Of course it is possible for an individual worker to reduce his real wage by accepting a lower money wage, for so long as other prices do not fall, a lower money wage means a lower real wage. But if all money wages were reduced a certain percentage, it does not follow that real wages would fall. Keynes' thesis is that a variation in money wage rates affects the demand for output and hence causes falling prices which leave real wages unchanged.[24]

This is apparently what Professor Boulding meant when he referred to the "modified 'wages fund' of the *General Theory*."[25] Thus real wages are determined in the capital market through decisions affecting investment, and not in the labor market. Keynes' theory involves an indirect type of wages fund theory since he argued that, given the propensity to consume, employment was a function of investment (defined as net additions to capital). The demand for labor depends on capital, and this is precisely the wages fund argument of the classical economists. In the Keynesian system, as in the classical economics, the wage bargain in the labor market affects distribution only indirectly, that is, as it affects the determinants of investment and future income.

The conclusion that distribution is largely independent of what happens in the labor market seems to be supported by the statistical

[24] It appears that Keynes was of two minds in this matter. By Chapter 19 of the *General Theory* he admitted that variations in money wages might affect real wages through what has come to be called the "Keynes Effect." Because a fall in wages will be accompanied by a fall in prices, the stock of money in "real" terms increases. Assuming the nominal amount of money is constant, this lowers the rate of interest and increases investment, thereby increasing employment and income. Notwithstanding this fact, Keynes rejected the manipulation of money wage rates as a policy for increasing employment. For one thing, in capitalistic economies there is no method available to bring about a universal reduction of the money wage rate. Employers must have sufficient bargaining power to push it downward, which they may not have. What is more, even if employers had such power, wages might be reduced in one industry before another, causing great labor unrest and dissatisfaction. What is still more, falling prices would have adverse effects on debtors. Keynes therefore felt that manipulation of the quantity of money would be a much sounder policy to follow in the event of unemployment. John Maynard Keynes, *The General Theory of Employment, Interest and Money* (New York: Harcourt, Brace and Company, 1936), pp. 257–271.

[25] Kenneth Boulding, *op. cit.*, p. 132.

evidence. If unions have the power to alter overall income distribution it should show up in an increase in labor's share of the national income.

The various studies on labor's share are of unequal reliability and not always comparable with one another. Despite differences of method and definition, however, most studies show that the secular shift in the employee's share has been relatively small.[26] Of course, we can never know what might have occurred in the absence of unionism, but the evidence does imply that unions *have not* improved the relative position of wage earners over time.

Although it is tempting to conclude that the impotence of unions to raise wages is a telling tribute to the accuracy of the wages fund theory, it must be noted that the validity of the doctrine does not depend on empirical evidence regarding the share of labor. Indeed, both Ricardo and Cairnes expected the share of labor to fall over time. The truth is that any fact about the progress of labor's share would be consistent with the wages fund doctrine, depending on certain arbitrary assumptions about changes in other variables. Thus Ricardo expected the share of labor and capital to decline at the expense of landlords. This followed from Ricardo's acceptance of the Malthusian theory of population and diminishing returns to land. If he found that labor's share had increased, the wages fund theorist would look for offsetting factors in technological development or in the willingness of workers to limit their numbers. The point is that the wages fund theory is not necessarily tied up with any definite trend in labor's share.

What is of interest is that unions are still seen as an obstacle to economic development. The same arguments raised against them by the classical economists are once again used to demonstrate that they act as an obstacle to economic growth because they hinder capital formation. Professor de Schweinitz, for example, has argued

[26] Harold M. Levinson, "Collective Bargaining and Income Distribution," *American Economic Review, Papers and Proceedings,* XLIV (May, 1954), pp. 315–316. E. H. Phelps Brown, and P. E. Hart, "The Share of Wages in National Income," *Economic Journal,* Vol. 52 (June, 1952), pp. 253–277. Joseph D. Phillips, "Labor's Share and 'Wage Parity,'" *The Review of Economics and Statistics,* XLII (May, 1960), pp. 164–174. George H. Hildebrand, "The Economic Effects of Unionism," *A Decade of Industrial Relations Research 1946–1956* (New York: Harper & Brothers, 1958), pp. 98–145.

that the labor movement in underdeveloped countries must be restricted in the interests of economic development. In drawing upon evidence from Western experience in the nineteenth and twentieth centuries, he shows that if unions are successful in redistributing income from profits to wages, savings decline, consumption rises, and investment is retarded. The rapid industrial growth of England, Germany, and the United States in the nineteenth century occurred when trade union action was either prohibited or restricted. In the Soviet Union trade unions have never been permitted to develop because low wage rates are necessary to provide high rates of saving.[27] As one writer has put the case:

> Unions fundamentally favor consumption; economic development requires keeping aggregate consumption down in order to free resources for investment. Effective unions, whatever else they may do, thus tend to delay, reduce, or prevent the growth of investment.[28]

Recognizing the incompatibility of trade unions and economic growth for newly industrializing societies, Professor Mehta suggests that unions should be given assignments that would interfere as little as possible with the requirements of rapid development.[29]

That the wages fund is still a crucial consideration for underdeveloped economies has been noted by Vakil and Brahmanand. They claim that in countries such as India the concept of capital as a stock of wage goods available to support labor in investment activities is realistic:

> We have maintained that unemployment in underdeveloped countries is due solely to the prevalence of the wage-goods gap. It is the inability of the economy to provide in the short-period the required surplus of wage-goods necessary in order that disguised unemployment can be employed in investment, that inhibits expansion in employment and investment. Countries like India have an abundance of population, but such an abundance does not necessarily connote a proportionate availability in employable units. If the economy can make good the wage-goods gap through a quick expansion in the output of

[27] Karl de Schweinitz, Jr., "Industrialization, Labor Controls, and Democracy," *Economic Development and Cultural Change,* Vol. VII (July, 1959), pp. 385–404.

[28] Adolph Sturmthal, "Unions and Economic Development," *Economic Development and Cultural Change,* Vol. VIII (January, 1960), p. 199.

[29] Asoka Metha, "The Mediating Role of Trade Unions in Underdeveloped Countries," *Economic Development and Cultural Change,* Vol. VI (October, 1957), p. 16–23.

wage goods, it is possible to raise the employment potential and thus initiate an upward cumulative process.[30]

The same argument is made by Professor Higgins. He notes that in some underdeveloped countries the labor cost of food determines the real wage rate and the spread between prices and real wages determines profit. When agricultural yields are good, wages are low and profit high. Investment thereby tends to increase. Once the new supplies of manufactured goods reach the market their prices fall. "If this drop in manufactures coincides with a bad harvest, so that wages and raw material costs are rising at the same time, investment may be cut off to a trickle."[31] Thus the experience of underdeveloped economies seems to support the proposition that the wages fund theory is not without validity.

In sum, the wages fund theory was not replaced by the development of marginal productivity analysis since the two theories are attempts to explain two quite different phenomena. In fact the attempt to use marginal analysis to explain wages has created a vacuum in wage theory, since marginal analysis can only explain specific wages and tells us nothing about the total wage bill or average wage rates, essentially macroeconomic issues. The Keynesian theory in a sense filled the void by returning to the problem of aggregates. It involves a wages fund theory in the sense that real wages are assumed to remain constant in the face of a change in money wages, and the latter proposition has many of the same consequences as the classical wages fund doctrine. In the Keynesian system, furthermore, the important consideration so far as income determination is concerned is net investment (or net additions to capital), and this is a fundamental part, indeed the essence, of the wages fund theory. The wages fund theory remained a part of capital theory because capital theory postulates a span of time, made possible through the advance of wages, between input and output. Moreover, experience in underdeveloped countries indicates that the wages fund doctrine is indispensable as a tool of analysis.

[30] C. N. Vakil and P. R. Brahmanand, *Planning for an Expanding Economy* (Bombay: Vora and Co., 1956), p. 23. See also pp. 211, 231–232, 397.

[31] Benjamin Higgins, *Economic Development* (New York: Norton & Co., 1959), p. 47. See also P. T. Bauer, "Regulated Wages in Under-developed Countries," *The Public Stake in Union Power* (University of Virginia Press, Charlottesville, 1959), Ed. by Phillip D. Bradley, pp. 324–329.

PAPERS OF THE MICHIGAN ACADEMY OF SCIENCE, ARTS AND LETTERS
VOL. XLVII, 1962 (1961 MEETING)

THE CASE FOR DEBT MANAGEMENT

ROBERT H. SCOTT
Michigan State University

AN increase in the volume of federal debt outstanding accompanies deficit expenditure by the government,[1] that is, it implies that fiscal policy is at work. Similarly, an increase in the volume of debt implies (*a*) an increase in the money supply to the extent that Federal Reserve banks purchase additional debt, and (*b*) an increase in the volume of liquid or "risk-free" assets if others purchase it. Thus inevitably fiscal policy and monetary policy operate concomitantly with variation in the volume of debt. It is useful, therefore, to abstract from changes in the *volume* of debt in order to view debt management as a stabilization tool of government that is separate and different from monetary and fiscal policy. For the purpose of this paper, then, debt management is defined as changes in the maturity mix of the outstanding volume of marketable federal debt in the hands of the public.[2]

There are always a number of terms and conditions attached to any debt instrument that determine its taxability, callability, eligibility, and saleability. But its maturity is the most important attribute for policy considerations because this factor is the principal determinant of what is called the "liquidity" of a debt obligation. Thus a Treasury bill is more "liquid" than a long-term bond, for the holder of a Treasury bill has greater certainty that he may conveniently obtain cash for his asset.[3]

[1] Minor exceptions to this rule occur whenever the Treasury varies its deposits and cash holdings.

[2] Unlike the *Treasury Bulletin* classification, in which "public" holdings of debt include holdings of *Federal Reserve banks,* here public holdings are thought to be exclusive of Federal Reserve holdings.

[3] For definitions of liquidity see: Abba P. Lerner, "Money," *Encyclopedia Britannica,* Vol. 15 (Chicago: Encyclopedia Britannica, Inc., 1946), pp. 693–696, reprinted in *Money and Economic Activity,* ed. Lawrence S. Ritter (New York: Houghton Mifflin Co., 1952), p. 15; and "Commercial Banks and the Problem of Liquidity," *Monthly Review,* Federal Reserve Bank of Kansas City (April, 1957), pp. 3–5.

Assume that the Treasury swaps long-term for short-term debt and thereby lengthens the maturity structure of the outstanding volume of debt. This reduces the liquidity of the public's asset holdings and implies a rightward shift in the usual Keynesian liquidity preference function. Fig. 1 shows two liquidity preference functions relating the rate of interest (R) to the money supply (M) for two debt structures (A_0) and (A_1) such that $A_0 < A_1$, that is, debt structure A_0, has shorter average maturity and is more liquid than debt structure A_1. One may use the average time to maturity of the debt as a rough index of A.

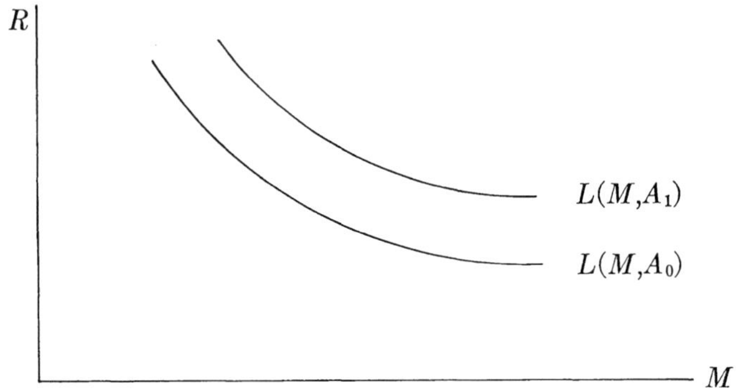

FIG. 1

For a given money supply (and implicitly a given level of income) the community will be willing to hold less liquidity of the type envisaged to be inherent in A only at a higher level of interest rates. The interest rate pertinent to the Keynesian framework is, of course, the long-term rate, and presumably higher long-term rates exert a restrictive influence on spending, *ceteris paribus*.

Counter-cyclical debt management, therefore, entails increasing the supply of long-term issues in periods of inflationary pressure and of short-term issues in recessionary periods.

Since the celebrated Treasury-Federal Reserve "Accord" of March 1951, interest rates on long-term Treasury debt have been free to fluctuate in response to conditions of demand and supply in the securities market. Data on the average length of time to maturity of the marketable public debt, found in the *Treasury Bulletin* for end of month dates, clearly show that the maturity structure of the federal debt has *increased* in periods of recession and *decreased* in periods of expansion—the exact opposite of the pattern dictated by the theory for prudent debt management.

Treasury authorities explain this perversity with respect to counter-cyclical objectives by reference to "other" considerations. Some of these are vague, such as the "need" to fund the debt and the "need" to avoid disrupting the "orderly" flow of savings into investment channels. These will be ignored here. One consideration, however, warrants discussion. It is the "interest cost" consideration.

In an inflationary period interest rates attain high levels both because of the great demand for capital and because of the restrictive policy of the monetary authorities in permitting a slower than usual rate of increase in the money supply. If the Treasury sells long-term debt in these periods, interest charges on the debt are higher than they otherwise would be. Prudent business management, of course, dictates that debt be funded in periods of low interest rates. But here, as in the case of balancing budgets, what is prudent for a business firm may be folly for government. Higher interest costs represent no additional burden (i.e., reduction in consumption) for the community as a whole.

Higher interest costs mean greater transfer payments from taxpayers to bondholders. On grounds of equity these transfers may be unjust. But the argument is less simple if alternatives appear in the picture. The Treasury might (Congress willing) monetize the entire debt and avoid interest charges altogether—an action certain to cause inflation and inequity in spite of extensive counteracting measures the Federal Reserve might take. No one seriously suggests that this be done, except perhaps as a very gradual long-run policy objective.

Another means of reducing interest charges would be available if the Treasury were willing to sell long-term debt when interest rates are low and buy it back when interest rates rise, that is, repurchase it when debt prices fall. The sheer size and overriding influence of the federal debt certainly give the Treasury authorities the power to "corner" the financial market to their own advantage, if they wished to do so, just as a speculator might, if strong enough, "corner" a commodity market. Again, considerations of equity erect the principal barrier to this procedure, and again, no one seriously suggests that this be done.

Finally, all measures, direct or indirect, taken in the interest of maintaining economic stability involve to a greater or lesser degree some distribution effects, although in the case of indirect controls these are thought to be more evenly dispersed among members of the community. For example, under tax policy it is pertinent to ask who pays the tax; under expenditure policy, which factors receive benefits; and under monetary policy, who gains from changing levels of interest rates.

It is clear, then, that special criticism of debt management on equity grounds is unwarranted. On the contrary, if each of the policy tools carries a fair share of the stabilization burden, the distribution effects among the individuals of the community will be diffused, while extensive reliance on one tool alone in the face of perverse behavior of the other tools will cause considerations of equity to loom large in the decision process.

To clarify this point assume (*a*) that the federal budget is balanced, and (*b*) that the Federal Reserve System undertakes to spur a recessed economy by reducing reserve requirements of member banks. With the maturity structure of the debt unchanged, the excess supply of lending power available to the banks pushes interest rates down. But if the Treasury then lengthens the debt, this pushes long-term interest rates back up, or at least keeps them from falling as much as they otherwise would, thus necessitating more extreme moves on the part of the Federal Reserve System—moves clearly discriminatory in that direct gifts of lendable cash accrue to banks. Concern over equity, therefore, dictates a positive counter-cyclical role for each policy tool.

On February 20, 1961, the Federal Reserve System abandoned its self-imposed eight-year-old "bills only" policy, under which it conducted open market operations only in the short-end of the market. Since then System authorities have engaged in swapping operations—buying bonds and selling bills in order to keep short-term interest rates high and slow the outflow of short-term capital that threatens our balance of payments position, while at the same time lowering long-term interest rates to spur the domestic economy. This is debt management as defined above. Of course, varying the maturity structure of the debt in the hands of the public is debt management no matter which agency of the government undertakes it.

The theory underlying these swaps rests on simple supply and demand analysis. Reducing the supply of long-term bonds raises their price and lowers their yield. Increasing the supply of short-term securities lowers their price and raises their yield. The markets for different maturities are separate and interdependent.

But how, one may ask, does this supply and demand theory fit with the neoclassical "expectations" theory developed by Hicks and Lutz of the determination of the maturity structure of interest rates, a theory that holds long-term rates of interest to be an average of intervening short-term rates?[4] Assume that the Federal Reserve reduces the supply of long-term issues and simultaneously increases the supply of short-term issues. If expectations are unchanged, then a sort of "arbitrage" portfolio manipulation will occur whereby marginal holders of long-terms will shift into intermediates, intermediate holders into shorts, and so on throughout the range of maturities. These shifts will push the long-term rate back up to its original level and the short-term rate back down to its original level. In other words, under constant expectations about the future pattern of rates the yield curve must remain unchanged.

It is this theory and this reasoning that lie behind current newspaper reports that many financial observers feel it will be impossible, or at least extremely difficult, for the Federal Reserve to achieve by debt management its avowed aim of restructuring the yield curve so as to discourage the outflow of short-term capital and stimulate the domestic economy.

The weakness of the neoclassical theory is found in the assumption of unchanging expectations. Why should expectations be insulated from obvious changes in the supplies of different debt maturities when, presumably, expectations regarding demand and supply factors determine expectations about future interest rate levels? Thus, if demand and supply factors are regarded as affecting expectations that in turn determine the shape of the yield curve, then a synthesis of the two theories emerges.

This synthesis indicates that our nation's monetary authorities should be able to induce a positively sloped yield curve to flatten out by debt management operations. Of course, the flatter the desired yield curve the more extensive must be the swapping maneuvers.

[4] Joseph W. Conard presents a concise statement of this theory in his *Introduction to the Theory of Interest* (Berkeley: University of California Press, 1959), Part Three.

And it would be extremely difficult to induce a negative slope into the yield curve and keep it there for any extended period of time. But modest operations should suffice to push short-term rates on government securities above the level of today's 2½ per cent and long-term rates below their present range near 4 per cent.

Furthermore, empirical evidence strongly supports the thesis that variation in the maturity structure of the debt has the prescribed effect both on the term structure of interest rates and on the level of long-term interest rates.[5] Both multiple regression and simultaneous equations experiments using monthly data covering the period from 1952 through 1959 to test the hypotheses indicate that as the debt is lengthened (*a*) the yield curve becomes more positively sloped, and (*b*) the long-term rate of interest rises, *ceteris paribus*.

In his study paper prepared for the Joint Economic Committee of Congress Warren L. Smith argues quite persuasively that little benefit can be expected from debt management policy employed for stabilization purposes.[6] In a journal article Herbert Stein argued, again persuasively, that Treasury authorities should attempt to minimize interest charges on the debt until it can be shown that countercyclical debt management is capable of something more than monetary policy alone can do.[7] This space is clearly too small to deal adequately with all facets of the question, but an attempt has been made above to show briefly the other side of the picture—to present an antagonist's case for implementing positive debt management policy.

[5] See the author's Ph.D. dissertation, "Debt Management for Economic Stability," Department of Economics, Harvard University, 1961.

[6] "Debt Management in the United States," Study Paper No. 19, Joint Economic Committee, 86th Congress, 2nd Session, January 20, 1960.

[7] "Managing the Public Debt," *Journal of Law and Economics*, I (October, 1958), pp. 97–104.

PAPERS OF THE MICHIGAN ACADEMY OF SCIENCE, ARTS, AND LETTERS
VOL. XLVII, 1962 (1961 MEETING)

LINGUISTICS

A GENERAL SYSTEMS MODEL IN LINGUISTICS — ENGLISH NOUN FORMS —

ARTHUR A. SAXE

The University of Michigan

INTRODUCTION

THE primary purpose of this paper is to introduce a General Systems model for consideration in linguistic analysis. I propose that such a model contains the potential for a unifying approach to linguistic problems.

Approaches to these problems have utilized certain conceptual tools which have seemed to be at odds with one another. As Joos has commented (with respect to Bloch's (1947) and Nida's (1948) positions), "These (ideas and modes of thought) are, between the two sides, as multiply different as could be imagined, or more so" (1957: 254). On the other hand, all of the analyses possess a common thread of approach, at least covertly. This modicum of agreement seems to lie in certain attributes of the data which impose themselves upon the analysts, given, of course, agreement on certain basic assumptions. These imposing attributes will strongly affect our systemic considerations.

Hockett analyzed two models for grammatical description, e.g., 'item and process' and 'item and arrangement,' and found one model as simple, useful, etc. as the other (1954: 399, *et passim*).[1] As we shall see, a major premise of this paper is that 'item and process' models are not inherently in opposition to 'item and arrangement' models, but rather are two sides of the same coin. The model presented here links these approaches in terms of analytic procedures applied to language considered as an ongoing communication system.

The various models that exist are responses to various foci of interest, the model chosen being a function of the questions we ask or

[1] If a cited source is reprinted in Joos, 1957, the page cited will refer to its position in that volume.

of the data we focus on. To illustrate the utility of the proposed model I shall therefore define the problem accordingly.

The problem is how best to represent the English noun forms, and in particular those forms which deviate from the "productive subclass."[2] Which model will most fruitfully represent the noun "paradigms," articulate them with the larger constructs of which they are a part and with the smaller constructs (e.g., morphemes) of which they are composed? Quite evidently, we are not interested solely in either structure or process, but in both. Our ideal solution should be a single model which will provide a context for all of our specific problems. If the model does no more than this, it will be satisfactory.

Hockett, however, (1954: 398–399) has established "metacriteria" for the evaluation of a model, which provide additional yardsticks. Although the limitations of this paper do not allow the full application of the model, the metacriteria are included so that the reader may judge whether or not it has the potential for fulfillment.

1. A model must be *general*: it must be applicable to any language, not just to languages of certain types.

2. A model must be *specific*: when applied to a given language, the results must be determined wholly by the nature of the model and the nature of the language, not at all by the whim of the analyst. It is not a lack of specificity if the model requires us to subsume certain facts more than once from different angles; it would be a lack of specificity if the model allowed us to take our choice in such cases, instead of forcing one choice or the other or their joint option.

3. A model must be *inclusive*: when applied to a given language, the results must cover all the observed data and, by implication, at least a very high percentage of all the not-yet-observed data.

4. A model must be *productive*: when applied to a given language, the results must make possible the creation of an indefinite number of valid new utterances. (The four so far presented are basic. Hockett is hesitant about the fifth.)

5. A model must be *efficient*: its application to any given language should achieve the necessary results with a minimum of machinery.

I propose that a satisfactory resolution of specific problems is to be found within the framework of a General Systems model as applied to English noun forms. Hopefully, this will provide a "single more nearly satisfactory model" which will subsume the common threads

[2] Usage of this term follows Gleason, 1955: 99.

of approach of different models and eliminate some of the bases of disagreement.[3]

The procedure will involve the following steps: first, state the assumptions and definitions of General Systems Theory; second, apply a General Systems model to the analysis of English noun forms; and third, discuss the controversy in the literature and present a view of the issues.

Assumptions and Definitions[4]

Scientists make certain assumptions about nature which are basic to the formulations of particular theories that attempt to explain nature. For our present purposes, there are four such basic assumptions:

1. Nature proceeds as events in continuous time-space.
2. Nature as a whole is a stable system of interlocking subsystems.
3. Nature is not understandable directly; we must isolate and conceptualize aspects of nature so as to make it intelligible.
4. Any level of systemic isolation is grounded in subordinate levels but has properties not derivable from these other levels (e.g., compounds and atoms).

Nature may be made intelligible in a number of ways. For present purposes I shall mention two. The Natural History approach attempts to make nature intelligible by dealing with nature as a time-space continuum; this is done with various techniques of conceptualization and isolation. Natural historians isolate events in the time-space continuum and conceptualize them in terms of a scheme which they conceive is pertinent to the understanding of another event also in the time-space continuum. In other words, they order the events with respect to a problem. The relevance of the ordering must stem from

[3] A note is in order as to the reticence with which I approach the problems in a field where I am a novice. In one sense, to attack the integrative problem is a reflection of humility in the shadow of the experts. Each analyst has powerfully represented his position within the context of his interests and assumptions. I shall not repeat in one way what they have said effectively in another. Rather I shall respond to Hockett's plea for more experimentation. A Systems approach has been found useful in other sciences. As Browning said, one's reach should exceed one's grasp. If its application here is found wanting, at least it is out in the open for revision or rejection.

[4] I am indebted to Lewis R. Binford of The University of Chicago for this particular approach to General Systems Theory. I assume full responsibility for any modifications or misapplications that may appear here.

the investigators' experience with the events themselves or from the findings of Systemic[5] studies which use the other assumptions about nature. The point is that the events themselves are not self-explanatory. If they were there would be no need for other types of explanation. Even the arrangement of the events merely relates what occurred and does not tell why or how the sequence was as it was. The explanation of the arrangement must always be referred to another type of analysis, the Systemic.

The Systemic approach attempts to make nature intelligible by isolating and conceptualizing nature as a system of interlocking subsystems. This approach makes it possible to deal with events, not with respect to their position in the time-space continuum, which necessarily makes each unique, but as repetitive events, i.e., the result of the operation of a stable system. Thus we may speak of the English language as a stable system regardless of its dialects or idiolects.

A *system* may be defined as a set of relationships between elements so that a change in one requires a change in others; in other words, a set of relationships between elements whose relation to one another is determined by their relation to the whole. The first definition focuses on what we have called function. The second on what we have called structure (below).

Because nature is considered to be a system of subsystems, any isolated system is considered to be a system of subsystems. Systems may be isolated by defining their parameters on the basis of intersystemic relationships. This involves the determination of whether:

1. The elements do actually articulate structurally and functionally (see definitions below),
2. The system as isolated can be considered as an element of a system more broadly conceived,
3. The elements of the isolated system can each be considered a subordinate system, or a closed system in its own right.

Closed systems are isolates in which a factor impinging from outside the defined parameters is a given, not relevant to the problem at hand.

Open systems are isolates in which a factor impinging from outside the defined parameters is relevant to the problem at hand.

The types of systems isolated depend upon the problem at hand, so that systems may be isolate on: (1) the basis of the stratification of

[5] Thus reserving 'systematic' for its usual connotation of methodical, orderly, etc.

nature, e.g., chemical isolate, atomic isolate, etc.; or (2) simply the basis of the interaction of two systems within one strata, e.g., the oxygen-carbon isolate and their many compounds; or (3) any number of other ways suitable to the question at hand. At times any of the above may be isolated with the addition of time-space continuum dimensions. These dimensions allow us to break up the strata, etc., into parts, so that even though we may speak of the English language as a stable system, we may also speak of it as it is actually spoken in different regions.

A system consists of elements related structurally and functionally:

An element of a system is any system that in another scheme of isolation could be considered as a closed system. For example, the earth is one element of the solar system, but it can also be isolated as a closed system consisting of such elements as geological, meteorological, or biological phenomena.

The structure of a system is the constant relationships between the elements of the system isolated. 'Constant' does not imply 'static.' For example, the earth and all the planets vary in their distance from each other and the sun at different points in time. Yet their relationship is constant and can be expressed in terms of mass, velocity, and distance. If this relationship were to change, the structure of the solar system would change.

The functions of a system are the variable relationships between elements within a structure at any given time.[6]

A systemic isolate will hereafter refer to any given isolated system wherein by the foregoing definitions the structure and functions are articulated. We shall therefore be in a position to discuss, for example, the morphophonemic systemic isolate on either a 'structure-level' or a 'functions-level.' The shorthand terms 'morphophonemic-structure' and 'morphophonemic-functions' will refer to these levels.

Assumptions Specific to Language Systems

Linguistic scientists make certain assumptions about the nature of language which are basic to their formulations.

[6] Anticipating, I may note that my definitions are similar to Hockett's mathematical analogs. "Structure" encompasses his *relation* type system which is correlated with IA constructions. "Function" encompasses his *operation* type system which is correlated with IP processes (1954: 394–395).

1. Language as a part of nature proceeds as events in continuous time-space.

2. Language as a whole may be considered as a stable system of interlocking subsystems. This assumption underlies all attempts to formulate, for example, a "theory of morphemic analysis" that would be valid no matter what particular language was being analyzed. The inclusion of cross-language data in all literature cited (except Bloch) is witness to this.

3. Language as a whole or languages in particular are not understandable directly. We must isolate and conceptualize aspects of them to make them intelligible.

4. Any level of systemic isolation is grounded in subordinate levels but has properties not derivable from them (e.g., syntax and phonemics).

There are two basic approaches to making language linguistically intelligible. The first is what we have called the Natural History approach and the second is the Systemic approach. Modern American linguistic science has placed a heavy emphasis on the first approach but has by no means excluded the second. The ordering of events by the natural historian[7] merely states what occurred; the explanation of the arrangement (e.g., the taxonomy) must always be referred to the other type of analysis which we have called Systemic. The arguments in favor of any particular taxonomic scheme, although phrased in terms of 'principles' and 'procedures' (Nida and Hockett) and 'more systematic arrangement of the known facts' (Bloch), have at least *implicitly* contained attempts to justify these taxonomies. Thus they have approached systemic formulations at a number of points based upon the structure and functions of languages. One example is the universally accepted 'principle No. 1' as phrased by Nida: "Forms which possess a common semantic distinctiveness and an identical form in all their occurrences constitute a single morpheme" (1948: 258).

Conflict arises over taxonomic schemes and techniques because there is no explicit unified systemic approach. This conflict is particularly evident in regard to those forms which do not fall into the 'productive subclasses.' It is here that we are allowed to 'take our choice,' which results in decisions seemingly "esthetic." It is here that confusion arises as to whether linguistics is a science or an art.

[7] Whether diachronic or synchronic is irrelevant to my formulation.

It is here that the different models fall beneath Hockett's meta-criterion No. 2.

Having stated the assumptions and definitions basic to an approach, I shall now consider the English language as a case in point, and particularly the noun forms. Because of the assumption and analytical isolation of systems of subsystems, however, I shall have to delimit cursively those systemic isolates that have a bearing on the English noun forms. For purposes of clarity I shall ignore the suprasegmentals. It will be helpful to the reader if reference is made to Fig. 1 throughout the following discussion.

The Discourse Isolate

The English language is here considered to be a stable system. That is, it possesses a structure. If this structure were to change (and therefore the systemic form) the result would not be the English language but something else. I shall begin the analysis by arbitrarily cutting into the English language at a systemic isolate level we may predefinitionally label *Discourse*.[8] The elements of this isolate may be given as a number of 'utterances' referred to in Fig. 1 as Utterance A, B, etc. Each *may be* what is thought of as a 'sentence' in traditional grammar. A may be a 'declarative sentence,' B may be an 'interrogative sentence,' etc. These utterances may include nonsentence constructions so that element C may be a 'construction class,' e.g., a prepositional phrase.[9] At any rate, the structure of the discourse isolate would be a statement of the constant relationship between these elements. Such a statement of structure would delimit the possible arrangements of the elements. These of course would be determined by the particular type of discourse under scrutiny.

A statement of functions would denote any particular variation in the combinations of elements as opposed to another combination. For example, if A refers to the 'declarative' class of utterances of which there are a number of different declarative-types, and if B refers to the 'interrogative' class of utterances of which there are a number of different types, and if a given type of B follows only a given type of A; then a statement of functions would be a statement of this 'condi-

[8] 'Predefinitionally' because the exact nature of these superordinate levels is peripheral to our noun focus; it is sufficient for our purposes merely to recognize their existence and to have some idea of their nature.

[9] Usage of this term follows Gleason, 1955: 137.

tioning.' It may be pointed out that the different potentially communicable meanings within a structure are explicitly denoted by the functions. An interesting corollary of this is that there is no 'separate semantic system.' Each systemic isolate level (above the phonemic) communicates its own level of meaning functionally. The problem is one of decoding the signals involved. Approaching in this way, we no longer have the fundamental problem of devising a new linguistics to deal with a separate semantic idiom. We have instead a technical problem of decoding in combination with the explicit restriction that we cannot make the functions of one level do the work of another.

The Syntax Isolate

Each of the elements of the Discourse isolate (wherein each was an open system) may be isolated and considered as a closed system on a subordinate level of analysis. This level is referred to as the *Syntax isolate* in Fig. 1, where for purposes of illustration I have chosen Utterance B as the element to be further analyzed as a closed system. Any of the other Utterances could just as well have been

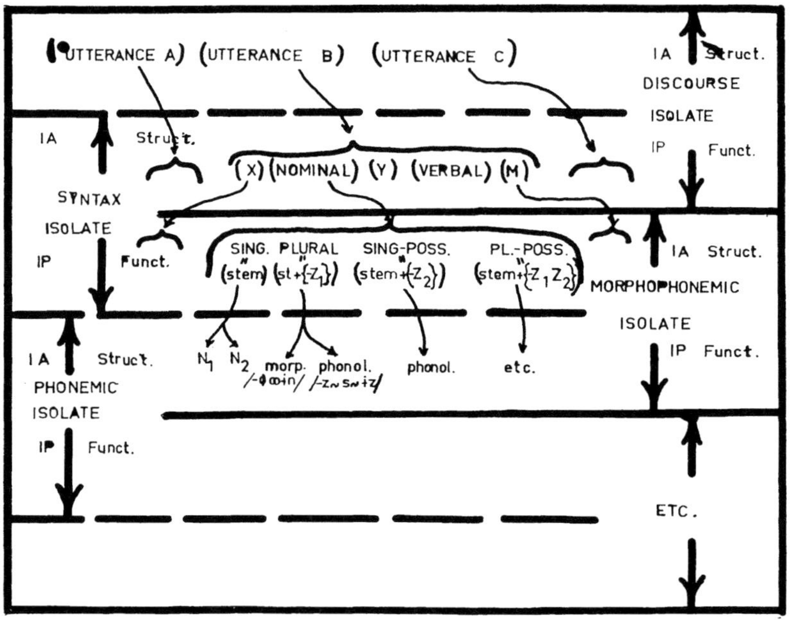

Fig. 1

chosen for syntax analysis. The only prescription is that it contain 'noun forms,' since they are the calculated destination. The type B Utterance as represented consists of the following elements: X, Y, M, VERBALS and NOMINALS. Other Utterances not dealt with here may contain the same elements, but what distinguishes them is the structure. The structure of any isolate is always the given. Presented here, the structure is prescribed as (X) – (NOMINAL) – (Y) – (VERBAL) – (M), which may represent an utterance of the type 'Did rats really go there?'[10].

Within this structure it should be recognized that each 'syntactic slot' may be filled by a number of morphemes or a number of combinations of morphemes. All 'constructions' (Gleason, 1955: 132) which fill the NOMINAL slot form one 'constituent class' (*ibid.*: 136) which we shall hereafter refer to as the 'noun class.' What is crucial is the syntactic function of the construction. This holds true whether for VERBALS, X-ALS or Y-ALS. It is only on this level that we can distinguish forms like /buk/, 'to book a game,' from /buk/, 'to read a book.' Furthermore, even though each of the constituent classes may be filled by a number of phonemic shapes, the occurrence of a particular form in one syntactic slot may require a particular form in another slot, e.g., 'Did rats really go there?' and 'Had rats really gone there?' A statement of functions on the Syntax isolate level would be a statement of these variable relationships between elements within the structure. When we examine the noun class we find that there are four functional variations of nouns determined by functional variations in the other elements. These are traditionally labeled singular, plural, singular possessive, and plural possessive. They may be represented as (stem),[11] (stem) + $\{\text{-}Z_1\}$, (stem) + $\{\text{-}Z_2\}$, and (stem) + $\{\text{-}Z_1Z_2\}$ respectively. Those forms that appear after the plus sign have been traditionally called 'inflectional suffix' morphemes.[12] The decision as to which morphs are to be joined as allomorphs of an i.s. morpheme is made on the basis of purely syntax-functional criteria.

[10] As previously noted, the suprasegmentals will be ignored. We may note that they are an important part of any utterance; if syntax, discourse, etc., were our concern, they could not be ignored.

[11] 'Stem' as used here may also cover those forms which are composed of stems plus derivational affixes; 'stem' therefore applies to any form which occurs in the noun construction class on the syntax-structure level, and specifically occurs in the singular slot on the syntax-function level.

[12] Hereafter 'inflectional suffix(es)' shall be abbreviated 'i.s.'

That is, a form like /boyz/, in itself, does not tell us whether it is plural, plural possessive, or singular possessive. We base our decision upon the nonnominal forms it appears with. Just as the quality of Nominal-ness was derived as a Discourse-functional variation, we find that the very quality of plural-ness as expressed in $\{-Z_1\}$ is determined by us as investigators from examination of the correlated functional variations within the syntax structure. The plural-ness is therefore overt. To put it another way, the statement that all noun-plural forms consist of a stem plus $\{-Z_1\}$ is a summary statement of a syntax isolate function. $\{-Z_1\}$ as an expression of meaning is derived and is only derivable from the syntax level, where it is an overt functional variation; it cannot be derived from the so-called 'morphological' level, where the differences upon which it is based are covert.[13] (On this level it can only be a given, a structural element of the 'morphophonemic' systemic isolate.) As an example we may consider the forms /boyz/ "boys" and /boyz/ "boy's," which are phonemically identical. Morphemically we express one as boy + $\{-Z_1\}$, and the other as boy + $\{-Z_2\}$. The first is the noun-plural and the latter is the noun singular possessive. These meanings are derivable only from a superordinate level of systemic analysis (more broadly conceived than morphemic). 'The boys ran away' and 'Each boy's hat had a ribbon' clearly distinguish between the two meanings of /boyz/, and do so on a syntactic level. There are some cases where we even have to go to the Discourse level to distinguish the function (the overt representation of meaning) of /boyz/. For example, 'The boy's hats fell.' and 'The boys' hats fell.'

We may further note that in English the 'paradigmatic classes' cannot always be defined on the overt basis of their inflectional suffixes. For example, /buk/ may be either a noun form or a verb form. We must go to the syntactic level in order to ascertain their function; then, on the basis of these functions, we may classify them morphologically. I doubt that the English language, as it exists today, is most fruitfully analyzed so as to contain a separate systemic level called 'morphological.' One may make a morphological taxonomy, but the justification for it must come from the syntax-functions level or the morphophonemic-structure level. (For further discussion of this point see pp. 363–64, *et passim.*) This is not meant to imply

[13] Hockett was vulnerable to Nida's critique (1948: 257) because he did not make this levels-distinction. See p. 364 of this paper.

a 'God's truth' position. The assumption is rather that each language has a structure independent of the investigator, and that that model which is most fruitful[14] in analysis is closest to expressing the reality. This in no way precludes the formulation of a more fruitful model.

It is a commonplace that Chinese and Latin as language systems are organized differently. The former has been characterized as 'isolating' and the latter as 'inflecting.' These are real differences. Using an inflective model for English (possibly tracing to the Latin grammarians), Gleason, who is not alone, comes up with a morphological level of analysis which consists of two kinds of item classes: paradigmatic classes and syntactic classes. The former are defined as 'parts of speech' composed of stem morphemes and i.s. morphemes and their allomorphs. There are four paradigm patterns which "define four paradigmatic classes." These are verbs, nouns, personal pronouns, and adjectives. The syntactic are defined as "classes of words which occur in the same or comparable environments in English utterances" (as the paradigmatic classes) but which are not inflected (1955: 95–96).

Gleason's procedure therefore is to delimit the paradigmatic classes on the basis of inflection, and then to delimit the syntactic classes on the basis of these. I feel this is backwards. We have already noted that the paradigmatic classes, e.g., noun and verb, cannot always be distinguished overtly on the 'morphological' level.[15] We have also noted that the i.s. morphemes similarly are not overtly distinguishable on the 'morphological' level, but are derived from functional alternation on the syntactic level. Operationally, this puts them in the same phenomenological class as the syntactic classes and specifically Gleason's 'nominals' (1955: 136). That is, they have no systemic reality, (no structure and function) apart from the syntactic level of isolation. Nouns do not articulate solely with nouns. The noun i.s. morphemes such as $\{-Z_1\}$, etc., therefore are abstractions from and summary statements of the functional variation on the syntactic level.

Gleason has noted that $\{-Z_2\}$ is rare in English. Why? Because there is an increasing trend towards the use of 'of'. English seems to

[14] As measured by Hockett's metacriteria, for example.

[15] We must note that Gleason is aware of this point; e.g., see pp. 106–107. It is only *after* a paradigmatic classification, however, that he resorts to syntax as a means of defining the residue.

be in the process of losing {-Z_2} "as it has already lost a large number of inflectional affixes during the last millenium" (1955: 97). If language is a system then elements cannot arise or drop out of the system without ramifications in the rest of the system. What is the nature of a system that drops noun paradigms as significant functional elements, and how does the system communicate meanings? I cannot hope to give a conclusive answer to this question here. I have already pointed, however, to the crucial role of syntax (and higher levels) in communication. In summary, this paper suggests that a "morphological level of analysis" (for English) be abandoned as an *analytical tool* for those phenomena called nouns. One may make a useful morphological taxonomy, (indeed this 'Natural History' approach has characterized American descriptive linguistics) but the justification for and the explanation of the taxa must be derived from levels of analysis wherein these phenomena function overtly, i.e., possess systemic reality. As Hockett's second metacriterion states, the results of analysis must be determined *wholly by the nature of the model and the nature of the language*, not at all by the whim of the analyst. Esthetic justifications for taxonomies, in the words of Anna Russell, leave us as befogged as before.

In the remainder of this paper I shall: (1) discuss the morphophonemic level of systemic analysis and present a morphological taxonomy consistent with my position on English noun forms; (2) examine some controversy in the literature in the light of my systems model.

The Morphophonemic Systemic Isolate

In the morphophonemic isolate I once again take an element (a type of noun) of a superordinate isolate (syntax) and isolate it as a closed system on a subordinate level. The elements of the system are those items which are variable within the noun slot on the syntax level. These items express syntactic-functional relations; singular, plural, etc.

What is constant in all of these forms within the syntax isolate level is the noun-ness which was derived on a discourse-function level and given on the syntax-structure level. What varies is (1) the particular i.s. morphemes, and (2) the particular stem morphemes filling the slots.[16] These then become the elements of the morphophonemic

[16] Note that I shall here include particular forms in defining the parameters

structure, e.g., the stem morpheme class and the i.s. morpheme class. These articulate structurally so that in any one construction, e.g., 'plural,' there is only one stem morpheme and one i.s. morpheme.[17]

Each class (stem or i.s.) internally exhibits variety when articulated with the other. These are the functions. They are expressed in the various phonemic shapes of the allomorphs of the elements and are considered to be a result of the interaction of these. Therefore, once again, functional statements can be made in terms of conditioning. My taxonomy of English noun forms (Fig. 2) is based upon such considerations. It *is not* a taxonomy of our systemic isolation. It *is* a morphological taxonomy based upon the results of that isolation. It is laid out along the two dimensions of stem and i.s.

For purposes of presentation I shall refer to the singular form of the stem allomorphs as the 'base form.' There are two classes of stem morphemes. The first, or N_1 in Fig. 2, are those which possess but one allomorph, i.e., those in which the base form articulates (without change in the phonemic pattern) with all of the i.s. morphemes. This class comprises the majority of stem forms in English. The second class, or N_2, possesses two allomorphs of the stem morpheme. The occurrence of each is regular and predictable. The base form allomorph occurs with no i.s. in the singular slots, e.g., /wayf/, and is 'inflected' (with $\{-Z_2\}$) in the singular-possessive slot, e.g., /wayf + s/. The other allomorph occurs in both plural slots (plural and plural-possessive) with $\{-Z_1\}$, $\{-Z_1Z_2\}$, e.g., /wayv + z/.

The i.s. morpheme dimension contains three morphemes and their allomorphs, as were determined on the syntax-functions level. These are $\{-Z_1\}$, $\{-Z_2\}$, and $\{-Z_1Z_2\}$. $\{-Z_1\}$ expresses plural-ness. $\{-Z_2\}$ expresses singular-possessiveness. $\{-Z_1Z_2\}$ expresses plural-possessiveness. Here on the morphophonemic level, these functions are expressed in the phonemic shape of the allomorphs when articulated with the phonemic shapes of the stem allomorphs. The processes undergone in this articulation are of two kinds, traditionally expressed as phonological conditioning and morphological conditioning. The phonemic shapes of the allomorphs involved are an expression of these processes,

of the system. I noted on p. 6 that, in addition to other criteria, a system may be isolated with respect to events in the time-space continuum. Because my concern is with English noun forms, the analysis at this point will encompass both systems and specific forms.

[17] *One*, because each i.s. is a representation of one syntax-function. The plural-possessive symbol, therefore, $\{-Z_1Z_2\}$ as used here, is one morpheme.

		gloss	Base Form	{-Z₁} /-z ~-s ~-ɨz/	{-Z₁} /-ø ∞ -ɨn/	{-Z₁} Loans	{-Z₂} /-z ~-s ~-ɨn/	{-Z₁Z₂}
N₁	A	'bee'	/biy/	/biyz/			/biyz/	/biyz/
	B	'ox'	/aks/		/aksɨn/		/aksɨz/	/aksɨnz/
	(C)	'crisis'	/kraysis/			/kray-siyz/	'of the crisis'	'of the crisis'
N₂	D	'wife'	/wayf/	/wayvz/			/wayfs/	/wayvz/
	E	'goose'	/gus/		/giysø/		/gusɨz/	/giysɨz/
	(F)	'index'	/indeks/			/indɨs-iyz/	/indeks-ɨz/	/indɨs-iyz/

Noun Subclasses ↑ (first two columns); Phonol. C'N'OT'D. ↑ (/-z ~-s ~-ɨz/); Morph. C'N'OT'D. ↑ (/-ø ∞ -ɨn/); Phonol. C'N'OT'D. ↑ ({-Z₂}, {-Z₁Z₂})

FIG. 2

just as the morphemes themselves are an expression of the role played by the allomorphs in the syntax and discourse processes. A statement of the type of conditioning therefore is a summary statement[18] of functions on the morphophonemic level.

The three i.s. morphemes are represented along the horizontal axis of Fig. 2. Of these, only the plural slot, {-Z_1}, is further subdivided. This subdivision represents the difference between phonological and morphological conditioning.

The majority of stems as 'modified' by {-Z_1} take /-s∼-z∼-iz/, the

[18] A summary statement is sufficient for our purposes as numerous statements of conditioning are available elsewhere (e.g., Gleason, 1955: 98).

particular phonemic shape being phonologically determined. The plurals thus formed comprise the great bulk of English noun forms, both N_1 and N_2.

The /-ø ∞-in/ allomorphs of {-Z_1} are morphologically conditioned. They are rare in English (no matter whether the stem is of N_1 or N_2 type). These are the controversial forms taxonomically, and especially so when combined with an N_2 type stem. I shall return to these later.

The third subdivision of {-Z_1} contains 'loan words' from other languages which have retained the formations used in the original tongue. These are intrusive into the language and, for purposes of clarity in presentation, will not concern us here. Suffice it to say, "There is a strong tendency to make these conform to the English pattern by changing the form of {-Z_1} to /-z∼-s∼-iz/," i.e., the productive subclass (Gleason: 1955: 99). We shall ignore such "words" as have not been so integrated.

The remaining two i.s. morphemes, {-Z_2} and {-Z_1Z_2}, can be dealt with rapidly. The allomorphs of {-Z_2}, /-z∼-s∼-iz/, are all phonologically conditioned. The allomorphs of the {-Z_2} part of {-Z_1Z_2} similarly are phonologically conditioned given the {-Z_1} form of the noun in question.

The taxonomy presented here, then, delimits six subclasses of nouns that occur in the English language. These are represented on the chart by the characters A, B, (C), D, E, (F). Subclasses (C) and (F) are concerned with the 'loan words' and have been excluded from present consideration.

Subclasses A and D comprise the majority of English noun forms. Subclass A is more frequent than D. This is due to the relative infrequency of type N_2 stem morphemes, which I have roughly estimated at 30 occurrences. Both A and D represent phonologically conditioned {-Z_1} suffix shapes. The differences are the occurrence of but one allomorph in A, and two allomorphs in D.

Both B and E possess morphologically conditioned {-Z_1} shapes, the difference again being the occurrence of but one allomorph of the stem in B, and two allomorphs in E.

The {-Z_2} and {-Z_1Z_2} conditioning is the same for all classes.

The taxonomy presented here is consistent with the systemic analysis and theoretically therefore with the structure of the English language as a whole considered as a system of subsystems. It results in a picture of the noun class that has but six (rigorously four) subclasses of nouns. It is not redundant in classification. It eschews special cases. That is, each problematical form or small group of forms does not have to be 'explained' in terms of its own peculiarities.

I have demonstrated that the stem and i.s. elements of noun forms are determined by reference to different systemic isolates where they are involved in functional variation. On the morphophonemic-structure level these become the structural elements, and their phonemic shapes are classified as expressions of morphophonemic functions.

I was obliged to consider the stems (as measured by the base forms) separately from the i.s. forms. To put it another way, I was *not* obliged to deal with 'words.' No linguist knows what a word is, and yet it seems that an implicit lay concept of 'word' has been the unit of focus in morphological analysis. Relationships and operations were just as important as phonemic shapes, in some cases more so. The model avoids the morphological machinations that go hand in hand with a word-type analysis.

Controversy in the Literature

Hockett (1947: 229) distinguished between morphemics and tactics, noting that both are necessarily involved in grammar. Nevertheless, he continued, we have a considerable range of choice in drawing the line between them. We have a choice which to emphasize and thereby complicate. The language is not disturbed by our choice but the resulting descriptions may be more or less clarifying in their depiction of the situation. He then states an a priori assumption that the greatest possible clarity "is achieved by the simplest possible tactics, whatever submorphemic complications may be necessitated" (1947: 229).[19] It is here that I take issue with him. If we assume a language structure independent of us as investigators then we must assume that the more fruitful descriptive analysis most closely approximates that reality. Hockett's a priori assumption of simple tactics/complicated morphemics implies a *completely* arbitrary choice on the part of the investigator. I take the position that the choice of approach *is* arbitrary but is also limited by the nature of the language, a position which Hockett supported in 1954. The different approaches naturally are determined by the questions we ask as investigators, but the most fruitful answers will be found to the degree that we

[19] He does this, though, "without steadfast conviction."

understand the structure of the language as it exists independently of us. Our systemic analysis led us to the conclusion that morphemics and tactics in relation to English noun forms are largely inseparable. In Hockett's terms we would say that we have emphasized tactics more than morphemics. This type of statement, in the light of intersystemic relations, is obviously oversimplified. Such oppositions are of questionable value when the language system allows no such distinction.

• • •

There are certain problems borne of strict adherence to Harris' general procedure for identifying utterances as morphemically the same (Hockett, 1947: 230). One is that alternants may be identical in meaning but almost (not quite) in complementary distribution, e.g., 'hoofs' and 'hooves.' "These would forbid the tactically desirable solution that there is but one noun plural morpheme in English." I agree with Hockett in disagreeing with this statement. My argument is on two different levels.

First, if they are identical in meaning, then the occurrence of the two different overt forms must either be due to dialect differences or free variation. A dialect difference does not alter the language; in fact a dialect is defined as such because it partakes of the same language system as other dialects but varies from them in the specific content of the slots that are filled. Free variation is linguistically nonsystemic and needs no more comment. If these examples are neither dialect nor free variation then there are connotative differences in meaning (minimally)[20] derivable from the syntactic or superordinate levels of the language system. This position is also applicable to the plural forms 'fish' and 'fishes' and 'brothers' and 'brethren.' Herein I would follow Hockett (1947: 234) and distinguish the first from the second in each case, postulating two distinct though homophonous morphemes for each; the different meanings would be derived from his tactics level (our syntax or discourse). I can utilize both distributional and semantic criteria in morphemic determinations inasmuch as the systems analysis suggested that both adhere in any one system and cannot justifiably be separated.

Second, no matter what we say about these 'words' (and we may note here that the gremlin of dealing with word-like things has crept into Hockett's analysis), in the present systemic analysis 'hoofs' and

[20] Nida, 1948, principle no. 7, pp. 265–266, with which I agree.

'hooves' do *not* forbid the conclusion that there is but one noun-plural morpheme in English. The stem morphemes are to be isolated from the i.s. morphemes. In the 'hoof' case the stem is of the N_1 type. The other is of the N_2 type. In each case the noun-plural allomorph (/-s/ and /-z/) is phonologically conditioned, expresses noun-plural meanings as derived on the syntax level, and does not contrast in identical environments with the other. Therefore they still form but one noun-plural morpheme, which I call {-Z_1}. It is at this point that I merge and agree with Hockett's solution of 'non-contrastive distribution' (1947: 233). Here is a case in point where it has proven fruitful to deal with stem and suffix separately.

• • •

Another problem of strict adherence to Harris' criteria is that "Sometimes a stretch of speech may be cut at either of two places, so as to produce equally satisfactory—and equally unsatisfactory—morphs" (Hockett, 1947: 230). Our analysis does not allow us to be completely arbitrary in our cutting. In English noun forms, when we are faced with a cutting problem which seems arbitrary, we are probably dealing with forms whose overt functions occur in more than one systemic isolate (see pp. 362–63). Noun forms such as 'oxen' /aksin/, in which the stem is of N_1 type, never present cutting problems even though the i.s. is morphologically conditioned. On the other hand, forms with an N_2 type stem may present cutting problems, especially when combined with a morphologically conditioned i.s., e.g., 'child' /čayld/ and 'children' /čildrin/. In English, with so few occurrences of morphological conditioning of N_2 noun forms, the cutting should be made so as to conform to the rest of the system. This is no more than an expression of the assumption of language as a system. I would therefore cut /aksin, breðrin, čildrin/ so as to isolate /-in/ as the {-Z_1} allomorph which would be applicable to all three. Thus 'children' and 'brethren' conform to the cutting of N_1 'oxen' on the one hand, and N_2 'geese' and 'wives' on the other.[21] Furthermore, I would eschew the concept of 'empty morphs' for English noun forms. That is, I would not cut 'children' as /čayld/ /r/ /in/ with /r/ as an empty morph on the assumption that every phonemic shape so cut should be totally accountable for on some level of systemic analysis.

[21] This in no way invalidates the portmanteau status of 'children' and 'brethren.' For a discussion of this point and an example of my cutting procedure see pp. 363–64.

In English the decision as to morphemes and their allomorphs is made above the morphophonemic level. Only the allomorphs are analyzed at this level. Therefore, because the phonemic system does not require an /r/, there is no justification for the sudden appearance of an empty morph.

• • •

Bloch's "English Verb Inflection" (1947), although specifically out of our focus on noun forms, raises an opposition of approaches which our systemic approach may help to clarify. This opposition may be summed up in Hockett's terminology (1954) as 'Item and Process' (IP) vs. 'Item and Arrangement' (IA). Hockett's conclusion is that neither model meets all the metacriteria for the evaluation of a grammatical description. Each in its own way comes as close as the other. Bloch prefers the IA model, which is understandable in terms of his problem orientation. IP statements "referring to processes of derivation, are useful for showing the relation of any inflected form to its base . . . (but) cannot be readily used for the description of specific forms, viewed as words in their own right." Bloch continues, "To describe the structure of a language as a whole . . . any sentence, phrase, or complex word can be described as consisting such-and-such morphemes in such-and-such an order; each morpheme has a meaning, and so also has the order in which they occur . . . " (1947: 243). I take the position that an opposition of IP to IA models is not justified in terms of fruitfulness (see Hockett, 1954: 399, where he makes this point). Furthermore, in the analysis of any given language system, because each model focuses on a different side of the coin, both are necessary for complete analysis.

Whatever reasons may exist in the history of linguistic science for the opposition, the two models are responses to two different questions. Hockett pointed this up nicely in his 'V. Mathematical Interlude' section (1954: 394–395). One question type focuses on elements of a system and their relations (our structure), while the other focuses on elements of a system "for which certain operations are defined" (our functions). Any system must contain both aspects, or it would not be a system. Hockett's view to the future includes the resolution of the opposition with an "eventual reintegration into a single more nearly satisfactory model". It is this model which I feel will have to be built along systemic lines, and have attempted to build here. My definitions of structure and functions were rigorous. A structure was defined as a constant relation between elements of a system, and it would seem therefore that an IA model would be most fruitful on any structure level. Function was defined as the variable relations between elements within a structure, and it would seem therefore that an IP model would best serve a discussion of functions. Furthermore, any system consists of subsystems, and as a matter of procedure, the empirical/operational test of whether the investigator has isolated a system is whether it can be further dissected into and analyzed in terms of subsystems. This means that what is a structure on one systemic isolate level, and is therefore most fruitfully dealt with in terms of an IA model, is involved in functional variation on a systemic isolate level more broadly conceived, and therefore would be more fruitfully dealt with in terms of an IP model. What I am now making explicit has already been done in the analysis. Statements of functions are statements of conditioning. Statements of structure are statements of items (elements) and their arrangement. What such an analysis makes clear is that what has traditionally been dealt with as 'morphology' (see Gleason: 1955: 96–97) does not stand alone in systemic analysis. The morphemes involved are at one and the same time a set of functional variations on the syntax-functions level and a set of items in arrangements on the morphophonemic-structure level. I suspect that much of the debate about IP vs. IA models for morphological analysis has revolved about the nonconscious systemic context within which the investigator was working.

• • •

The non-base form of N_2 type stems are portmanteau items (Hockett, 1947: 239–241). They simultaneously express base form meaning and the syntax-function of "plural." In other words, /giys/ is an allomorph of both {gus} and {-Z_1}. Pluralness, and therefore {-Z_1}, is overt and derived on the syntax-function level. /giys/ is overt as the functional variation of {gus} on the morphophonemic level but only occurs in that syntax relationship called "plural." Therefore it is a portmanteau, but because there are two levels of overt plural functions represented (e.g., syntactic and morphophonemic), it is more consistent to represent /giys/ as /giys + ø/. The /ø/ is the overt phonemic representation of {-Z_1}. This picture is recognition of {gus + -Z_1}→ /giys + ø/,[22] which is recognition of the fact that the English language does communicate "pluralness" at different levels.

[22] See p. 364 for further discussion relevant to this point, e.g., "replacives."

The base forms of N_2 may similarly be considered as portmanteaus of "singular-ness." However, for purposes of taxonomy, the most economical statement is achieved by merely treating the nonbase, N_2 type stem forms.

• • •

"An overt formal difference among related forms... constitutes a morpheme, if in any of these forms this difference, together with a zero tactical difference, is the only significant feature for establishing a minimal unit of phonetic-semantic distinctiveness." This was Nida's fourth principle (1948: 262). On the basis of this fourth principle he treated "noun plurals in English such as *men, feet, mice, teeth* as occurring with 'replacives' (i.e., replacements which are morphemic). Whether we are to say that both a replacive morpheme and a zero alternant of the plural suffix occur in these words depends on how much significance we attach to that part of the structural design which provides the basis for zero" (pp. 263–264).

As I have stated previously, I do attach equal significance to the "structural design," as 'equal' as to the 'overt' stem. I do not do so, however, because of an 'esthetic feeling' or to be democratically fair about overt and covert status, both of which enter into Nida's conclusions. I do so because the objective criteria of the model, structure, function, and system, constrained me to do so.

On the morphophonemic isolate level (structural), which has been traditionally designated as the 'morphological' or 'morphemic' system, the stem and i.s. types are presented simultaneously as elements of the structure. The IA model expresses noun-plural as stem + $\{\text{-}Z_1\}$. We must remember that any form that appears here is simultaneously both a functional variation within the syntactic isolate and a structural element within the morphophonemic isolate. This means that the noun-ness of the stem is not determined by its paradigm but by its structural position on the syntax level and its functional position within the discourse isolate. I have already pointed to the difficulty of distinguishing nouns from non-nouns except on these levels. I have also pointed out that operationally the noun i.s. morphemes and their allomorphs can only be derived on the syntax-functions levels. Pluralness is not a function of the noun form on the morphophonemic level; that is, a stem form is not plural because it takes $\{\text{-}Z_1\}$. Rather it is said to take $\{\text{-}Z_1\}$ on this level because of its functional position on the syntax level. There is nothing less overt about including an /ø/ in {-Z_1} than a /-z/ at the level at which {-Z_1} is defined. It is not, as Nida said, dependent "upon how much significance we attach to that part of the structural design which provides the basis for zero." The structural design is crucial in English, not only for zero and for our i.s. morphemes, but for the very quality of nounness itself.

Is the "replacive" solution in opposition to the /-ø/ solution? This problem does not arise in relation to N_1 type stems. For these, the /-ø/ solution is the only possible way to represent a form such as /šiyp/ in a plural context. However, N_2 plural stems (which are also portmanteaus) have a "built-in" phonemic change to represent the syntax-plural-function. The replacive model is therefore a descriptive statement of the change in phonemic shape of an N_2 stem when it is functioning in a plural context, e.g., articulates with {-Z_1}. This is supported by the fact that these changes are perfectly correlated with the singular-plural dichotomy. Plural stem forms of subclass D, however, e.g., /wayv-/ are bound forms, unlike those of subclass E, e.g., /giys/ (Fig. 2). If we apply the "replacive solution" to class E, we must also apply it to class D. This would not violate the definition of a replacive as 'a replacement that is morphemic.' In other words, in class D there are two phonemic changes, one in the stem, and one in the i.s. Therefore, because the stem changes in both D and E are syntax-functionally identical, we are *constrained* by our model to represent both subclass D and E plurals as a stem (containing a replacive) plus some form of {-Z_1}: /stem/ + /e ← (R)/ + /ø/.

Nida recognized that the linguistic meaning of a morpheme can be defined in terms of the types of constructions in which it occurs (1948: 264). Why, then, is our solution to the last argument so different from his? Why was Hockett vulnerable to Nida's criticism (see below)? The answer to both questions is the same. That is, both investigators lack a unified systemic technique for analyzing the language systems they are dealing with. Let us briefly examine the arguments.

In arguing against Hockett's portmanteau proposal, Nida suggests that using Hockett's own logic one could combine all "plural affixes of nouns in one morpheme... with the partly homophonous third singular suffix of verbs" (1948: 257). Nida's point is well taken. But if Hockett had stated the level of systemic isolation for which his generalization had validity no such criticism could have been made. As previously mentioned, verbs and nouns are distinguished on the syntax-structure level and the discourse-functions level. Both are thereafter to be dealt with as separate systems on subordinate levels of systemic analysis. Even if homophonous forms occur in both verb and noun slots, e.g., "I fish with worms" and "I eat fish," the meaning of each is distinguished syntactically; the two forms therefore are different morphemes.

• • •

Nida, Hockett, and most other descriptive linguists recognize and define operational procedures for analysis. These have proven useful and have provided a basis for a modicum of agreement among many investigators. These 'principles' are applicable to the analysis of any language. Many of them are interrelated so that a statement of one implies an aspect of another. They have been formulated so as to encompass, ideally, all the problems one encounters in linguistic analysis. A good example is Nida's aforementioned principle No. 5, which states that the linguistic meaning of a form can be defined in terms of several criteria (one being the types of construction in which it occurs). This implies a recognition of the various ways a language can operate as a system of communication. A language operates differently not only at one level of isolation but also at different hierarchical levels. Such principles have proven very useful. In the analysis of a specific language system, however, these principles must be integrated and applied within the limits set by that system. We need a unified theory of application and analysis which is now lacking. To this end I have proposed a General Systems Theory.

This paper has focused upon the implications of a General Systems Theory as applied to English noun forms. I have contented myself with its most general applications except for specific problems discussed in the literature. In these cases I have attempted to present and defend such particular viewpoints as stem from my systemic assumptions and techniques. Except for these points of disagreement I support the positions of the authors cited and the basic assumptions of descriptive linguistics. The orientation presented here should be viewed as a proposed addition to these or, in Hockett's terms, an approach "looking towards an eventual reintegration into a single more nearly satisfactory model."

It is my opinion that a General Systems model may satisfy Hockett's metacriteria for the evaluation of models. We might remember the adage: "Behold the turtle: he makes progress when he sticks his neck out."

LITERATURE CITED

BLOCH, BERNARD. 1947. English Verb Inflection. Language, 23: 399–418. (Reprinted in Joos, 1957: 243–254.)

GLEASON, H. A., JR. 1955. An Introduction to Descriptive Linguistics. New York: Henry Holt and Co.

HOCKETT, C. F. 1947. Problems of Morphemic Analysis. Language, 23: 321–343. (Reprinted in Joos, 1957: 229–242.)

——. 1954. Two Models of Grammatical Description. Word, 10: 219–231. (Reprinted in Joos, 1957: 386–399.)

JOOS, MARTIN, ed. 1957. Readings in Linguistics. New York: American Council of Learned Societies.

NIDA, EUGENE A. 1948. The Identification of Morphemes. Language, 24: 414–441. (Reprinted in Joos, 1957: 255–271.)

PAPERS OF THE MICHIGAN ACADEMY OF SCIENCE, ARTS, AND LETTERS
VOL. XLVII, 1962 (1961 MEETING)

GEOGRAPHY. A. GENERAL

PRECIPITATION REGIME IN THE LOWER PENINSULA OF MICHIGAN

DIETER BRUNNSCHWEILER

Michigan State University

IN spite of its rather precarious nature and its high preference as a conversation topic, the climate of Michigan has received surprisingly little scholarly attention from the geographer's side. In order to obtain any detailed analysis of our climate we have to go back to such classics as Leighly's study (1) on the often-mentioned, little-understood "Lake Effect," or even look along the shelves of forlorn theses for Seeley's 1917 master's thesis (2) on the climate of Michigan.

The present study deals with one element of Michigan's climate and with the Lower Peninsula only. Precipitation in its varying aggregates of moisture is very much in evidence in our state throughout the year. We have become aware of droughts as well as of rainy seasons, we have heard of the snow belt, we are familiar with freezing rain, and we believe in Indian summer. The major purpose of this paper is to present the climatological background to some of the popular concepts related to Michigan's precipitation regime.

AVAILABLE DATA AND PERIOD OF OBSERVATION

The records (3) of the period 1931–52 of 75 stations situated in the Lower Peninsula are reliable enough to be used in a regionally and temporally comparative analysis of precipitation. It is of paramount importance in precipitation studies that the length of the period be uniform. Even though a period of 22 years is not sufficiently long to incorporate the potential extremes it does include perhaps 90 per cent of the range which is apparent at stations with records dating back to the last century. The density of stations with homogeneous records leaves much to be desired but has proved to be high enough to exclude subjective interpolations. All except four counties of the Lower Peninsula have at least one station with a homogeneous 1931–52 record.

Regional Distribution of Monthly, Seasonal, and Annual Precipitation

As a first step in the analysis of Lower Peninsula precipitation characteristics I have drawn 12 monthly isohyetal maps, based on the mean monthly values of the previously mentioned period (see Figs. 1–4). The total amount of precipitation (including the water equivalent of snow) is comparatively low in all winter months (December, January, February), normal values being below 3 inches throughout the peninsula. A clearly noticeable west-east decrease is evident, however, which is even more conspicuous on the map representing total annual snowfall, 50%–70% of which falls in the three winter months (Fig. 5). A certain concentration of winter precipitation is visible in the triangle formed by Allegan, Paw Paw, and Kalamazoo, and in the extreme northwest. It should be noted that areas immediately adjacent to Lake Michigan receive a considerably lower amount than stations some distance inland. Two snow tongues in the south central part and one in the northern "thumb" are conspicuous in the otherwise snow-deficient southeast.

Spring, hydrometerologically speaking, begins in March in lower Michigan. A marked contrast between the wet south and the dry north of the peninsula is immediately obvious from the inspection of the first and second spring months (March, April). The three-inch isohyet migrates northwards parallel to itself through the course of spring until it reaches the latitude of Saginaw Bay. The meridional gradient of rainfall has reached a maximum in May with most of the southern tier receiving close to or above 4 inches of rain. An independent cell of higher rainfall has formed in the north-central part by this time. While the same general arrangement still prevails in the first summer month (June), scattered rain cells, indicative of increased convective activity, have replaced the former simple bipartite zonation by July. A conspicuous feature of the midsummer month is the "dry belt" rooting in Saginaw Bay and connecting northwestward with Grand Traverse Bay and southeastward with Lake St. Clair. In the fall months (September, October, November) the northern and northwestern portions of the peninsula experience a strong regional maximum of precipitation which then migrates

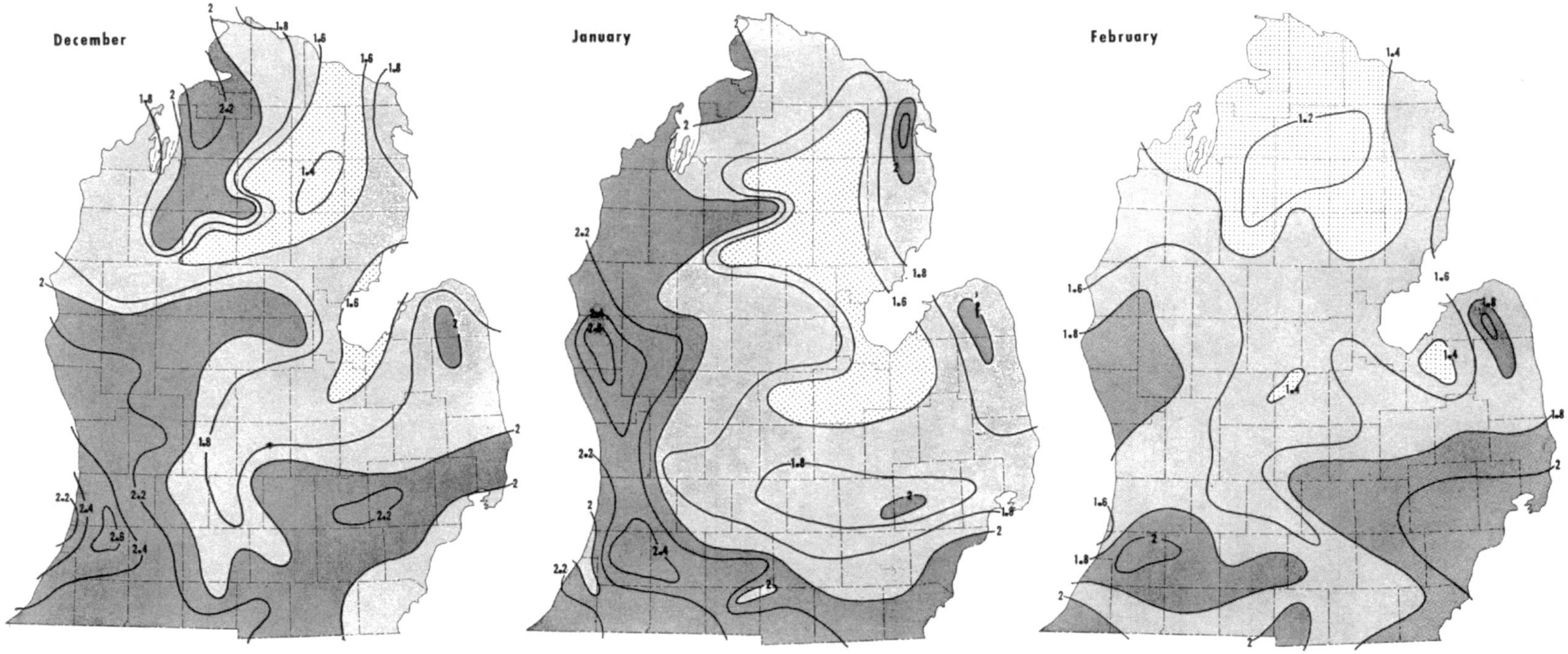

FIG. 1. Mean monthly precipitation, winter months (inches).

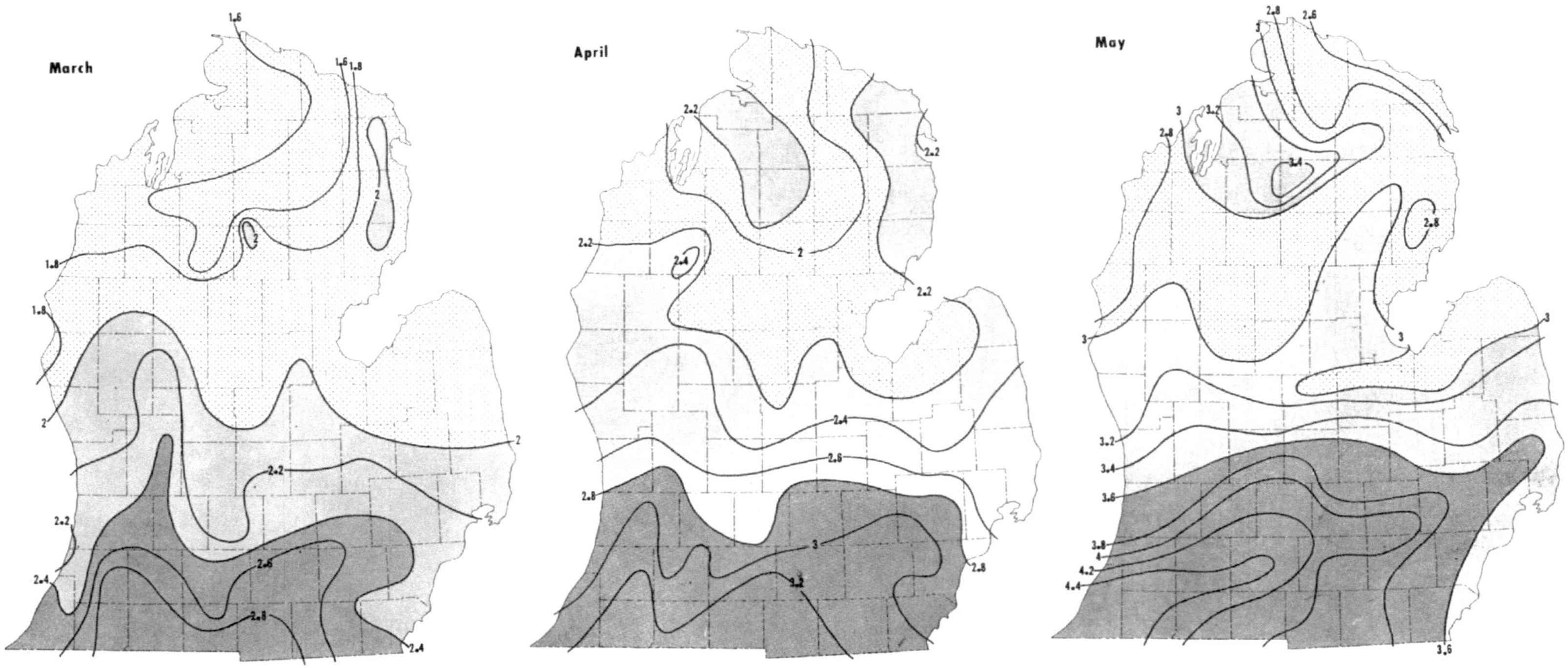

FIG. 2. Mean monthly precipitation, spring months (inches).

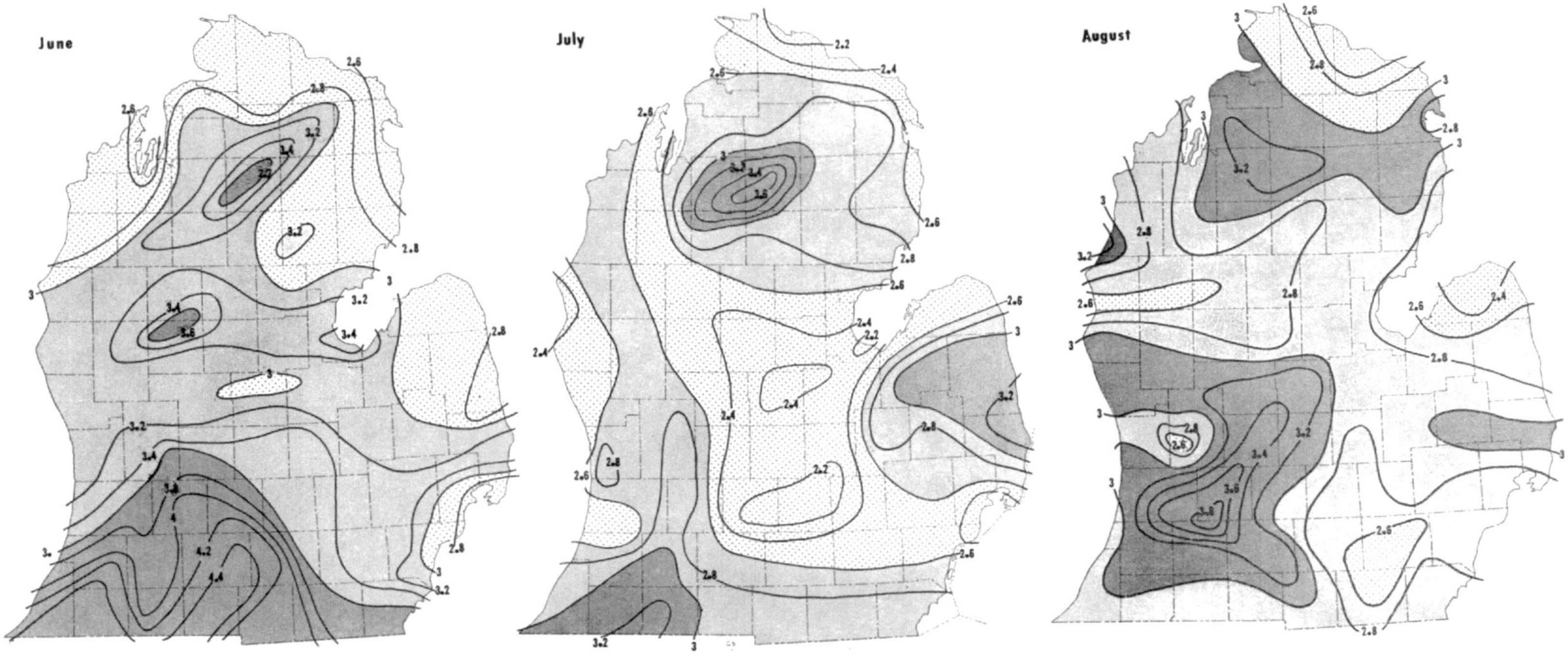

Fig. 3. Mean monthly precipitation, summer months (inches).

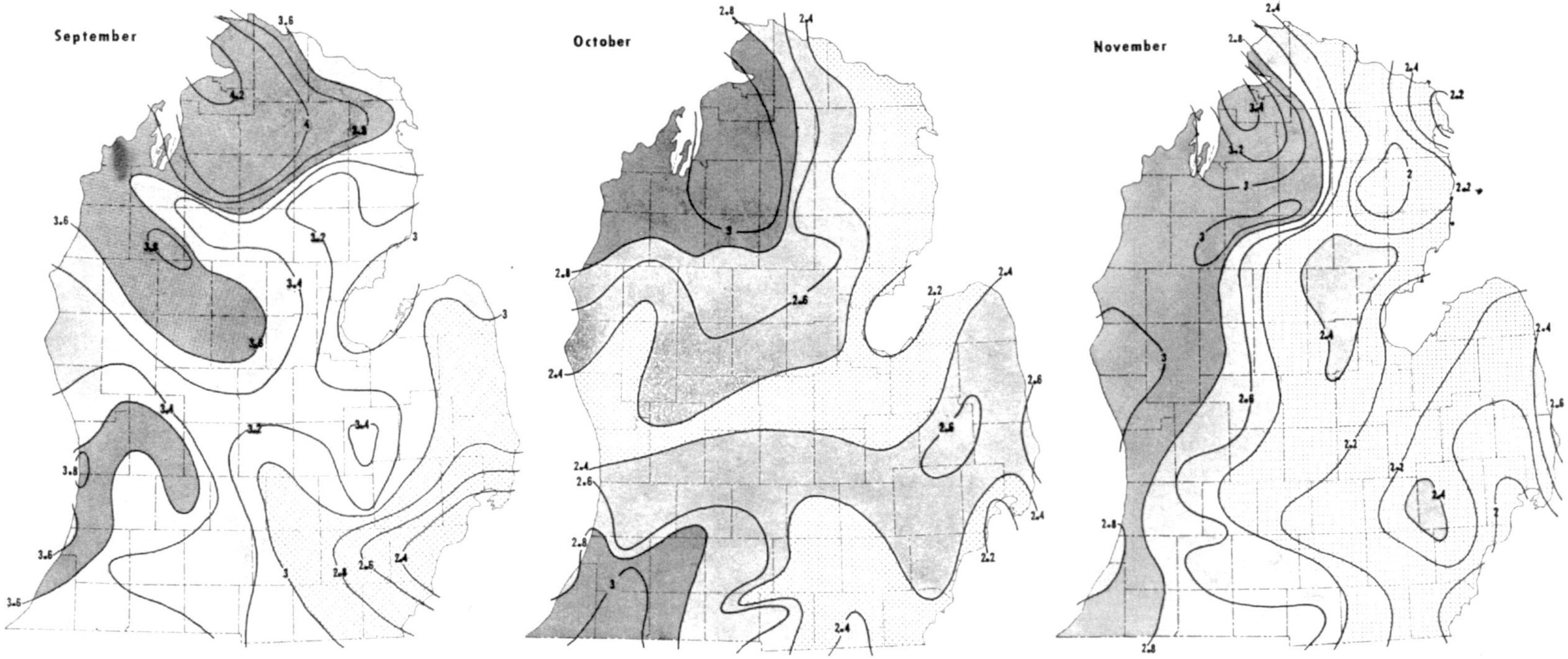

FIG. 4. Mean monthly precipitation, fall months (inches).

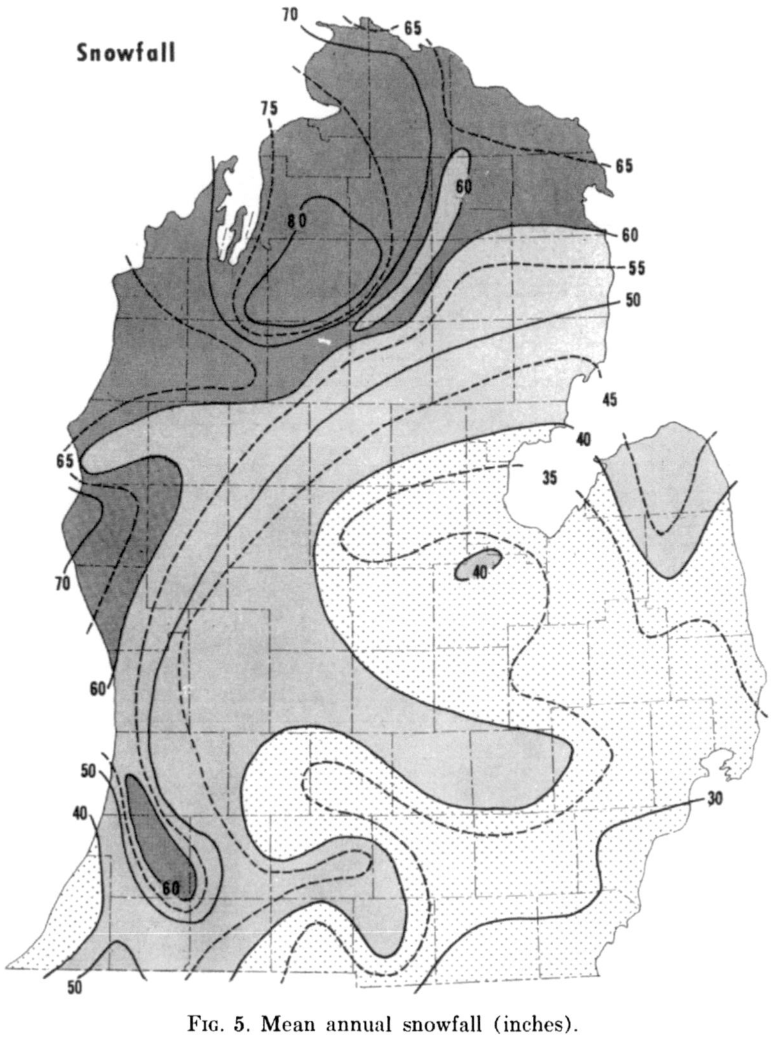

FIG. 5. Mean annual snowfall (inches).

southward along the lake to reach its winter position in the extreme southwest by December.

The monthly maps are integrated into four seasonal maps in Fig. 6. In order to accentuate the regional differences in the seasonal rainfall I have again filled in a three-toned pattern between isohyets

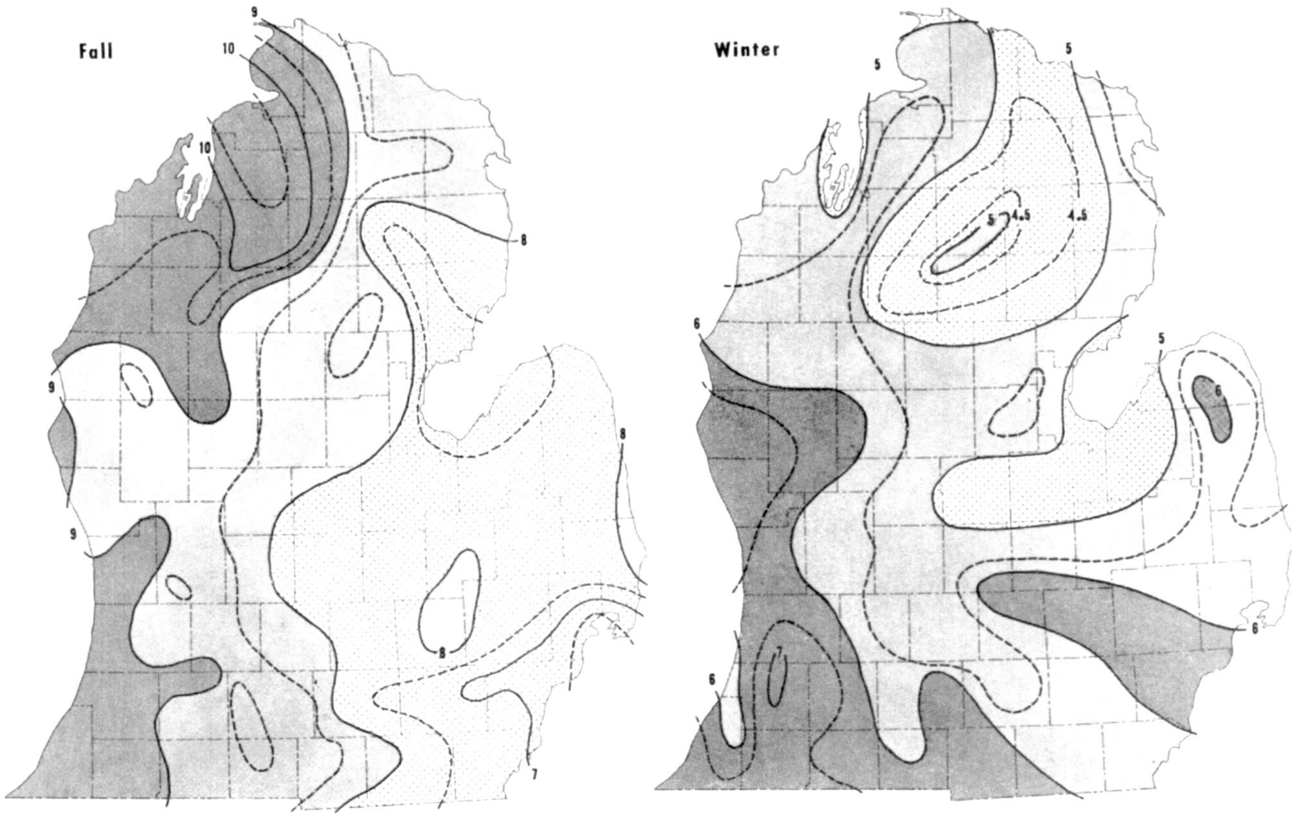

FIG. 6. Composite seasonal precipitation (inches).

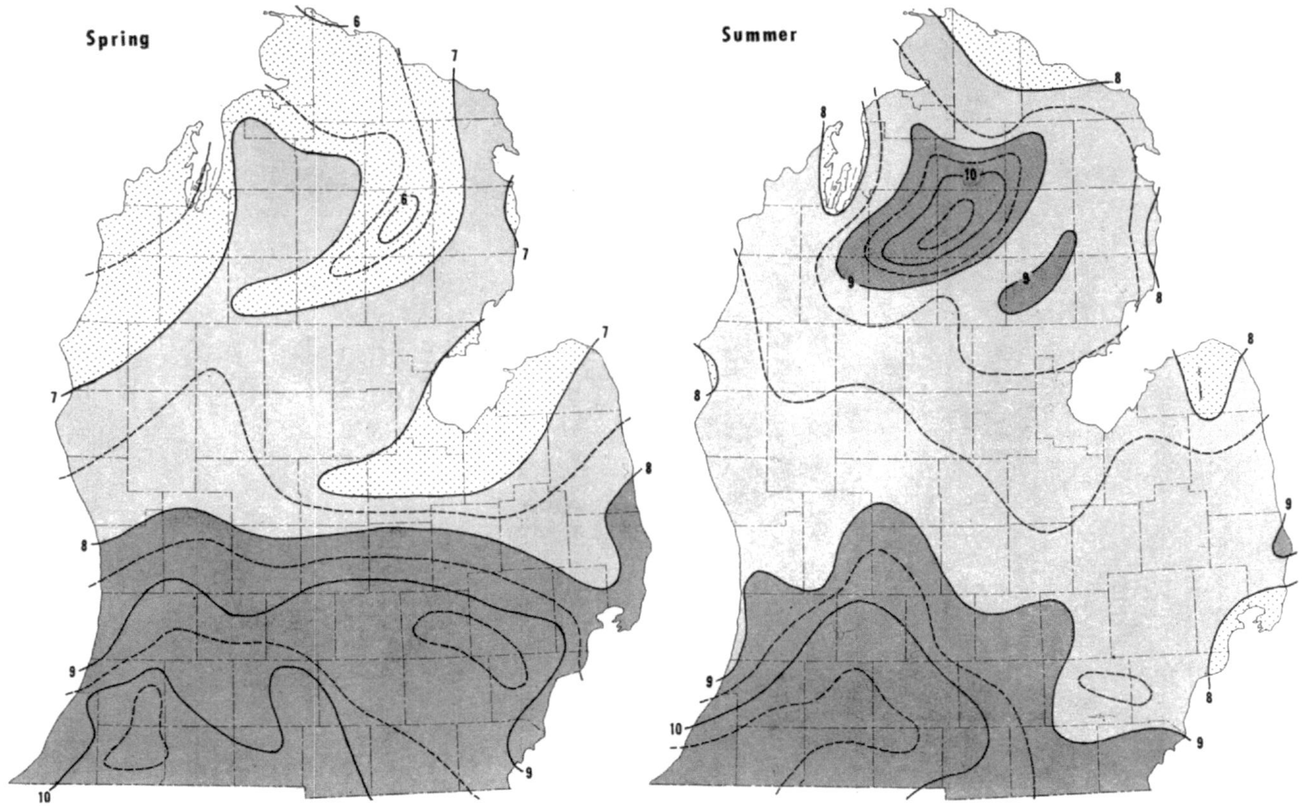

Fig. 6 *Continued*

delimiting areas of high, medium, and low precipitation. An "axis" of heaviest precipitation seems to rotate in a counterclockwise direction during the annual cycle. Thus the wettest/driest portion is the southwest/northwest in winter, the south/north in spring, and the northwest/southeast in fall, while the distribution can be called even in summer.

The annual map (Fig. 7) clearly defines the southwestern/north-

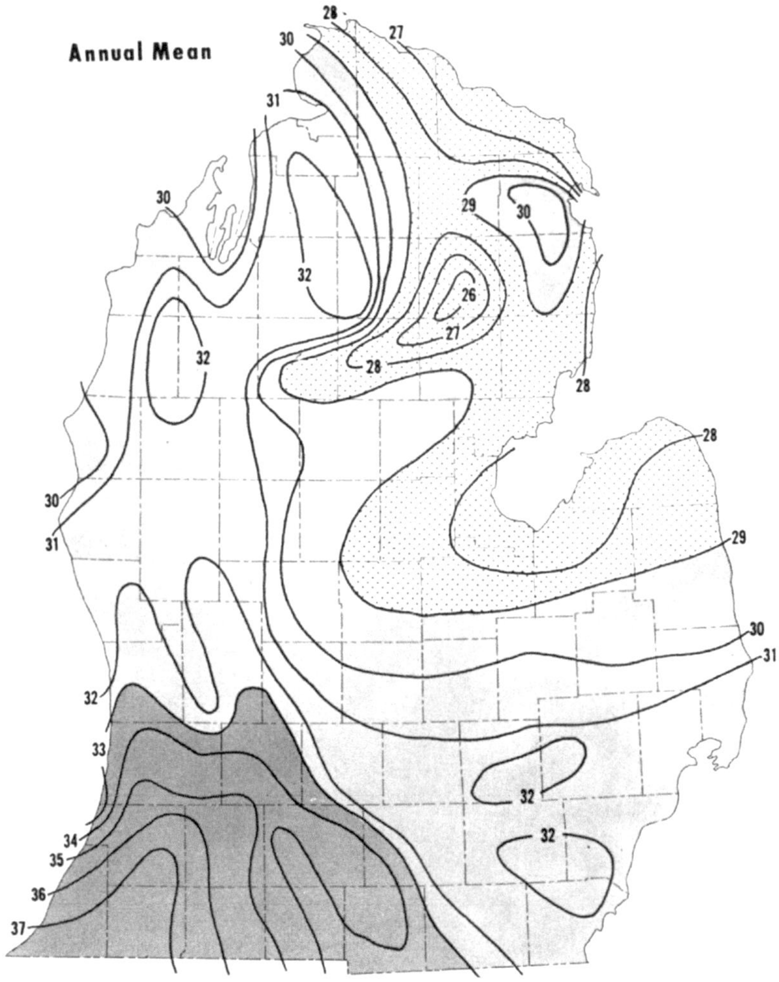

FIG. 7. Mean annual precipitation (inches), 1931–52 period.

eastern quarter as the wettest/driest part of the Lower Peninsula. The isohyets of 33 and 29 inches, respectively, enclose most of these extreme areas, which outline between them a belt of intermediately high precipitation crossing the peninsula diagonally from northwest to southeast.

The Temporal Precipitation Regime

The determination of regime types on the basis of mean values is always imaginary to a certain extent in areas of high monthly precipitation variability. The deviations from the mean monthly values observed in individual years reaches well above 100% in most months and at most stations. It is therefore necessary to establish whether the periodicity indicated by the means is statistically valid. This may be achieved by studying the scattering of the individual values of monthly series. After plotting the dispersion of the monthly means during the observation period I found that the elimination of the extreme quartiles (5 years each) yielded a relatively small mean deviation of the remaining variate, which then consisted of the 55%-fractile (12 years). This central 55% of the total variate is represented on Fig. 8 in the form of black bars in 15 type station histograms. The white dots within the bars are the monthly means of the observation period. For the majority of the stations the periodicity of the means is more or less approximated by the upper and lower ends of the 55%-fractile. With some unavoidable subjectivity we might say that a particular monthly value can be called a maximum if it exceeds that of the highest amount of the 50%-fractile of the preceding and succeeding month. Working strictly with means I would suggest that an arbitrary but meaningful single-month maximum can be said to exist if the highest monthly amount exceeds that of the next highest by at least 20% (of the maximum amount). A double-month maximum exists if the values of the two rainiest months lie within 10% of each other, both being 20% above the amount received in the next-highest month.

If these methods are applied to the 1931–52 records, the precipitation regimes of the 75 Lower Peninsula stations fall into three distinct types.

1. South of a line connecting Benton Harbor, Lansing, and Port Huron the maximum of the annual precipitation curve lies either in May ("acute" single-month spring maximum, type station: Ann Arbor, AA on Fig. 8) or both in May and June ("extended" double-

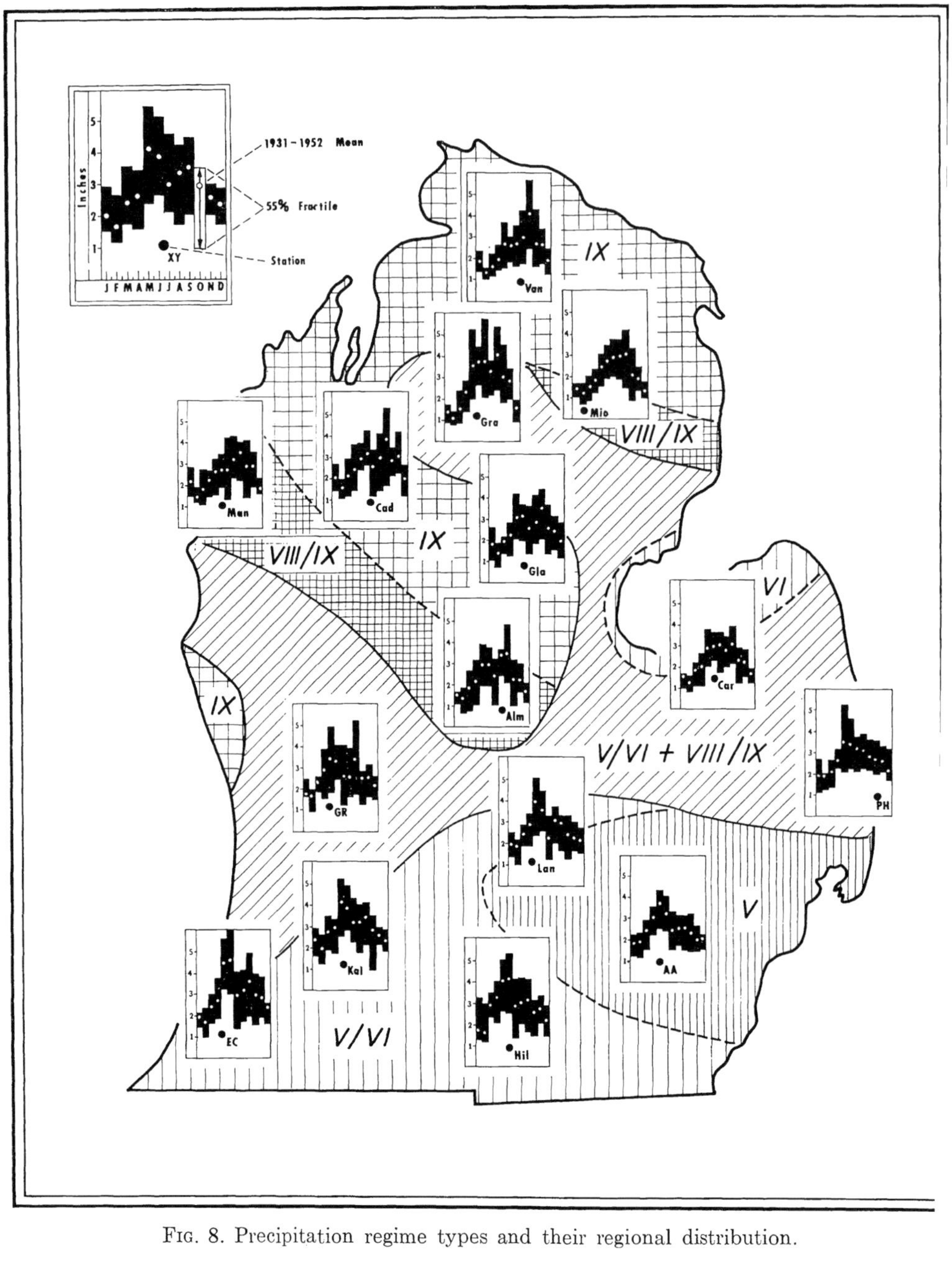

FIG. 8. Precipitation regime types and their regional distribution.

month spring maximum, type stations: Eau Claire, EC; Kalamazoo, Kal; Hillsdale, Hil; Lansing, Lan). An extension of this type fringes Saginaw Bay.

2. The northern section, with a peninsular extension to the center portion of lower Michigan, experiences a marked fall maximum, occurring either in September (acute fall maximum, type stations: Vanderbilt, Van; Cadillac, Cad; Gladwin, Gla) or both in August and September (extended fall maximum, type stations: Manistee, Man; Alma, Alm; Mio). A small outlier of this type is centered in Muskegon.

3. Most of the central third and a considerable portion of the north central interior takes part in both the spring and fall rainy seasons of the adjacent areas. The peaks are extended (May/June and August/September). The northern stations in this zone all have higher rainfall in fall than in spring while the reverse is true for the southerly locations (Type stations: Grand Rapids, GR; Caro, Car; Port Huron, PH; Grayling, Gra).

Common to most stations of the Lower Peninsula is the location of the lowest monthly mean in February—this in contrast to other areas marginal to the Great Lakes where winter maxima of precipitation are quite normal (Buffalo!). Another characteristic common to practically all stations in lower Michigan is a pronounced midsummer "slump" of rainfall. There is not a single station with a mean July maximum. At Lansing only 10 times has July been the wettest month of the year since observation began 97 years ago. (Curiously enough four of those ten wet Julys have been recorded within the last six years!)

Even though the previous analysis seems to have yielded acceptable results, another and perhaps more objective method to describe the temporal distribution of precipitation has been investigated. If we subject the monthly series to harmonic analysis and synthesis it is possible to express the observed curve of the means as the algebraic sum of sine curves. The first three harmonics of the sine curve have yielded phase angles which indicate a pronounced late spring/early summer maximum in the south, a late summer and fall peak in the north. This method also stands or falls, of course, with the "meaning of the mean," the latter being only a convenient but not always reliable statistical measure.

Conclusion

It is beyond the scope of this paper to penetrate into the intricacies which cause the anomalous behavior of Michigan's precipitation regime. Be it mentioned, however, that a corollary to this study has shed some light on the curious midsummer decline of rainfall, which is dynamically the most puzzling anomaly in our "continental" climate. The magnitude and exact mechanism of the "Lake Effect" on precipitation has never been satisfactorily demonstrated. Whoever wants to cope with this problem has to rely on the synoptic-dynamic basis of our climate, studying in particular the stabilizing and labilizing effect of the lakes in individual seasons upon traversing air masses and pressure systems. Hodgins (4) has already proved that the high frequency of anticyclonic control in midsummer has a damping effect upon convective activity. On the other hand, the relief effect along the east shore of Lake Michigan seems to be just strong enough to trigger the release of precipitation in conditionally unstable air masses. But the major genetic factor behind the march of seasonal precipitation over Michigan is undoubtedly the vigorous frontal activity along the polar front as it traverses the state in spring and fall in its northward and southward migration, respectively.

In conclusion it may be said that the temporal and regional distribution of precipitation in our area forms clearly recognizable patterns in spite of its marked variability from year to year. Further investigation both as to cause and effect of Michigan's precipitation should prove to be a highly important and rewarding activity for anyone seriously interested in the local physical geography.

Acknowledgments

The author is indebted to the following persons for help in some phase of this investigation: A. H. Eichmeier, Head, East Lansing Office, U. S. Weather Bureau; A. Stephen Messenger, Graduate Assistant, Michigan State University; Donald A. Pontti, Graduate Student, Michigan State University; and John F. Chandik, Special Assistant, Michigan State University.

LITERATURE CITED

1. Leighly, J. "Effect of the Great Lakes on the Annual March of Temperature in the Vicinity." PMASAL, 27 (1941), pp. 377–414.

2. Seeley, D. A. "The Climate of Michigan and Its Relation to Agriculture." Thesis for the degree of Master of Science, Michigan State University, 1917.
3. Weather Bureau, U. S. Department of Commerce, "Climatic Summary of the United States-Supplement for 1931 through 1952." Climatography of the United States No. 11-16. Washington, 1956.
4. Hodgins, L. E., "Genetic Climatology of the Great Lakes Region." Thesis for the degree of Master of Arts, Michigan State University, 1960.

PAPERS OF THE MICHIGAN ACADEMY OF SCIENCE, ARTS AND LETTERS
VOL. XLVII, 1962 (1961 MEETING)

THE WATER BALANCE OF THE LOWER PENINSULA OF MICHIGAN

A. STEPHEN MESSENGER

Michigan State University

INTRODUCTION

Concepts.—Water is an essential nutrient for plant growth and is needed in much larger quantities than any other; but whereas a large proportion of every other nutrient absorbed by the plant is retained, the outstanding characteristic of the water is its continuous one-way flow from the soil through the roots up the stems into the leaf surface, where it is evaporated mainly inside the stomata, through which it diffuses into the air.

The rate at which water is transferred from the soil into the air by the plant is controlled by four separate processes: the transfer of water from the soil through the cell protoplasm of the absorbent rootlets into the vascular system; the transfer from the vascular system through the protoplasm of the leaf cells bounding the stomata; the energy supply falling on the leaves that can be used for evaporating the water in the stomata and from the surface of the leaf; and the transfer of the water vapor from inside the stomata into the main body of the air by diffusion and convection.

When moisture supplies are unlimited the rate of evapotranspiration from a large vegetation-covered surface is controlled primarily by atmospheric factors. This climatically controlled rate of evapotranspiration has been termed "potential evapotranspiration" by C. W. Thornthwaite.

Wind, humidity, and solar radiation all affect this potential rate. Since these factors vary with temperature, however, the latter, in addition to average day length, has been used to compute the potential rate of evapotranspiration.

When moisture supplies are limited, on the other hand, this potential rate will not be realized, the actual evapotranspiration rate being controlled by the additional factors of soil moisture tension and specific plant characteristics.

On medium and fine textured soils, soil moisture tension increases as the soil water supply drops below field capacity, thereby making it impossible for the potential evapotranspiration rate to be maintained. The computed (and measured) difference between the reduced amount of evapotranspiration and the computed potential amount is termed the *water deficit.*

One of the specific plant characteristics mentioned above is depth of rooting, a factor which can be easily included in calculations of the water deficit. The leaf surface albedo or reflectivity is another factor which affects the actual rate of evapotranspiration, but the differences between species of trees are apparently not great, nor are the differences between grass and many green-leaved vegetables. The differences between major groups of plants, though important, are not considered in the current computational procedure, probably because the albedo also varies with leaf moisture content.

When soils are filled to their capacity with water, any additional precipitation not offset by evapotranspiration induces percolation of the excess into the ground water system. In water balance computational procedures this excess is termed as the *water surplus.*

Objectives of the present study.—The present study was primarily designed to characterize the Lower Peninsula's climate by a method or methods which would best portray the role of present climate in phytogeography, plant growth, and soil genesis. An additional purpose was to establish qualitative relationships between the water balance parameters and the more permanent features of the climate in an attempt to estimate climatic conditions of the past which have affected both plants and soils. Correlations and relationships have not yet been studied, except in a very general way, and consequently only the climatic analysis is reported here.

Methods.—Mean annual potential evapotranspiration (hereafter referred to as PE) and water deficits were computed from (1) mean values of precipitation and temperature for the period 1931–52, (2) mean values for the 10 driest years (based on annual precipitation) out of the first 20 years of complete records since 1930, and (3) the accumulation of monthly deficits through the two consecutive years (1929–50) having the highest cumulative sum. Mean annual water surplus values were computed only from (1) above.

Acknowledgments.—The sincere interest and coöperation of Dr. Dieter Brunnschweiler and Dr. E. P. Whiteside of Michigan State

University is hereby gratefully acknowledged. Thanks is also due to Mr. A. H. Eichmeier, the state climatologist, for his excellent coöperation.

Presentation of Data

Mean annual PE, 1931–52, (Fig. 1).—The results of this analysis bring out the lower PE values of the northern one-third of the peninsula as would be expected, the absolute lowest value being 546 mm. or 21.5 inches at Vanderbilt Trout Station and the highest, 689 mm. (27.1 inches) at Adrian.

Patterns that were not expected were: (1) the isolation of the southern part of Grand Traverse Bay, where the value of 606 mm. (23.9 inches) is comparable to areas 20–25 miles to the southwest, and (2) the peninsular position of Lansing and Milford, which have PE values as much as 48 mm. or 1.9 inches *lower* than stations to the north in the Saginaw Valley area.

Mean annual water deficit, 1931–52, (Fig. 2).—In all computations of deficits and surpluses here presented, the soil available moisture-holding capacity of 14 inches was used. This value approximates the maximum found in Michigan for a soil profile 4 feet deep.

The mean annual water deficit varies from a low of 8 mm. (0.3 inches) at Grayling to a high of 62 mm. (2.4 inches) at Mount Clemens, both of these values fitting well into the respective regional patterns.

Mean annual water deficit of the dry 50% of twenty years, 1931–58, (Fig. 3).—The pattern exhibited by this analysis indicates that nearly all of the northern one-half of the peninsula is relatively free from *frequent* high-deficit years, as is an area in the southwestern section of the peninsula. The area of high deficits has a distribution similar to that of the normal; Flint and Milford, however, are subject to periodic dryness of such a magnitude that they connect the extreme southeastern dry area with the Saginaw Valley dry area. The disjunction between the two low deficit areas is related to the frequency of low July precipitation values, a belt between Holland and Bay City being characterized by a 16–24 per cent frequency of July precipitation values of less than one inch (1909–58).

The extremes range from a low of 24 mm. (0.9 inches) at Alpena Camp to a high of 102 mm. (4.0 inches) at Mount Clemens.

Mean annual water deficit of the driest two consecutive years,

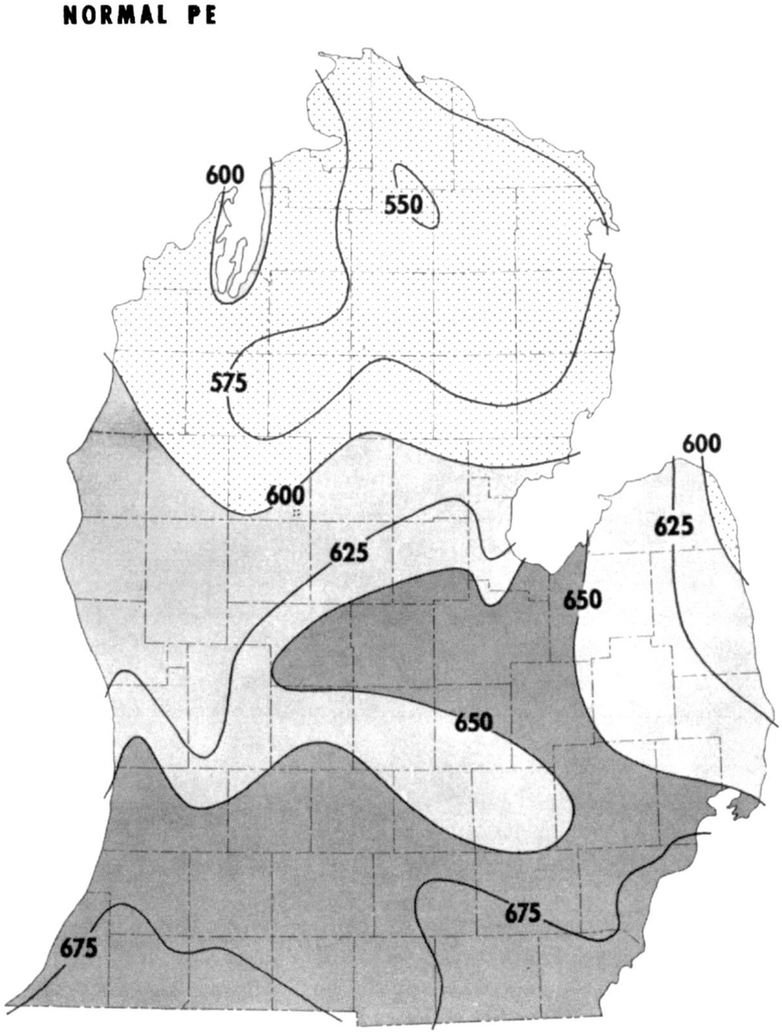

FIG. 1

1929–50, (Fig. **4**).—Results of this analysis show a further reduction of the low-deficit area in southwestern Michigan, nearly all stations in the northern one-half of the peninsula thus exhibiting lower deficits than those in the southern one-half.

Traverse City and Alpena (city) exhibit higher deficits than

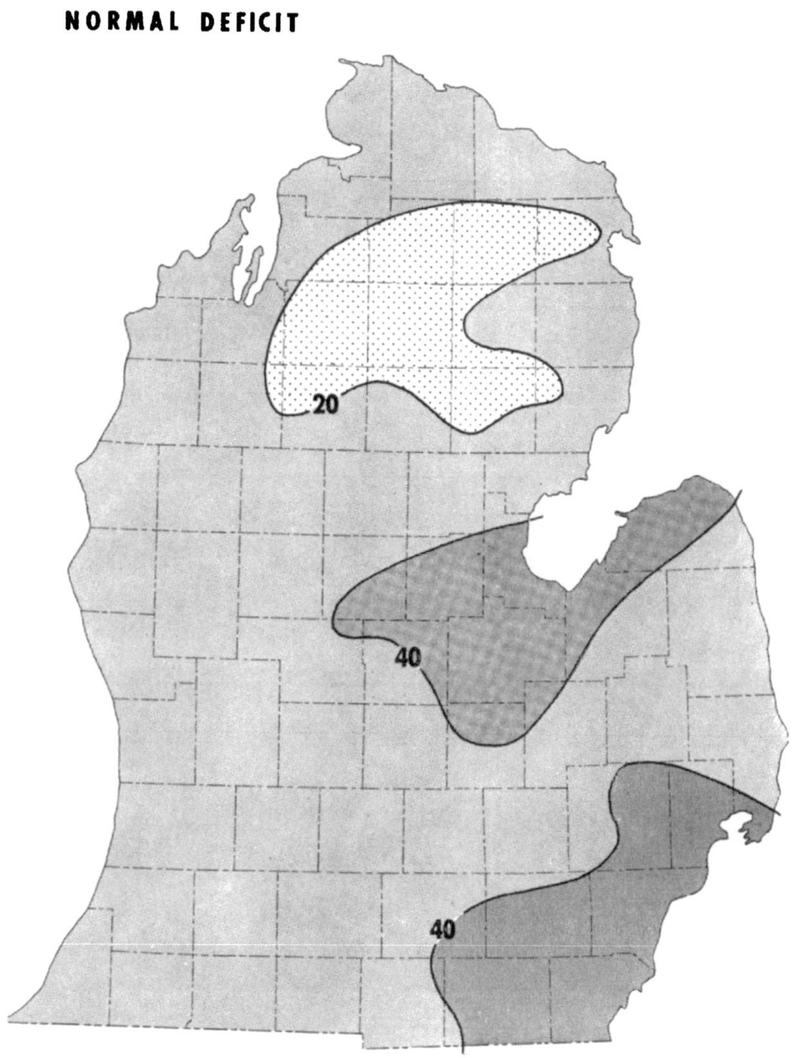

FIG. 2

surrounding areas. Zones with comparable deficits lie 50 to 80 miles south of these stations.

A value of 99 mm. (4.0 inches) isolates Lansing, but the stations at Charlotte and Battle Creek have values of 106 and 104 mm.

MEAN DEFICIT, DRY 50% OF YEARS

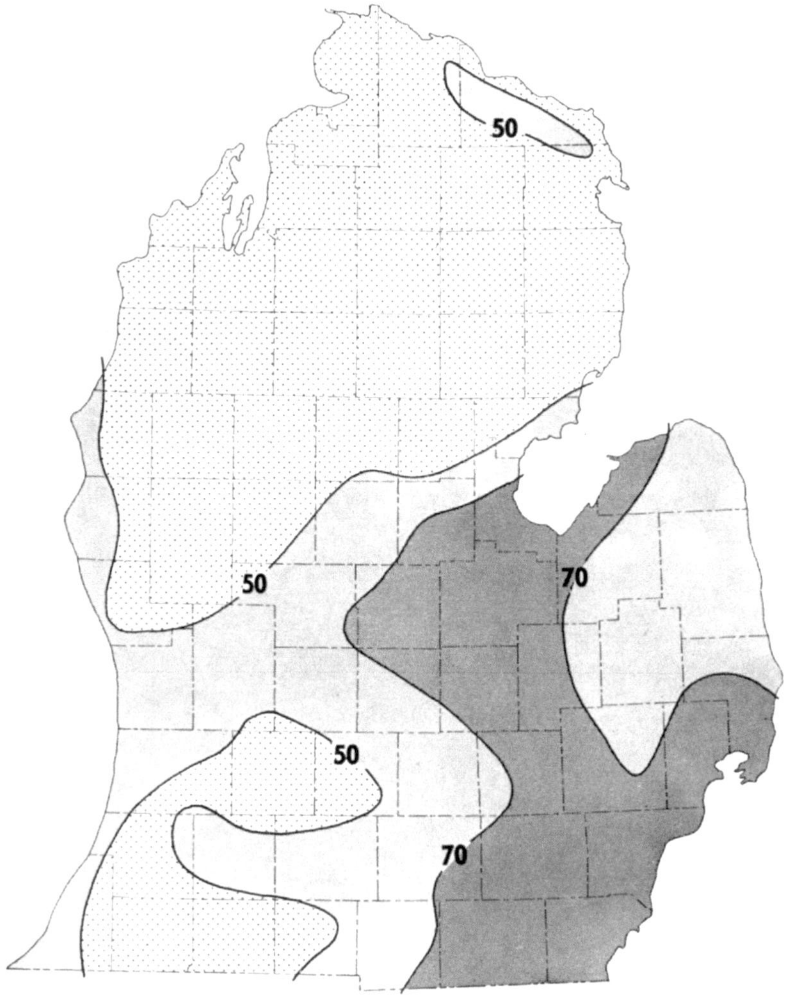

Fig. 3

respectively, indicating a peninsula-like connection between Lansing and the Cassopolis-Dowagiac area.

The extremes range from a low of 69 mm. (2.7 inches) at Alpena Camp to a high of 176 mm. (6.9 inches) at Midland.

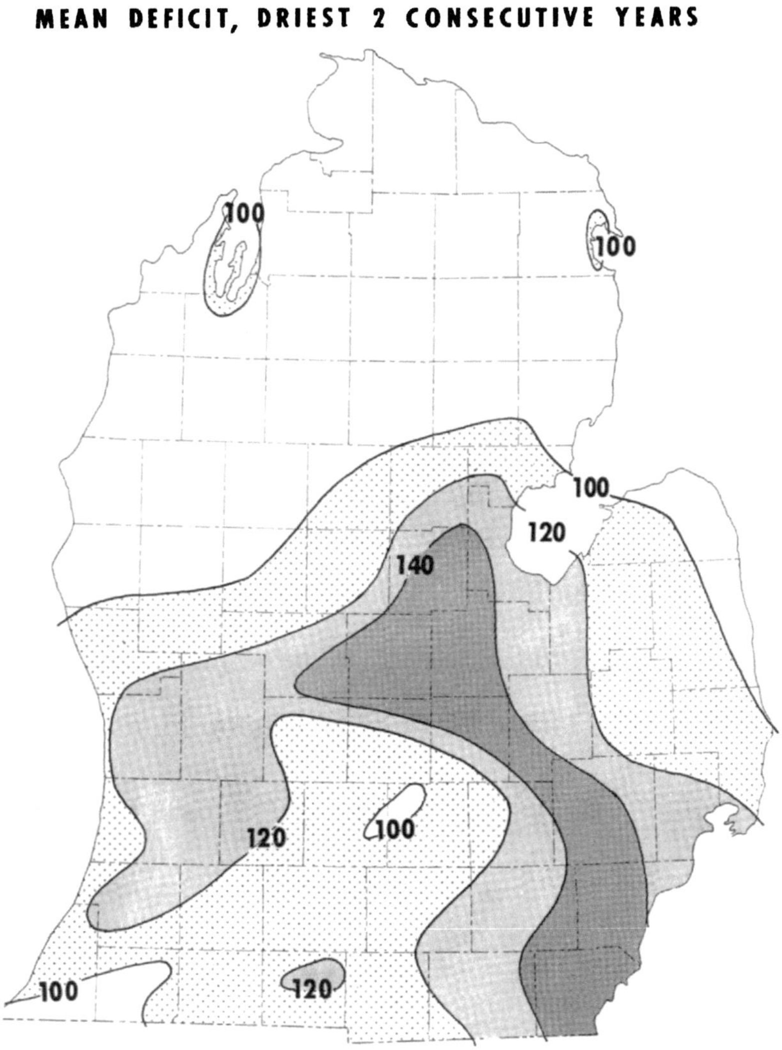

FIG. 4

Mean annual water surplus, 1931–52, (Fig. 5).—The belt of relatively high surpluses (above 200 mm. or 7.9 inches), extending from northwest to southwest, is strongly contributed to by snowfall. The northern section (above 43°N), however, is also affected by fall pre-

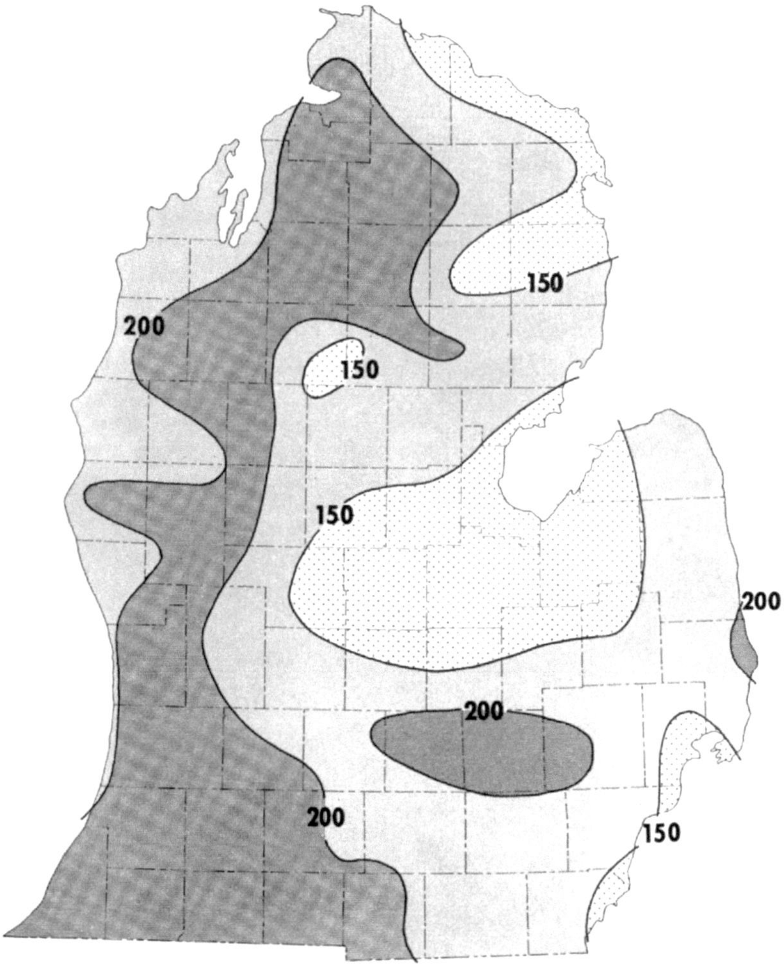

FIG. 5

cipitation, while the southern section is influenced mainly by early spring precipitation. This seasonal distribution of surplus may be important to soil processes because of the differences in soil temperature profiles between these two seasons. An additional analysis, not presented here, has indicated that sands in the northern section normally are brought back to field capacity before December and before soil temperatures are too low for chemical activity to take place.

The highest surplus value computed was 304 mm. (12.0 inches) at Paw Paw. In the northern section of the high surplus belt, the maximum is 269 mm. (10.6 inches) and is shared by Grayling and East Jordan. The minimum amount of surplus is exhibited by Bay City, with a value of only 74 mm. or 2.9 inches.

Discussion

The regional differences in seasonal precipitation brought out by Brunnschweiler and the regional differences in the water balance are of such a magnitude that some climate-dependent natural phenomena might be expected to vary accordingly.

Thornthwaite found that growth of a particular plant increased with increasing PE. Hare found that the southern limit of the boreal forest in North America corresponded closely to the 520 mm. (20.5 inches) iso-PE line.

Water deficit values, if used properly, should give an indication of the relative severity of drought.

Water surplus values should be related to the relative degree of leaching where similar-textured soils are involved and temperatures are not greatly different.

Conclusions

Regional differences in the water balance of the Lower Peninsula are apparent in all the analyses thus far completed. The high normal or periodic deficits in the southeastern section warrant particular attention. The strong gradient of west-to-east decrease in surplus values seems also noteworthy from the standpoint of soil genesis.

Many geographic anomalies exist in Michigan. Some may be products of bygone conditions, but how is this to be determined unless an exhaustive analysis of the present environment is made?

LITERATURE CITED

Brunnschweiler, Dieter. 1961. Precipitation Regime in the Lower Peninsula of Michigan. Prog. and Abstr. of Pap., Geog. Sec., MASAL.

Mather, J. R. 1954. The Measurements of Potential Evapotranspiration. Publ. Climatology, 7 (1).

RUSSELL, E. JOHN. 1950. Soil Conditions and Plant Growth. Longmans, Green and Co., London.

THORNTHWAITE, C. W., AND J. R. MATHER. 1955. The Water Balance. Publ. Climatology, 8 (1).

——. 1957. Instructions and Tables for Computing Potential Evapotranspiration and the Water Balance. Publ. Climatology, 10 (3).

——, AND DOUGLAS B. CARTER. 1958. Three Water Balance Maps of Eastern North America. Resources for the Future, Inc., Washington, D. C.

WEATHER BUREAU, U. S. DEPT. OF COMMERCE. Climatic Summary of the United States—Supplement for 1931 Through 1952. Michigan.

WHITE, DONALD P. 1958. Available Water: The Key to Forest Site Evaluation. First North American Forest Soils Conference. Agric. Exp. Sta., Mich. State Univ., East Lansing.

Papers of the Michigan Academy of Science, Arts, and Letters
Vol. XLVII, 1962 (1961 Meeting)

EFFECTS OF VEGETATIVE COVER ON FROST PENETRATION

F. M. HELMREICH, *Bangkok, Thailand*

O. H. CLARK, *Michigan Conservation Department*

Plan and Purpose of Field Study

In 1953 a co-operative agreement between the Rural Rehabilitation Corporation and the Michigan Department of Conservation was initiated to resolve certain problems relating to water in the glacial drift soils in the northern section of lower Michigan. A small watershed, approximately 600 acres in size, was selected near the town of Hoxeyville, Wexford County, Michigan. This selection was made because it encompassed an area of overgrazed pasture, fields cultivated in small grains and hay, and wood lots typical of many parts of northern Michigan. The drainage system consists of two tributaries with intermittent flow combining to form a single channel at a point about 300 feet upstream of its juncture with Poplar Grove Creek, a tributary to the south branch of the Manistee River. The soils are uniformly coarse in texture and at the upper end are designated as Kalkaska sand. This is the area significant to this paper. The upper part is slight rolling with minor changes in elevation. The extreme upper part of one tributary originates in a Norway pine (*Pinus resinosa*) and jack pine (*P. banksiana*) plantation, planted in 1936 at 6-foot centers in the bottoms of the plowed furrows. To the south of these plantations lies a natural mixed hardwood area of about the same age, made up of aspen (*Populus tremuloides*) and oak (*Quercus alba*). These plantations are the areas upon which the research in this paper is based. It developed in the process of the investigations that the frost penetration and time of melt under different vegetative cover vary significantly and might affect materially the surface runoff in the spring. This paper will discuss and analyze the data collected relating to these frost conditions.

Procedures and Results

In the spring and summer of 1955 Bouyoucos moisture blocks were installed at depths of 6, 12, 24, and 36 inches and in a few places

48 inches in various locations on the watershed. From observations during the winter of 1955–56 it was decided to establish special frost study locations with blocks placed at depths of 2, 4, 6, 8, 10, and 12 inches. Two of these special study sites were located in jack pine, Norway, and hardwood stands, and one site in an open pasture immediately adjacent to these stands. In addition there was one site each in the Norway and jack pines with blocks at depths of 6, 12, 24, 36, and 48 inches, and one site in the jack pines with blocks at 6, 12, 24, and 36 inches. After having been calibrated for uniformity, the blocks were installed in June and July 1956. Installations were made with a three-inch auger and care was taken not to mix the different horizons in back-filling. Considerable effort was made to obtain good contact between the blocks and the soil.

The soil at these installations was classified as Kalkaska sand, a soil of high permeability with depth to ground water of 40 to 50 feet. The study sites in the pine plantations were located in the bottoms of the planting furrows, with 100 feet separating the sites. Readings on the resin-treated plaster of paris blocks were taken with a Bouyoucos soil moisture meter, which reads directly the percentage of available moisture. No corrections for soil temperature were made. When freezing occurs the resistance in the block rises rapidly, as if the block were dry. When the block is completely frozen the meter will read zero per cent available moisture. Observation showed that readings between 5 and 40 per cent available moisture were an indication that partial freezing or thawing was in progress, provided the soil was at or near field capacity before freezing started.

The electrical resistance method for measuring frost has the advantage that no disturbance of the soil, or even the snow cover over the area to be checked, is necessary. Care was taken during the readings that nobody tramped down the snow near the blocks. Readings were taken at least once a week and more frequently during the breakup periods. A temperature recorder was located in the watershed. It developed mechanical difficulties, however, and the mean temperatures therefore were taken from the official U. S. Weather Bureau station at Cadillac, Michigan, about 15 miles distant. This station had shown good correlation with the watershed during the previous winter.

Precipitation, temperature, and frost penetration.—All precipitation measurements were reduced to water content. In October, 1956, precipitation over the watershed amounted to 0.75 inch, all in the form of rain. Rain amounting to 1.0 inch continued until November 15, followed by 0.15 inch of snow, followed on November 20 by 0.25 inch of rain. On November 21 and 22 there was altogether 0.85 inch of snow, resulting in a snow cover of 8 to 12 inches over the watershed. Another 0.25 inch of snow fell by November 30. As a result of temperatures above 32° F. together with 0.25 inch of rain by December 6, most of the snow disappeared from the open ground but a few inches were left in the pine and hardwood plantations. Another 0.15 inch of rain on December 12 cleared the rest of the snow from the open ground, but snow remained under forest cover. The following low temperatures, while having no effect on the forest-covered soil, caused the soil in the open ground to freeze to a depth of 8 inches (Fig. 2). Because of temperatures above 32° F. and 0.45 inch of rain on December 23 followed by 0.3 inch of rain with snow, this frost disappeared again at Christmas. Another 0.4 inch of snow fell by the end of December and 0.35 inch up to January 5, with daily mean temperatures around 20° F. Snow cover over the watershed was between 8 and 14 inches. Fig. 1 shows that by January 5 frost site F2 showed frost at the 2-inch, 4-inch, and 6-inch levels and L12 frost at the 6-inch level. Both are locations in the Norway pine plantation. F4, also in the Norway pine plantation, showed no frost. All other locations were free of frost except for F1, in the jack pines, which showed frost at 2 inches.

With mean temperatures below 20° F. and another 0.45 inch of snow on January 12, all blocks in the Norway pine plantation showed frost down to the 8-inch level; the other locations remained unchanged. On January 19, after another 0.25 inch of snow and mean temperatures below 10° F., all three locations in the Norway pines, F2, F4, and L12, showed frost down to the 1-foot level. F3, in the jack pines, also showed frost to the 1-foot level, while F1 showed frost only to the 8-inch level. L11 and L13, also in the jack pines, showed frost at the 6-inch level but no frost at the 12-inch level. F7, in the open ground, showed frost only at the 2-inch level. There was no frost in the hardwoods.

It should be mentioned here that the readings for F7 do not appear to show the worst condition for the open ground, since there was usually a greater cover of drifted snow over F7 than in some more

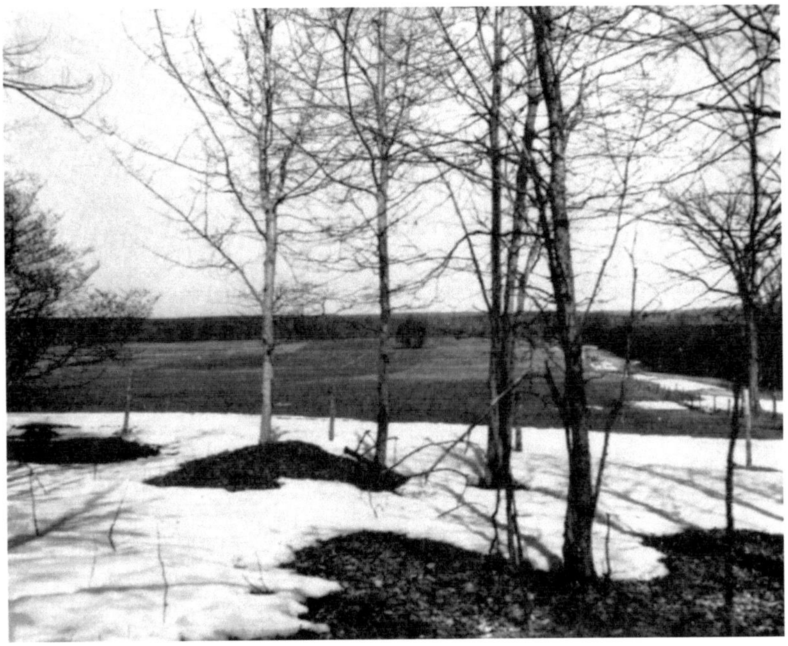

Fig. 1. Hardwood stand (foreground), jack pine and Norway pine (at right), and open field.

exposed spots of the watershed. L8 and L9 give a better picture of the exposed places and agree closely with other locations there, not incorporated in this paper.

A mean temperature of 38° F., 0.2 inch of rain on January 21, and 0.6 on January 22 was the cause for a general reduction of frost depths when readings were taken on January 26. Again it should be noted that F2, F4, and L12, in the Norway pines, show frost down to the 6-inch level, with L12 showing frost even at the 1-foot level, while the frost depth in the jack pine and in the open ground, which was depleted of most of its snow cover, is much shallower. Still there is no indication of frost under hardwood cover. By January 31 another 0.4 inch of snow fell, with mean temperatures around 15° F., and by February 2 all except the hardwood locations show frost down to the 6-inch level and some to the 1-foot level.

Up to February 22, with mean temperatures around 15° F. and a total of 0.6 inch of snow, freezing progressed in all locations except in the hardwoods. F2 and F4 in the Norway pines have all installations (greatest depth 12 inches) frozen, and L12 shows frost at the 6-inch, 1-foot, and 2-foot levels, which is the greatest frost depth recorded anywhere in the watershed.

Less severe were the conditions in the jack pines, where F3 showed a frost depth of only 8 inches while F1, L11, and L13 showed frost down to the 1-foot level, but not at the 2-foot level. The same applies to the open ground, where, though F7 showed frost only to the 6-inch level for reasons mentioned before, L8 and L9, which have less snow cover, showed frost at the 6-inch and 1-foot levels but no frost at the 2-foot level. Hardwoods were still frost-free even at the 2-inch level.

A warm spell on February 24 and 25 together with 0.15 inch of rain on February 25 caused a general reduction in frost depth but also left the open ground bare of snow cover except in drifted places. Again it will be noted that measurements on March 1 indicate that the Norway pine plantation was least affected by the thaw. Only the 12-inch block of F4 temporarily thawed, while F2 showed frost at all blocks and L12 at the 6-inch, 1-foot, and 2-foot levels. The jack pine plantation seems to have thawed in places completely, and the 2 and 4 inches of frost recorded at F3 and F1 (Fig. 2) were probably caused by the low temperatures after February 25. The open ground stayed frozen at the 6-inch and 1-foot levels, and the hardwoods were still unaffected by frost.

The daily mean temperatures stayed below 32° F. up to March 10, and there were only traces of snow. Data taken on March 8 (Tab. I) give the following picture. Norway pine: F2 and F4, frost at all installations; L12, frost down to the 2-foot level. Jack pine: F1, L11, and L13, frost down to the 1-foot level; F3, frost to the 10-inch level. Open ground: F7, frost to the 6-inch level; L8 and L9, frost to the 1-foot level, with the 2-foot block of L8 showing signs of freezing. Hardwoods: For the first time this winter, one block at the 8-inch level of F5 and one block at the 2-inch level of F6 show frost. All other installations are still frost-free.

March 11 brought a rise in temperature together with 0.5 inch of rain. From March 12 to 15 the daily mean temperature was well above 32° F., with another 0.5 inch of rain in the evening of March 14. This breakup period had no effect whatsoever on the soil in the

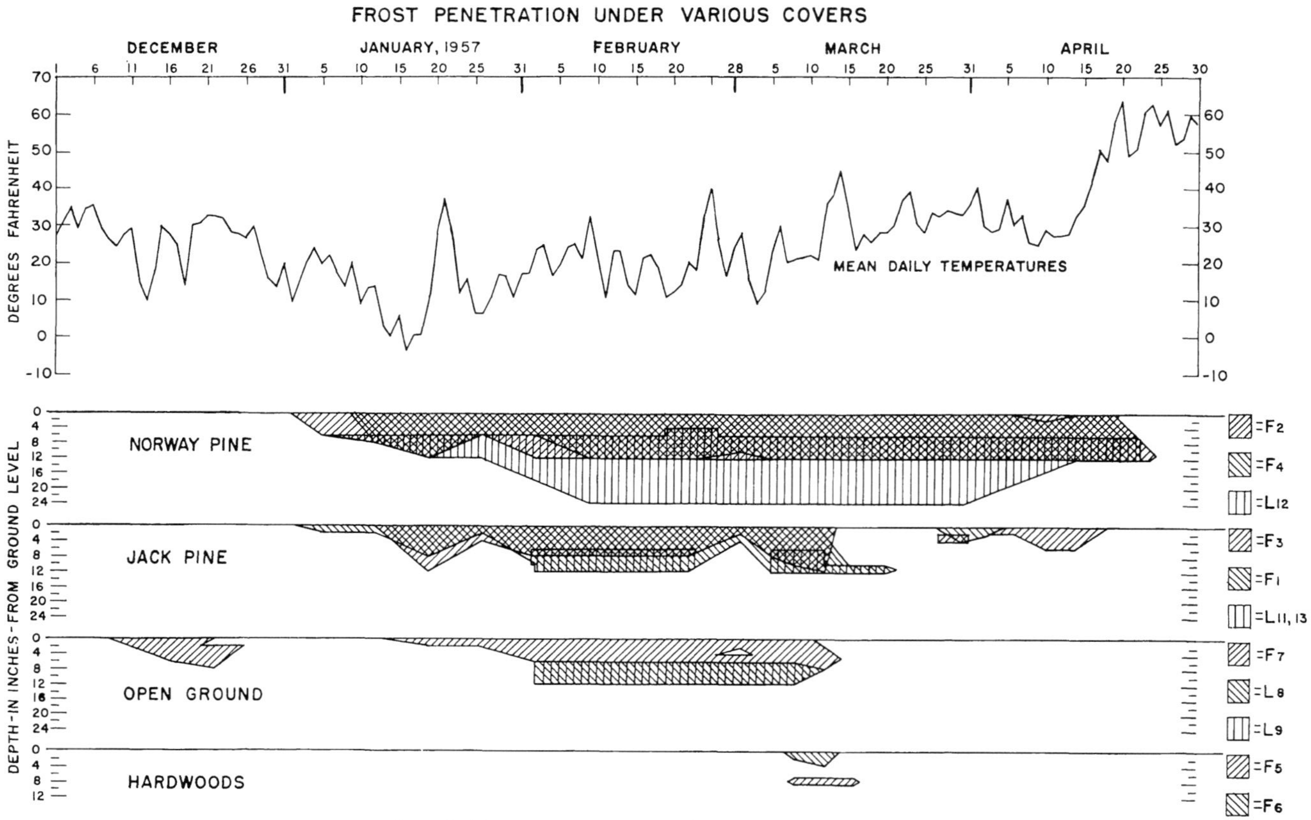

Fig. 2. Frost penetration under various covers.

TABLE I

Soil Moisture Percentages at Various Depths

Zero = frozen block

March 8, 1957

	Norway pine			Jack pine				Open ground			Hardwoods	
	F2	F4	#12	F1	F3	#11	#13	F7	#8	#9	F5	F6
2″	0	0	—	0	0	—	—	0	—	—	64	0
4″	0	0	—	0	0	—	—	0	—	—	67	48
6″	0	0	0	0	0	0	0	0	0	0	70	68
8″	0	0	—	0	0	—	—	66	—	—	0	70
10″	0	0	—	0	0	—	—	66	—	—	64	68
12″	0	0	0	0	66	0	0	24	0	0	67	75
2 ft	—	—	0	—	—	60	62	—	18	62	—	—
3 ft	—	—	60	—	—	57	44	—	62	64	—	—

March 12, 1957

	Norway pine			Jack pine				Open ground			Hardwoods	
	F2	F4	#12	F1	F3	#11	#13	F7	#8	#9	F5	F6
2″	0	0	—	20	30	—	—	54	—	—	65	20
4″	0	0	—	10	15	—	—	15	—	—	67	20
6″	0	0	0	54	10	35	42	0	—	—	64	68
8″	0	0	—	20	—	—	—	15	—	—	0	70
10″	0	0	—	0	58	—	—	66	—	—	64	68
12″	0	0	0	0	0	0	0	74	—	—	67	75
2 ft	—	—	0	—	—	54	58	—	—	—	—	—
3 ft	—	—	62	—	—	58	48	—	—	—	—	—

March 16, 1957

	Norway pine			Jack pine				Open ground			Hardwoods	
	F2	F4	#12	F1	F3	#11	#13	F7	#8	#9	F5	F6
2″	0	0	—	65	70	—	—	64	—	—	65	65
4″	0	0	—	58	65	—	—	66	—	—	67	58
6″	0	0	0	65	65	62	65	68	—	—	70	68
8″	0	0	—	63	42	—	—	68	—	—	32	70
10″	0	0	—	68	63	—	—	68	—	—	64	68
12″	0	0	0	15	48	38	15	76	—	—	67	75
2 ft	—	—	0	—	—	62	60	—	—	—	—	—
3 ft	—	—	62	—	—	58	48	—	—	—	—	—

March 23, 1957

	Norway pine			Jack pine				Open ground			Hardwoods	
	F2	F4	#12	F1	F3	#11	#13	F7	#8	±9	F5	F6
2″	0	0	—	65	72	—	—	80	—	—	66	75
4″	0	0	—	58	65	—	—	78	—	—	68	74
6″	0	0	0	65	66	64	35	78	67	58	70	68
8″	0	0	—	64	66	—	—	76	—	—	64	72
10″	0	0	—	68	64	—	—	72	—	—	64	68
12″	0	0	0	54	68	66	20	78	62	62	67	75
2 ft	—	—	0	—	—	62	62	—	62	62	—	—
3 ft	—	—	62	—	—	58	58	—	62	64	—	—

March 27, 1957

	Norway pine			Jack pine				Open ground			Hardwoods	
	F2	F4	#12	F1	F3	#11	#13	F7	#8	#9	F5	F6
2″	0	0	—	62	72	—	—	72	—	—	66	75
4″	0	0	—	58	20	—	—	75	—	—	68	78
6″	0	0	—	65	66	—	—	78	—	—	71	68
8″	0	0	—	64	66	—	—	78	—	—	66	72
10″	0	0	—	68	64	—	—	74	—	—	64	68
12″	0	15	—	60	68	—	—	82	—	—	67	76
2 ft	—	—	—	—	—	—	—	—	—	—	—	—
3 ft	—	—	—	—	—	—	—	—	—	—	—	—

March 30, 1957

	Norway pine			Jack pine				Open ground			Hardwoods	
	F2	F4	#12	F1	F3	#11	#13	F7	#8	#9	F5	F6
2″	0	0	—	0	70	—	—	57	—	—	65	75
4″	0	0	—	35	0	—	—	64	—	—	67	74
6″	0	0	0	65	40	64	28	67	63	55	70	68
8″	0	0	—	64	65	—	—	68	—	—	64	69
10″	0	0	—	68	63	—	—	66	—	—	63	66
12″	0	15	0	60	67	58	40	75	64	63	66	67
2 ft	—	—	15	—	—	63	62	—	64	65	—	—
3 ft	—	—	62	—	—	58	60	—	64	66	—	—

TABLE I—*Continued*

March 20, 1957

	Norway pine			Jack pine				Open ground			Hard-woods	
	F2	F4	#12	F1	F3	#11	#13	F7	#8	#9	F5	F6
2″	0	0	—	63	68	—	—	64	—	—	65	72
4″	0	0	—	58	64	—	—	66	—	—	67	70
6″	0	0	—	63	65	—	—	68	—	—	70	68
8″	0	0	—	63	64	—	—	68	—	—	64	70
10″	0	0	—	68	63	—	—	68	—	—	64	68
12″	0	0	—	38	66	—	—	76	—	—	67	75
2 ft	—	—	—	—	—	—	—	—	—	—	—	—
3 ft	—	—	—	—	—	—	—	—	—	—	—	—

Norway pine plantation, which stayed frozen down to the 2-foot level (Tab. I). Readings on March 12 in the jack pines indicated that, though all 1-foot blocks were still frozen, thawing was in progress at most other locations, with some blocks already completely thawed out. On March 16 even the 1-foot blocks were nearly thawed out. F7 in the open ground showed thawing on March 12 and was completely thawed out by March 16. The same was true for the isolated spots in the hardwoods. From March 15 to March 21 the mean daily temperature dropped again below 32° F. and a total of 0.5 inch of snow fell on the watershed. Readings taken on March 23 showed practically all installations free of frost with the exception of those in the Norway pines, which were still frozen solidly down to the 2-foot level. Though the daily mean temperatures stayed mostly above 32° F. until the end of March, the Norway pine plantation stayed frozen solidly to the 10-inch level while signs of thawing began to show at the 1-foot level of F4 and the 2-foot level of L12 on March 27 and 30. All other locations were frost-free with the exception of a few shallow isolated blocks in the jack pines that started freezing due to low night temperatures. The total precipitation for March 21 to 31 was 0.1 inch of snow on March 29. With the exception of April 1, 5, and 7, the daily mean temperatures were below 32° F. until April 14. There was 0.1 inch of rain on April 1, 0.25 inch of snow and rain on April 4, and 1.1 inch of rain, changing into wet snow, on April 5 and 6. There was 0.1 inch on April 11 and on April 13.

In this period F2 and L12 in the Norway pines stayed frozen solidly to the 1-foot level, while F4 showed signs of thawing at the 2-inch, 4-inch, and 1-foot level, with the 6- to 10-inch layer frozen at all times. By April 14, even F4 was completely frozen to the 1-foot level. Frost also came back into the jack pine area, and by April 14 it was frozen to the 6-inch level. The open ground and the hardwoods showed no frost.

After April 14 the daily mean temperature rose rapidly. By April 18 frost had disappeared from the jack pines, and F4 in the Norway pine stand showed signs of thawing at the 2-inch and 1-foot levels, while F2 in the same stand was still frozen solidly. On April 20, F2 showed signs of thawing at the 2-inch, 4-inch, and 1-foot levels, as did F4, with the 6- to 10-inch blocks still completely frozen. It was not until April 24 that all frost had disappeared from the Norway pine plantation.

Analysis of data.—It is apparent from the data in Fig. 2 and Table I not only that the soil under the Norway pine plantation froze one week earlier and to a greater depth than the soil under the jack pines, but also that it stayed frozen all winter. Rain and a prolonged period of thawing temperatures in March and the beginning of April did not remove this frost. The frost developed was of the concrete type, which is the most unfavorable for water infiltration, and could not be penetrated by an auger. Water was standing in the furrows during the warm periods. The depth of 2 feet of frost was the deepest observed in the watershed during the winter. Most important is the fact that during the March breakup period the soil under the Norway pines did not thaw out, and infiltration was negligible. This would suggest that reforestation of large areas with one species of trees such as Norway pine could increase the flood potential of a watershed unless precautions were taken. A sudden rise in temperature together with a warm rain may well have the same or worse effects in a Norway pine plantation as on open ground. This would be true especially if the pine were situated on sloping ground and on heavier soils, which tend to freeze deeper than sand.

The jack pine plantation, as did the Norway pines, provided protection against frost in December, when frost depths up to 8 inches were recorded in the open ground. This was due to a snow cover which could sustain itself in the forests during that period and thus protected the ground. It froze a little later than the Norway pine plantation and to a shallower depth. It also thawed during the thawing periods. This was probably due mainly to a lesser crown density in comparison with the Norway pines. Thus less snow was intercepted and the depth of protective snow cover in form of loose, freshly fallen snow was increased, which accounts for shallower frost depths. Furthermore, sunlight can penetrate crowns of less density more freely, which accounts in part for the response to thawing periods. The frost was not of the concrete type throughout the plantation, and only rarely and in isolated spots was water standing in the furrows.

The freezing in the open ground was very spotty and was influenced almost entirely by the amount of snow cover. Nowhere, however, did the frost reach the 2-foot level as it did in the Norway pines.

The best behavior in relation to frost was shown by the hardwoods. In spite of the low temperatures during the winter the soil never froze, except for a few isolated places which showed stalactite type of frost for short periods. It appears not only that the trees provided protection against wind drifting without restricting the air movement and good light penetration, but also that the fallen foliage was a far better insulator than pine needles.

From this study it is obvious that for good watershed management, which is one of the main purposes of the large scale reforestation programs in Michigan, close attention will have to be paid to the species of the trees to be planted. Strip cropping has proven very successful in the agricultural field, and there is no reason why it should not be successful in forestry, especially on sloping ground. Water running from frozen areas can then penetrate into the ground in the unfrozen strips, thus decreasing the flood peaks and increasing the ground water level.

Conclusions

1. The most severe soil frost with the greatest depth and longest duration was observed in the Norway pine (*Pinus resinosa*) plantation.

2. Frost in the jack pine (*P. banksiana*) plantation and in the open ground was less severe.

3. The hardwood showed practically no frost at all during the observation period.

4. Thawing took place from the surface downwards as well as from the lower frost line upwards.

5. It would seem that for large-scale watershed management close attention will have to be paid to the species of the trees to be planted.

6. The advantage of strip planting of different tree species is indicated.

Acknowledgments

This work was made possible by the financial assistance granted by the Rural Rehabilitation Corporation of Michigan and the cooperation of the staff and field personnel of the Fish Division of the Michigan Department of Conservation. Special mention should be made of Mr. Leon Bigelow, on whose property part of the experiment was run.

LITERATURE CITED

Bouyoucos, G. J., and A. H. Mick. 1940. An Electrical Resistance Method for the Continuous Measurement of Soil Moisture under Field Conditions. Mich. State Univ., Expt. Sta., Tech. Bull. 172.

——. 1954. Electrical Resistance Methods as Finally Perfected for Making Continuous Measurement of Soil Moisture Content under Field Conditions. Mich. State Univ., Expt. Sta., Quart. Bull. 37 (1): 132–149.

Helmreich, F. W. 1956. Report on Watershed Research at the Experimental Watershed near Hoxeyville, Michigan. Mich. Rural Rehabilitation Corp., Res. Rep. (unpublished).

Lutz, H. J., and R. F. Chandler, Jr. 1946. Forest Soils. John Wiley & Sons, Inc., New York. pp. 269–276.

Post, F. A., and F. R. Dreibelbis. 1942. Some Influences of Frost Penetration and Microclimate on the Water Relationships of Woodland, Pasture, and Cultivated Soils. Proc. Soil Science Soc. Am., 7: 95–102.

Rowland, E. F., L. H. Stolzy, and G. A. Crabb, Jr. 1954. Frost Determination by Electrical Resistance. Highway Res. B., Wash. D. C., Bull. 100, 17–21.

Russel, E. J. 1956. Soil Conditions and Plant Growth. Longmans, Green and Co., London. pp. 329–332.

Storey, H. C. 1955. Frozen Soil and Spring and Winter Floods. Yearbook of Ag., U.S.D.A., pp. 179–184.

Papers of the Michigan Academy of Science, Arts, and Letters
Vol. XLVII, 1962 (1961 Meeting)

THE MONSOONS OF ASIA AND THEIR SIGNIFICANCE

STANLEY D. DODGE

The University of Michigan

THE windbelts of the world have provided up to now the base for the orthodox system of climates. There is, however, a major difficulty. The monsoons have had to be considered as irregular and as an exception. The purpose of the present paper is to present the monsoons in a new light, so that they appear, not as an exception, but perhaps as a clue to what the wind systems of the world are really like.

On pilot charts of the North Pacific the line showing the northern limit of the northeast trade winds cuts off in the winter months near Midway Island, in longitude 177° 21′ East. In summer the line extends 33° farther west, to the vicinity of Guam in longitude 144° 45′ East. The trade winds should be continuous across the Pacific, but there is much uncertainty about them in the western part. West of the points indicated, wind in the winter is from the north or north-northeast, whereas the trades are nearly easterly. In the summer, the winds west of the trades are from the southwest, almost at right angles to the trade winds themselves. The foregoing are facts requiring interpretation.

In the ordinary interpretation of the monsoons the weather conditions of India are taken as the classic examples. In the summer, as is well known, winds blow from the southwest, bringing moisture off the Arabian Sea for the drenching monsoon rains. In the winter, on the other hand, the wind blows from the northeast, bringing prolonged drought because they are then from the land, except along the Coromandel Coast, where the northeast winds, blowing across the Bay of Bengal, bring copious rains to Madras and stations north and south. The usual explanation is that in the summer India is under the influence of a low-pressure area centered over the Thar or Indian Desert in the northwest of the peninsula. Indeed, though the winds of the Deccan blow from the southwest and the Arabian Sea, those of the Ganges Basin and the foothills of the Himalayas are shown on Indian charts as blowing from east to west, that is, toward the Thar Desert

TABLE I
RAINFALL[1] IN INCHES IN BOMBAY
Monthly Means, 1921–30

January,	0.36 inches	July	24.02 inches
February,	0.01	August	10.56
March,	0.01	September,	12.76
April,	0.01	October,	1.74
May,	0.29	November,	0.65
June,	21.43	December,	0.11

Total, 71.95 inches

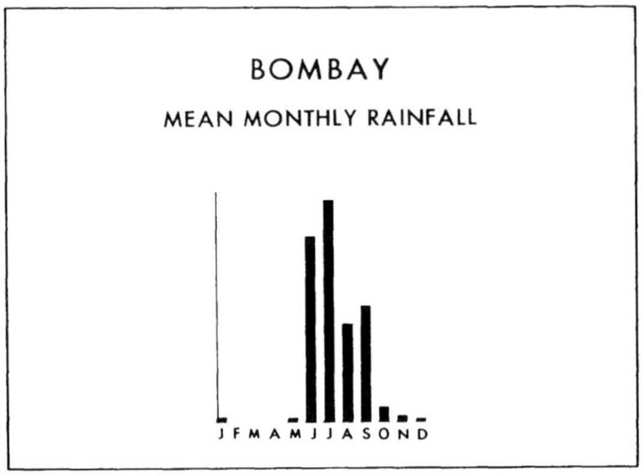

FIG. 1

area of low pressure. All these facts, however, need to be reconciled with those noted in the western part of the Pacific Ocean if what happens in southeastern Asia is to be sufficiently understood.

The nature of the beginning of the monsoon rains in India is obscured when monthly means alone are considered. Data for Bombay will illustrate the matter (Tab. I and Fig. 1).

The sudden great increase in rainfall from May to June is characteristic, but again, taking monthly averages obscures the facts. In the decade 1921–30 rain fell in May in only three years, whereas there was rain in every June. In 1921 there was no rain in May at all, but in June, 26.74 inches. Such an abrupt change in monthly quan-

[1] H. Helm Clayton, *World Weather Records, 1921–1930,* Washington, Smithsonian Institution, 1934, p. 117.

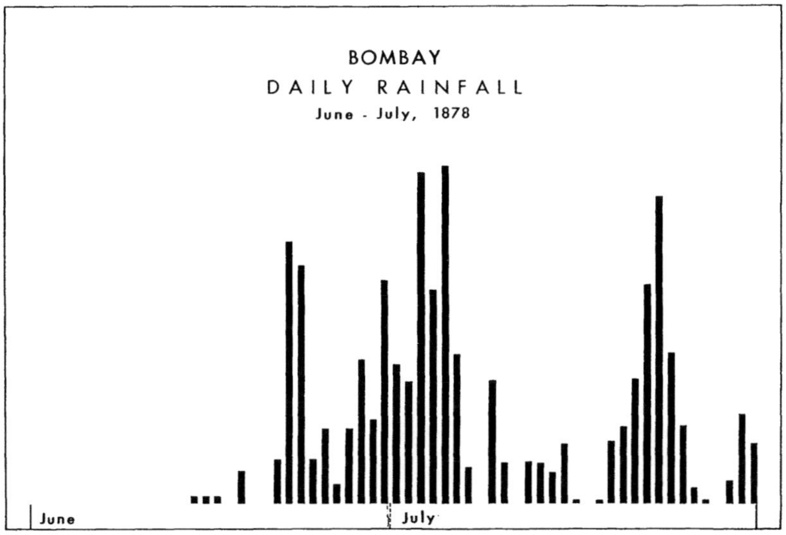

FIG. 2

tity is typical, for in only three years was the June rainfall less than 10 inches. But even that obscures the facts of the beginning of the monsoon. Day-by-day records are needed. In 1878, 0.03 inch fell in May, 0.01 on the 19th and 0.02 on the 21st.[2] There was no rain again till the 14th of June, when 0.08 inch fell. From then on there was rain every day till toward the 1st of September, except that there was none on the 18th, 26th, 27th, 30th, and 31st of August. There were four rainless days in September, but in October the whole month was dry except for eight rainy days. November and December were entirely rainless, except that November had 2.63 inches falling over three days in the first week. The point is that even during the rainy season rain is not continuous. The graph for June and July 1878 shows the situation (Fig. 2). The real rains began on the 14th of June. On the 22d, there were more than 4 inches. Rainfall dropped off to about half an inch on the 26th but rose again to 3½ inches on the 30th. And so on through July.

Although rainfall throughout the summer monsoon fluctuates,

[2] Blanford, Henry F., *Report on the Meteorology of India in 1878*, Calcutta, Superintendent of Government Printing, 1880.

the rains do begin rather suddenly. Fig. 3 shows the "date" of the beginning of the monsoon rains in southeastern Asia. The isochrons represent the first day of the month in which the rain is twice that of the preceding month. Thus at Saigon the mean rainfall in February is 1.4 mm. and in March, 30.3 mm. March, therefore, with about twenty-two times as much rain as February is considered as being the first "monsoon" month. However, Hyderabad, near the mouth of the Indus River, does not have the defining increase in monthly rainfall till July. June rain is only 0.09 inch, and July rain is 3.83 inches. Between Saigon and Hyderabad the beginning of the monsoon rains advances westward. The rains begin, also in March, at Colombo, Ceylon, and at Cochin, south of latitude 10° on the Malabar Coast, but not till April at Belgaum, near Goa, 6° farther north, and not till May at Bombay, near latitude 20°.[3]

Such "beginnings" for "monsoon rains" may also be traced in eastern Siberia and Korea, well north of what is ordinarily considered as being "Monsoon Asia." At Chemulpo, on the Korean west coast, the beginning is in April; at Vladivostok, 10° farther north, rains fitting the definition begin in May.[4] A May "beginning" may be traced as far north and west as Yakutsk, Irkutsk, and Kirensk.[5]

The fluctuations in the quantity of rain from day to day and from week to week in Bombay and elsewhere in India are comparable to those in the eastern United States. They suggest the passage of fronts. The isochrons of the beginning of the rainy season in southern and eastern Asia suggest a front between opposing wind systems. Since the front may be traced from Malaya to eastern Siberia, from Saigon to Yakutsk, it is more than a separation between local air masses. The mass or anticyclone, of which the front marks the edge, involves probably the whole of Asia, even of Eurasia. It confronts an Arctic anticyclone on the north, a North Pacific one on the east, and others in other directions.

From the center of the Asiatic anticyclone, air moves always outward. Because of the Coriolis force it bears to the right, setting the whole mass in motion clockwise.

The center of the anticyclone is a center of cold. It is more intense in the winter than in the summer. Hence, in winter, the whole Asiatic

[3] Clayton, *op. cit.*, pp. 103, 117, 120, 122, 135, and 142.
[4] Clayton, *op. cit.*, pp. 154 and 191.
[5] Clayton, *op. cit.*, pp. 168, 170, and 193.

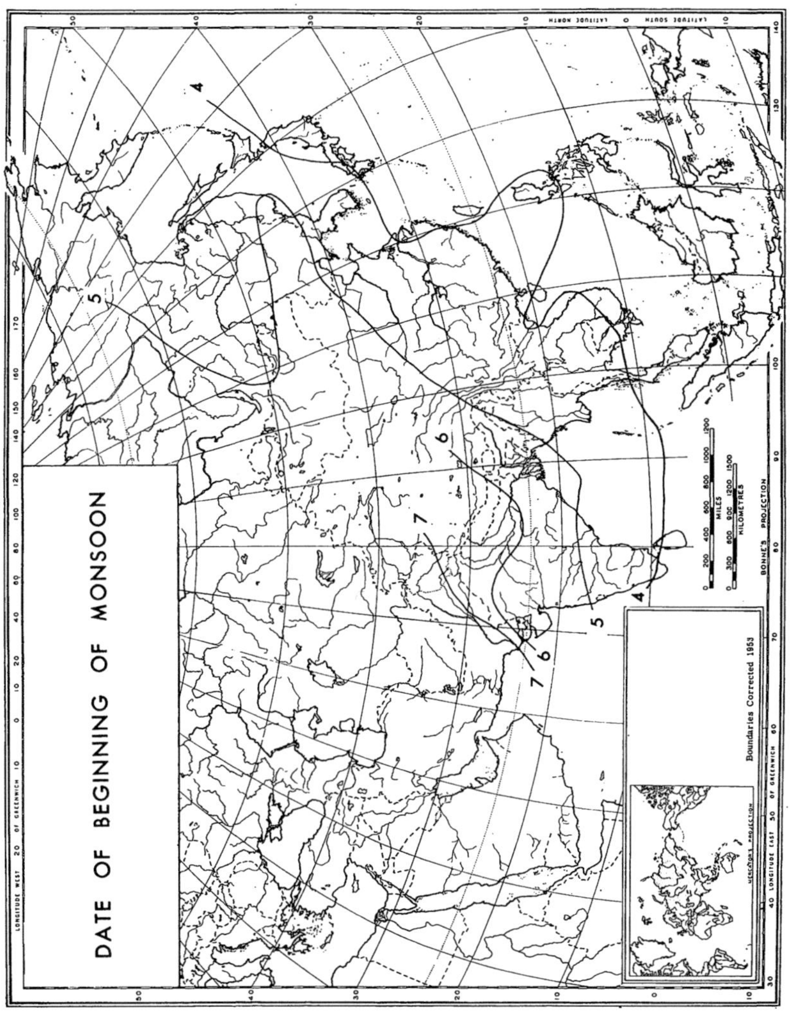
DATE OF BEGINNING OF MONSOON
ASIA—500 M
Boundaries Corrected 1953
BONNE'S PROJECTION
MILES
KILOMETRES
4
5
6
7

Fig. 3

anticyclone is greatly expanded. It spreads far out over the western Pacific. It is for this reason that, as noted earlier, the trade winds of the North Pacific Ocean are cut off near Midway Island, some 30° east of their summer position. The winds of the western North Pacific are then mainly from the north and north-northeast, an eastern limb of the Asiatic anticyclone. On the southern limb, the winds are from the northeast, the familiar winter winds of India, bringing drought to the Deccan and the Malabar Coast, but rain to the greater part of the Coromandel Coast.

In summer, because of heat, the Asiatic system is weaker. The North Pacific system expands westward into the Indian Ocean. The southwest winds on the Malabar Coast are part of the western limb of the North Pacific anticyclone, which has its center somewhere near the Hawaiian Islands. By July the outer margin of the North Pacific anticyclonic winds has normally reached Pakistan. In some years it reaches still farther west. In 1930 Jask, on the north shore of the Gulf of Oman, had 0.71 inch of rain, the only summer rain at Jask in the decade 1921–30.[6]

Northwest and north of the winds of the southwest limb of the Pacific system, when it spreads over India, the easterly winds of the Asiatic system still prevail. It is for this reason that there are easterly winds in the Ganges Basin, giving rise to the probable error of attributing the monsoons to a low pressure area over the Thar Desert.

So much for some of the facts, and so much for a start on the interpretation of them. There are many more facts which may be elucidated by the theory here proposed and which tend strongly to support it. If it be true that the continent of Eurasia is dominated by an anticyclone with a clockwise movement of air and the North Pacific by another, then North America and the North Atlantic Ocean ought to have similar systems. If they do not, there is little general significance in the interpretation of the monsoons here given; if they do, then the present interpretation of the Asiatic monsoon may be significant for climatic theory in general.

First, the west coast of Central America and of Mexico shows a northward movement of the beginning of the summer rainy season

[6] Clayton, *op. cit.*, p. 122.

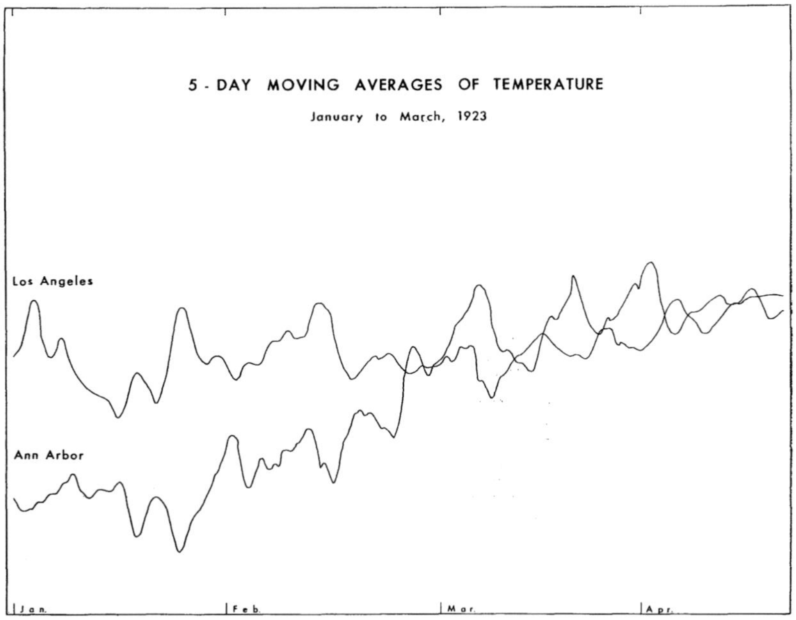

FIG. 4

similar to that of the Malabar Coast of India.[7] It moves northward to the limits of the Sonora Desert, which is thus for North America like the Thar Desert of India and Pakistan. The cause would appear to be the summer westward extension of the North Atlantic anticyclone, bringing southwest winds to the part of the Pacific Ocean south of Lower California similar to those of the Arabian Sea brought by the westward expansion of the North Pacific anticyclone. The westward expansion of the Atlantic system is allowed because of the summer weakening of the North American anticyclone.

Second, this part of the significance of the Asiatic monsoon, then, depends on whether there is indeed a North American anticyclone.

[7] Rains, on an average, begin as follows on the west coasts of Central America and Mexico: San Salvador (lat. 13° 42′ N.), February; Salina Cruz (lat. 16° 12′ N.), April; Mazatlan (lat. 23° 11′ N.), May. The month indicated is the one in the first six months of the year in which the rainfall is one and one half times that of the preceding month. Ref. Clayton, *op. cit.*, pp. 306, 308, and 315.

If there is one, then as it expands and contracts a-periodically temperatures should rise and fall in the eastern United States and those in the southwest should behave in opposite fashion. When the North American anticyclone expands, places in the northeast have northerly winds, and so falling temperatures, whereas those in the southwest have southerly winds, and so rising temperatures. When the anticyclone contracts, the east has southerly winds of the western limb of the North Atlantic anticyclone, and the west northerly winds of the eastern limb of the North Pacific system. Fig. 4 shows five-day moving averages of temperatures for Ann Arbor and Los Angeles. When the temperature falls in Ann Arbor, it rises in Los Angeles, and *vice versa*, and so generally throughout the year. It will be noted that the fact is shown by the opposition of peaks on one curve to troughs on the other. Such opposition seems to confirm the existence of a North American anticyclone. There are other facts which fit into and help strengthen the idea. The whole, then, seems to confirm the suspicion that the Asiatic monsoon is not an exception to normal circulation in the atmosphere, but instead a clear indication of atmospheric truth the world over.

Papers of the Michigan Academy of Science, Arts, and Letters
Vol. XLVII, 1962 (1961 Meeting)

THE RURAL SETTLEMENT OF THE SAVANNAH COUNTRY

SHERWIN H. COOPER

University of North Carolina

The Savannah Country is the Georgia side of the drainage basin of the Savannah River below the fall line. Augusta is at its upper end and Savannah, 120 miles away by road, at its lower (Fig. 1). Even a cursory reading of the history of the basin would suggest the proposition that social and economic forces control not only kinds of settlement, but also their overall density and distribution. Further study, on the other hand, reveals that particulars of settlement such as the sites and situations of individual dwellings and agglomerated settlements are more closely related to such factors as soil drainage and fertility, modes of transportation, and roads. That this should be true is obvious but at best has been only casually mentioned thus far in published works on the region. My main purpose is to call attention to some of these particular relationships, and to sketch in those features of the social and economic milieu of the past and present which have strongly influenced the choices of land made by the main body of settlers.

The Region

The Lower Savannah River Basin in Georgia is a wedge-shaped section of the Atlantic Coastal Plain oriented roughly northwest to southeast, abutting at its broad end on the Piedmont and narrowing to a few miles near the sea at Savannah (Fig. 2). In its upper or northern part it lies at the heart of the decadent Cotton Belt—a region stretching from North Carolina to Alabama along the lower Piedmont and Inner Coastal Plain—while in its lower or southern part it provides a passageway from the remnants of that belt to the sea. Near Savannah the region is moist and subtropical, while near Augusta, some one hundred miles inland and four hundred feet higher, continental effects appear in the form of colder and longer winters and reduced rainfall. Although the climatic contrast is statistically slight, it nevertheless is manifest, along with contrasting soils, in a difference in natural vegetation between the interior and seaward parts of the basin.

Flowing past both the upper and lower parts of the basin is the Lower Savannah River. In viewing it today, one could hardly suspect it of having once played a vital role in the settlement of Georgia. Its broad floodplain—unbroken from Augusta to Savannah—plays host to a swampy wilderness, and with it the river forms one of the greatest natural barriers of the eastern seaboard. The whole bottom land

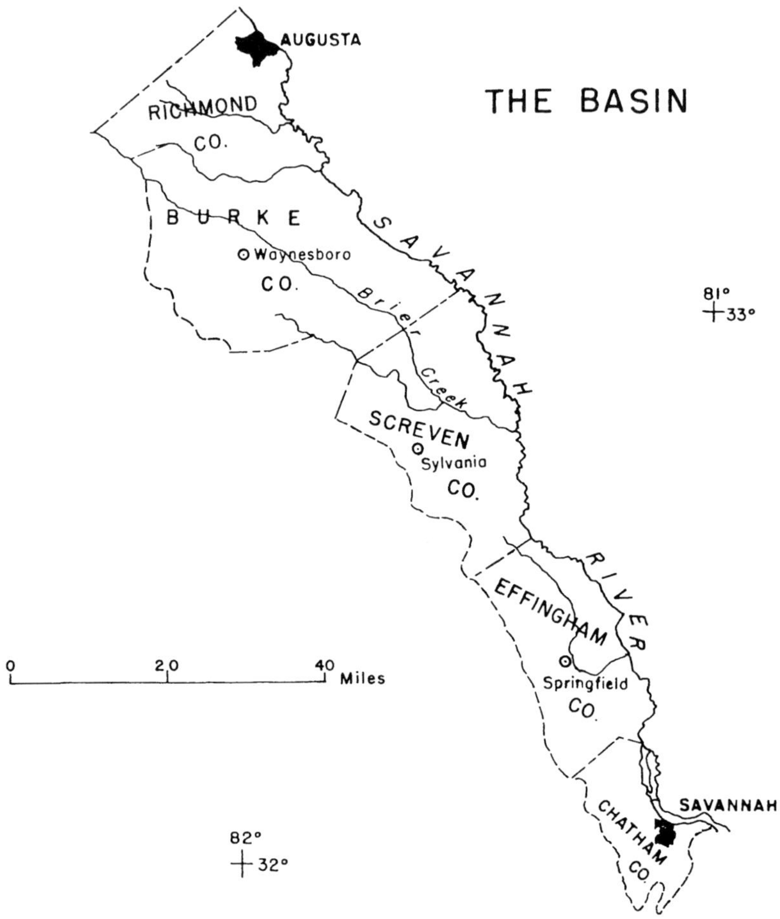

Fig. 1

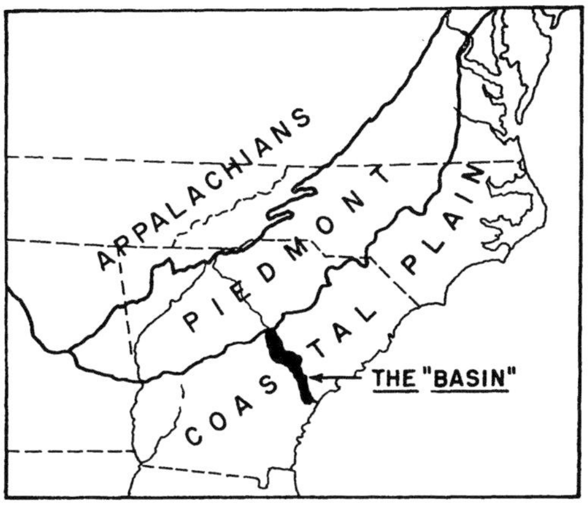

FIG. 2. Physiographic divisions of the southeastern United States.

area is a broad zone of little or no settlement interposed between the more or less continuous fields and farms of the opposing uplands in South Carolina and Georgia (Fig. 3). The river remains important to the basin, however, in that it drains almost all of its well-watered area of some 2,000 square miles.

Agriculture, fundamentally the life of the basin, is limited for the most part to the uplands. There spindly, ephemeral rows of cotton and corn replace the everlasting oaks and cypresses of the swamp below, and in place of a sluggish, meandering river flow swifter streams of highway and rail traffic. Everywhere rural dwellings punctuate the harmony of woodland and field and lend a human quality to the scene that is unknown in the river lowland below.

PERIODIC SHIFTS OF SETTLEMENT

Settled upland and unoccupied lowland were not always the rule, however. In fact, the arrangement was at one time the opposite and has shifted often in the 200-year history of the basin. This constant reweaving of the fabric of rural settlement is attributable to social, economic, and technological revolution within the basin itself and in the nation at large.

It is clear enough that after the initial settlements had been made by Oglethorpe and his followers at Savannah and Augusta in the 1730's the introduction of rice planting produced the first major

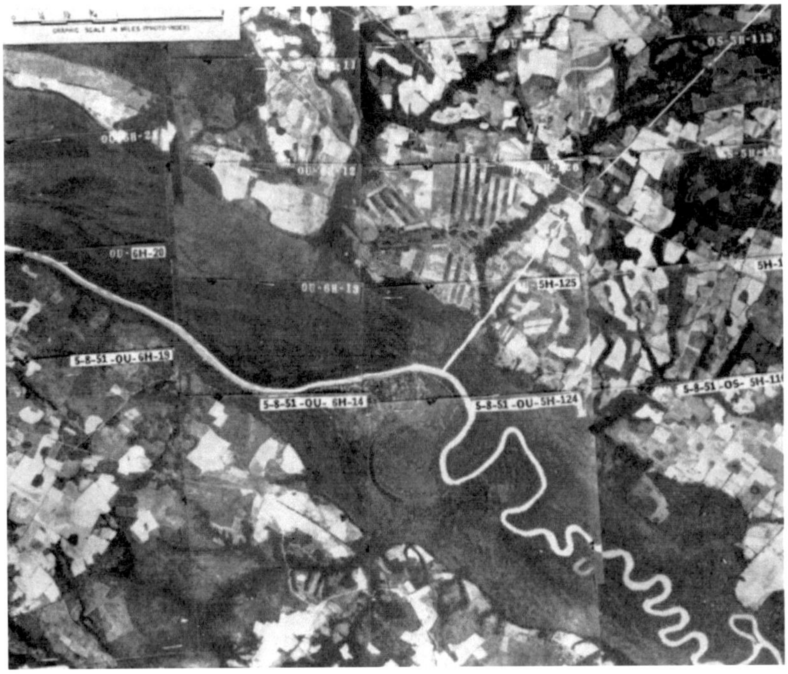

FIG. 3. A section of the Lower Savannah River and its floodplain. The lighter areas are cultivated fields on the uplands. South Carolina is on the far side, and Georgia on the near side. (From U.S.D.A. Commodity Stabilization Service aerial photo index of Allendale County, S. C.)

change. At the time American population was concentrated largely on the eastern seaboard, and rice, which commanded a vast market, could be grown easily on the river floodplain. Rural settlement, until then claiming only scattered houses in the backwoods, soon boasted great clusters on the river flats.

The cotton gin was invented in **1793**, and before long the thinly settled uplands of the basin became widely cultivated, densely settled agricultural areas. The upland farmers borrowed the plantation system from the lowland rice growers and with it the lowland form of settlement—the plantation cluster or "village."

A third great change in settlement came with the Civil War. The clusters of plantation settlement were broken up, the rice lowlands were largely abandoned, and dispersed individual houses became the dominant form wherever rural settlement survived. When railroads entered the picture numerous new hamlets and villages appeared. The dominance of the dispersed dwelling and the small farm supply center persists today but is being weakened by new social and economic forces which are causing an overall attrition of rural settlement in the basin.

With changing conditions, then, many lands which had been supporting substantial settlement lost their power to continue doing so, while other lands, previously unused, gained that power. The lost power was almost always a matter of economics, not agronomy. The basin has always been agricultural, and in some instances land was simply abandoned when the crop to which it was best or perhaps solely adapted lost its market. Over broad areas, however, what usually happened was that settlements changed in form, pattern, density, or distribution rather than appearing or disappearing altogether.

Colonial Rice Planting and Pioneer Farming

The first activity to draw people to the vast interior between Augusta and Savannah was skin trading.[1] It ushered in rural settlement, for the traders, by their regular use of old Indian trails, established the main lines of overland travel in the basin. The epoch of skin trading was followed by two contemporaneous phases of colonial settlement, each stemming from a distinct set of social, economic, and land factors. One, under a riparian plantation rice economy which was based on the demands of the world market, owed its particular location to topography and soil drainage. The other, under an upland pioneer farm economy which was largely the product of extrinsic population pressure, owed its location to the marginal nature of the land available at the time.

The plantation rice phase grew out of the lucrative plantation systems farther north—in particular the contemporary rice economy of the Carolina coastal lowlands. The nature of rice culture limited it to the swamps of the Lower Savannah and its tributaries when it was finally introduced to the basin in 1749.[2] Owing to its proximity

[1] Georgia, *The Colonial Records of the State of Georgia* (Atlanta: "Printers for the State," 1904–1915), XXXV, p. 307.

[2] Savannah Writers' Project, *Savannah River Plantations* (Savannah: Georgia Historical Society, 1947), XVIII, p. 98.

to the great natural waterway of the Savannah, to difficulties of road-building and travel on the poorly drained adjacent uplands, and to the bulkiness of the harvested crop, settlement in the lower basin clung close to navigable waterways for most of the colonial period.

Negro slaves supplied the large amount of heavy labor which rice and the clearing of swamps required.[3] The best lands thus fell into the hands of the relatively few planters who had capital and slaves enough to develop and work the rich bottoms of the Savannah River. The adjoining sandy pine uplands of the lower basin were either infertile or poorly drained, so that the combination of already committed rich lowlands and undesirable uplands discouraged large scale European immigration to the lower basin. For some time past the Revolutionary War settlement was still largely confined to river sites.

But this was true only of the lower basin, for the upper basin was well endowed with accessible and well-drained fertile land. Once Indians no longer barred their path there was a great influx of settlers to the southern Piedmont from crowded or worn-out lands of the northern Piedmont in Virginia and Maryland. Although they at first avoided the formidable sand hills near the fall line, they ultimately pushed the farm frontier well to the south of Augusta[4] into the well-drained, fertile lands of the Louisville Plateau and the Tifton Upland (Fig. 4).

The upland settlers naturally followed the easiest paths into the virgin wilderness—Indian trails, public roads, and navigable creeks. Most farmers cleared land as close to the roads as possible for convenience to markets and for the security of ready contact with neighbors. The absence of hostile Indians and the openness of the pine forest enticed a few to seek good land away from the roads, so that, although settlement in general was closely articulated to the main roads, small clearings were to be found here and there on fertile islands in an otherwise unoccupied sea of poor lands.

The close of the colonial period found the basin densely settled along the lower part of the river and sparsely settled in the uplands. In both the overall pattern of settlement was linear—in the lowlands

[3] Ulrich B. Phillips, ed., *Plantation and Frontier Documents: 1649–1863* (Cleveland: A. H. Clark Co., 1909), Vol. I, pp. 259–265.

[4] Wilbur Zelinsky, "An Isochronic Map of Georgia Settlement, 1750–1850," *Georgia Historical Quarterly*, 35, No. 3 (September, 1951), p. 192.

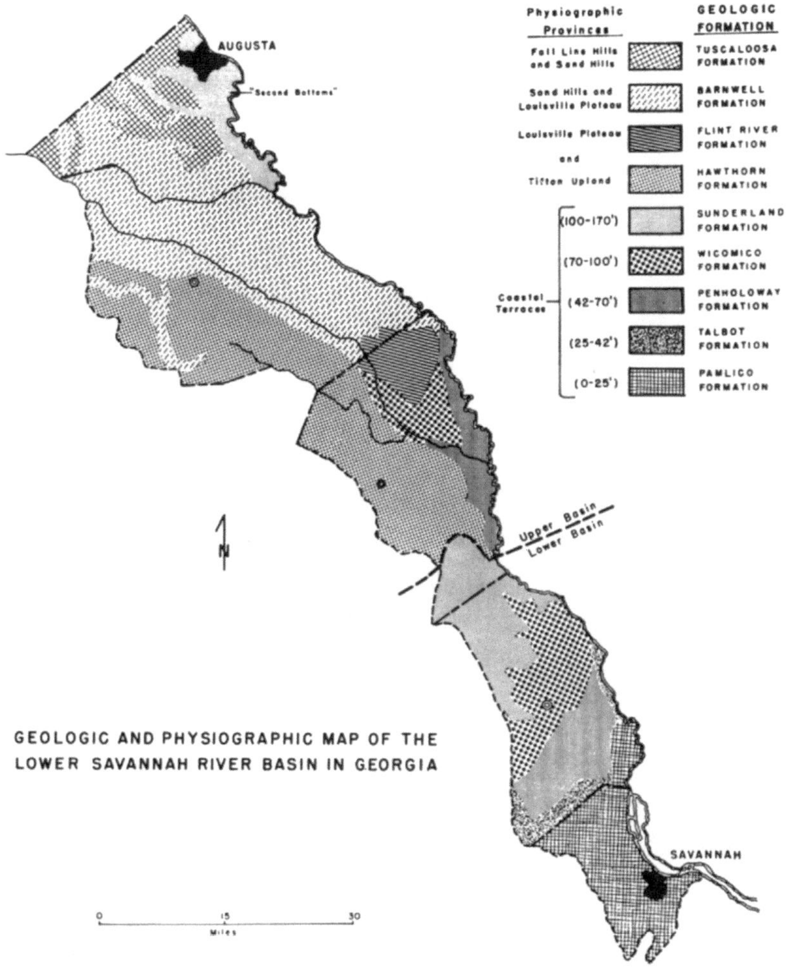

FIG. 4

because of the narrowness of the floodplain, in the uplands because of adhesion to roads and to larger creeks. The uplands supported typical pioneer farmers and backwoodsmen, while the lowlands built the plantation system that was to dominate the entire basin in the next stage of occupance.

COTTON PLANTING AND THE REVOLUTION IN TRANSPORTATION

The invention of the cotton gin opened the well-drained uplands to the plantation system, which had been confined to the river lowlands. Rural population increased rapidly—with Negro slaves outnumbering white farmers for the first time in many sections[5]—while vast acreages on the uplands of the upper basin were cleared and planted to cotton and corn.

Early in the prewar period settlement was generally strung out along the main routes of migration—roads, the railroad, and the Savannah River. The pattern is familiar when one reviews the history of the American frontier. Peculiar to the basin and the South, however, was the later form of occupance growing out of the plantation system. There slaves in clusters the size of hamlets and villages dwelled on the best of river lowlands and uplands. Such compact settlements not only contained most of the people but also occupied the greater part of the cultivated land of the region. They were the truly distinctive form of settlement of the period.

A heavy increase of population, together with the new commercial agriculture of the uplands, created a need for new and better transportation,[6] and before long improved overland roads and the steamboat on the Savannah fulfilled that need. The river became alive with rafts, canoes, cotton boxes, and steamboats, whereas before it had provided little more than a port of entry for settlers at Savannah and a frontier staging point at the fall line town of Augusta. When Augusta became a great cotton market for a vast segment of both the Piedmont and the Inner Coastal Plain, the river strengthened its role by linking that town with its seaboard outlet, Savannah. In so doing it provided an improved and longer avenue of penetration of the basin by settlers. The application of steam power to river boats led to a similar attempt with overland transportation, and before long the railroad made its debut in the basin. The revolution in transport was the key to the origin, location, and development of many an agglomerated settlement. Steamboats strengthened the role of such

[5] W. S. Rossiter, *A Century of Population Growth in the United States, 1790–1900,* U. S. Bureau of the Census (Washington: U. S. Government Printing Office, 1909), p. 132.

[6] Ulrich B. Phillips, *A History of Transportation in the Eastern Cotton Belt to 1860* (New York: Columbia University Press, 1908), p. 103.

river ports as Augusta and Savannah, while the railroad, which ran down the middle of each county from Augusta to Savannah, gave rise to a string of depots, some of which grew into hamlets and villages, and stimulated the growth and importance of such central places along it as Waynesboro.

By 1860 then, on the eve of the war, the basin had not only laid out and developed most of its main lines of travel, but had also completed its transformation from a region of upland subsistence farms and lowland rice plantations to one of plantation agriculture on both upland and lowland. The steamboat had replaced the cotton box, the railroad the stagecoach, and the cluster of slave quarters the isolated farm dwelling.

The Dispersion of Settlement

The great agglomerations of ante-bellum plantation settlement disintegrated in the aftermath of the Civil War. The former slaves dispersed, and new, scattered dwellings popped up all over the countryside. Little virgin soil had to be broken and few new paths cut through any wilderness, for forest and soil alike had already been widely exploited under the plantation system. Add to these factors the survival of a network of old farm roads, and one must conclude that dispersal to settle the land anew was not too awesome a venture.

Soon after the war the old roads, lying dusty with disuse, were rejuvenated, and, in the human traffic they bore, so was the landscape rejuvenated. Where cotton and corn had grown without interruption over vast acreages there were now independent dwellings and farmsteads every thirty or forty acres.[7] For a brief period at least, cotton was no longer king.

When the turmoil of transition had subsided and cotton had reasserted its supremacy over all other crops, rural settlement had once again rooted itself firmly into a definite pattern. The rudiments of that pattern—an upland landscape blanketed with individual dwellings, profusely mottled with hamlets, and sprinkled with villages and towns—survive even today, at a time when the inexorable tides of urbanization and farm decline threaten to engulf the region.

The old rice flats near Savannah, however, long the most densely

[7] Merle Prunty, Jr., "The Renaissance of the Southern Plantation," *Geographical Review*, 45, No. 4 (October, 1955), p. 469.

settled lands of the basin, never recovered. These lands, richest of all in the basin, proved themselves fit for little but plantation rice culture. Inasmuch as rice planting depended on large gangs of labor and that kind of labor was no longer available after Emancipation, the land was generally abandoned (Fig. 5). The saturated floodplain soils were ill-suited to cotton, and, no matter what the substitute for rice, without elaborate drainage works which required huge investments of capital and labor to maintain, the rich alluvium was of no use to farmers. To compound the difficulty the relict ditches and dikes of the ante-bellum rice flats were so constructed as to prevent the use of modern machinery.

In the ninety-odd years since the breakup of the plantations and the dispersal of dwellings the settlement process has been, first, "trial and error" occupance of whatever land could be acquired, then finer and finer adjustment to the better-drained and more fertile soils of the uplands. By 1920 rural settlement had reached an all-time high, but since then it has steadily declined. In fact, except for the expansion of the urbanized areas centered in Augusta and Savannah and the growth of rural nonfarm population nearby, the early colonial pattern of scattered individual farmsteads and large wooded areas is beginning to recapture the basin. The regression of the rural areas is evident in that the rural population numbers the same today as it did in 1890 and in that many parts of the basin have been returning to lumbering and cattle-raising—activities that had more or less been forsaken for two hundred years.

Rural nonfarm settlement, on the other hand, has been growing proportionately and is often found occupying such poor agricultural lands as the sand hills and even some of the better lands near the cities. For the most part the growth of rural nonfarm population has been taking place in the city-dominated counties of Richmond and Chatham, where the overall population has been increasing. The almost wholly rural counties of Burke, Screven, and Effingham, while counting many nonfarm workers among their people, have all been losing population of late. There has been a gradual but unmistakable emptying of the countryside. Growth in these counties has occurred only in and around the two largest towns, both on main roads and the railroad. These towns—Waynesboro (4,461) and Sylvania (2,939)—and their militia districts (townships) were the only areas showing a gain in population from 1920 to 1950. With the increasingly rapid transport of recent years they have become the only important trans-

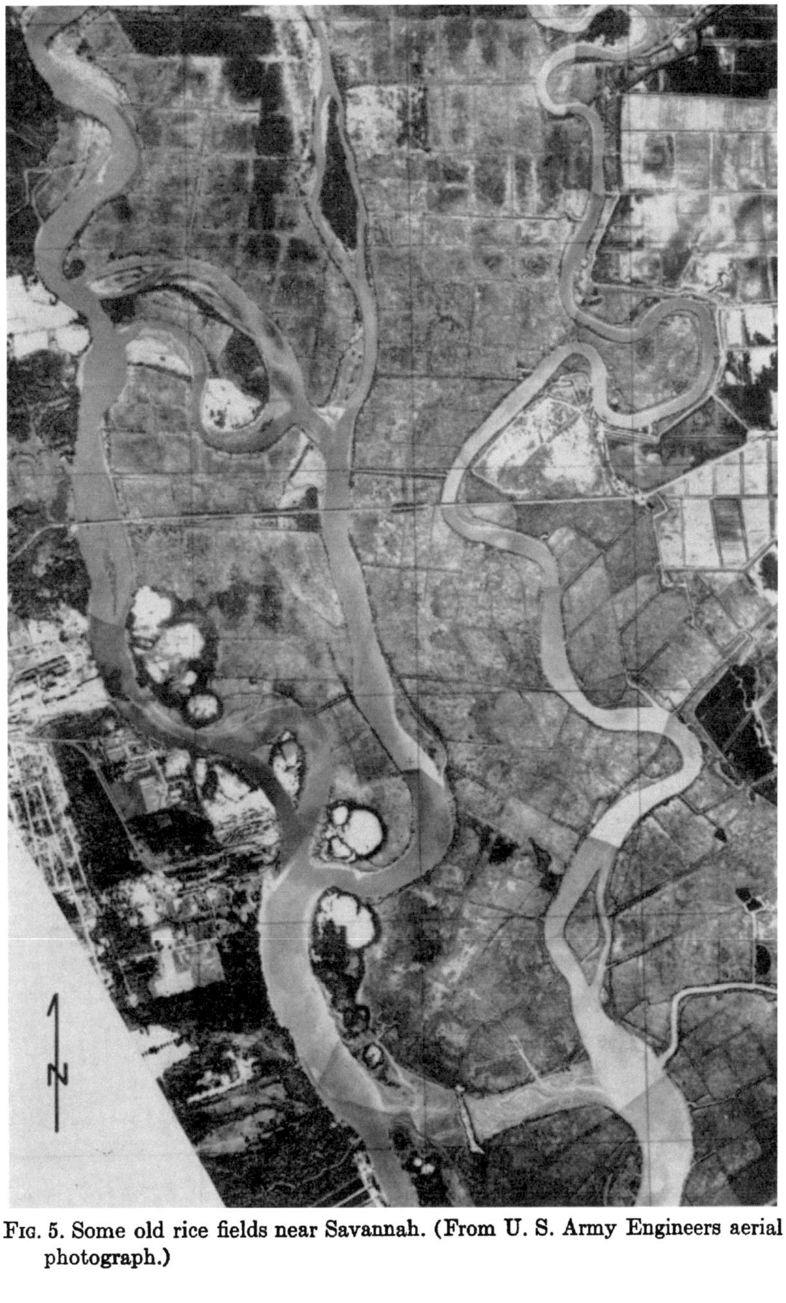

Fig. 5. Some old rice fields near Savannah. (From U. S. Army Engineers aerial photograph.)

portation hubs and supply centers within the rural basin. Without exception the other towns, villages, and militia districts in the farming areas have either declined in population or stagnated.

Rural Settlement Today

There is today a strong accordance of farm dwellings with well-drained loamy soils on the uplands and a few restricted areas of artificially drained heavier types on the lowlands near Augusta and Savannah. Some farmers still cling to the marginal sands of Burke and southern Richmond counties because of the low cost of getting fertilizer from Augusta, but by and large the marginal soils of the basin support a rather tenuous sort of settlement, if any at all.

Actually, a truly dispersed rural population is found only in the upper part of the basin, whereas in the lower basin the pattern is linear. The difference is almost wholly a matter of topography and soils.

The upper basin is a broad, well-drained upland capped with soils weathered out of Appalachian and Piedmont material deposited in an ancient sea. The floor of that sea was subsequently uplifted and exposed in Upper Cretaceous and Tertiary times.[8] The lower basin, on the other hand, consists of very low-lying marine terraces of Quaternary Age (Fig. 4), and what well-drained uplands there are are for the most part the exposed sandy tops of old beach ridges and therefore sterile and droughty. The rest of the area lies soaked in flat swales which dip almost imperceptibly below the general level of the beach ridges.

Except for the sterile and rough Fall Line Hills and sand hills of Richmond County and some wet depression in Screven, the widespread occurrence of level but well-drained soils in the upper basin has permitted rural settlement there to be spread out more or less evenly. The lower basin, highly limited in such arable land, has not only fewer settlements but also, because of their confinement to the old beach ridges, ones that are far less dispersed than those in the upper basin.

The contrast in topography and drainage of the two parts of the basin is equally significant in its effects on the road pattern, which in

[8] Laurence LaForge, et al., *Physical Geography of Georgia*, Geological Survey of Georgia, Bulletin 42 (Atlanta: Stein Printing Co., 1925), p. 5.

turn affects settlement. Owing to the great breadth of the well-drained uplands of the upper basin, the main highways and railroads are able to maintain fairly straight courses without crossing very many streams or wet depressions, and at the same time to run through the heart of rural settlement (Fig. 6). As a result there is a great thickening of farm service centers along these transportation lines, where they are in the best position to receive and distribute goods. In fact, the number of hamlets and villages along the paved roads and railroads is about the same as that on all the dirt roads put together, and dirt roads exceed by far the total mileage of paved roads and railroads in the basin.

In the lower basin, however, the main highway and railroad bypass the greater part of rural settlement. They follow almost perfectly straight paths and in so doing cut across, rather than follow, the grain of settlement (Fig. 7). Following the circuitous routes of

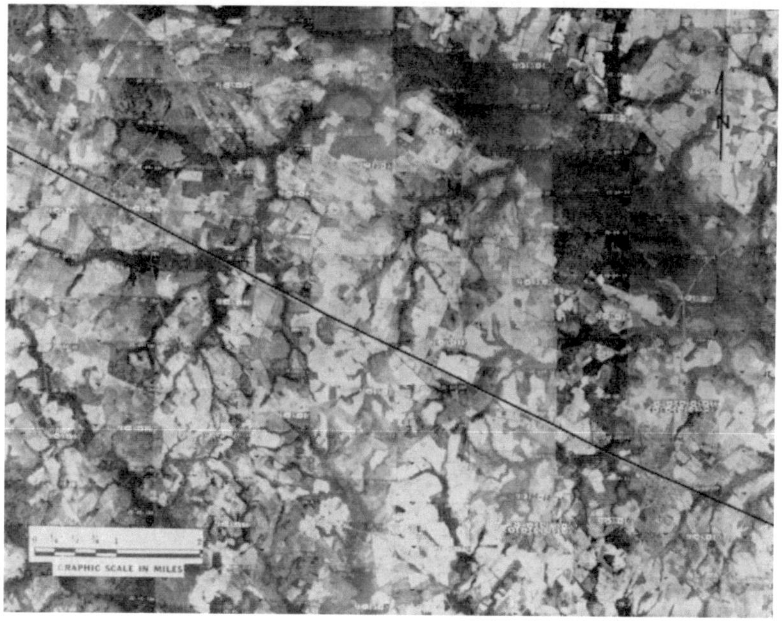

FIG. 6. A section of broad upland in Burke County, Georgia. The black line is the main highway of the area. (From U.S.D.A. Commodity Stabilization Service aerial photo index.)

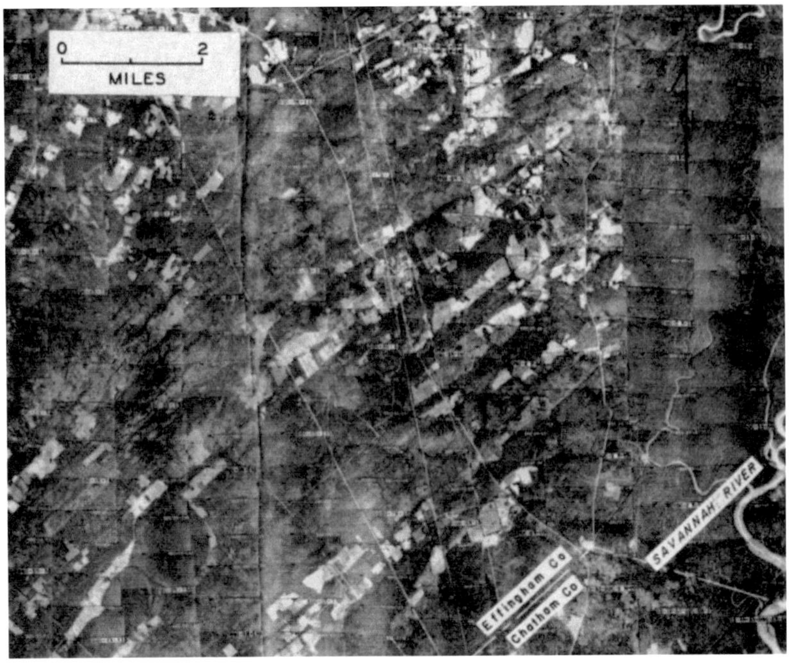

FIG. 7. Old beach ridges in Effingham County, Georgia. (From U.S.D.A. Commodity Stabilization Service aerial photo index.)

the old beach ridges would have been too costly and unrewarding. There are therefore relatively few settlements along great stretches of the main roads, in sharp contrast to the upper basin. The close accordance of the main body of rural settlement with secondary rather than primary roads is characteristic of the relative backwardness of the lower basin.

CONCLUSION

While it is true that the pattern of settlement typifying each stage of occupance was born, for the most part, of a unique set of social and economic conditions interacting with the land, many features of each stage were passed on to the next, and some even beyond. Such were, among others, the homestead of many a pioneer farmer in the uplands, the old rectangular rice flats near Savannah, and most of the old roads. Today's landscape is thus not entirely unique but rather the product of society and the land, past and present.

A boat trip down the Savannah from Augusta today would wind round meander after meander of shallow, muddy water and through seemingly endless rows of swamp hardwoods before reaching Savannah. Only occasionally would the dark monotony of the swamp be relieved by a bare bluff topped with a clump of pines, a pasture, a corn or cotton patch, and now and then a ramshackle cottage. The sole suggestions of modern commerce and industry in the two hundred-odd miles of river between the two cities are the crossing of a highway bridge about midway and of a railroad just before the smoke of Savannah's industrial towers comes into view.

The quietness of the river scene is perhaps symbolic, for the basin of the Lower Savannah River in Georgia has, insofar as its rural landscape is concerned, passed the acme of its growth and vitality as a region of settlement. Even in the uplands, where rich virgin stands of southern pines long ago fell to the axe, an unmistakable lethargy seems to have set in. There are few signs within the rural basin that it is keeping pace with the positive, progressive adjustment of man to land found elsewhere in the United States and, for that matter, in its own past. It seems, rather, to be engulfed by the general decline of the rural Old South.

PAPERS OF THE MICHIGAN ACADEMY OF SCIENCE, ARTS, AND LETTERS
VOL. XLVII, 1962 (1961 MEETING)

THE RELATIONSHIP OF EARLY SETTLEMENT ON THE GULF COASTAL PLAIN TO THE USE OF RIVERS AS COUNTY BOUNDARIES IN TEXAS

DAN DILLMAN

Western Illinois University

A POLITICAL map of Texas reveals an interesting arrangement of county boundaries with three distinct regional patterns appearing in (1) the Gulf Coastal Plain, (2) the far west-El Paso-Big Bend area, and (3) the central part from the Panhandle to the Rio Grande, with an extension along the Red River Valley. Of these regions, the first, i.e., the Gulf Coastal Plain, fails completely to conform to the rectangular arrangement common to most western states. Not only are the county boundaries irregularly shaped, but their orientation is totally different from the rectangularly delimited counties, being northwest to southeast instead of north to south (Fig. 1).

If the stream pattern is superimposed upon the county boundary pattern, it is apparent that the orientation of drainage runs counter to boundary lines in the north, west, and south of the state. Also it can be seen that on the Coastal Plain, rivers act as borders for some counties, but in others they run through the middle of the political unit (Fig. 2).

Often the causes of such irregular distribution of cultural phenomena in a region can be traced to the early stages of settlement when streams were intimately related to man's movement. In some areas the rivers would form hindrances to cross-country movement and would act as effective barriers to communication. In other places, however, rivers would serve as regional bonds when the routes of communication could easily parallel their gradients or when local topography did not restrict crossing. In the first instance, population would tend to concentrate away from the water courses, which then acted as boundaries through a predominantly empty area. In the latter case, population would tend to cluster along the level or fertile valleys with border lines arbitrarily drawn through the interstream area.

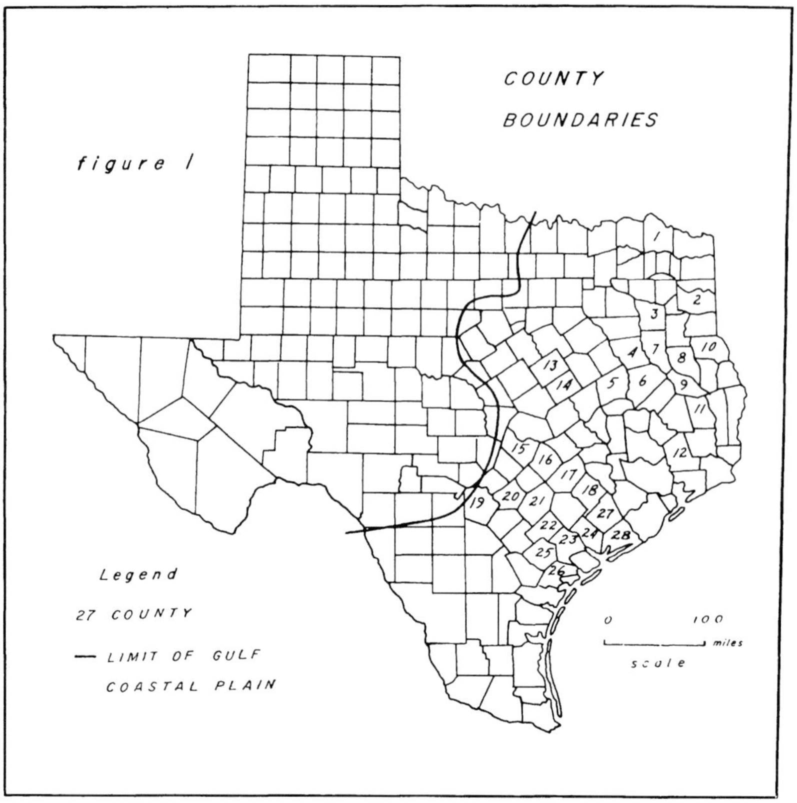

FIG. 1

With the preceding idea in mind, i. e., rivers acting either as unifying or disunifying influences, the question might be asked: What geographic factors of this sort during the early stages of settlement on the Gulf Coastal Plain could have contributed to the use of streams as boundaries in some counties but not in others, where streams appear to flow through the middle of counties?

An initial impression might be that settlers could have utilized the southeasterly flowing rivers as routes to the interior, subsequently spreading out from nuclei settlements in the valleys as the number of people increased. This in fact did not happen in most areas on the

LOCATIONAL REFERENCE OF COUNTIES AND COUNTY SEATS FOR FIGS. 1 AND 3. COUNTY NUMBERS REFER TO FIG. 1.

Fig. 1—County	*Fig. 3—County Seat*
1—Red River County	
2 —Harrison County	2 —Marshall
3 —Smith County	3 —Tyler
4 —Anderson County	4 —Palestine
5 —Leon County	5 —Centerville
6 —Houston County	6 —Crockett
7 —Cherokee County	7 —Rusk
8 —Nacogdoches County	8 —Nacogdoches
9 —Angelina County	9 —Lufkin
10 —Shelby County	10 —Center
11 —Tyler County	11 —Woodville
12—Liberty County	
13 —McLennan County	13 —Waco
14—Falls County	
15 —Travis County	15 —Austin
16 —Bastrop County	16 —Bastrop
17 —Fayette County	17 —LaGrange
18 —Colorado County	18 —Columbus
19 —Bexar County	19 —San Antonio
20—Guadalupe County	
21 —Gonzales County	21 —Gonzales
22 —DeWitt County	22 —Cuero
23 —Victoria County	23—Victoria
24—Jackson County	
25 —Goliad County	25 —Goliad
26—Refugio County	
27—Wharton County	
28—Matagorda County	

Coastal Plain.[1] Even to approach the coast from the Gulf was a hazardous undertaking, for barrier beaches extend for hundreds of miles from the Rio Grande Delta almost to the Sabine River (Fig. 2). Texas rivers on the Coastal Plain could not be used like the tidewater in New England or Virginia because the diminished seasonal flow restricted navigation even for shallow draft vessels. The importance of the northwest to southeast flowing streams in the populating of east-

[1] Ellis W. Shuler, "The Influence of the Shoreline, Rivers, and Springs on the Settlement and Early Development of Texas," *The Texas Geographic Magazine,* Vol. IV, No. 2, 1940, pp. 26–31.

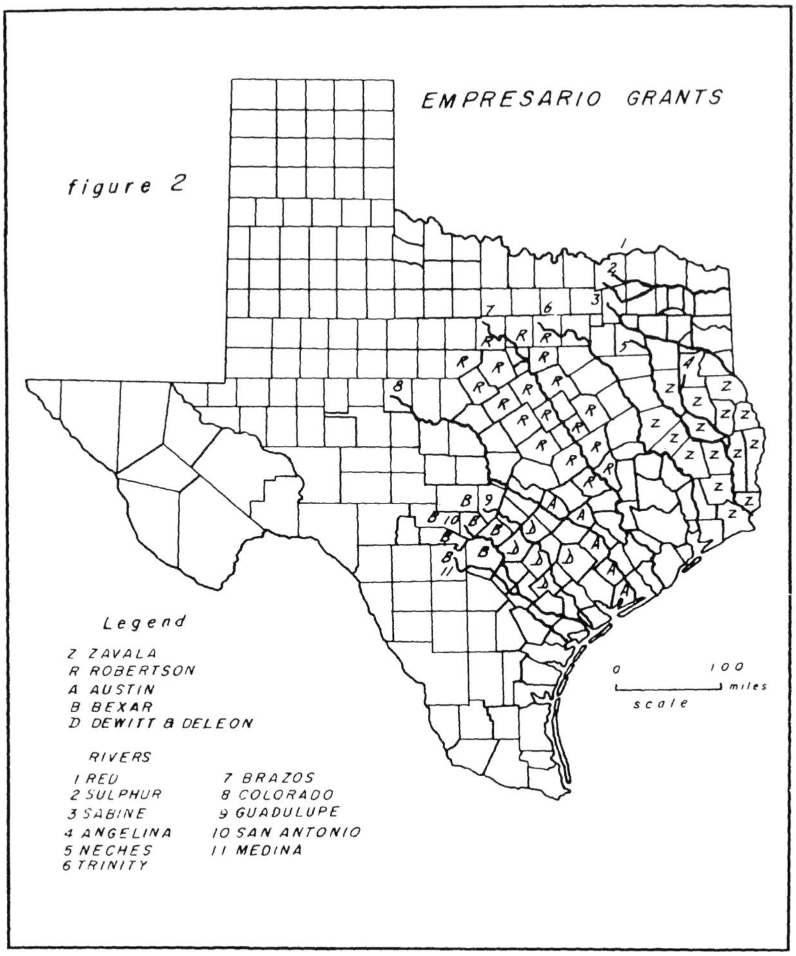

FIG. 2

ern Texas appears to have been minor.[2] For these reasons, routes of settlement were not oriented from the coast inland, but overland from the northeast where the gateways to Texas immigration were situated

[2] R. N. Richardson, *Texas: The Lone Star State*, New York: Prentice Hall, Inc., 1943, pp. 18–83, 187–212.

at Pecan Point and Jonesborough in the present Red River County (Fig. 1).[3]

As the influx of new Texans poured across the Coastal Plain from the northeast, the numerous southeasterly flowing streams were encountered. Soon it was discovered that most of the streams were underlain with mud or quicksand which precluded mass crossing except at a few places.[4] These rivers acted as effective barriers to local cross-country movement, and even where permanent ferries could be established, sites were chosen usually where streams cut across harder beds, because at these points meandering was slight, e.g., Moseley's Ferry across the Brazos River on the Old San Antonio Road where the river crossed a harder clay belt in sandy terrain.[5] Thus, river bottoms with solid rock bases were at a premium as fording sites, e.g., Austin Chalk outcrops at the present site of Waco.[6] The location of the few early fordable places where routes converged became all-important as centers of settlement. The stream crossings would then have acted as nodal points for the local regions, and when these areas were populous enough to assert their county status, boundaries were placed away from the most densely settled places making the rivers flow approximately through the middle of the new county, e.g., McLennan County and the Brazos River; Liberty County and the Trinity River (Fig. 1).

Still other factors, in addition to the nature of the stream beds, can be realized as evidence of early river valley unity on the Coastal Plain, e.g., (1) the tendency of rivers to bisect the areas of Empresario Grants, and (2) the preference of colonists for bottom lands instead of prairies.

The Empresario Grants originated from the efforts of the Mexican government to settle the empty land between the Rio Grande and Louisiana. The Grants accomplished this task by allowing aliens to obtain land contracts and bring in colonists—something Spain had not been able to do.

Sometimes a large stream occupied the central area of an Em-

[3] *Ibid.*, pp. 187–212.

[4] Shuler, *op. cit.*, pp. 26–31.

[5] *Ibid.*, pp. 26–31.

[6] F. W. Johnson, *Texas and Texans*, Chicago and New York: The American Historical Society, 1914, Vol. II, pp. 613–1037.

presario Grant, e.g., (1) the Medina River in the Bexar District, (2) the Guadalupe River in the grants of DeWitt and DeLeon, (3) the Brazos River in the Robertson Grant, (4) the Colorado River in the Austin Colony, and (5) the Neches River in the Zavala Grant (Fig. 2).[7] When county boundaries were formed, it was through sparsely populated territory with the rivers remaining near the center of the county: e.g., in Robertson's Grant, the Brazos River flows through Falls and McLennan counties; in Austin's Colony, the Colorado River flows through Colorado, Fayette, Bastrop, Travis, Wharton, and Matagorda counties; in the DeWitt and DeLeon grants, the Guadalupe River flows through DeWitt, Gonzales, and Guadalupe counties.

The preferential attitude of pioneers on the Coastal Plain toward settling bottom lands, where local conditions permitted, arose from their belief in the inherent infertility of treeless prairie soils. Besides, the prairies could not provide the required wood for shelter, fuel, and fences; and these early pioneers were accustomed to an abundance of resources in the Eastern forests.[8] These three causes of river valley unity, (1) streams flowing across former Empresario Grant areas, (2) streams possessing choice fording sites, and (3) preference for bottom lands by early settlers, during the early stages of settlement, would seem to persist today. The location of many county seats near the center of certain counties, on a river which previously flowed across an Empresario Grant, or where settlement first developed in bottom lands is evidence of this: e.g., in Robertson's Grant, Waco in McLennan County; in Austin's Colony, Columbus in Colorado County, LaGrange in Fayette County, Bastrop in Bastrop County, and Austin in Travis County; in the DeWitt and DeLeon grants, Victoria in Victoria County, Cuero in DeWitt County, Gonzales in Gonzales County, and Goliad in Goliad County (Fig. 3).

Where rivers are utilized as county boundaries on the Coastal Plain, the disunity of the early stages of settlement, i.e., the rivers did not serve as original unifying regional bonds drawing people together, still appears to persist in the county line arrangement to the south, e.g., river boundaries of former Empresario Grants continue to

[7] Shuler, *op. cit.*, pp. 26–31.
[8] Richardson, *op. cit.*, pp. 187–212.

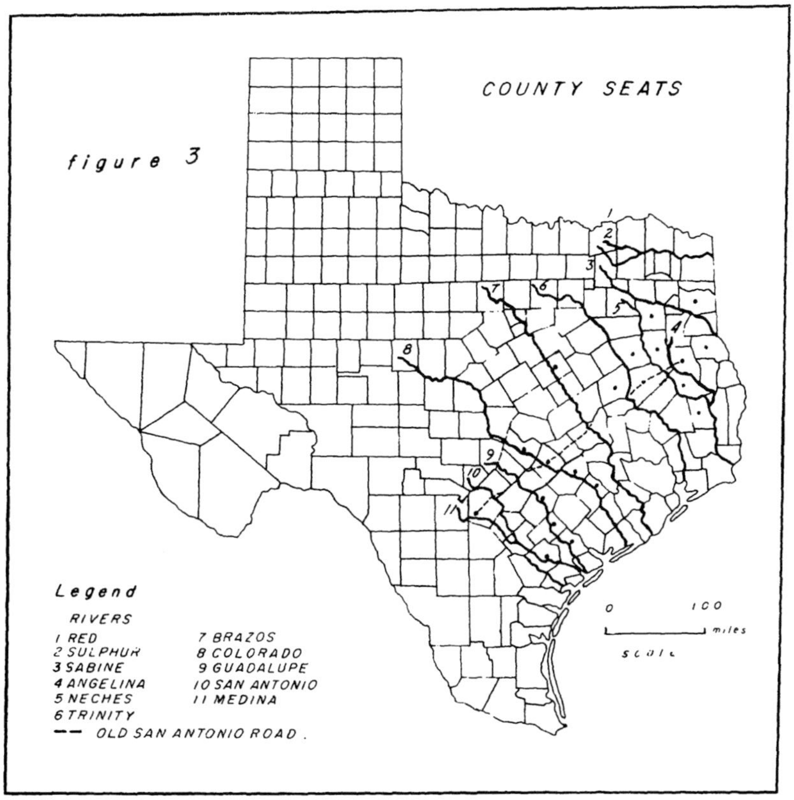

FIG. 3

act as county borders in some of the oldest counties of Goliad, Refugio, and Jackson (Fig. 1).

The northeastern part of the Coastal Plain is a region which possessed streams hindering cross-country movement. As a result, settlement undoubtedly did not gravitate to their banks. Ten county seats can be found away from the water courses, e.g., Palestine, Woodville, Lufkin, Rusk, Nacogdoches, Center, Marshall, Tyler, Crockett, and Centerville.[9] With the exception of Shelby County, where the county seat was moved from Shelbyville to nearby Center, these original

[9] Johnson, *op. cit.*, pp. 613–1037.

county seats remain the focus of land routes in their local region. It would seem, therefore, that rivers bordering these counties have never been a unifying influence (Fig. 3).

Of the principal Coastal Plain rivers, the Colorado, Guadalupe, and Brazos have exerted the greatest unifying force upon riparian settlement, while the other major streams have fostered disunity in this respect.

Papers of the Michigan Academy of Science, Arts, and Letters
Vol. XLVII, 1962 (1961 Meeting)

POLITICAL POTENTIALS IN THE HORN OF AFRICA

EDWIN E. MEADER

Ann Arbor, Michigan

JUTTING into the Indian Ocean and backed by a complex hinterland, the Horn of Africa presents a striking feature of the African map (Fig. 1). On a map of the world it represents an isolated corner distinctive for its shape, stimulating in its remoteness, and perhaps repugnant for its barrenness and heat. For thousands of years men have roamed this territory. It is adjacent to the ancient Land of Cush told of in the Bible. It is part of the country called Punt visited by the expeditions Queen Hatshepsut depicted on her tomb at Thebes. Greeks and Romans called it "Regio Cinnamomifera" and thought it the most southerly part of the habitable world.[1] Strabo and Herodotus wrote of the Icthyophagi living on its shores, feeding on the dead fish cast up by the sea. In the sixth century A.D., the torrid coasts were known to Arab sailors as Bar Ajam, Land of Heat.[2] Inland, Prester John is more than a myth in the isolated upland valleys, and the legend is still told of the descent of Abyssinian kings from the son King Solomon tricked from the Queen of Sheba.

From ancient times migrations have moved from the Arabian Peninsula across the Red Sea and Gulf of Aden into the Horn. From Yemen came the Hamitic and Semitic peoples who established themselves on the Abyssinian plateau. From the Hadhramaut and region of Dhofar successive waves of Sabaeans, Gallas, and Somalis moved through the lowlands of the Horn, resulting in a great fanlike distribution of Cushite peoples. Intermixtures of the invaders with residents blurred ethnic lines. The Arabic word for Abyssinia is "habash," meaning "mixture," but distinctly Hamitic, Semitic, and Negro traits remain, and the region has been called "a museum of peoples."

For the present purpose the area included within the Horn is

[1] Ralph E. Drake-Brockman, *British Somaliland* (London: Hurst and Blackett, Ltd., 1912), p. 2.

[2] *Ibid.*, p. 13.

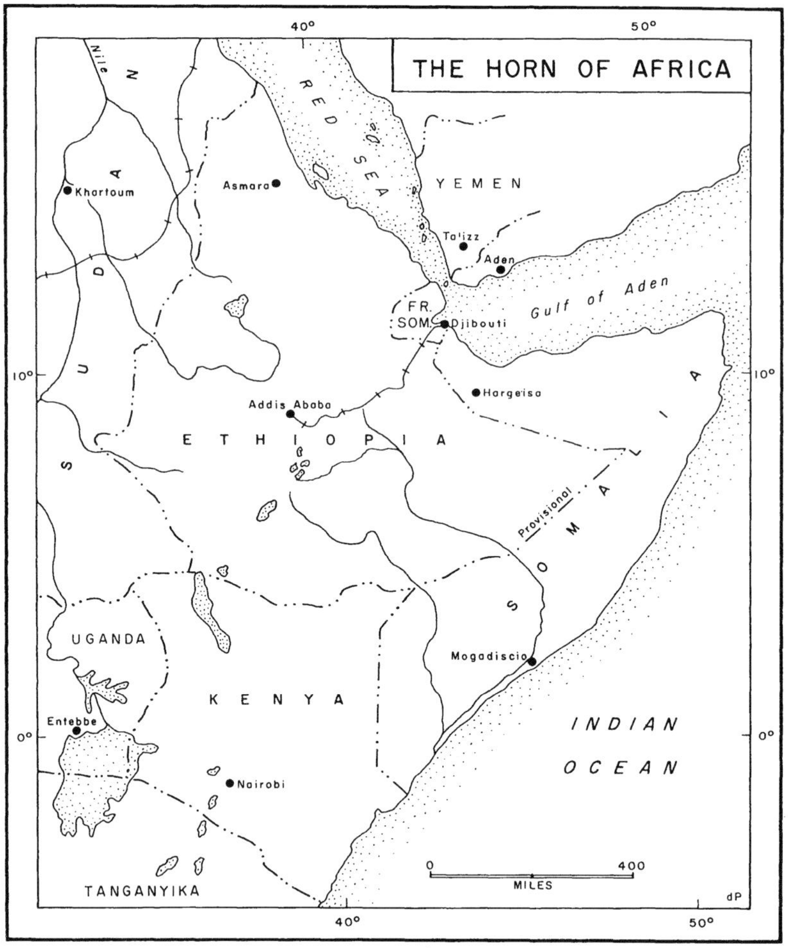

Fig. 1

Ethiopia with Eritrea, the Somali Republic, French Somaliland, and that part of Kenya lying within approximately a hundred miles of the Somali border. Excluding the latter, this totals a little more than 703,000 square miles. The total population is estimated variously from 9 million to 17 million.

The contrasts in this region are spectacular. Heights of 14,000 feet overlook sinks below sea level. Cool plateaus are adjacent to broiling coasts. Violent and heavy rainfall in the west subsides sharply to almost absolute aridity in the east. Steppes and deserts of the eastern lower lands allow easy movement to roaming people. The high lands are gouged with deep chasms enforcing isolation and encouraging provincialism, independence, and suspicion. The economic life and population distribution come into focus easily against this background. Nomadism dominates the lower country with settlement only in the south, where there is irrigable land and limited dry farming. In this semidesert the camel is the prime economic factor. Herds and flocks follow the interminable search for enough water and pasture. On the plateaus pastoralism and settled agriculture support life. The determined volume of natural resources other than agricultural is hardly significant.

How natural that in this land of lofty, exclusive heights surrounded by vast, open plains the contrast should continue into the two principal political structures. One is centuries old, fragmented like the land in its local aspects; a challenged autarchy supported by a moribund culture. The other became a nation 1 July 1960; a parliamentary government with political parties energetically organized. It had a full complement of national and international problems, including crises.

From the standpoint of political geography the Empire of Ethiopia and the Republic of Somalia each present an interesting prospect for study and analysis. But, taking them together with the French enclave and considering them as components of a geographic region is almost inevitable. This is true primarily because of the strategic aspects of the region deriving from the geographic function of location.

From several viewpoints Ethiopia may be regarded as an association of vertical factors. Physically there is a strong sense of the vertical. The Abyssinian plateau, extending from Eritrea to Kenya at elevations of from four to eleven thousand feet, is approximately half of the country. The Somali plateau, to the east, is about as high.[3] The Ethiopians themselves divide their country vertically into cold, temperate, and hot zones. Almost as vertical as the plateau escarpments are the precipitous river gorges, some more than a mile deep, which fractionate the country. These huge gashes in the earth

[3] L. T. C. Kuo, "Ethiopia," *focus,* V, No. 10 (June, 1955), 1.

offer an amazing variety of flora and fauna in short horizontal space, from high altitude plants at the top to tropical malaria at the bottom.

There are vertical aspects among the Ethiopians in their social organization as well. The accepted superiority of the Christian Amharas, occupying the most favored parts of the Abyssinian plateau, contrasts sharply with the inferiority of the pagan Gallas and Danakils of the lowest plains and deserts. The two million Amharas' consistent political consciousness and genius for organization have made them the ruling power. Yet with this power they combine a vast ignorance, self-satisfaction, vanity, and indolence. But even their contempt for things foreign does not exclude humor, generosity, and hospitality to strangers.[4] The topography in which they live inhibits trade and communication, engenders a sense of adequacy and indifference to what is unfamiliar. Ethiopian bases of value are so alien to Americans as to be hardly recognizable—and vice versa.[5] The source of inspiration to revolution and change is difficult to imagine, so great is the weight of conservative unconcern.

The Ethiopian Church, monolithic and ultraconservative, also represents a vertical aspect in that its archbishop sits at the pinnacle of a hierarchy committed to the maintenance of the status quo. It absorbs perhaps a fifth or sixth of the male population as priests, parasites in a partly obligatory system of personal "advisors."

Similarly, the government is a vertical organization. The Emperor, at least, is vertical and in respect of him the populace and all Ethiopian officialdom including the Crown Prince are horizontal, literally, with faces in the dirt. The government has all the incredibly lugubrious characteristics of a one-man show too big for one man. The Emperor makes all of the important decisions. Advancement is sought by regular showings of face at the palace.

Local government is in the hands of village chiefs. They are traditional figures and are the principal representation of government to most of the people in a land historically ravaged by wars, uprisings, and roving bands.[6] To quote an authority, "the mass of

[4] Ernest W. Luther, *Ethiopia Today* (Stanford: Stanford University Press, 1958), pp. 36–37.

[5] *Ibid.*, p. 26.

[6] Margery Perham, *The Government of Ethiopia* (London: Faber and Faber, 1948), p. 275.

the people are hardly aware even of their own existence in a larger and more unified whole."[7]

The Emperor, so often lauded as the patron of his people, laboring in their behalf for education, for health, and for general improvement, must have been singularly galled to hear, half way around the world, of his being deposed by his Crown Prince. This was "... to end three thousand years of injustice to the Ethiopian people.... concrete action aimed at improving the standard of living... etc."[8] The Prince promised a constitutional monarchy and a system for permitting political parties, but ultraconservatism won. Still, the unrest of the educated minority had shown itself, and some remained to grow.[9]

The background of Ethiopian foreign policy is well expressed in the following quotation from historians of the country. "For long periods, especially since the rise of Islam, the Abyssinians have lived cut off from the outside world, and during these periods, in which they were surrounded by barbarians and foreigners and enemies of their faith, they developed a distrust of foreigners."[10] This sounds like an understatement and it is. Ethiopians still think they are surrounded by barbarians, foreigners, and enemies of their faith, but they are no longer cut off. Suspicion and distrust are not unique to Ethiopia but they are more active in its policy, just as they are in the nature of its people.

In the United Arab Republic Ethiopia finds its ancient enemy, largely because of the vulnerability Egypt feels concerning Ethiopian control of the source of the Blue Nile, which provides most of Egypt's water. Probably the Egyptians will never be wholly satisfied until they control Lake Tana and the whole course of the Blue Nile. A more practical and limited objective is some form of international control for riparian purposes. Meanwhile it appears to be UAR policy to inhibit Ethiopian development in every possible way and to encourage and support Ethiopian opposition.

While Islam is not yet much of a political force it is probably the most potent threat Nasser has for Ethiopia. The Christians, cen-

[7] Luther, *op. cit.*, p. 1.

[8] *Newsweek*, 56: 27, December 26, 1960.

[9] *Time*, 76: 18, December 26, 1960.

[10] A. H. M. Jones and Elizabeth Monroe, *A History of Ethiopia* (Oxford: Clarendon Press, 1935), p. 6.

tered in the Abyssinian highlands, find Muslim majorities almost encircling them. There is fear of being surrounded completely. If Islam continues its current conquests among the central and east Africans the encirclement may soon be a reality. The Abyssinian homeland, which successfully resisted the spread of Islam into eastern Africa for centuries, now faces a developing religious potential with inescapable political overtones.[11]

The United States and Soviet Union positions vis-a-vis Ethiopia follow a familiar pattern. The U. S. has poured more than $104 million into development work and in support of the army of about 28 thousand. Soon after the U. S. associated itself with the 1959 British proposal to unite British Somaliland with Somalia the Emperor accepted a Soviet invitation of five years' standing and went off to Moscow. What may accompany the $100 million loan with which he returned is not specified, but there is no reason to believe that the familiar Soviet pattern of exporting communism with economic assistance is being abandoned. Not long after the Emperor was returned to his throne by the British during World War II the Soviet Embassy in Addis Ababa opened with an inordinately large staff, which has continued to grow. Subsequently came a Soviet hospital and staff. So far neither the U. S. nor the Soviet Union has taken a position with Ethiopia in regard to a touchy and currently important foreign problem—the territorial and boundary differences with neighboring Somalia. An American diplomat is quoted as saying, "On this issue there is no compromise with the Emperor. We are with him or against him."[12]

A British colonial administrator writing in Berbera in 1912 made this comment on Somali character: "It must be remembered that the man who will twist a lion's ears to pull it off his master when he is being mauled cares little for death and less for threats."[13] Forty years later another Englishman commented, "The Somalis are independent people unused to any sort of central government, unaccustomed to any form of direct taxation, and intensely suspicious of

[11] J. Spencer Trimingham, *Islam in Ethiopia* (London: Oxford University Press, 1952), p. 15 *et seq.*

[12] Edward A. Bayne, "Somalia on the Horn," *American Universities Field Staff Reports Service, Northwest Africa Series,* VII, No. 9 (March 11, 1960), 6.

[13] Drake-Brockman, *op. cit.*, p. ix.

the foreigner, the white man, the Christian. The only forms of national sport are raiding one's neighbor's camels, seizing his water supplies, and emasculating his males."[14] These are the people who comprise the new nation formed in July 1960 when the former British and Italian administered territories joined and entered the United Nations.

Physically Somalia needs almost everything a nation could need except land and people. It is already trying to get more of what it needs least. The economy has had essential assistance since World War II from British, Italian, and U. S. grants. Only massive further assistance will enable the government to maintain fiscal responsibility. Schools, roads, water development, communications are all vitally needed even after what is generally regarded as good Italian trusteeship under the United Nations since 1950.[15]

Even the assets of land and people are so limited in some respects that one is led to wonder what may be the minimum requirements for a nation. The land is extensive, about 250 thousand square miles, and nearly all steppe and semidesert. Settled agriculture can be supported almost exclusively in the area near and between the Juba and Webi Shibelli rivers. The remainder is for nomads. The people share a vast love for a free-roaming life, a disdain for village and urban dwellers, and a sense of confident superiority founded upon abysmal ignorance of affairs and people outside their desert world. A few Somalis have become aware of their lack of a native written language, but the remedies have been unacceptable.

These seemingly monumental disadvantages for nationhood have not inhibited an extraordinary, rapid political development. Political parties first took root under British tutelage and continued under Italian trusteeship. A general enthusiasm developed for political organization, but it developed along tribal lines. For the Somali politician the urgent and immediate need is political stability. The government must successfully and politically manage, on a daily basis, a large group of illiterate, nomadic, tribalized people.[16] Somali politicians must negotiate a difficult path. They must keep their state

[14] J. Darrel Bates, "The Somaliland Protectorate," *Corona* (May, 1953).

[15] A United Press International dispatch printed in the *Detroit Free Press*, February 6, 1961, indicated an oil strike near Mogadiscio. Subsequent information does not permit an assessment of this event, which could be momentous.

[16] Bayne, *op. cit.*, No. 6 (March 4, 1960), 7.

alive with infusions of foreign grants or loans and secure foreign assistants for all kinds of skills from artisans to teachers and executives to operate the state. They must jockey for political advantage among people whose tribal loyalties are their primary, if not only, acknowledged responsibilities.

The most important external concern for Somalis is Greater Somalia, a rearrangement of borders to include French Somaliland, the Ogaden Province of Ethiopia, and eastern Kenya. "Greater Somalia" is the rallying cry. It is the focus of active Somali nationalism. The suggested generosity on the part of Somalia's neighbors would relieve an incipient irredentism. It would resolve the border dispute which Ethiopia chooses to ignore at present. It would eliminate the serious difficulties which have arisen from nomads crossing international boundaries in pursuing their centuries-old practice of changing pasture with the seasons. Britain has been supporting this solution since 1941. Egypt supports it, of course. Ethiopia is not inclined to relinquish territory. On the contrary, the Emperor has urged the return of all Somalis to their homeland in a federation similar to that shared with Eritrea.

Previously it was noted that the United States and the Soviet Union have not taken official positions regarding Somali expansionism. The U. S., Italy, and Great Britain have expressed an intent to continue economic aid, but Greater Somalia will be the crucial problem in the immediate years ahead. In January 1961 Somali nomads making their customary crossing into Ethiopia were strafed by Ethiopian planes and attacked by Ethiopian infantry bearing automatic weapons. Twenty Ethiopians and a hundred Somalis were killed. Hundreds of Somalis were wounded, and there was a large loss in camels, goats, and cattle dead or wounded. The Emperor had ordered an example made of the "marauding" Somali.[17]

Of the others concerned directly with Greater Somalia, the French are not likely to offend Ethiopia or to give way except for some definite political advantage. The 80 thousand Somalis of Kenya are the least politically conscious group of all and present no current problem.[18]

Chinese communist activity in the Horn came to light in the crash of an Egyptian airliner near Elba in September 1960. There

[17] *Newsweek,* 57: 39–40, January 16, 1961.
[18] Bayne, *op. cit.,* No. 8 (March 8, 1960), 8.

appeared evidence of the financing of ultranationalist, antiwestern Somali leaders to the extent of $4 million in Swiss banks. Additional evidence involves an illegal shipment of arms to the Somali coast near Berbera. The location of Somalia gives the Red Chinese a remarkable opportunity to attack the United States through Ethiopia, the French through Djibouti, and the British through Kenya.[19]

Speaking of Somalia's external problems, the Commandant of the Somalia National Police commented, "... two elements seem the most difficult of all to me. The first is Ethiopia. We are never certain what they will do and we must be prepared. Second, the Russians. We have never had to deal with them before but it is certain they will be here."[20] That was early in 1960. Now the Soviets and the Satellites are there.

Solutions to the troublesome problems of the Horn of Africa will continue to be offered. There has been serious consideration of Somalia's joining the British Commonwealth. In addition to Emperor Haile Selassie's proposed federation others have been suggested. One is a federation of Greater Somalia, including all Somali lands, with Ethiopia under United Nations supervision. Another "would include parts of Somalia, the Sudan, the Congo, Nyasaland and Northern Rhodesia."[21]

In this odd corner of the world an ancient and conservative way of life is being challenged by the dynamics of modern international politics. Arnold J. Toynbee, in *Civilization on Trial,* wrote, "Future historians will say, I think, that the great event of the twentieth century was the impact of Western civilization upon all the other living societies of the world of that day. They will say of this impact that it was so powerful and so pervasive that it turned the lives of all its victims upside down and inside out."[22] The political potentials in the Horn of Africa uniquely illustrate this process.

[19] Robert Counts, "Chinese Footprints in Somalia," *Reporter,* 24: 32–34, February 2, 1961.

[20] Bayne, *op. cit.,* No. 5 (March 1, 1960), 8.

[21] Drew Middleton, *New York Times,* March 12, 1961.

[22] Arnold J. Toynbee, *Civilization on Trial* (Oxford: Oxford University Press, 1948), p. 214.

Papers of the Michigan Academy of Science, Arts, and Letters
Vol. XLVII, 1962 (1961 Meeting)

B. RECREATIONAL GEOGRAPHY

THE TERMINOLOGY OF RECREATIONAL GEOGRAPHY

ROSS N. PEARSON
The University of Michigan

During the years immediately following World War II and up until the publication of *American Geography: Inventory and Prospect,*[1] in 1954, the study of recreational phenomena was considered to be one of the newly emerging fields of geography. At that time it was generally thought of as a subdivision of economic geography. Since then the amount of research dealing with recreational subjects indicates that recreational geography no longer can be considered on the fringe of geography or an infant branch of the subject. Further evidence indicates that it should not be confined within the realm of economic geography. Recreational land use is important in urban areas, as well as in forested regions and along lake shores, and many aspects of recreational study have implications in the field of cultural geography. During the time that this growth has occurred it seems apparent that there has been considerable confusion, which has handicapped the development of recreational geography, in the terminology used to describe recreational features. It is hoped that what is presented here will serve as a basis for clarification.

If we start with the basic word *recreation,* we find it defined in most dictionaries as "refreshment of body or mind." Thus it follows that any person seeking or participating in such activity is a *recreationist* and that the numerous occupations, businesses, and economic enterprises which provide the means for individuals to secure "refreshment" collectively make up the recreation industry. The interest of geographers in such matters is twofold: first, the areal distribution of recreational activities is based in part upon natural environment, which with cultural additions to the landscape causes differences from place to place; secondly the existence of recreational activities

[1] P. E. James and C. F. Jones (Editors), *American Geography: Inventory and Prospect* (Syracuse: Syracuse University Press, 1954), p. 251.

in many areas makes possible a livelihood for local residents who provide visitors with entertainment, goods, and services. In this discussion we shall examine first the various groups, or recreationists, which provide the market for the recreation industry, and then attempt to describe the several forms of recreational land use.

The Recreationists

In the past it has been common practice to refer to all persons who travel away from their place of residence for pleasure as *tourists* and to the various activities that provide them with entertainment, food, and lodging as *tourism*. Since the root word *tour* implies travel it is suggested that these two terms be applied only to those aspects of recreation in which movement occupies a large part of the time and where travel, with sight-seeing, is the factor which brings refreshment of body or mind. *Tourists*, then, are transients. They spend most of their time in travel and usually do not stay in any locality more than one day and night. Tourists may or may not have a previously planned itinerary, lodging is often whatever is available as evening approaches, and meals are usually taken wherever food can be secured at the time it is desired.

Most reports dealing with the tourist business imply that the people being served are away from their residence overnight. It may be noted here that a large part of the recreation activity of Michigan and many other regions is made up of persons who leave home during the daylight hours and return during the evening. Some 65 per cent of the 20 million visitors to the Michigan State parks are accounted for by those located in the southern four tiers of counties,[2] most of whom are day tourists. Among the most important individual tourist attractions of Michigan, in terms of number of visitors, are places like Greenfield Village, Tiger Stadium, and the sports fields of the universities. They attract nonresidents for short periods of time, mostly those who do not stay overnight, but nevertheless provide considerable income for the area visited.

In many parts of the world recreational development is based upon attracting visitors who do relatively little traveling other than that involved in the journey to and from their home, and who have a single temporary residence while engaged in leisure-time activities. It is suggested that recreationists of this type be referred to as *vaca-*

[2] Automobile Club of Michigan, *Michigan State Parks*, 1957–1958 edition.

tionists. A vacationist has a fixed abode, takes most of his meals there, and whatever his daily activities may be returns to it each evening. Some are attracted to commercial establishments that can lodge and feed a number of people for an extended period of time. Others own or rent individual cottages and prepare their own meals.

On the basis of length-of-stay there are several types of vacationists. Among the most common are *resorters*—families, couples, or individuals who take up residence at a commercial resort or cottage for a one-, two-, or three-week period. Many areas attract recreationists who stay for several months, or *seasonal residents.* Most of the latter live in a single housing unit, rented or owned, and a few take accommodations in a hotel or inn. A third special class of vacationist is the type who visits an area for two or three days, usually on weekends. These *weekenders,* like tourists, spend much of their time in travel, but unlike them may return to a single area several times during a year and carry on most of their activities from a single base. One of the most rapidly increasing types of vacationist in the United States is made up of persons who carry their shelter and cooking equipment with them, or *campers.* Some campers are undoubtedly tourists, but because of the amount of gear they haul and the time involved in setting up and taking down the camp most of them stay for several days in a single locality.

Although numerous attempts have been made to classify recreationists on the basis of activity, this seems to be a rather futile line of approach. Most recreationists participate in a variety of pursuits and some in none at all. Sightseeing, hunting, fishing, swimming, boating, skiing, and spectator events are important for both tourists and vacationists, but for many relaxation is gained by a change of routine, rest, and doing nothing. Change of scenery and climatic differences, real or psychological, are always important.

Recreational Land Use

The many ways in which people seek refreshment of body or mind and the several types of recreationists make it impractical to recognize a single category of recreational land use. There are at least four types of land utilization, however, that can be classified as primarily recreational. Strangely enough, three of the four are intensive in nature and have many characteristics of urban areas.

The term recreational land is most often applied to forested areas, lake shores, mountainous regions, and other places with scenic attractiveness and few inhabitants. Except for tracts that have been designated state and national parks or public recreation areas, these are seldom purely recreational and it is suggested that they be classified as *wild lands.* Such lands even in state and national forests have some use for lumbering and grazing, and those in private ownership may also be held for petroleum exploitation, watershed protection, or speculative purposes. In a great many instances they are recreational by default; that is, there is no other use to which they might at present be placed. These lands provide a habitat for game, have streams which support a fish population, add to the scenic beauty, and may have scattered cabins or cottages. Although they are primarily recreational, they are not entirely so; hence the designation wild land.

The development of recreation in the northern lake states and in many other parts of the world has brought about an intensive type of land use which consists of a large number of structures clustered about a particular feature, often a lake shore. These can best be designated as *resort areas.* The most prominent structure is the single-family seasonal dwelling or cottage. In many the development has been preceded or followed by the construction of a commercial resort. Such agglomerations have grocery stores, gas stations, taverns, overnight cabins, permanent homes, and other features which cause the settlement to be urban in nature. Business places and areas of congestion tend to be concentrated near points of greatest access, and it is not unusual for such conditions to be accompanied by what might be called recreational slums. On a small scale, resort areas have most of the problems that are associated with large cities: trash and garbage disposal, road maintenance, law enforcement, fire protection, even zoning regulations. Resort areas are recreational in nature, but they are quite different from the broad extensively used wild lands and should be so designated.

With the increase in travel that has taken place in recent years and with the improvement of highways there has emerged a third form of land use that is essentially recreational. At especially favored spots along major routes of travel there are clusters of gas stations, restaurants, taverns, motels, gift shops, grocery stores, and residences which might be referred to as *highway service hamlets.* Especially concentrated at major highway junctions and expressway interchanges, these clusters also cater to the commercial traveler, but they derive much of their business from the tourist and vacation trade. In spite of their relatively recent development, they bear a striking resemblance to Trewartha's unincorporated hamlet.[3] The major differences are their orientation toward routes of travel and the fact that their services are specialized to fit the needs of a nonresident population.

The fourth classification of recreational land use is that found in urban areas. All cities have private recreational facilities, and Bartholomew[4] estimates that 7 per cent of the developed area of the average American city is devoted to public parks and playgrounds. Obviously most of these are for the use of local people, but in many cities there are attractions which bring a large influx of nonresidents each year, creating sources of employment and income for the city population. Among such attractions are sports arenas, theaters and music halls, convention buildings, historical structures and museums, large department stores, and night clubs. Many hotels, shops, and restaurants depend heavily upon this trade. We have no way of determining what percentage of the patronage of the attractions is made up of out-of-towners, but it is certain that the attractions could not exist in such numbers and sizes if they had to depend entirely upon the residents of the city in which they are situated.

Summary

The study of recreational phenomena by geographers has now reached the point where there is need to recognize that there are several types of recreationists and that there is no one clear-cut type of recreational land use. Tourists are attracted to certain kinds of areas, and recreationists who stay for longer periods of time are attracted to others. The land use types that have developed to accommodate vacationists are quite different from those that serve the needs of over-night or day-time visitors. It would seem that past concern with "north country" recreational development has caused types which are essentially urban to be overlooked.

[3] G. T. Trewartha, "The Unincorporated Hamlet," *Annals of the Association of American Geographers*, Vol. 33 (1943), pp. 32–81.

[4] H. Bartholomew *Land Uses in American Cities* (Cambridge: Harvard University Press, 1955).

PAPERS OF THE MICHIGAN ACADEMY OF SCIENCE, ARTS, AND LETTERS
VOL. XLVII, 1962 (1961 MEETING)

SOME NOTES ON THE USE OF SALES TAX STATISTICS IN THE STUDY OF THE RECREATION INDUSTRY OF MICHIGAN

WILLIAM R. BRUECKHEIMER

Western Michigan University

FROM the time of its adoption in 1933 until the tax increase of January 1, 1961, the Michigan sales tax was imposed upon retailers of tangible personal property at the rate of 3 per cent of their receipts from retail sales.[1] Each month the Michigan Department of Revenue issues a bulletin giving a summary of tax collections by counties and kind of tax.[2] This information along with unpublished sales tax statistics secured from the tax offices in Lansing has proved to be a rich source of material for the student of recreation; a source largely untapped as of the year 1952.

In the spring of 1952, both Ross N. Pearson of The University of Michigan and I found these tax summaries useful in the development of our dissertations on the recreation industries of Ogemaw and Alger counties, respectively. I owe much to Pearson for ideas for using and organizing sales tax statistics in the study of recreation.[3] This paper presents some of the findings which developed from a recent study of 1960 sales tax statistics as well as from research done on 1951 statistics for my dissertation.[4]

[1] Retail Sales Tax—Act 167, P. S. 1933. The short-lived 4 per cent sales tax in the fall of 1959 has no bearing upon this paper.

[2] Research and Statistical Bulletin, Michigan Department of Revenue, Lansing, Michigan.

[3] Ross N. Pearson, "Recreation and Its Significance in the Economy of Ogemaw County, Michigan" (Unpublished Ph.D. dissertation, Department of Geography, University of Michigan, 1953).

[4] William R. Brueckheimer, "The Significance of the Recreation Industry in Alger County, Michigan" (Unpublished Ph.D. dissertation, Department of Geography, University of Michigan, 1953).

A Comparison of Sales Tax Collections for Selected Counties

By plotting the monthly per capita collections for selected counties on graphs, one quickly perceives very significant differences in the seasonal pattern of retail sales between tourist counties (Mackinac and Alger) and other counties which one might characterize as agricultural (Lenawee) or urban-industrial (Kent). Fig. 1 shows sharp rises beginning with May for the tourist counties until the peaks in sales are reached in the months of July and August, the peak tourist and vacationing months. In both Alger and Mackinac counties, the November sales tax collections are higher than the December collections—the reverse of collections in the other counties pictured and for the state as a whole. It is interesting to note that whereas March sales in Mackinac County are approximately 50 per cent of those of Kent County ($2.08 vs. $4.15), July sales are approximately double those of Kent County ($7.43 vs. $3.74) (Appendix I).

Kent County, on the other hand, shows a very different curve; a curve with a significant drop in sales during the months of July and August, the same months which were the peak months in Mackinac and Alger counties. Although there are many factors involved in the economy of Kent County which will not be considered here, certainly one important explanation of this drop in sales is the fact that many thousands of people normally living in Grand Rapids are vacationing and therefore spending elsewhere during this period.

The agricultural county of Lenawee, Michigan's number one corn county, has a curve much like that of the urban-industrial Kent County although flatter and with less variation from month to month.

The Per Cent of Retail Sales Attributable to Tourists and Vacationers

I have made some attempt to determine the per cent of retail sales tax collections which can be attributed to the spending of tourists and vacationers in Mackinac County. Total sales tax collections for the year 1960 were $469,283. Using March ($22,709) as a base month for the reason that March comes closest to the average for the eleven months excluding December in Michigan, the total

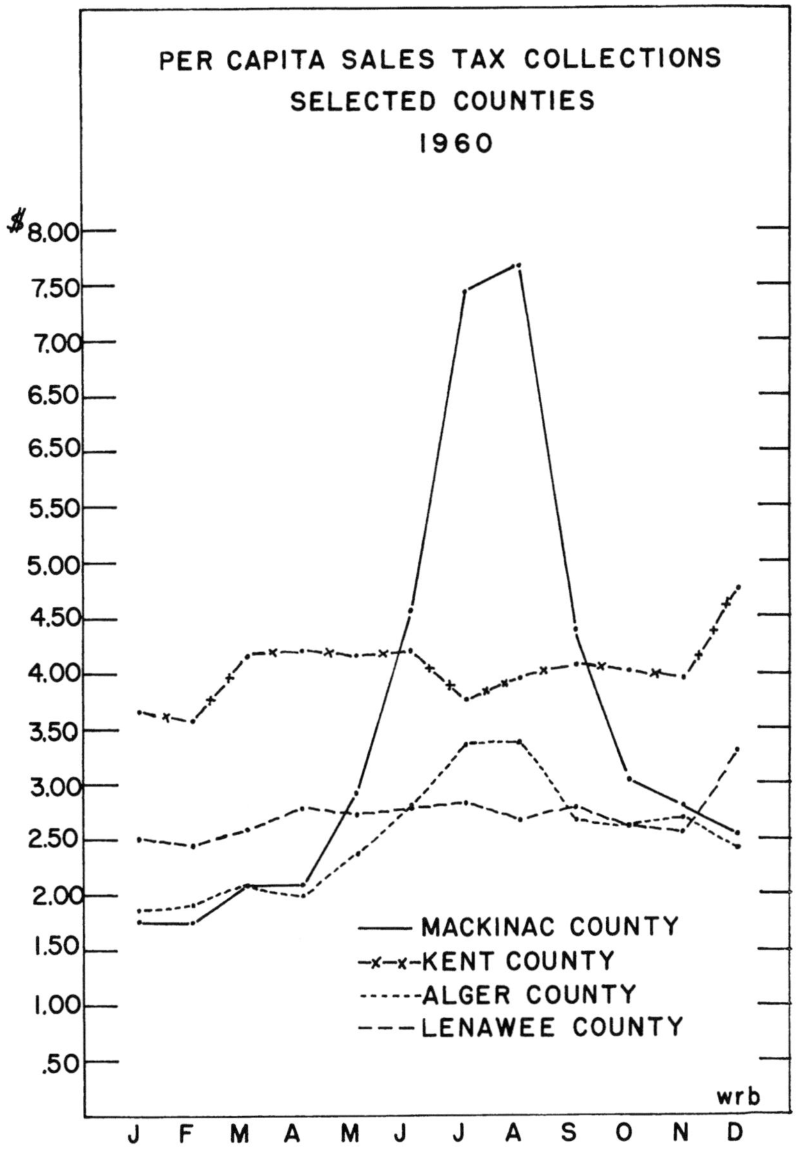

Fig. 1. Per capita sales tax collections, selected counties, 1960.

sales tax collections might have been expected to be $272,508 in Mackinac County. The difference ($196,775) represents 42 per cent of the total. By capitalizing this figure at 3 per cent, the estimated retail sales attributed to tourists and vacationers would be $6,559,-166.

As far as I can determine, this great increase in retail sales during the recreation season is almost entirely the result of the influx of tourists and recreationists into and through the county. Rent is not taxed; therefore the very large sum of money which represents rent is not entirely accounted for by the above figure, although an unknown share is represented indirectly through purchases by the rentees. Many resort and motel owners and others serving tourists during the recreation season spend the winter months after hunting season as their own vacations in other areas of the country, so that a significant portion of the tourist and vacation dollars is spent outside the environs of the county. In addition, a portion of the rent dollars collected in tourist season is spent during the winter months within the county.

It seems evident, therefore, that actual tourist and recreational expenditures in the county far exceeded the estimated figure of $6,559,166 in retail sales. This is substantiated in part at least by Robert McIntosh of Michigan State University, who estimated the total 1957 tourist and resort business in Mackinac County at $12,-840,575 by capitalizing estimated tourist food expenditures at 28 per cent[5]—that per cent of the vacationist's dollar estimated by the American Automobile Association as spent for food purchases.[6]

Indirectly, the economic significance of the tourist and recreation industry on retail sales is much greater than indicated above. Many of the inhabitants of Mackinac County could not live in the county if it were not that a portion of their income came from the tourist and recreation industry; their existence in the county, therefore, as well as that of the service establishments meeting their needs, is to a considerable degree dependent upon the tourist and recreation industry.

[5] Robert Woodrow McIntosh, "A Measure of the Volume and Geographic Distribution of the 1957 Tourist and Resort Business in the Upper Peninsula of Michigan," PMASAL, 44 (1959), 323–326.

[6] *Americans on the Highway* (Washington, D. C.: American Automobile Association, 1956), p. 4.

The Effect of the Tourist and Recreation Industry on Selected Types of Business Activity

Monthly sales tax collections for various kinds of business activity can be plotted on graphs, and from such graphs, Figs. 2 and 3, one can readily see the influence which the tourist and recreation industry exerts on these types of businesses.[5]

It can be inferred from the graphs that many eating establishments in Alger County owe their existence to tourists and vacationers. There is a striking increase in business during the summer tourist season and also during deer season in November.

Many other types of business activity are greatly affected by the influx of tourists and vacationers. Notable among these are drugstores, liquor stores, sporting goods stores, grocery stores, and filling stations as shown in Figs. 2 and 3.

These businesses supply the immediate needs or desires of the tourist and vacationer and, therefore, reflect the influx of tourists and vacationers more than businesses which satisfy long-time needs. It is interesting to note that November, deer season, is the peak month for liquor sales in Alger County. Certain types of businesses such as department stores and general merchandise stores are similarly affected during the tourist season, but only moderately so as shown in Figs. 2 and 3.

Another approach using sales tax statistics was used in Table I. Four types of businesses were considered: filling stations, taverns, restaurants, and grocery stores. The number of establishments in the state of Michigan was divided into the 1950 population of the state, and a figure for the average number of people per establishment in each group was secured. The same was done for Alger County. By dividing these state averages into the population of Alger County, one arrives at the number of establishments in each group which might have been found in the county if Alger were characteristic of the state as a whole.

By subtracting the latter figures from the actual number of establishments in each group found in Alger County, I was able to arrive at the number of excess establishments in each group. By this process, I found that there were 64 excess establishments out of a total of 126 establishments in the four groups or, in other words, slightly over 50 per cent of the establishments in these groups might have been said to owe their existence to the recreation industry.

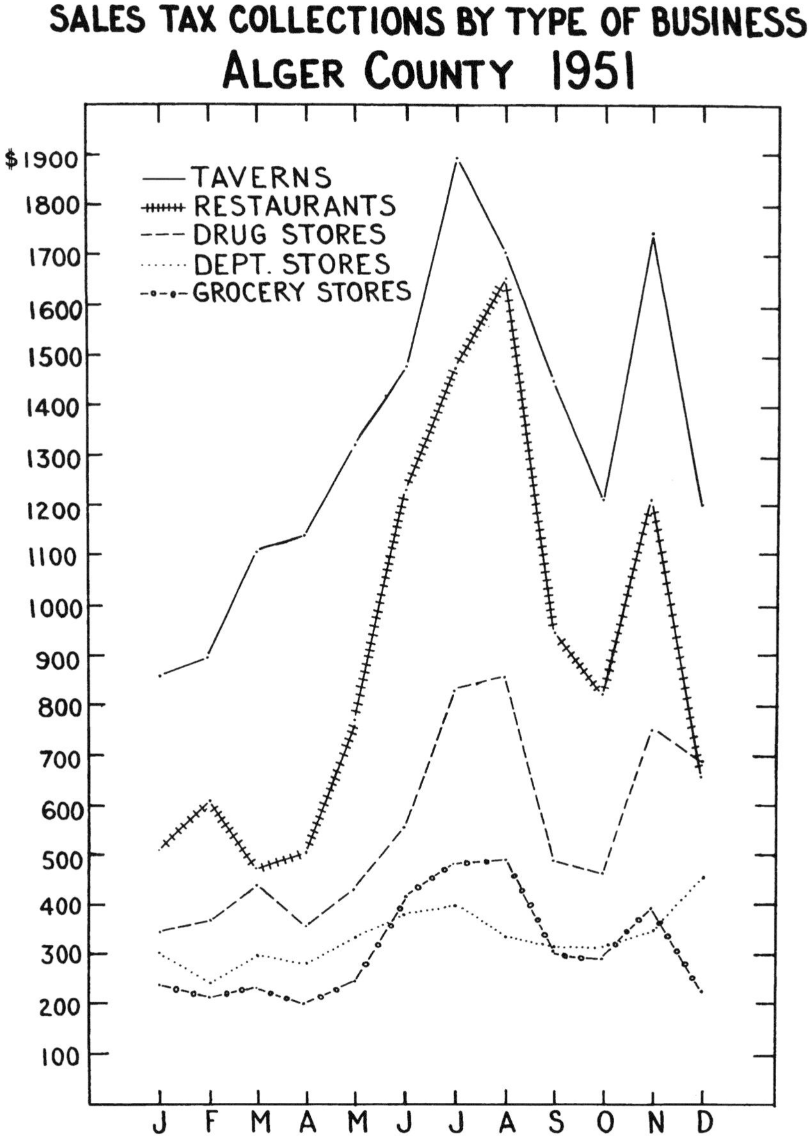

Fig. 2. Sales tax collections by types of business, Alger County, 1951.

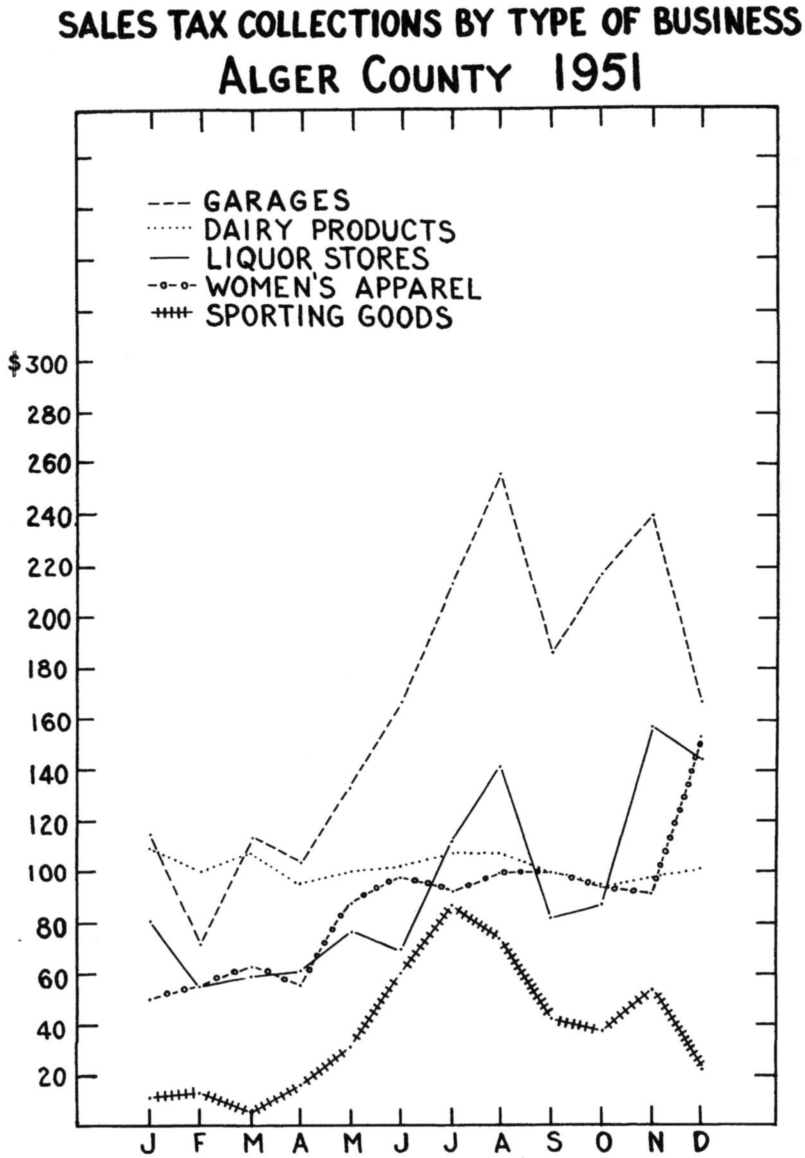

FIG. 3. Sales tax collections by types of business, Alger County, 1951.

TABLE I

EXCESS ESTABLISHMENTS (SELECTED KINDS OF BUSINESSES) IN ALGER COUNTY

Type of Business	People per Establish-ment (Michigan)	People per Establish-ment (Alger County)	Establish-ments (Alger County)	Establish-ments in Alger County on Basis of State Averages	Excess Establish-ments in Alger County
Filling stations.....	617	357	28	16	12
Taverns...........	931	360	27	11	16
Restaurants........	656	286	35	15	20
Grocery stores......	513	278	36	20	16
					Total 64

DISTRIBUTION OF THE TOURIST AND RECREATION INDUSTRY AS REVEALED BY SALES TAX STATISTICS

Sales tax statistics tell a great deal about the distribution of the recreation industry over the state as a whole. Two maps were prepared using 1960 sales tax statistics. The first map (Fig. 4) shows the per cent increase in sales tax collections for July 1960 over March 1960 by counties. The change in collections range from actual decreases in such urban-industrial counties as Ingham, Kalamazoo, Kent, Oakland, Saginaw, and Wayne to increases of over 200 per cent in Keweenaw, Lake, Leelanau, Mackinac, Montmorency, and Roscommon counties. In three additional urban counties—Calhoun, Genesee, and Washtenaw—the per cent increase was less than one per cent while the remaining large urban-industrial counties with the exception of Berrien County (St. Joseph-Benton Harbor) had increases of less than 10 per cent (Appendix II). The highly developed recreational industry of Berrien County must account in part at least for the 20 per cent increase in retail sales. In addition, the presence of thousands of migrant workers in this number one fruit and truck farming county would account in part for increased retail sales.

The greater share of the farm counties in the southern part of the state below the Muskegon-Bay City line had increases of less than 25 per cent—the exceptions being Allegan, Barry, Cass, Huron, Livingstone, Sanilac, and Van Buren counties. For the most part

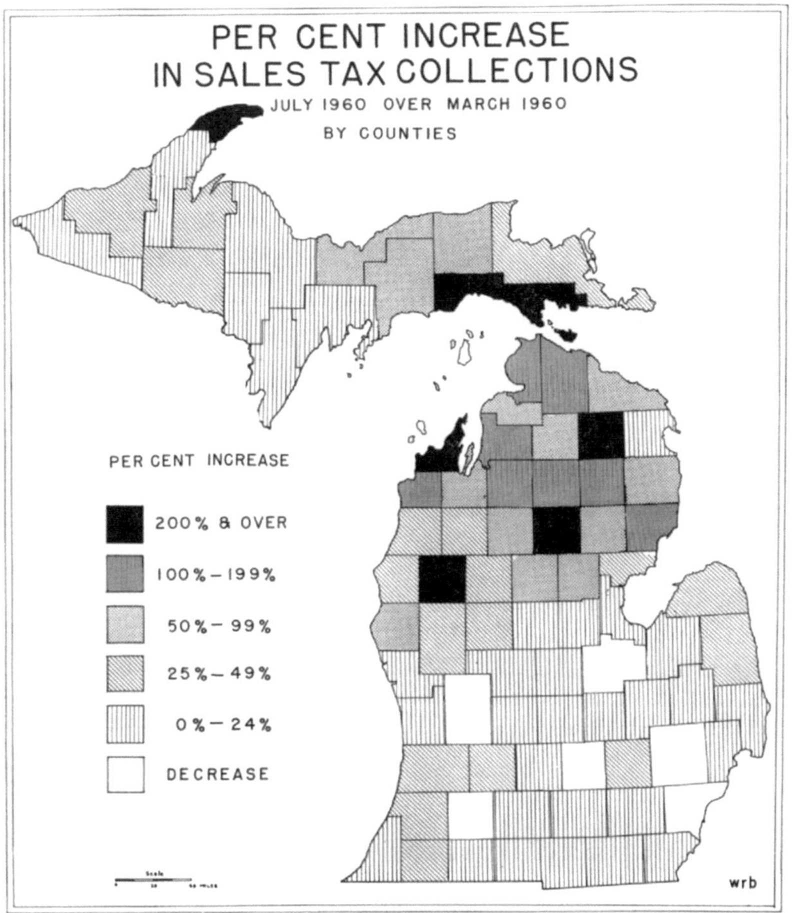

FIG. 4. The per cent increase in sales tax collections for July 1960 over March 1960 by counties.

these counties have large and well-known recreational areas which may account in part for the larger increases in retail sales in these counties.

The map clearly points up those counties in which tourist and recreational activity account for major increases in retail sales. Twenty-seven counties benefit from increases in retail sales of more than 50 per cent for July over March. Thirteen of these have increases of more than 100 per cent. It is interesting to note that none of these 27 counties is situated below the Muskegon-Bay City line mentioned so frequently in geographical literature in discussions of forest, climates, and soils.

Northern Michigan, however, does not benefit evenly from the influx of tourists and vacationers. In the western part of the Upper Peninsula, that is in the ten-county area west of Alger and Schoolcraft counties, only one county, Keweenaw, had an increase of more than 50 per cent in retail sales. In fact, in six of the ten counties, the increase was less than 25 per cent.

Statistical conclusions based upon per cent increases, however, can be very misleading. A relatively small increase dollar-wise can look sensational on a percentage basis if the original base is small. For example, while Keweenaw County had the largest per cent increase (264 per cent) in retail sales in the state, the actual dollar increase was only $5,983 and was exceeded by most of the counties in the state (Appendix II).

With this in mind, a second map, Fig. 5, was made showing the actual dollar increase in sales tax collections for July 1960 over March 1960 for each county. This second map reveals that certain counties which did not stand out on the per cent increase basis as tourist and recreation counties do rank importantly in the recreation industry. Notable is the case of Berrien County where the per cent increase for July was only 20 per cent over March. But the dollar increase for Berrien County was $95,272, highest in the state of Michigan. In addition, Macomb County with only a 5 per cent increase in collections had an actual dollar increase of $61,542, while St. Clair with a 17 per cent increase had an actual increase of $52,462. It seems evident that the beaches and parks of Macomb and St. Clair counties entice the urban dwellers of Wayne County in large numbers. The popularity of the Lake Michigan shoreline is also more evident on the second map than on the first.

Although there is some variation in the intensity between the two maps, the northern part of the Lower Peninsula and the eastern part of the Upper Peninsula still show up as outstanding tourist and recreation areas.

A great deal more could be said about many of the counties which stand out on both maps as outstanding recreation counties. Roscommon County, with a population of only 7,075, deserves men-

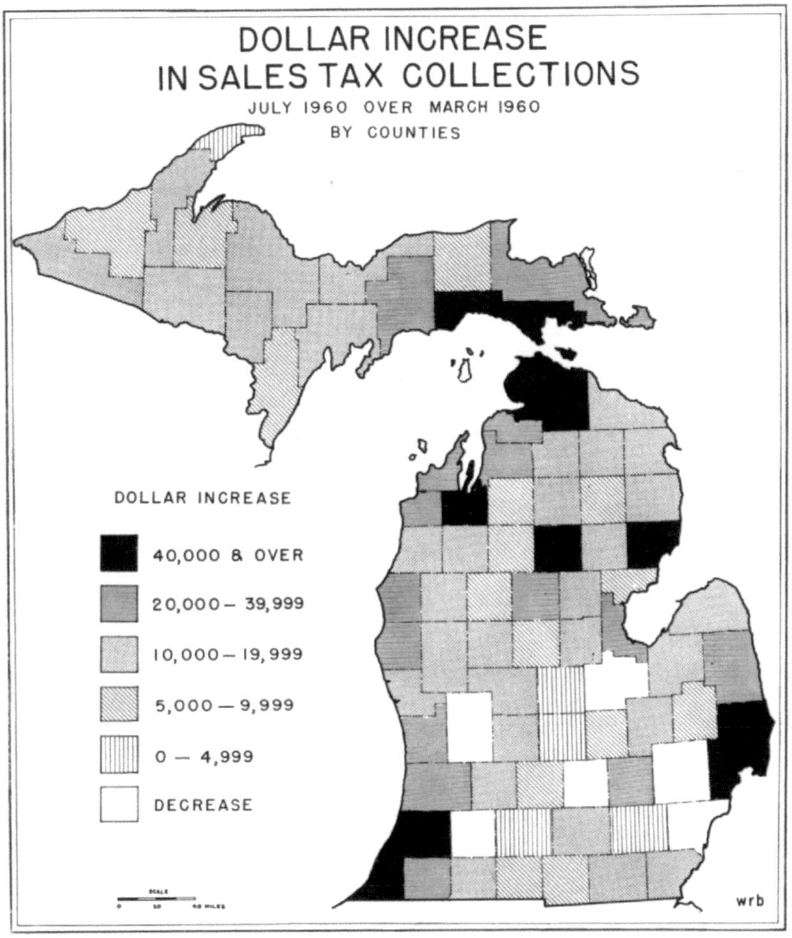

FIG. 5. The actual increase in sales tax collections for July 1960 over March 1960 by counties.

tion along with Emmet, Grand Traverse, Leelanau, Charlevoix, Cheboygan, Antrim, Benzie, and many others. The facts presented, however, will not permit a final determination of just which counties or businesses benefit the most from the recreation industry or of the degree to which they are affected. This paper had for its purpose the presentation of some ideas and techniques which should help to make the recreational industry of Michigan better understood in both its spatial dimensions and intensity in the future.

Certain questions have still to be answered. For example, to what extent does agriculture bring about an increase in retail sales during the summer months? The answer to this question is necessary if we are to determine the true recreation picture in those counties which are outstanding agriculturally as well as recreationally. It would, therefore, be presumptuous at this stage to attempt a definitive description of the recreation industry of Michigan on the basis of present knowledge.

APPENDIX I

SALES TAX COLLECTIONS FOR SELECTED COUNTIES, 1960*

Month	Alger County		Mackinac County		Lenawee County		Kent County	
	Total Tax Collected	Per Capita Tax Collected	Total Tax Collected	Per Capita Tax Collected	Total Tax Collected	Per Capita Tax Collected	Total Tax Collected	Per Capita Tax Collected
Jan.	$17,038	$1.85	$19,087	$1.75	$194,328	$2.51	$1,323,822	$3.67
Feb.	17,487	1.90	19,214	1.76	189,094	2.45	1,289,982	3.58
March	19,241	2.09	22,709	2.08	200,356	2.59	1,495,042	4.15
April	18,253	1.99	22,722	2.08	214,963	2.78	1,515,144	4.20
May	21,773	2.37	31,738	2.91	210,115	2.72	1,499,563	4.16
June	25,615	2.79	49,569	4.54	214,429	2.77	1,514,520	4.20
July	30,714	3.34	81,110	7.43	218,283	2.82	1,349,796	3.74
Aug.	30,981	3.37	83,891	7.68	206,285	2.67	1,416,436	3.93
Sept.	24,437	2.66	47,923	4.39	215,583	2.79	1,466,943	4.07
Oct.	23,875	2.60	33,122	3.03	202,118	2.61	1,448,394	4.02
Nov.	24,761	2.69	30,572	2.80	198,644	2.57	1,423,685	3.95
Dec.	22,209	2.42	27,626	2.53	254,849	3.30	1,707,852	4.74

* Source: *Research and Statistical Bulletin*, Michigan Department of Revenue, Lansing, Michigan, 1960.

APPENDIX II

1960 MARCH AND JULY SALES TAX COLLECTIONS COMPARED BY COUNTIES

County	March Sales Tax Collections	July Sales Tax Collections	Per Cent Increase July over March	Actual Dollar Increase July over March	Population of County
Alcona	$ 11,139	$ 21,213	90%	$ 10,074	6,260
Alger	19,241	30,714	60	11,473	9,190
Allegan	124,654	156,897	26	32,243	57,406
Alpena	86,830	105,015	21	18,185	28,757
Antrim	16,488	42,552	158	26,064	10,328
Arenac	27,424	36,899	35	9,475	9,799
Baraga	15,553	20,773	34	5,220	6,959
Barry	61,205	78,360	28	17,155	31,588
Bay	361,199	382,231	6	21,032	105,594
Benzie	17,124	39,315	129	22,191	7,659
Berrien	477,210	572,482	20	95,272	148,813
Branch	86,575	95,533	10	8,958	34,843
Calhoun	511,474	514,635	.6	3,161	138,378
Cass	65,849	94,904	44	29,055	36,610
Charlevoix	37,000	66,706	80	29,706	13,216
Cheboygan	33,923	84,866	150	50,943	14,218
Chippewa	87,910	127,089	44	39,179	32,487
Clare	32,497	57,387	77	24,890	11,559
Clinton	70,633	74,350	5	3,717	37,865
Crawford	13,512	32,164	138	18,652	4,917
Delta	101,470	120,585	19	19,115	34,061
Dickinson	71,334	81,532	14	10,198	23,679
Eaton	109,841	115,110	5	5,269	49,887
Emmet	48,976	109,662	124	60,686	15,646
Genesee	1,292,975	1,304,407	.9	11,432	370,303
Gladwin	25,711	41,133	60	15,422	10,650
Gogebic	64,615	76,558	18	11,943	24,234
Grand Traverse	124,595	199,455	60	74,860	33,144
Gratiot	116,645	116,950	.3	305	36,936
Hillsdale	88,899	96,742	9	7,843	34,605
Houghton	94,450	111,179	18	16,729	35,579
Huron	91,100	118,055	30	26,955	33,812
Ingham	889,648	821,597	decrease	−68,051	211,634
Ionia	105,805	120,783	14	14,978	42,849
Iosco	39,734	85,380	115	45,646	16,213
Iron	45,297	56,338	25	11,041	17,079
Isabella	88,373	98,367	11	9,994	35,250
Jackson	456,372	474,847	4	18,475	130,948
Kalamazoo	640,201	624,337	decrease	−15,864	169,151
Kalkaska	7,633	15,513	103	7,880	4,343
Kent	1,495,042	1,349,796	decrease	−145,246	360,574

APPENDIX II (*Cont.*)

County	March Sales Tax Collections	July Sales Tax Collections	Per Cent Increase July over March	Actual Dollar Increase July over March	Population of County
Keweenaw.......	$ 2,279	$ 8,262	264%	$ 5,983	2,387
Lake............	5,790	19,518	236	13,728	5,280
Lapeer..........	98,956	106,915	8	7,959	42,075
Leelanau........	9,838	32,280	227	22,442	9,210
Lenawee........	200,356	218,283	9	17,927	77,301
Livingston......	88,841	111,249	26	22,408	37,899
Luce...........	16,888	25,520	51	8,632	7,784
Mackinac.......	22,709	81,110	257	58,401	10,923
Macomb.........	1,129,935	1,191,477	5	61,542	406,553
Manistee........	48,549	67,430	39	18,881	18,876
Marquette.......	145,510	170,557	17	15,047	55,604
Mason..........	57,571	83,355	45	25,784	21,699
Mecosta.........	46,109	57,544	25	11,435	21,001
Menominee......	45,424	53,001	17	7,577	24,655
Midland.........	138,938	149,522	7	10,584	51,154
Missaukee.......	10,063	17,002	68	6,939	6,683
Monroe.........	223,430	239,042	7	15,612	100,980
Montcalm.......	105,855	118,971	12	13,116	35,380
Montmorency....	6,221	21,856	252	15,635	4,402
Muskegon.......	490,460	504,673	3	14,213	148,950
Newaygo.......	49,139	66,272	35	17,133	23,941
Oakland.........	2,526,055	2,500,555	decrease	−25,500	686,893
Oceana..........	34,257	55,154	61	20,897	16,413
Ogemaw.........	26,441	44,194	67	17,753	9,591
Ontonagon......	22,373	28,376	27	6,003	10,447
Osceola.........	30,135	39,311	31	9,176	13,572
Oscoda.........	6,215	15,075	143	8,860	3,388
Otsego..........	23,033	41,234	79	18,201	7,511
Ottawa..........	259,959	300,998	16	41,039	8,362
Presque Isle.....	28,304	42,362	50	14,058	12,855
Roscommon......	19,411	73,774	244	54,363	7,075
Saginaw.........	653,894	645,492	decrease	−8,402	189,237
St. Clair.........	313,161	365,623	17	52,462	106,396
St. Joseph.......	123,651	142,665	15	19,014	42,190
Sanilac..........	68,826	91,706	34	22,880	32,103
Schoolcraft......	23,487	43,568	86	20,081	9,200
Shiawassee.......	152,838	159,529	6	6,691	53,194
Tuscola.........	97,025	108,027	12	11,002	41,764
Van Buren......	119,725	170,308	42	50,583	48,060
Washtenaw......	545,938	548,132	.4	2,194	172,016
Wayne..........	10,135,613	9,620,045	decrease	−515,568	2,650,001
Wexford.........	56,853	75,439	33	18,586	18,182
Totals.........	26,366,209	27,227,549		851,606	

* Source: *Research and Statistical Bulletin*, Michigan Department of Revenue, Lansing, Michigan, 1960.

Papers of the Michigan Academy of Science, Arts, and Letters
Vol. XLVII, 1962 (1961 Meeting)

COMPETITION FOR PARK AND RECREATION LANDS IN SOUTHEAST MICHIGAN

ARTHUR C. ELMER

Division of Parks and Recreation, Michigan Department of Conservation

Land—hilly land; flat land; tree-covered or barren; land which naturally or by development can provide outdoor recreation opportunities for people—land is our greatest need. Land with open spaces, woods, waters, and places to park, picnic, and camp doesn't just happen; it must be found and acquired. It must fit into a pattern of use, need, and accessibility. We must acquire the land we need—not for today but for the years 1970 and 2000. It is still available, but it won't be on those dates. After land acquisition, we need to plan the development of these areas to serve the recreational needs of the people of Michigan and its visitors, and lastly, we need to do the actual development.

This discussion is designed to cover only the first phase of this problem, and then only in a small part of the state—6 counties out of the 83, but counties which now contain approximately 50 per cent of the total population of the state and are very likely to retain this high percentage. These 6 counties in southeast Michigan are Livingston, Wayne, Oakland, Washtenaw, Monroe, and Macomb. In these counties villages such as Novi and Wixom, which not so long ago supported populations of a few thousand, are now six-mile-square incorporated cities "urban sprawl," brought about by the population explosion which threatens to increase our population by one-third by 1970.

Between 1944 and 1950 the Department of Conservation set up as desirable units for recreation some 13 areas lying largely in a parabolic curve around Detroit. These areas were to encompass some 150,000 acres, of which we hoped to purchase 100,000. Some 50,000 acres have been acquired to date at an average cost of $85 to $100 per acre. Lands acquired in the early 40's were the easy lands—lands where the landowners in most cases were willing to sell. To acquire the remainder will be costly. Some of it is lost—totally outside the realm of possibility for purchase for recreation purposes. Many of the suitable lakes within these purchase areas are now highly developed resorts or year-long residential properties. Our acquisition for recreation will be in direct competition with the urban developers, the subdivision contractors and builders, and the people who want a "place in the country."

Acquisition of land for parks and recreation will be difficult, too, because we must "block in" southeast Michigan recreation areas. "Blocking in" means purchasing those areas within existing project boundaries so as to permit orderly development. They are not lands we want but lands we need, and "must" lands are always difficult to purchase and high-priced.

Present legislation provides for a budget of $1,000,000 to acquire park and recreation lands in 1961. Our studies reveal that to acquire the lands we need in the next ten years, most of which are still unencumbered and available, we will need a minimum of $58,000,000.

The competition for lands for park and recreation uses and activities is keen. Homes, industrial sites, platted subdivisions, shopping centers, roads, power lines are all competing for these same lands. Land is normally sold to the highest bidder, rather than on the basis of the greatest use for the most people, and the State must be prepared to pay the "high dollar" for the land it needs.

As city populations grow, open lands adjacent disappear, and unless planned for through zoning by agencies such as county and state zoning boards, the Detroit Regional Planning Commission, Huron-Clinton Metropolitan Authority, and the Department of Conservation, there will be no opportunity for outdoor recreation activities near cities of large population.

The advent of the limited access roads, both free and toll, has further complicated the problem of acquisition of lands for recreation. Such roads are another competitive use for land. Ease of transportation makes moving out of the cities more justifiable—it takes 15 minutes instead of an hour to get from home to the plant. This road program does other things, too—it removes from private use thousands of acres of land. Whether these roads cross private or public lands, they complicate development.

In order to justify the limited access four-lane highways which are designed for speed and comfort in driving, we must provide the places—lands and facilities—for outdoor enjoyment. People can use roads only to drive on, and even stopping is hazardous; we need, therefore, to acquire and develop land for parks—land and water where people can enjoy outdoor leisure and outdoor activities. A highway is just a means of getting to where one wants to go, not a place of quiet, rest, or enjoyment.

Michigan in 1950 had a population of 6,371,766. In 1960 the population was 7,778,220, an increase of 22 per cent. The number of persons per square mile in Michigan is 136.4, as compared to 347.4 for New York. This comparative figure indicates the possibilities which still exist for acquiring lands and waters for recreation in Michigan *if we act now.*

The population of the city of Detroit, Wayne County, in 1950 was 1,849,568. In 1960 it was 1,655,681, a loss of 10.5 per cent. Where these people (as well as many more) went is evident in the following chart, which shows the tremendous increase in the population of the counties in and around Wayne.

	1950	*1960*	*Per Cent Increase*
Oakland	396,001	686,893	73.5
Washtenaw	134,606	172,016	27.8
Monroe	75,666	100,980	33.5
Livingston	26,725	37,899	41.8
Macomb	184,961	406,553	119.8

(Wayne gained only 8.8 per cent in this ten-year period.)

According to Marion Clawson, Director of the Land Use and Management Program of Resources for the Future, Inc., it is estimated that overall outdoor recreation will increase tenfold between the years 1956 and 2000. In his report to the National Conference on State Parks in Maine in 1960 he stated, "Three acres of state park for 100 persons of the total state population is not an unreasonable standard for the present level of demand for state parks." This would require a total state park acreage for Michigan of approximately 235,000 acres, or about a 30 per cent increase over our present park acreage. If we adopt an "eight acres per 100 population" figure as a modest but still defensible future standard for the year 2000, however, and assume again the State's population will increase to 12 million persons by that time, then we will require nearly 1 million acres of park land for our Michigan residents, without considering the tremendous increase in use by residents of neighboring states.

The public demand for outdoor recreation has been tremendous—

and actually, it has only started. Park use in Michigan is reflected in these figures: in 1950, state park attendance was 11,667,793; in 1960, more than 19,000,000, an increase of 62.8 per cent. Camping uses of state parks increased even more dramatically—from 41,046 in 1950 to 148,681 in 1960; an increase of 250 per cent.

As I said earlier, land is our greatest need; if we don't get it now it will be lost forever, and even now the cost will be high and the problems of acquisition many. Although the State Conservation Department has the power of eminent domain, we use it reluctantly, and only after all other reasonable means of direct negotiation have failed. In order to compete for these lands which people need and want without realizing it until it is too late, we must start now and use every means at our disposal.

The land exists, but the real job before us is to acquire it before it is pre-empted for other uses. We can't create another Porcupine Mountains in Michigan, but we can and must acquire the private holdings in the existing southeast Michigan recreation areas, acquire potentially suitable park sites in the lower half of the state, acquire the Great Lakes shoreline areas, the sand dunes, and the forested areas that have park and recreation significance and potential.

Outdoor recreational opportunities are as valuable and important to us as our health, our homes, and our jobs, and it is imperative in our overall budgeting for the welfare of our people that parks and recreation areas receive their fair share of the funds available.

People can be the "forgotten element" in our planning; we think of highways, buildings, shopping centers, airports, and other similar interests when what would be in the peoples' interest is a park with trees and grass and sky. Southeast Michigan, with its tremendous need for open areas for outdoor recreation, is first on our list of areas to be acquired.

Papers of the Michigan Academy of Science, Arts, and Letters
Vol. XLVII, 1962 (1961 Meeting)

AN OPERATIONAL MODEL FOR PREDICTING RESERVOIR ATTENDANCE AND BENEFITS: IMPLICATIONS OF A LOCATION APPROACH TO WATER RECREATION

EDWARD L. ULLMAN AND DONALD J. VOLK
Meramec Basin Research Project, Washington University

This is a paper in geographical prediction, applied to a practical problem. Antecedents of the approach however can be found in the work of Hotelling, Clawson, and other economists, and can even stretch back as far as von Thünen in 1826. The problem is twofold: to predict attendance for a potential, large new reservoir far from other reservoirs, and secondly to estimate some of the minimum benefits from the intangible activity of recreation. Formidable data problems beset any work in this field, but we feel that we have acquired enough to illustrate our procedure concretely, and carry our analysis at least beyond the "heroic assumption" stage necessarily characteristic of some earlier work. We shall not only present a model of wide application but also operate it with real although less than perfect data on a real problem, to produce a real answer, for a real place. This in turn has further implications for other places and for national policy.

Both the attendance prediction and benefit assessment problems are difficult. There is no past trend to use for a brand new facility or flow pattern, so that the standard method of prediction by simply projecting the past or making elaborate time series calculations and the like cannot be employed. By way of contrast our method calls for obtaining data from other localities and might be termed the "geographical analog" or comparative method. Here the problem is to find comparability or substitutability between places, or clearly to adjust for differences. Another feature of our solution is the use of a spatial interaction approach by obtaining actual flow data and using some versions of a gravity model (or simply calculating distance regression lines) to order the data into form suitable for prediction. We rely on interaction or data on connections, rather than on static description. This method provides an illustration of some of the exciting advances that can be made with this newer approach. It also indicates what data to search out since data are neither found nor created unless one knows what to look for. The data available and the nature of this problem, however, do not warrant elaborate statistical or mathematical treatment, which would tend to produce a spurious precision and needless refinement on many aspects, although other approaches might also be attempted.

The Problem and Method of Predicting Reservoir Attendance

In connection with our studies of the Meramec River Basin, much of it forested and thinly populated extending from the St. Louis suburbs down into the Ozarks, we needed to assess the demand for water recreation at a potential multiple purpose reservoir. St. Louis, with a metropolitan population of more than 2,000,000, is a long way from water. The nearest large impoundments are shown in Fig. 1, with circles proportionate to total visitor-day attendance and

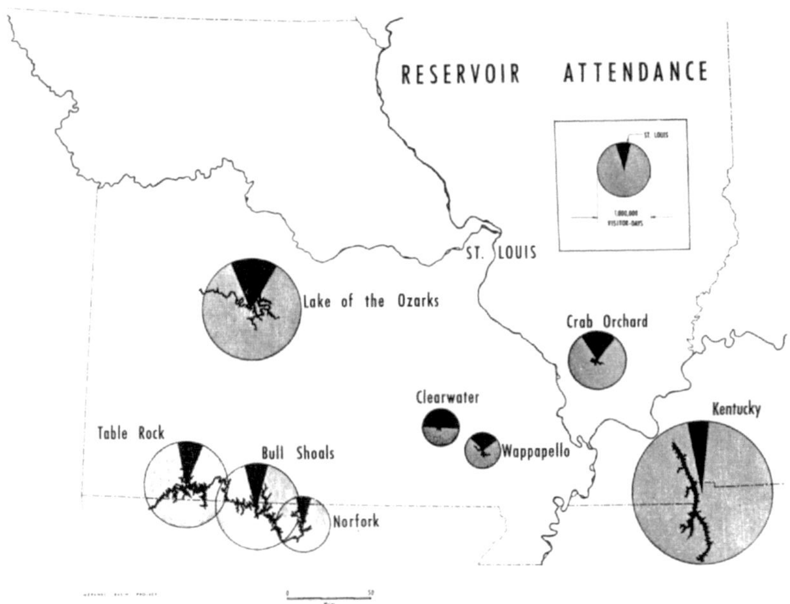

Fig. 1

pie sectors proportionate to estimated St. Louis visitor-days, ranging from 4 per cent at Kentucky Lake on the southeast to 50 per cent at Clearwater in the center.[1] Generally speaking, the nearer the lake, the greater the St. Louis attendance, as would be expected. Some variation occurs because of size and quality of the lakes, as will be noted later.

In order to determine the drawing power by distance zones of the various lakes we sought out data on origin of visitors by small geographic divisions. Our biggest block of data came from unpublished creel census reports of the Missouri Conservation Commission which show origin of visitors by counties.[2] These were based on several thousand sample interviews over a period of years on various arms of lakes or for whole impoundments representing the smaller lakes of Clearwater, Taneycomo, and Wappapello. These data were then converted into county per capita indices and mapped. Fig. 2 is one representative map. Note generally how attendance decreases with distance. These indices were then plotted on logarithmic paper, and regression or trend lines fitted by inspection. The fit was fairly close, but there was also considerable scatter, in large part occasioned by smallness of sample when applied to small counties. The slope of the lines fell off however in an essentially linear fashion and varied from a drop-off ranging from about the square of the distance to the fourth power.

The fact that the decline was so relatively rapid with distance might be construed as an indicator of the similarity and substitutability of one impoundment for another. Similar calculations which we and others have made for more nearly unique attractions such as Grand Canyon or the Grand Tetons show a more gradual decline at a rate less than distance squared.[3] However these more gradual declines represent the effect of combined visits to several

[1] Alton Lake nearby on the Mississippi is not considered because it has different characteristics from the others in the universe, although it does figure in St. Louis recreation and has some attractive features.

[2] See also: Joseph W. Kathrein, *An Intensive Creel Census on Clearwater Lake, Missouri During Its First Four Years of Impoundment, 1949–1952*. Transactions of the Eighteenth North American Wildlife Conference; Washington: Wildlife Management Institute, 1953.

[3] Marion Clawson, *Methods of Measuring the Demand for and Value of Outdoor Recreation*. Resources for the Future, Reprint No. 10. Washington, 1959. Our calculations made from National Park and other surveys.

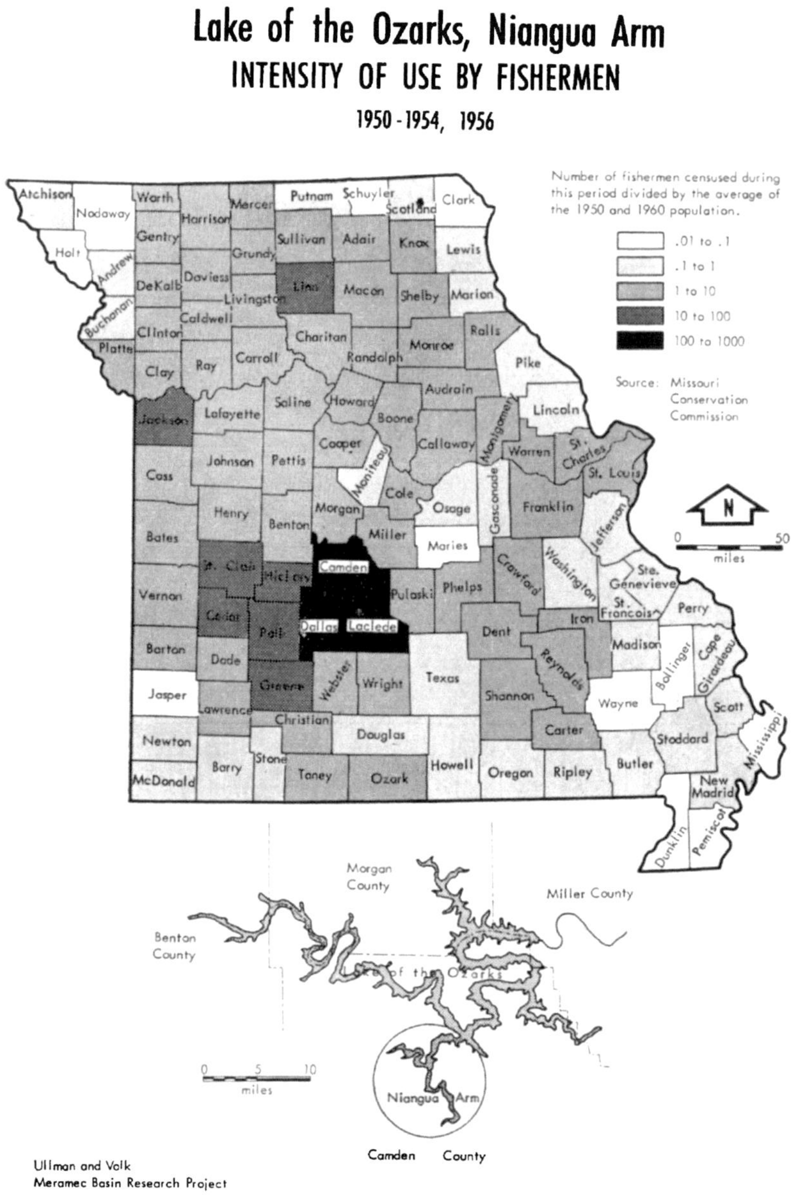
Lake of the Ozarks, Niangua Arm
INTENSITY OF USE BY FISHERMEN
1950-1954, 1956
Number of fishermen censused during this period divided by the average of the 1950 and 1960 population.
.01 to .1
.1 to 1
1 to 10
10 to 100
100 to 1000
Source: Missouri Conservation Commission
N
0
50
miles
Camden
Hickory
Dallas
Laclede
Morgan County
Miller County
Benton County
Lake of the Ozarks
0
5
10
miles
Niangua Arm
Camden County
Ullman and Volk
Meramec Basin Research Project

FIG. 2

western attractions pulling simultaneously, whereas the average trip length to impoundments in our universe was less than four days. The exponent also declined for the western attractions in the more recent years as compared to the immediate post-war period, reflecting apparently increased mobility and ability to travel longer distances.

In addition to the creel census data we have gathered other data from other sources including an intensive survey at five Missouri state parks conducted jointly by ourselves and the Missouri State Park Commission in 1960. We are still processing some of these, but thus far they appear to conform to the others. We have therefore taken the median drop-off by the cube of the distance and plotted this on the logarithmic graph, Fig. 3. The horizontal axis shows distance from reservoirs, the vertical axis is per capita attendance by counties derived from the several thousand origins sampled.

On this graph we have also indicated two other regression lines: (1) a *high* representing generally high per capita attendance expectation from urban, high income counties, and/or to reservoir of high quality, and/or lack of intervening opportunities from closer im-

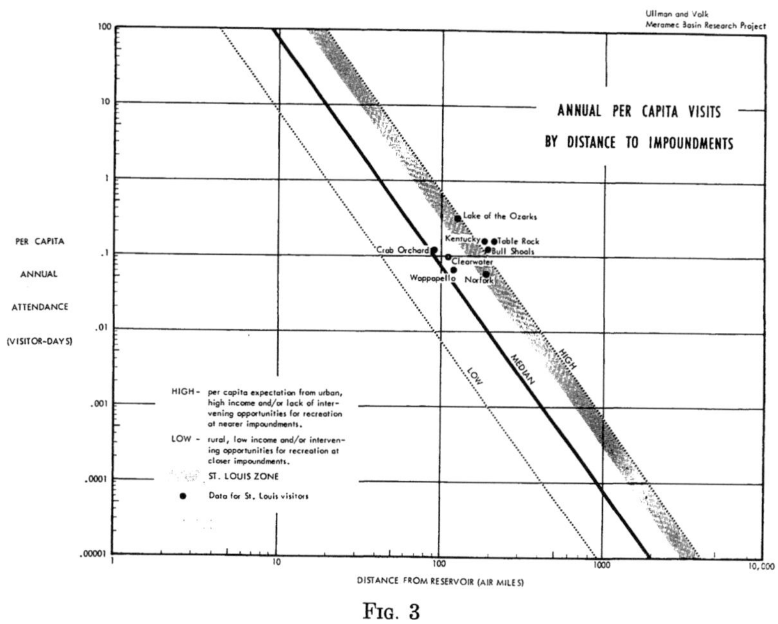

FIG. 3

poundments suitable for recreation, (2) a *low* representing rural, low income counties and/or lower quality impoundments and/or presence of intervening opportunities for recreation at closer impoundments. The shaded zone represents generally the category in which St. Louis visitors fell.

To find a starting point for the lines on this graph, since our initial data simply gave us an index of rate of decline with distance, it was necessary to make further calculations which gave us an estimate of 10 visits per capita for the median line at a distance of 20 miles. This was based on more than a dozen sample calculations for counties near reservoirs. An example will illustrate the method. Wayne County, a rural county in the Ozarks, whose midpoint is about 20 miles from Clearwater Lake, a 1,500 acre, relatively small reservoir, had 11.8 per cent of the creel census fishermen visitors sampled at the lake; multiplying this times the total attendance of 414,000, reported by the Corps of Engineers, gives a figure of 49,183, which, when divided by the population of the county, gave a figure of 5.2 visitor-days per capita per year from that county. Other calculations for other counties at similar distances and at other impoundments showed a range from 1.6 up to approximately 30, with 10 apparently a reasonable median. In many reservoirs it was possible only to compute a still higher upper limit because the creel samples applied only to one arm of the lake and total attendance figures were available only for the whole lake.

The procedure thus for making the estimate of 10 was not entirely satisfactory, but can be improved as we process more data from our later surveys, although we feel that it is unlikely to alter the figures significantly.

As an illustration and additional check on the regression lines, the St. Louis per capita visitor-days at the eight major impoundments have been plotted on the graph. These St. Louis visitor-days have been estimated from a variety of sources, and are only approximations, yet are somewhat independent and are isolated from the general calculations which went into the regression lines and anchor points. Note that five of the impoundments are in the high zone or just beyond, and three in the median zone. The five high ones are all larger, better impoundments; the three lower ones, Crab Orchard, Clearwater, and Wappapello are all smaller and generally poorer either in setting or development. This may explain the difference.

Fig. 3 then represents our attendance prediction model. From it we can estimate attendance by zones around any reservoir essentially similar in size and character to the average of the other reservoirs. For a reservoir 50 miles from St. Louis, for example, we read off four visits per capita per year in the middle of the gray band at the 50-mile distance. (Remember the graph scales are logarithmic.)

The actual calculation of predicted attendance is shown in Table I, where the per capita annual visits anticipated from each zone's characteristics are multiplied by the zone's population to give predicted total annual visitor-days in the last column. Note the overwhelming contribution of metropolitan St. Louis with its 2,000,000 population, accounting for ⅘ of the visitor-days alone.

TABLE I

Estimated Attendance at Hypothetical Major Impoundment 50 Air Line (60 Road) Miles from St. Louis

Distance Zones from Hypothetical Reservoir	Population in Zone	Per Capita Category*	Annual Visits Number	Total Annual Visitor-days
0–10	20,000	median	15+	300,000
10–20	40,000	median	10+	400,000
20–40	90,000	high	10+	900,000
40–60	2,000,000‡	high	4	8,000,000
60–100	300,000	low	.4	120,000
100–200	2,000,000	median	.03	60,000
200–300	10,000,000	median	.005	50,000
300–500	30,000,000	median	.001	30,000
Beyond 500	(Rest of U. S.)	Estimate		50,000
			Total	9,910,000

* Categories refer to per capita annual visits expected based on numerous origin surveys in 1960, at large impoundments in Missouri, Illinois, and Kentucky. *High* refers to per capita expectation from urban, high income and/or lack of intervening opportunities for recreation at nearer impoundments; *low* refers to rural, low income and/or intervening opportunities for recreation at closer impoundments. *Median* is average between two extremes.

‡ Includes most of St. Louis metropolitan area.

A Method of Calculating Benefits

The next step after predicting attendance is to calculate benefits. Intangible recreation benefits are difficult to calculate, with the result that recreation tends either not to be counted or arbitrarily made low. Rules of thumb figures of $1.00 to $1.60 per visitor-day have also been used. On these bases the recreation benefits from the 9,900,000 estimated would be enormous, although the greatest part would not be reimbursable by the Federal Government under current rules of procedure. What is needed is another, more persuasive method of calculating benefits. A basis for attempting such a method was provided by Harold Hotelling on a distance-zone basis some years ago and further examined by Clawson, Milstein, and others recently.[4]

Table II presents another, and conservative method for calculating travel benefits—represented by diverting visitors from more distant impoundments to a closer one. The principle implied is that cutting down travel by locating activities closer is just as legitimate a benefit as building a highway to cut down travel costs.

Paragraph I of Table II details the location and attendance at the eight nearest large reservoirs drawing St. Louis visitors. These are the same reservoirs shown on the map and graph.

Paragraph II indicates the visitor miles saved by locating a hypothetical impoundment 60 road miles (50 air line miles) from St. Louis. The critical figure here is the estimate of ⅓ diversion from more distant impoundments. This is based on some questions asked at four Missouri State Parks in our joint sample survey in 1960 and particularly the question: "If a lake similar to this one were built half as far away from your home, would this decrease your visits to this lake?" to the extent of "eliminate completely, reduce greatly, reduce slightly, no effect, don't know." These were scored 100 per cent, 75 per cent, 25 per cent, 0 and the last, "don't know" (responded by about 20%) allocated accordingly. The combined score was ⅓ diversion. Since a lake 60 miles from St. Louis is much less than half the distance to most of the other impoundments, ⅓ is undoubtedly too low, and ½ or some other figure might be better. However asking people what they might do before they do it is uncertain, so the conservative diversion of ⅓ was used. The other item, average trip length of four days, came from the same survey.

[4] Clawson, *op. cit.*

TABLE II

TRAVEL BENEFITS TO POTENTIAL NEW RESERVOIR CLOSER TO ST. LOUIS THAN EXISTING IMPOUNDMENTS

	Distance from St. Louis		Total Visitor-days (1960)[1]	St. Louis Visitor-days[2]	
	Air Line	Road		% of total	Number
I. *Impoundment*					
a) Lake of the Ozarks	135	180	4,000,000 est.	15	600,000
b) Clearwater Reservoir	105	130	410,000	50	200,000
c) Wappapello Reservoir	115	140	440,000 (1959)	30	130,000
d) Bull Shoals Reservoir	200	270	2,580,000	10	250,000
e) Norfork Reservoir	190	270	1,220,000	10	120,000
f) Table Rock, Taneycomo	220	270	3,000,000 est.	10	300,000
g) Kentucky Lake	180	230	7,500,000	4	300,000
h) Crab Orchard Lake	90	120	1,200,000	20	240,000
Total			20,350,000		2,140,000

II. *Number of annual round trips and visitor miles saved represented by:*

1) Diversion of ⅓ St. Louis visitor-days to new impoundment about 60 road miles (50 air line miles) from St. Louis (app. 725,000 visitor-days)[3]

2) Average trip length of four days.[4]

Impoundment	Annual Round Trips	Calculation	Visitor Miles Saved
a) Lake of the Ozarks	50,000	(180 − 60 = 120 miles × 2 = 240 × 50,000)	12,000,000
b) Clearwater Reservoir	15,000	(130 − 60 = 70 miles × 2 = 140 × 15,000)	2,390,000
c) Wappapello Reservoir	11,000	(140 − 60 = 80 miles × 2 = 160 × 11,000)	1,760,000
d) Bull Shoals Reservoir	21,000	(270 − 60 = 210 miles × 2 = 420 × 21,000)	9,030,000
e) Norfork Reservoir	10,000	(270 − 60 = 210 miles × 2 = 420 × 10, 00)	4,270,000
f) Table Rock and Taneycomo	25,000	(270 − 60 = 210 miles × 2 = 420 × 25,000)	10,500,000
g) Kentucky Lake	25,000	(230 − 60 = 170 miles × 2 = 340 × 25,000)	8,500,000
h) Crab Orchard Lake	20,000	(120 − 60 = 60 miles × 2 = 120 × 20,000)	2,400,000
Total	177,000		50,850,000

III. *Annual travel savings:*

a) At 2.3¢ per visitor mile saved (8¢ per car mile ÷ 3.5 passengers)[5]	$1,170,000
b) At .57¢ per visitor mile saved (2¢ per car mile covering only variable cost for gas, oil ÷ 3.5 passengers)	290,000
c) At 85¢ per hour representing time saved ($2.00 per hour for driver, plus $1.00 for one passenger, 0 for other passengers; at average speed 40 m.p.h. = 1,270,000 visitor hours × .85	1,080,000
d) At .57¢ per visitor mile plus .85 per visitor hour (b + c)	1,370,000
e) At 2.3¢ per visitor mile plus .85 per visitor hour (a + c)	2,250,000

Sources:

[1] Lake of the Ozarks—local estimates; Crab Orchard—Fish and Wildlife Service; Kentucky Lake—TVA. All others—U. S. Corps of Engineers, 1960 (except Wappapello, 1959).

[2] Based on Missouri State Conservation Creel Census Data, varying years; Meramec Basin—Missouri State Park sample surveys, 1960; Fish and Wildlife Service, 1960 (Crab Orchard); Meramec Basin and Kentucky State Parks, 1960 (Kentucky Lake).

[3] Based on estimates derived from Meramec Basin—Missouri State Park Joint Survey, 1960.

[4] *Ibid.* [5] 3.5 passengers derived from *ibid.*

The results of this procedure are summed up in the last column "Total visitor miles saved," 50,850,000. (This figure should probably be increased more than 10 per cent to take account of other visitors in still closer zones who would also save miles [see Table I].)

In paragraph III the visitor miles saved are converted into dollars under varying assumptions which are self-explanatory. (These again should be increased more than 10 per cent to count savings for non-St. Louis visitors.) Which assumptions should be used is difficult to determine. What is vacation or particularly weekend driving time worth? It appears reasonable that people would value their time more than the gas and oil saved in paying an attendance fee, if a straight market economy approach were taken. In any case the calculations presented here are conservative; we count *no* benefits for *generated* traffic, as well as conservatively estimating diverted traffic.

No account is taken of possible loss of business diverted from competing impoundments. These impoundments were built primarily on the basis of other benefits (power, flood protection, etc.) and only the recreation facilities per se need to be considered. The overriding practical consideration however, is that recreation is growing so rapidly that there would be only a very short term loss, if any, from the small diversions, so that the net disbenefit would be minute.

One reduction however might be made to take account of inflation in the official attendance figures (counting some ten minute stops as visitor-days, etc.). If they are inflated by a factor of two, about the maximum reasonable to expect, then the various savings would drop one-half. However we have been conservative, as noted, in our other calculations, so that much, if any, total drop should not be expected. Finally, water recreation is rapidly growing so that even if our figures were too high now based on 1960 conditions, they would probably not be by the time a reservoir were completed and attendance had gone up to predicted levels. We have deliberately taken what economists term an *a fortiori* approach—that is to see whether we get enough benefits to justify an expenditure, not to try to get all the benefits.

Translating the various annual savings into capital, results in some large figures as follows:

Annual Benefits =	Capital over 50-Year Period	
	at 2½%	at 5%
$ 500,000	$14,180,000	$ 9,130,000
1,000,000	28,360,000	18,259,000
2,000,000	56,720,000	36,420,000

From these amounts capital costs of both reservoir and recreation facilities (which have a shorter life) should be amortized together with annual operation and maintenance, deducted from the annual benefits, but a sizeable amount apparently would be left over on the larger figures to pay a good part, or most of the cost of a competitive impoundment.

Implications and Conclusions

The large attendance prediction and the conservative estimate of recreation benefits from a reservoir located close to large urban markets, not otherwise served, has implications both for location of reservoirs and national water policy. Many of the reservoirs constructed throughout the country for various purposes have served to satisfy urban recreation demand to some extent. However, these reservoirs were constructed primarily to serve purposes other than recreation—and their location has been dictated primarily by the requirements of these other purposes. For those reservoirs to be constructed near large markets increased recognition of recreation benefits should be included in their planning, although existing federal reimbursement procedures make this difficult. Furthermore as recreation has become more important, more attention should be paid to the location of reservoirs to satisfy recreation demand. The benefits to be derived from developing such recreation-oriented sites might well be large enough to offset any disbenefits (e.g., reduced hydro-electric power or flood control) resulting from developing sites which have less than optimum location for these other, non-recreational purposes. From a national standpoint the net recreation benefits could well be so high as to override lower benefits from other accepted purposes.

The area of the United States with conditions of both urban market and lack of recreation water might extend from Omaha-Kansas City through St. Louis to Cincinnati and Pittsburgh. North of this zone are the Great Lakes and the numerous lakes of the Glacial moraine area; south of it are the Oklahoma-Arkansas-Ozark impoundments and the TVA system, while to the east is the ocean, a not strictly competitive facility. To the west population is sparse as far as the Pacific littoral. However many other urban areas in the country might also show large recreation benefits.

From inspection it appears that the St. Louis area particularly fulfills these requirements. The prediction of 9,900,000 visitor-days in 1960 would give a lake near-by an attendance greater than at any other impoundment in the country, including the giant lakes of Kentucky Lake with 7,500,000 in 1960, and Texoma with 6,700,000 in 1959. Furthermore the combination of suitable stream and valley for impoundments, plus reasonably good scenery and warm weather in the Ozarks, argues for the same conclusion.

The prediction model presented represents the first, but critical step in assessing the potential. With these data as a starter, other calculations can readily be made—dollars spent, per cent visitors from outside the state, per cent from outside the basin, economic impact of reservoir recreation on the local area, et cetera.

To be truly effective, however, such models call for far better data on recreation than are now generally available. Attendance figures need to be improved, reliable, detailed origin data for small areas need to be collected, and some before and after surveys of behavior made in order to have firmer bases for prediction. The increasing importance of recreation calls for some hard facts on which to base hard analysis.

Papers of the Michigan Academy of Science, Arts, and Letters
Vol. XLVII, 1962 (1961 Meeting)

C. THE GEOGRAPHY OF ASIA

MARKET GARDENING IN KOCHI PREFECTURE, JAPAN[1]

JOHN D. EYRE

University of North Carolina

Recent geographical literature has established how urban growth in Japan has stimulated commercial vegetable production in both truck gardening zones immediately outside large cities and in a large number of small, widely scattered pockets of horticultural activity.[2] The former have the decided advantage of proximity to market, while the latter try to maximize some local advantage to offset their greater transportation costs. Both types are engaged in a continuous struggle for a share of the large city market, wherever possible scheduling their vegetable harvests and shipments for the period when vegetables are in greatest demand, supplies from competing areas smallest, and promise of profit brightest.

The main characteristics and problems of the more specialized and prosperous vegetable-growing areas are exemplified by southern Kochi Prefecture. Although situated in southern Shikoku far from its primary markets in Osaka and Tokyo, it flourishes on the basis of offseason (late fall to early summer) vegetable specialization. It is without peer in Japan in the quality of its produce, the skills used in cultivation and marketing, and the astuteness with which planting schedules have been fitted into the climatic patterns of the winter months.

[1] Based upon field work in Japan during summer, 1958, with financial assistance from the Social Science Research Council and the University Research Fund of the University of North Carolina.

[2] John D. Eyre, "Sources of Tokyo's Fresh Food Supply," *Geographical Review*, XLIX (1959), pp. 455–74; George H. Kakiuchi, "Stone Wall Strawberry Industry on Kuno Mountain, Japan," *Economic Geography*, 36 (1960), pp. 171–84; Yoshikatsu Ogasawara, "The Distribution Pattern of Cash Crop Fields in Japan," *Proceedings of the IGU Regional Conference in Japan, 1957* (Tokyo, 1959), pp. 440–45; and John H. Thompson, "Urban Agriculture in Southern Japan," *Economic Geography*, 33 (1957), pp. 224–37.

Interest in offseason vegetables dates back to 1799, when a feudal lord returning from Edo (Tokyo) obtained early cucumber seeds from farmers south of Osaka and had them planted near Kochi city. Remaining little more than a local curiosity for more than a century, offseason vegetable growing moved into a commercial stage in the late 1800's through the stimulus of early sales in Osaka and the creation (1906) of a prefectural experiment station to develop new vegetable strains and improved management techniques. Commercial foundations erected arduously over the course of a half-century were virtually wiped out during and immediately after World War II by food production restrictions which favored subsistence crops. Once the restrictions were removed in 1949, however, Kochi vegetable growers rapidly expanded their commercial acreage until in 1952 it was larger than in the prewar period. Since then, the acreage, volume, and value of commercial vegetables have moved upward to a record high plateau.

Location Factors

Offseason harvests are made possible by southern Kochi's mild winter climate, which is noted for its relatively high temperatures, long hours of continuous sunshine, and well-distributed rainfall. Contributing factors are a southern location, the protection from cold north winds provided by the high mountains of central Shikoku, and marine influences from the adjoining Pacific Ocean. Average monthly temperatures for the six coldest months are: November, 57°; December, 48°; January, 44°; February, 45°; March, 50°; and April, 59°.[3] Similar or slightly higher temperatures for the same period can be found in Japan only in the southern Kii Peninsula (Wakayama Prefecture) and southern Kyushu (Miyazaki and Kagoshima Prefectures). The long frost-free period is 240 days at a minimum.

Among its other natural attributes, southern Kochi has most of the prefecture's lowlands of agricultural value. As shown in Fig. 1, a series of depositional plains dot the margins of Tosa Bay, formed by short rivers pouring from the mountainous interior. The largest of these lowlands, the centrally located Kochi Plain, is the key agricultural area and has Kochi city, the prefectural capital and main

[3] Kochi Prefecture, *Kōchi-ken no sokusei engei* (Offseason Horticulture in Kochi Prefecture) (Kochi, 1958), pp. 2–3. Temperatures are Fahrenheit.

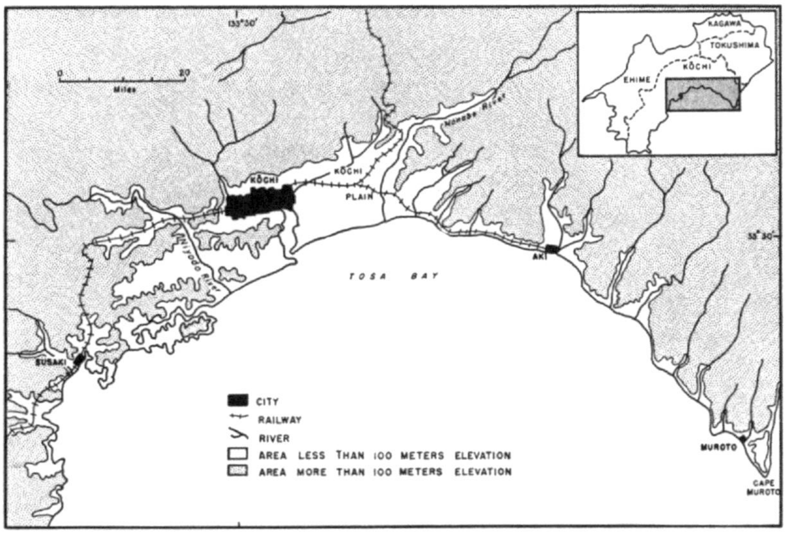

FIG. 1. General location map of southern Kochi Prefecture.

agricultural collection and shipment center. Most of the coast except for the southwestern part, where rocky cliffs rise precipitously from the sea in many places, is bordered by a narrow sandy strip.

Within this physical setting, commercial vegetables are grown in two general localities: (1) along the coast, where they are the farming mainstay, and (2) in the interior of the plains, where they are planted as a second crop on drained paddies after the fall rice harvest. The coastal section is the more important for crops maturing in midwinter because it has daytime temperatures at that time as much as 3° warmer than points one-half mile inland (Pl. I, Fig. 1). The higher temperatures are caused by rapid daytime heating of the sandy soil and the presence offshore of a branch of the warm Japan Current, which leaves the main stream at Cape Muroto and flows westward along the Tosa Bay coast.

The coastal section is most heavily planted in vegetables from Susaki eastward to Cape Muroto. Production is small-scale, one farm household operating only an average one-half to three-fourths acres, and is intensive, with at least two crops planted on the same land in one year. It is also expensive, involving large outlays of money for special equipment to protect crops from the lowest winter temperatures and to ease the periodic water shortages (Pl. I, Fig. 2). Labor needs during the busy winter and spring months are filled with young women recruited from nearby mountain villages.

In contrast, vegetable growing on the inner plains or on terraces on bordering slopes is usually only one phase of a broader paddy cultivation system. Farmers raising rice have larger average land holdings than along the coast, but normally devote only a small part of it, an average one-half acre, to vegetables because of the time demands of vegetable growing in competition with rice-growing activities. For the same reason, they tend to favor plantings which will mature in late spring or early summer before rice planting time (Pl. II, Fig. 1). About 6,500 acres of paddy in the plains east of Kochi city bear two crops of rice each year, the only place in Japan where this is done. Although present double-cropped acreage is smaller than the maximum size, 15,000 acres, reached in 1932, it is apparently stabilized and thus discourages more widespread interest in commercial vegetables in these areas.

Production Techniques

Southern Kochi's winter weather is normally warm enough so that most crops can be raised without artificial heat. However, sharp drops in nighttime and occasional daytime temperatures make it necessary to provide some protection for crops during the coldest months. Covers made of oiled native paper, an important handicraft product from Kochi's mountain villages, was used originally, but it has been replaced in large part by vinyl plastic sheeting since the early 1950's.

Protective structures covered with vinyl range in size from small hot caps used for seedlings to large greenhouses. The larger units, made with bamboo or metal framing, are erected in late September and remain standing until summer, when they are dismantled in time for the typhoon season. The cost of the vinyl, which must be renewed every two years, is such a financial burden that the prefectural government has had to provide loan funds through the agricultural cooperatives for its purchase. A rule of thumb is that the first year's profits pay for the vinyl and other equipment, leaving the second year's profits clear, provided vegetable markets remain steady and the growing areas free of serious typhoon damage. Growers prefer to erect their vinyl houses on different ground each year to avoid plant disease hazards. However, this practice is ruled out along the coast because of land shortages, so that farmers there must invest in fumigating equipment.

Distinctive vinyl structures and cultivation techniques are used for crops maturing in the three major harvest periods: late fall-early winter, winter, and late spring-early summer. The geographical distribution of the three seasonal production types is shown in Fig. 2.

Winter production calls for the heaviest investment in equipment. A few permanent greenhouses using glass, located in the suburbs of Kochi city, are too expensive for widespread adoption. Instead, farmers use vinyl-covered houses, either with or without artificial heat. Steam-heated units, first used in 1931, are restricted to a coastal strip of 150 acres just east of the Kochi Plain (Pl. II, Fig. 2). The heavy cost of boilers, pipes, and fuel can be borne only by farmers specializing in such profitable offseason specialties as eggplants, which are very susceptible to cold damage. Although unheated vinyl units used for other crops are operated intact all winter, the most widespread practice is to use the structures only for the coldest part of winter, then open their tops and sides to expose the mature plants to the warm sun (Pl. III, Figs. 1 and 2). This method

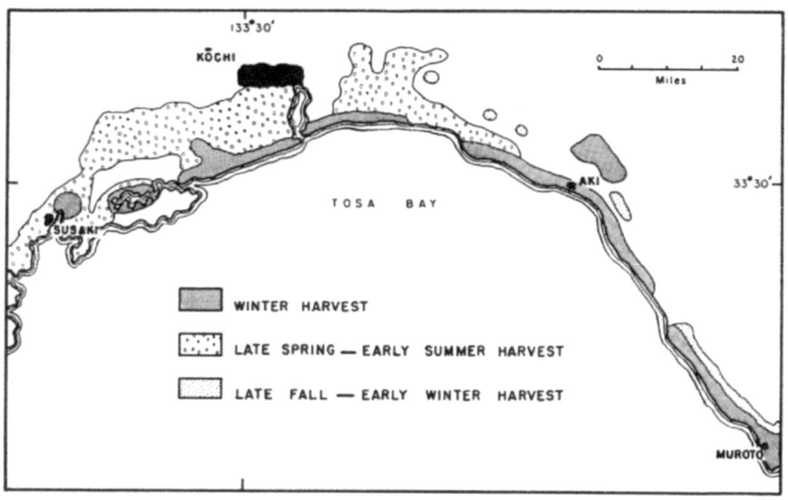

FIG. 2. Distribution of commercial vegetable production in southern Kochi Prefecture by harvest period.

is used for growing most of the vegetables harvested in midwinter and one-fourth of the total offseason production. It is the prevalent method along the coast and in a few scattered inland locations.

Crops maturing at the extremes of winter do not require such elaborate protective structures. Crops harvested in late spring or early summer are commonly set out under oiled paper, vinyl caps, or vinyl tunnels shaped over curved bamboo ribs. By the time seedlings outgrow the covers they are sturdy enough to withstand weather extremes. This simpler method, which involves only a small cash outlay for equipment, is used for more than one-half the total offseason vegetable planting, and it has widest adoption on drained paddies and interior areas of dry field (unirrigated) cultivation.

The first part of the harvest of offseason vegetables, in late fall and early winter, is achieved without protective covers. This method involves planting seeds in late summer and harvesting crops until they are killed by frosts. A relatively new technique, it has come into vogue rapidly along the southeastern coasts where frosts are of later occurrence than farther west. Plantings there are concentrated on narrow terraces overlooking Tosa Bay. Similar favorable climatic conditions are found along the southwestern coast, but it suffers from inadequate transportation links with Kochi city and a general lack of agricultural development, as shown by the general absence of commercial vegetable growing. Late fall production is still of minor importance, only 10 per cent of the total offseason vegetable acreage. A few remaining vegetables, such as cabbage and ginger, do not fit neatly into any one of the three cultivation methods described and are raised outdoors without protection by conventional methods.

Main Crops

Southern Kochi markets a wide range of offseason vegetables, yet it concentrates on the following items (with their 1957 acreage): cucumber, 878 acres; tomato, 217 acres; eggplant, 97 acres; pumpkin (*kabocha*), 476 acres; taro, 210 acres; cabbage (*kanran*), 225 acres; and watermelon, 1,070 acres. The harvest period of each is shown in Fig. 3.

Cucumbers are in continuous production from mid-October until mid-June. Plantings that mature in October are relatively new and although rewarding because grown without protection costs they

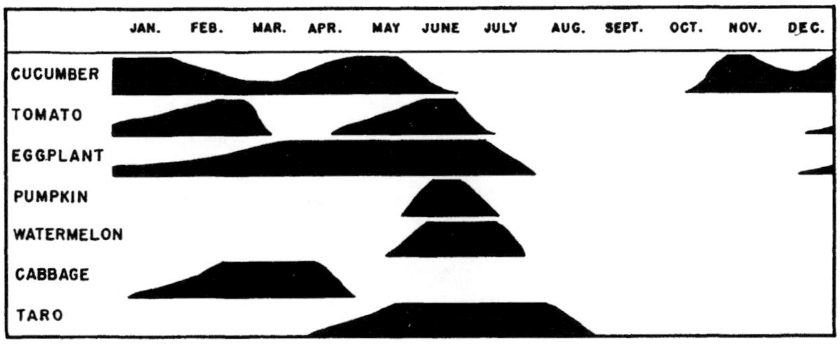

FIG. 3. Seasonal harvest schedule for main commercial vegetables raised in southern Kochi Prefecture.

must survive the annual typhoon threat. Heated vinyl houses can yield two crops spread over the winter months, but the bulk of the cucumber acreage is grown without artificial heat for late spring and early summer markets. Among the most dependable income earners, cucumbers are very popular as a winter crop for drained paddies, especially in the vicinity of Susaki and Aki.

Tomato production has two peaks, one in late February-early March and the other in mid-June. Of these, the latter is by far the larger. The most popular cultivation method is to start the plants in unheated vinyl houses or frames, later opening the structures to the open air as the weather warms. A small volume is raised as a second crop after cucumbers in heated vinyl houses along the coast. Tomatoes have been one of the real boom crops since World War II due to Allied Occupation purchases and, since 1952, the increasing consumption of salad by the Japanese public. The latter trend has touched off expanded tomato production in many other prefectures, so that southern Kochi shippers face sterner competition in urban markets. Tomato acreage is noticeably large around Susaki and from Aki eastward along the coast.

The fact that eggplants are easily hurt by low temperatures discourages their winter cultivation in most places in Japan. Hence, those raised in southern Kochi during December to March in heated vinyl houses can command sufficiently high prices to justify the heavy production costs. However, most of the eggplants are started in the winter under vinyl covering without heat and reach market in late spring or early summer (Pl. IV, Fig. 1). They are one of the best-known Kochi specialties, famous for their shape, color, and taste. Production tends to concentrate in the coastal strip between Susaki and Aki.

Pumpkins, watermelons, and taro hit their peak in June, thus reaching market before the flood of July-August production from areas farther north (Pl. IV, Fig. 2). All three are planted mostly on unirrigated fields because they thrive there and their relatively late harvest date makes them unsuitable as a second crop on drained paddies. One of the unique and most highly specialized watermelon-growing areas is the dry stream bed in the lower reaches of the Niyodo River (Pl. V, Fig. 1). Planted in winter, the watermelons mature rapidly under the influence of warmth from the stones and gravel in the stream bed. Harvested unusually early, in mid-June, they fetch premium prices. The harvest is complete by August, when the river rises and puts the growing area under water temporarily.

Cabbage is a strictly winter crop that is popular as a second crop on drained paddies. A new addition of great promise is an early sweet potato that matures in June, far in advance of the late summer harvest elsewhere. Small and sweet, it has registered steady sales and increases in acreage. Peas, a popular crop along the southeastern coast, green peppers, ginger, lettuce, and strawberries are among the many minor offseason crops.

The selection of crops to be raised is not left to chance. Every July, prefectural marketing officials review the past crop year and market conditions. Best prospects for the following year are then determined and their suggestions are circulated among southern Kochi growers through the agricultural cooperatives. Farmers tend to favor one particular vegetable but diversify in order to benefit from good markets for some items and to avoid ruinous losses if an oversupply for others does develop. There is also some local interest for such horticultural specialties as flowers (carnations), bulbs (tulips), and small-scale orcharding (grapes and peaches) in and around the vegetable-growing areas.

The emphasis on offseason vegetables is so great that, paradoxically, there is a shortage of vegetables in southern Kochi local markets in late summer, so that supplies must be brought in from other parts of the prefecture. To offset these periodic shortages and to help improve the economic lot of depressed mountain villages, the prefecture is trying to promote commercial vegetable growing in a few selected mountain locations. The cool mountain climate is ideal for summer vegetables.

TRANSPORTATION TO MARKET[4]

Southern Kochi's fringe location in southern Japan poses a serious transportation problem in terms of the time and cost required to get fresh produce to Honshu markets. Earlier, produce moved solely by boat eastward around Cape Muroto to Osaka, but in November 1935 the Sanuki Line, a part of the national railway system, was completed through Kochi city and westward to Susaki. Subsequently extended farther westward to Kubogawa and eastward to Aki, it provides a convenient cross-island link between southern Kochi and the northeastern Shikoku port of Takamatsu (Kagawa Prefecture). From Takamatsu, rail ferries transport freight cars across a narrow neck of the Inland Sea to the Honshu port of Uno (Okayama Prefecture) and rail connections with Kobe-Osaka, Tokyo, and other major Honshu cities.

Steep increases in rail freight rates that would have jeopardized southern Kochi's market position were proposed by the national government in 1950 and were modified only after vigorous local protests. Partial compensation for the increased rates were gained by shippers in the form of special rates for bulk and long distance shipments and accelerated rail service. While the rail trip formerly took three days, fresh produce now moves from Kochi city to Osaka in a single day. The time required for shipment to Tokyo has been cut from four to three days, an important factor in expanded Kochi vegetable sales on the Tokyo market. Time in transit is important for two reasons. First, there are no refrigerated freight cars, although cars of special design allow a maximum circulation of air around the shipment. The fact that southern Kochi ships most of its produce during the colder half of the year thus helps keep spoilage to a minimum. Second, produce is sold in city markets by the auction method, so that the appearance and condition of vegetables after the long rail haul has a critical bearing upon prices realized.

Improved rail service has reduced the importance of ocean carriers, yet one privately owned ship company still operates on the Kochi-Osaka run, taking two days each way for the trip. Shipping

[4] Based in part on Kochi Prefecture Horticultural Cooperative League, *Engei no Tosa* (Tosa Horticulture) (Kochi, 1958), pp. 48–57.

suffered a bad setback in April 1956 when a ship bound for Osaka with a cargo of vegetables sank in a sudden storm. That the rail service is not invulnerable to such disaster was shown by the tragic sinking of a Takamatsu-Uno ferry in May 1955, bringing a heavy loss in life and serious disruption of the tight southern Kochi shipping schedule.

Local produce movements in southern Kochi are by rail, truck, or ship. The prefecture operates a pool system with agricultural cooperatives which equalizes the cost of vegetable shipment from points near and far to the nearest rail head. This system has been a boon to vegetable growers along the southeastern coast.

City Markets

With the exception of a small volume sold locally, southern Kochi vegetables are for sale in distant city markets. Osaka-Kobe, Kyoto, and other cities in the massive urban cluster at the eastern end of the Inland Sea have long been the most reliable outlets, and during the postwar period until 1951 dominated Kochi sales. Since 1952, however, the great size and purchasing power of the Tokyo market has manifested itself, and in spite of the longer shipping distance involved Tokyo has replaced Osaka as the leading customer.

Shipping statistics for 1957 show that Tokyo received 40 per cent of the southern Kochi shipments by volume and 45 per cent by value, compared to 40 and 32 per cent, respectively, for Osaka. The smaller Osaka market receives large amounts of offseason produce from nearby parts of Honshu, which helps glut the winter market; Tokyo is larger and with its farther north location does not have the same amount of local vegetable production during the winter months. The specific points of consumption of much of the produce sold in Osaka and Tokyo is difficult to establish because it moves to nearby cities or into broader regional marketing systems. Minor vegetable flows from southern Kochi go to Nagoya, Kanazawa, the only large Sea of Japan coast customer, and the cluster of cities in northern Kyushu. Sales in Nagoya, which has one of Japan's best-developed truck gardening zones in its outskirts, is convincing evidence of southern Kochi's ability to command a substantial slice of the city market in winter in the face of spirited competition from Honshu growers.

Southern Kochi marketing authorities pay attention to such

fine details as the special market appeal of certain crops or varieties in various markets. For example, the smallest and best-shaped eggplants find their best sales in Kyoto, where such qualities are esteemed. In similar fashion, early sweet potatoes, tender and sweet when boiled, have their warmest reception in Osaka, where boiling is favored as a preparation.

Marketing efficiency has been a keystone to southern Kochi's success. Prefectural authorities keep a close watch on market conditions through branch offices in Tokyo and Osaka, the only such organizations maintained by any prefecture except Aomori (for apple sales). Daily reports on prices and suggested timing of vegetable shipments is relayed back to southern Kochi, where they are circulated among growers. Realizing that present markets can be held only by quality control, the prefecture has a corps of inspectors who subject shipments to rigorous checks. Vegetables are carefully boxed or wrapped to withstand the long trip to market without damage (Pl. V, Fig. 2).

The greatest concern of southern Kochi growers is the rapid rise in offseason vegetable production in other parts of Japan. Coastal Miyazaki Prefecture in southeastern Kyushu also has fine winter weather similar to that of southern Kochi and has been expanding its winter crops under the supervision of a former Kochi Prefecture official. Concentrating heavily on cucumbers and pumpkins, Miyazaki growers have the cities of northern Kyushu as their immediate market but are sending larger shipments to Osaka and Tokyo. Suffering from higher transportation costs and lower quality, Miyazaki produce is as yet unable to offer serious threat to southern Kochi's sales of quality vegetables.

Competition is also growing rapidly around Osaka, Tokyo, and several other larger cities where farmers in the truck gardening zones are adopting the vinyl house production methods pioneered in southern Kochi. One noticeable result has been a gradual shrinkage of Osaka purchases of early June cucumbers from southern Kochi. In Tokyo's case, farmers in the Izu Peninsula of Shizuoka Prefecture and the Boso Peninsula of Chiba Prefecture are especially active in their delivery of offseason melons and cucumbers. Both areas have a warm winter climate and are within easy truck-shipping distance of Tokyo.

Even with the rising tide of production and market competition, however, southern Kochi vegetable growers are confident of their ability to maintain their strength in Honshu markets. In the expanding national market for fresh vegetables, southern Kochi has the combined strengths of a suitable winter climate, scientific cultivation and marketing operations, and an established reputation for quality with which to bolster its present rank as Japan's premier off-seasonal vegetable producing area.

PLATE I

FIG. 1. June vegetable-growing landscape on the narrow coastal plain just east of the Kochi Plain. Tosa Bay is beyond the line of trees to the right. Main crops are cucumbers (foreground), eggplants, and taro.

FIG. 2. Vegetable plantings and concrete water storage tank just west of Aki.

PLATE I

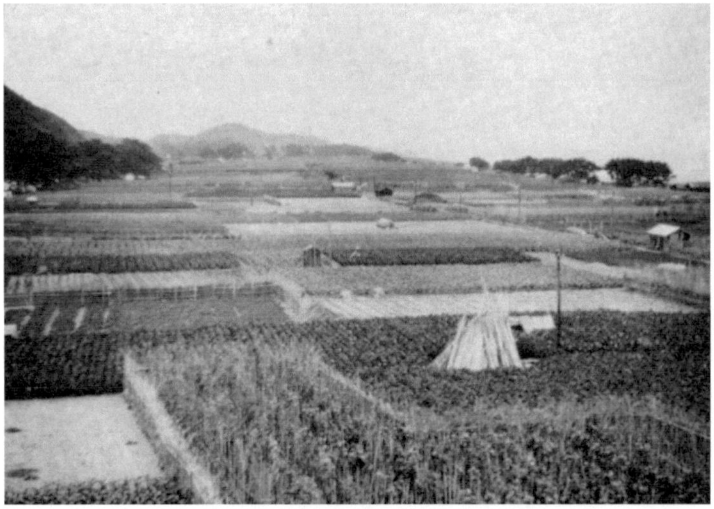

FIG. 1

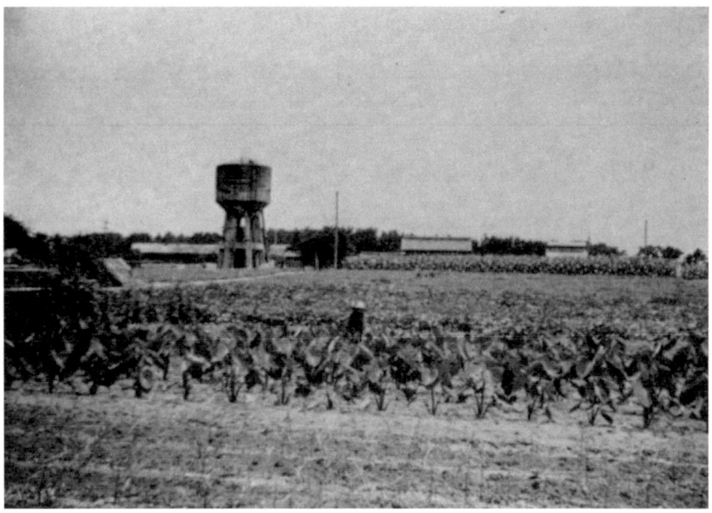

FIG. 2

PLATE II

FIG. 1. June preparation of paddy for rice planting following winter cultivation of vegetables. Poled cucumbers and a remaining vinyl house are in the adjoining field to the right.

FIG. 2. Main concentration of heated vinyl houses along the coast of Tosa Bay just east of the Kochi Plain. A typical sunny, warm December day. (Courtesy of Iwanami Film Co.)

PLATE II

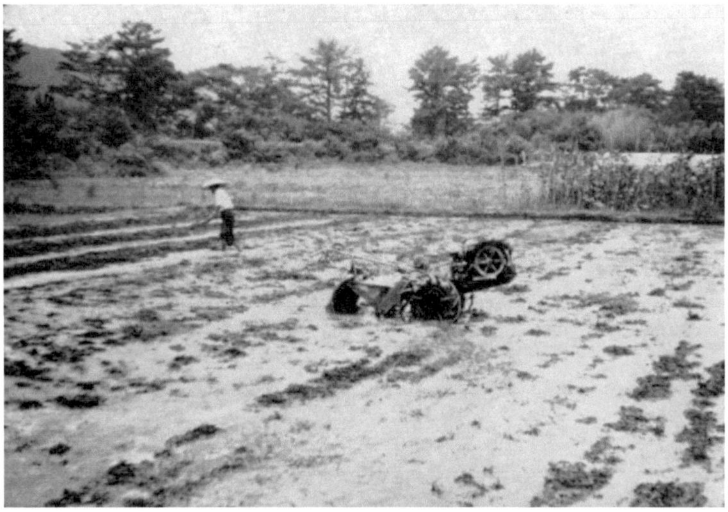

FIG. 1

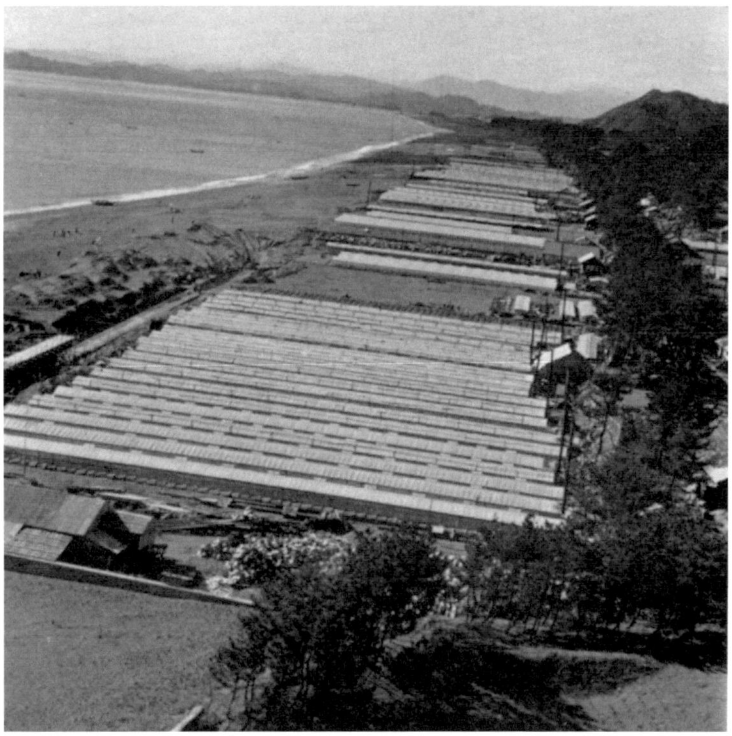

FIG. 2

PLATE III

FIG. 1. Tomatoes planted in vinyl frames (December). Opened to the warm sun during the day, they must be covered with vinyl sashes and straw mats at night to protect against low temperatures. (Courtesy of Iwanami Film Co.)

FIG. 2. Tattered vinyl houses at the end of the winter growing season (June). Frames will be dismantled and reused the next year. Typical landscape in lowlands west of Kochi city.

PLATE III

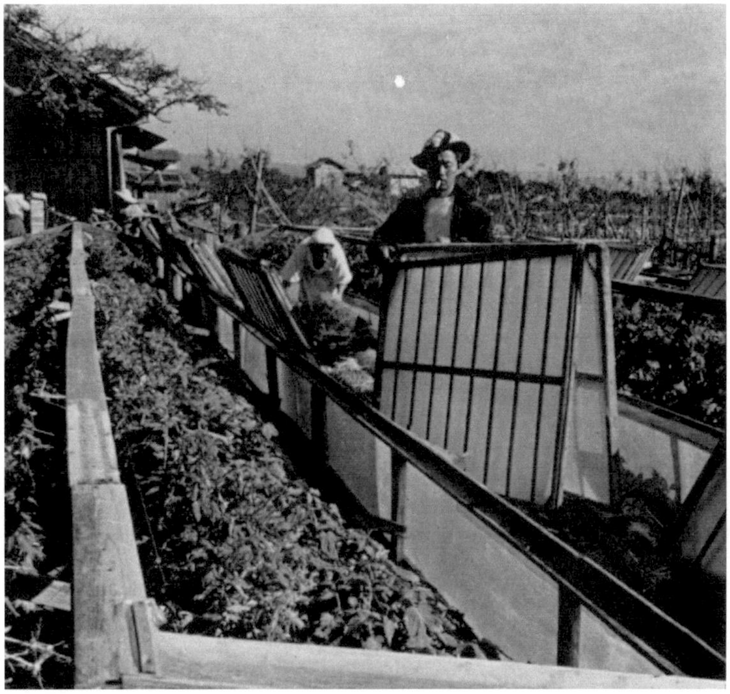

FIG. 1

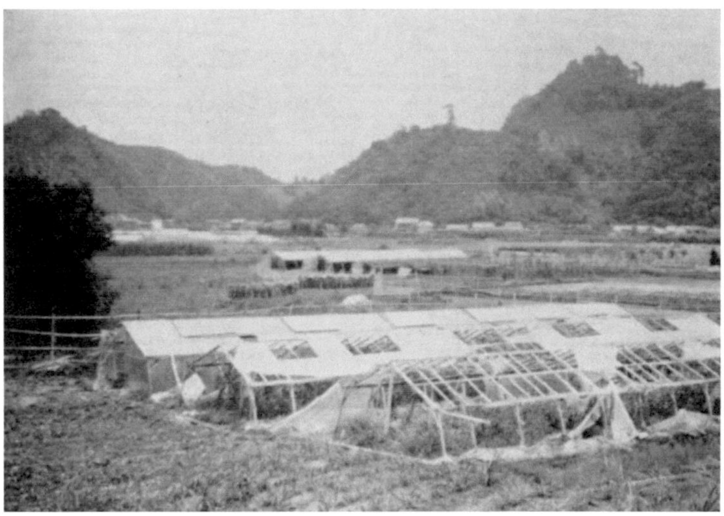

FIG. 2

PLATE IV

FIG. 1. Planting of eggplants near the fruiting stage (June). These were started under vinyl hot caps in early spring.

FIG. 2. Field of mature taro on the sandy coastal plain east of the Kochi Plain. Poles stacked in background are for supporting cucumbers.

PLATE IV

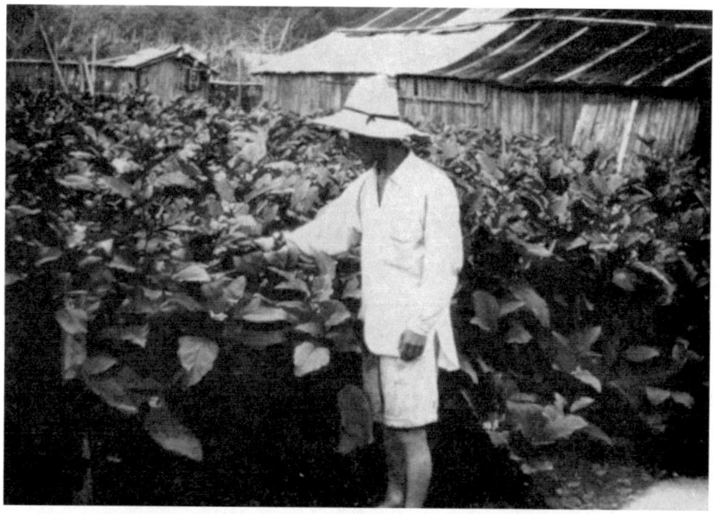

FIG. 1

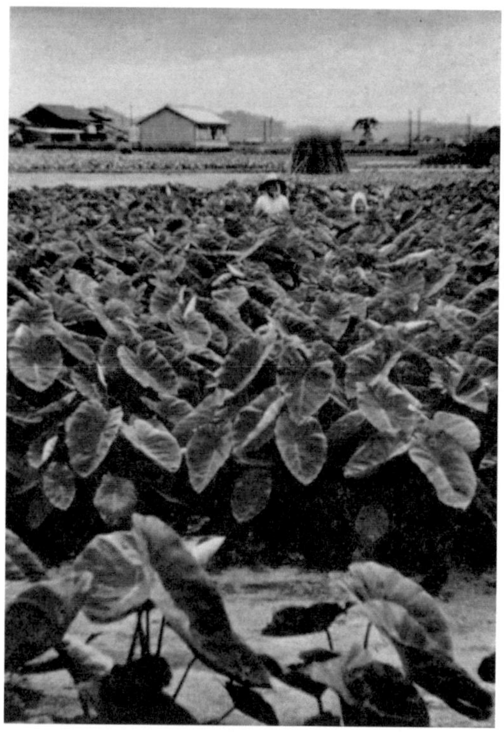

FIG. 2

PLATE V

Fig. 1. Bed of the lower Niyodo River planted to watermelons in June. Scattered pumps tap the plentiful ground water. The river bed is under water by late fall.

Fig. 2. Carefully wrapped watermelons grown in the lower Niyodo River basin. Thick straw matting prevents bruises while en route to market.

PLATE V

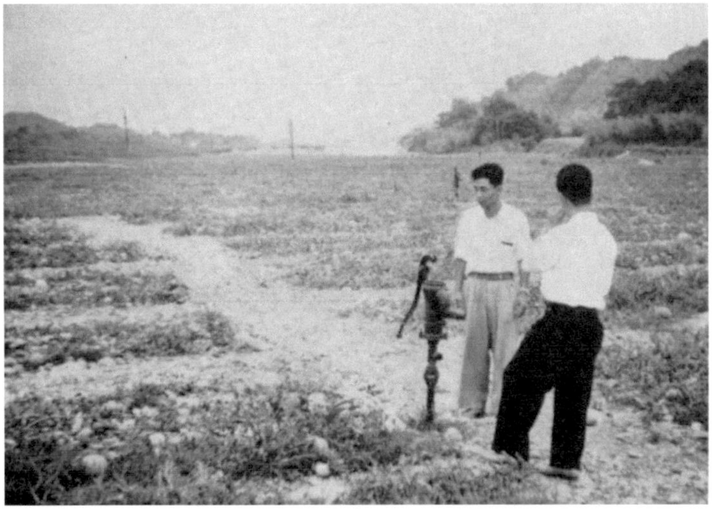

FIG. 1

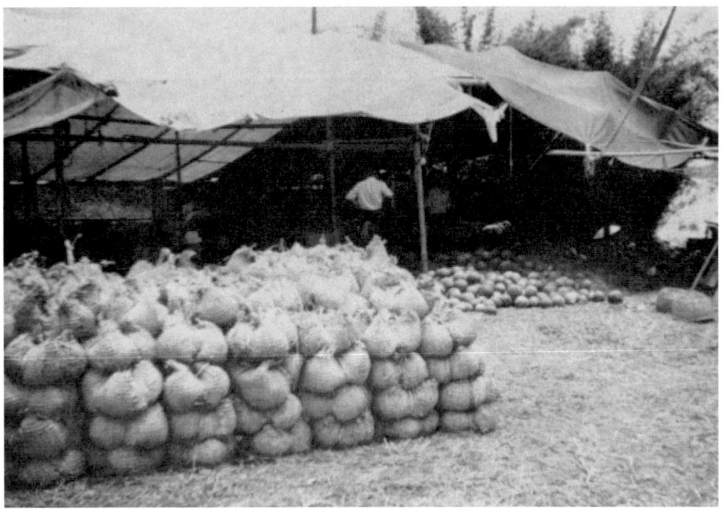

FIG. 2

Papers of the Michigan Academy of Science, Arts, and Letters
Vol. XLVII, 1962 (1961 Meeting)

TOJI: A STUDY OF A SEMIAGRICULTURAL COMMUNITY EXPLOITING MARINE RESOURCES[1]

GEORGE H. KAKIUCHI

University of Washington

Introduction

Tōji is one of three hamlets (ōaza) making up the village (mura) of Asahi and is located in the southeastern part of Izu peninsula, approximately 114 miles southwest of Tokyo. The settlement itself is situated on a narrow beach, which is hemmed in on three sides by steep slopes coming down to the sea. It is therefore physically isolated from the surrounding areas. In addition, it is poorly connected with the town of Shimoda (Fig. 1), the only important urban center in southern Izu. In fact, in 1954 there was only one small bus which came in daily at noon. Furthermore, being near the tip of a peninsula, Tōji is adversely located in respect to the mainstream of traffic flow along the highly urbanized Pacific coast of southwestern Honshū. Such isolation has been detrimental to cultural and economic development, which is reflected in the strong communal attitude of the people and the apparent poverty of the community.

Assuming that the situation described above is to be found in other similar coastal communities, then the study of Tōji could give us clues for further intensive and comparative studies, which would provide a basis for generalization. It is with this in mind that the settlement of Tōji is examined.

Historical Evolution of Tōji

Being hemmed in on one side by the sea and sandy beach and enclosed on the other three sides by mountains, Tōji is very poorly endowed with arable land. Nevertheless, historical records indicate that the original settlers did not engage in the exploitation of marine

[1] Much of the information and statistics for this paper were obtained during a field trip in December 1953, while the author was in Japan on a Fulbright Fellowship.

resources but in farming.[2] This factor is important in explaining the physical structure of the present-day settlement and also the attitudes of the people toward the exploitation of the products of the sea.

In time the number of households and population grew, so that farming alone could no longer adequately support the community, even though a number of the farmers rented land outside the hamlet, and other means of supplementary income had to be found. In this effort, the people first turned to other resources of the land, namely the exploitation of the products of the forests. The indications are that this occurred sometime before the end of the seventeenth century. For example, the tax records for 1688 show taxes on deer skins, bamboo, and other similar products.[3] But in time, the combined sources from the farmland and forest of Tōji, too, became inadequate, and it was only then that the farmers reluctantly began to turn to the resources of the sea.[4] This historical evolution seems to be characteristic of many of the land-scarce seaside hamlets, unless by traditional vocation the community is composed of fisher-folk.

Available information indicates that the people of Tōji began to exploit marine products around the turn of the nineteenth century as a source of additional income. By 1849, various fish and shellfish catches were being taxed. The same records show that by this time, 90 per cent of the farmers were cultivating less than 8/10 of an acre and fully 50 per cent were cultivating less than half an acre.[5] Today, the average farm size is about half an acre even though the surrounding slopes have been carefully terraced. This lack of arable land is reflected in that only 11 per cent of all the householders of Tōji are listed as full-time farmers, while the percentages for the neighbor-

[2] Ashikari Takashi, *Rinkai-son ni okeru nō-gyo seisan ketsugō no katei ni tsuite: Izu Tōji-son no baai* (On the process of the fusion of agricultural and fisheries production in a seaside village: case study of Tōji on the Izu peninsula) (Tokyo: Department of Geography, Tokyo University of Education, 1952), p. 133. (Unpublished B. A. manuscript).

[3] Between 1590 and 1688 the cultivated acreage increased by about 2 per cent while the number of farm households increased by 76 per cent. By 1848, the number of farm households had again nearly doubled, with no appreciable increase in cultivated acreage. *Ibid.*, p. 132.

[4] In Japan the people consider the exploitation of marine resources, especially the small-scale traditional type of fishing, as rather low in prestige and therefore turn to the sea only reluctantly.

[5] Ashikari Takashi, *op. cit.*, p. 144.

ing inland hamlets of Kisami and Ōgamo are 75 per cent and 47.3 per cent.[6]

Although the trend toward exploiting marine resources was evident before the Meiji Restoration in 1868, the date usually taken as the beginning of Japan's modernization, essentially the present socioeconomic pattern of the community took form after 1868 with the rise of a modern cash economy. The transition from a feudal subsistence-oriented economy to a cash economy gave new significance to and need for sources of cash income throughout Japan. For Tōji, however, handicapped by the lack of arable land and access to urban markets, the opportunities for cash cropping were very limited. Specialized cash cropping, namely out-of-season vegetables and flowers, might have been a possibility, for the coastal areas of southern Izu peninsula rarely experience frosts. In fact, it is said that some of the more enterprising farmers had tried, but their endeavors had not been profitable. Tōji, isolated as it was, could not compete with other areas similarly endowed but better located in terms of access to transportation facilities and markets.

Contemporary Scene of Tōji

The consequences of isolation can most effectively be brought out by comparing Tōji with its two neighboring hamlets of Ōgamo and Kisami. Both of the latter are located close to and on the two main roads leading into the port town of Shimoda, whereas Tōji is situated at the end of a narrow and winding road that must cross over the mountains which separate the settlement from the interior valleys (Fig. 1). The distance from Tōji is a bit too far and arduous to take the cash crops by man-drawn "rear-carts" or in bamboo baskets, and as mentioned earlier only one bus comes in daily at noon. As a result, while cash cropping, dairying, and working in Shimoda are the main sources of supplementary cash income in Ōgamo and Kisami, these sources are practically nonexistent in Tōji.

Being thus limited in economic opportunities related to land, the farmers had to seek alternative sources of supplementary income, especially after the rapid expansion of cash economy after the

[6] There were 95 households in Tōji, of which 91 had some farmland. The total population was 567. The total cultivated acreage was 46.5 acres, of which 18.3 acres were in paddy. Statistics taken from unpublished basic data collected by the village for the *1950 World Agricultural Census*.

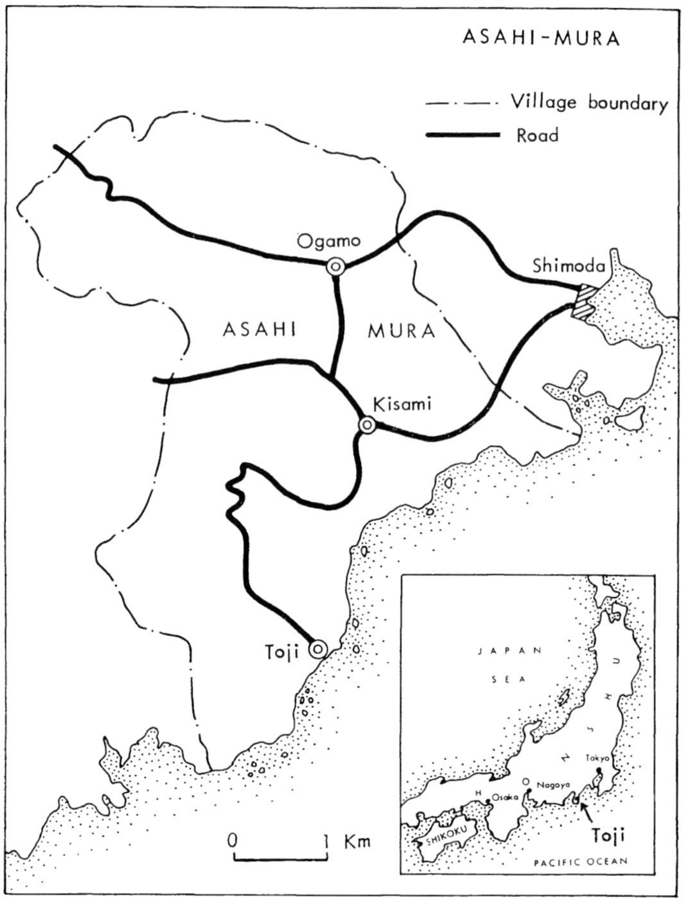

FIG. 1

Meiji Restoration. The immediate alternative lay in the adjacent coastal waters of the hamlet, to which the community had traditional legal rights. Initially, fishing had been much more important, but today the fish catch accounts for only a small percentage of the total production of marine products. The decline in fishing can probably be attributed to the rise of modern commercial fishing companies operating out of larger and better located fishing ports. The small coastal hamlets such as Tōji found it increasingly difficult to compete and had to make further adjustment to the changing national economic structure. The change that occurred was that the people increasingly turned to the gathering of seaweeds, especially agar-agar. In 1938, seaweeds accounted for 49 per cent of the total marine production, but by 1948, the percentage had risen to 89.6 per cent, with agar-agar alone accounting for 42 per cent.[7]

One of the prime reasons why the gathering of seaweeds became so important as a source of supplementary income was that it could be done with hardly any capital investment. Furthermore, the products could be dried and bundled into compact bales or packages and, being nonperishable, could be easily shipped out at the convenience of the local cooperatives or buyers. Such an activity, much less competitive than cash cropping or dairying, is therefore well suited to an isolated and poor community like Tōji and is quite common in comparable coastal hamlets elsewhere, as in the southern part of Bōsō peninsula in Chiba prefecture.

Socio-Economic Characteristics

Under the conditions outlined, the situation which evolved was that although the farms continued to provide the people of Tōji with some degree of self-sufficiency, the exploitation of marine resources became gradually more important in their livelihood as the principal sources of cash income. The result is that while the outward appearance of the settlement and the attitudes of the people still retain certain characteristics of its agricultural background, many of the socio-economic aspects associated with communities gathering seaweeds have been superimposed upon the scene. For example, it is rather common in hamlets where the gathering of seaweeds is important to have a higher than average predominance of the female over the male population. In Tōji, the females outnumber the males 313 to 254.[8] In contrast, in many of the agricultural villages in the interior of Izu peninsula the males outnumber the females.[9] The same high predominance of females is also seen in the coastal villages of southern Bōsō peninsula where seaweeds are gathered. The ratio

[7] 1938 and 1948 records of the Tōji Marine Products Association.

[8] Local 1948 population records of Asahi-*mura*.

[9] Bureau of Statistics, Office of the Prime Minister, 1935. *Kokusei chōsa hōkoku, Shizuoka-ken* (Population census of 1935, Shizuoka prefecture) (Tokyo: Bureau of Publications, Ministry of Finance, 1935), p. 5.

in a place called Nagao is 146 to 100 and in Shirahama 119 to 100.[10] There is a good reason for this phenomenon, and this is that the diving for seaweeds and also for shellfish is done by the women and girls of these communities. The divers, called *ama* (women of the sea), therefore, are distinct assets and are kept at home instead of being sent elsewhere to seek work, as is often the case in agricultural villages. The work of the *ama* is considered so important that the community subsidizes their training. According to the local people, the diving work began sometime during the early years of the Meiji Period.

This rather unusual role of the women in the economy of these coastal communities also has had interesting repercussions in the social customs. For instance, there is a higher incidence of new husbands establishing residence with their brides' families, which is very uncommon in agricultural settlements. Also there is a definite tendency for the families to keep the girls at home as long as possible. A family with a daughter who has reached marriageable age is less interested in her marriage prospects than it is in her abilities as a diver. The result is that the girls marry at an older age, which is contrary to situations elsewhere, and many wives are older than their husbands.[11] Although these customs have become less prominent, in 1949, for every three couples where the husbands were older than the wives, there was one where the wife was older than the husband.[12] Nationally the ratio was seven to one in 1935.[13]

On the other hand, residence lots and house construction in Tōji are similar to those of agricultural hamlets, where conditions are more spacious than they are in fishing villages. For example, farmers in Japan usually devote about 10 to 17 per cent of their floor space to various work activities, while in the fishing hamlets the percent-

[10] Aono Hisao, "Suisan chiri-gaku yori mitaru gaibō engan gyoson," (A seaside fishing village of outer Bōsō examined from the viewpoint of fisheries geography), *Otsuka chiri-gakkai rombunshū* (Collected papers of the Ōtsuka Geographical Society, Vol. 2) (Tokyo: Ōtsuka Geographical Society, 1933), pp. 195–196.

[11] Yanagida Kunio, *Kaison seikatsu no kenkyū* (Studies of life in seashore villages) (Tokyo: Japan Society of Anthropology, 1949), pp. 118–119.

[12] Local 1949 population records of Asahi-*mura*.

[13] Bureau of Statistics, Office of the Prime Minister, 1935. *Kokusei chōsa hōkoku, Nihon* (Population census of 1935, Japan) (Tokyo: Bureau of Publications, Ministry of Finance, 1935).

age is only about 5 to 13 per cent.[14] In Tōji the percentage devoted to work space is about 10 per cent and so occupies a median position.[15] And although the exploitation of marine products is very significant in the economy of Tōji, most of the people still consider themselves primarily farmers. Thus Tōji reflects aspects both of settlements exploiting the land and of settlements exploiting the sea.

It is very difficult, however, to say which is more important in contributing to the livelihood of the community. Village statistics which should clarify the situation do not agree with each other. For instance, the *Agricultural Census* of 1950 shows that approximately 80 per cent of the households derive their livelihood mainly from farming, but the records in the local Marine Products Association office indicate that about 50 per cent of the households derive their livelihood principally from the exploitation of marine products. The best that can be said, therefore, is that both are important.

Exploitation Rights of Marine Products

Except for fishing, the people of Tōji, like so many other similar small coastal hamlets which began as agricultural settlements, are not directly engaged in the exploitation of the local marine resources. Instead larger operators, usually from nearby villages, come and bid for the exploitation rights to the coastal waters which belong to the settlements. The situation is the same in southern Bōsō peninsula.[16] The system in Tōji seems to have originated sometime between 1867 and 1877, when agriculture was much more important than it is now.

The contractor's rights extend from April through October. From the money the community receives, deductions are first taken out for such community responsibilities as the maintenance of roads, schools, hospitals, and the like. The money that is left is then divided equally among the 66 households which were in existence at the time the first contracts were let out.[17] Although the remaining 29 families do not share in the income from the rental payments, they do have the right to fish, as do all the other people of Tōji. In 1948, each of

[14] Shōji Hikaru, *Jūkyo no Eisei-gaku* (Sanitation science of residential homes) (Osaka: Kosēikan, 1938), p. 24.

[15] Information obtained through interviews.

[16] Yanagida Kunio, *op. cit.*, p. 144.

[17] Information obtained from the offices of the local Marine Products Association.

the 66 families received ¥10,000 ($28.00) and in 1949 ¥5,000 ($14.00).

In addition to the income from the rental payments, the people also benefit from being employed by the contractor in the gathering and processing of seaweeds and shellfish. There is no doubt, then, as to the importance of the exploitation of marine resources in the economy of the community as the principal source of cash income.

Summary

In summary it can be said that although 80.5 per cent of the households in Tōji are mainly engaged in farming, the exploitation of marine products is certainly very important as a source of supplementary income. At least two-thirds of the households receive cash payments from the leasing of the community coastal water rights, and most of the households have members working for the contracting company in the gathering or processing of seaweeds and shellfish. Still others are engaged in fishing. For at least 35 per cent of the households, the exploitation of the marine resources constitutes, in one way or another, the principal source of livelihood.

But because the settlement initially began not as a fishing community but as a farming community, the general outward appearances and attitudes are those of an agricultural settlement. As population increased within an area severely limited in arable land and as a cash economy became increasingly important, especially after the Meiji Restoration, the people were forced to turn to the sea as a supplementary source of livelihood. As a result, certain socio-economic characteristics associated with communities exploiting marine products became part of the Tōji scene. Thus a study and analysis of the community's history and evolution can provide many clues to the present situation.

The fact that the people of Tōji have continued to exploit marine products, rather than going into what might be more lucrative endeavors, can be explained in terms of the relative isolation of the settlement from urban markets. Lacking good transportation connections with urban markets, the people have been at a competitive disadvantage in the growing of cash crops, although climatically they are just as well-situated as some of the better located settlements nearby which have come to specialize in out-of-season vegetables and flowers. Also, the increasing importance of nonperishable marine products, mainly agar-agar, is again a reflection of isolation and increasing competition from commercial fishing companies operating out of better-located ports.

Unless the connections of Tōji with the urban markets are improved, the situation in this community will continue substantially in its present form. The limited amount of arable land will be used essentially for subsistence crops. With the average landholding being about half an acre, the community as a whole must continue to buy part of its basic food and other daily needs. Thus it is that a secondary or supplementary source of income is necessary, and for Tōji, it has been the sea. But even when both sources are combined, the standard of living is low, which is characteristic of all the small semi-agricultural communities engaged in the exploitation of marine resources. This is reflected in the fact that, although the labor demand is high during the summer and fall months when farming and the gathering of seaweeds and shellfish coincide, most of the work is done by family labor.

In the final analysis, it is rather difficult to classify the community of Tōji clearly. If we look at the community from the socioeconomic standpoint, it resembles a community exploiting marine resources, but if we look at the outward appearance of the settlement and the attitudes of the people, then it resembles an agricultural community. And conflicting statistical information does not aid attempts at analysis.

PAPERS OF THE MICHIGAN ACADEMY OF SCIENCE, ARTS, AND LETTERS
VOL. XLVII, 1962 (1961 MEETING)

AGRICULTURAL PROBLEMS IN TOHOKU (NORTHEASTERN JAPAN)

TOSHIO NOH

Tohoku University

TOHOKU has long and rightfully been considered as the most backward and poverty-stricken region in Japan. It lay far distant from the old core area of Kinai. It was regarded as a frontier zone, and its relatively long and cold winters and frequent cool summers made it hazardous and unattractive. Even after the center of actual government was moved to Edo (present Tokyo) in the early seventeenth century, Tohoku still retained and deserved its unfavorable reputation. The circumstances were not improved after the Restoration in 1868. This was largely due to the historical coincidence that the majority of the Tohoku feudal lords had staunchly opposed the movement that brought in the new government. Tohoku was treated, very much as was Hokkaido, as a colonial area in the land development policies of the New Japan. The region retains its reputation for backwardness, which is still valid as far as culture is concerned and especially deserved because of the retarded development of its manufacturing industries.

The Tohoku district comprises the six northernmost prefectures of the main island of Honshu. It covers an area of 29,740 square miles and had a population of 9,393,880 by the 1955 census. This makes Tohoku, except for the north island of Hokkaido, the most sparsely populated region in Japan, having, after Hokkaido, the second, third, and sixth least populated prefectures of the nation.

Agriculturally, however, Tohoku presents an interesting anomaly. In the yield per unit area of Japan's most important crop, rice, Tohoku today leads the nation, even though at one time it was the district of lowest yields. The two maps in Fig. 1 show the sharp contrast between present yields and those of former days. Even in 1920 and later Tohoku was the low yield region. Traditionally the district was backward in rice cultivation as in everything else. It suffered from frequent bad crops and consequent famine caused by low summer temperatures. The radical change that we see today was

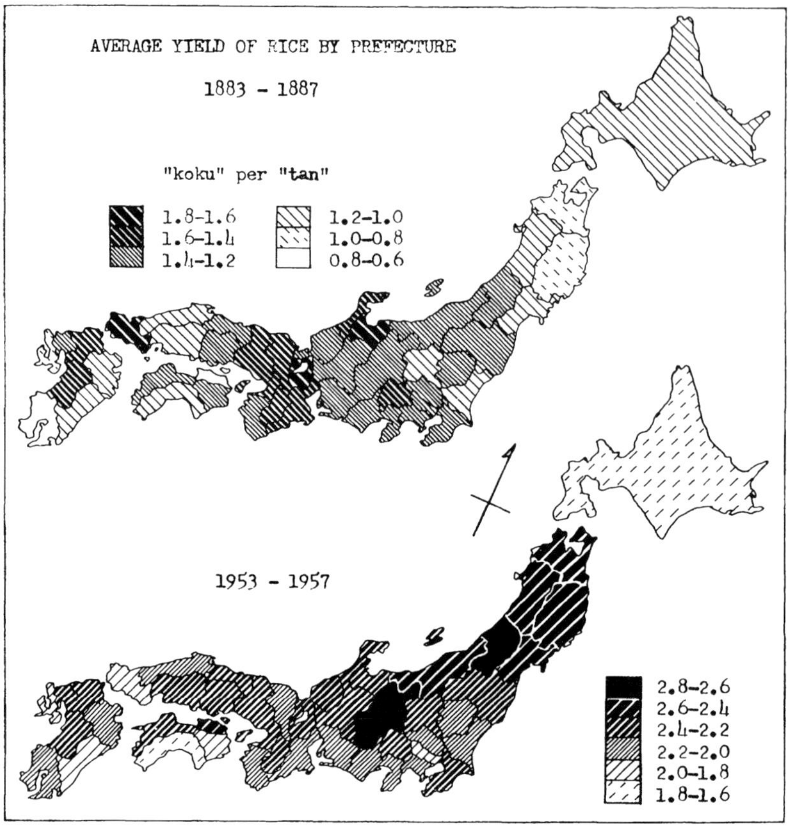

FIG. 1

brought about by the introduction of new varieties of rice adapted to low temperature conditions, such as Rikuu 132, Fujisaka 5, and Sasashigure.

In a country like Japan, where rice takes such a predominant place in the farm economy, it would seem that Tohoku is now agriculturally advanced rather than backward. This is but part of the story, however. It is the main objective of this paper to show the relationships between rice production and other aspects of the total agricultural economy.

Situated as it is in the far north of Honshu, Tohoku is characterized by relatively low temperatures, particularly in winter, and relatively deep snow accumulation. Winter coldness is more pronounced on the Pacific side of the region, while the accumulation of snow is much heavier on the Japan Sea side. Differences in temperature within the region are far less in summer than they are in winter, but once in every several years there is a summer when air temperatures for June–July are considerably lower than average. This being a period which is very important for the growth of rice, yields are hazardously decreased. In the history of Tohoku there are many records of famine caused primarily by low summer temperatures. The recent introduction of new varieties of rice has almost solved this menace. In fact, the farmers of the region are almost convinced that they are permanently free from cool summer damage.

There are other climatic handicaps in Tohoku which are not solved. Cold, snowy winters prevent the raising of a second crop, following the rice harvest, as is practiced in most other parts of Japan. The approximate northern limit of double cropping runs through the southern part of Fukushima, which is the southernmost prefecture of the region. This does not mean that all second cropping stops at this line. There is some north of it, especially near the larger cities, but on the whole such production is negligible. Tohoku farmers are thus compelled to emphasize their summer crop, rice, and here they have made highly important progress in recent years.

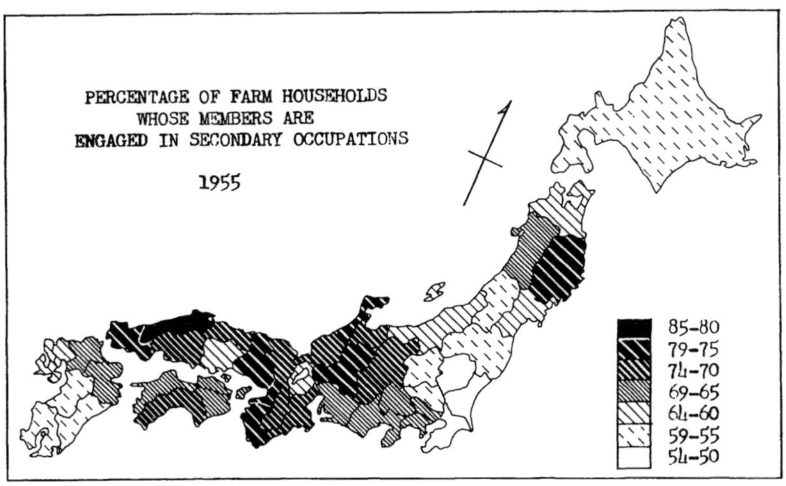

Fig. 2

In the winter when the fields are frozen or covered with snow, many Tohoku farmers migrate to other parts of Japan seeking employment on fishing boats, in sake breweries, and in other jobs (Fig. 2). In a sense, then, the high yields that have been achieved for rice are in themselves an indicator of the handicaps of Tohoku in comparison with other regions where important income can be expected from winter cropping.

It is never easy to compare the agricultural standards of different areas where not only the patterns of farm management but also the ways of life differ. Certain statistics from the Agricultural Census of 1957 are of interest here. In average gross income per farm Tohoku is shown to be far above the average for the nation, second only to Hokkaido. This high figure results mainly from the income from crops, of which rice accounts for more than 80 per cent. This places Tohoku in first place both in total crop income and in income from rice.

Hokkaido's supremacy in gross income results mainly from extra-farm sources, e.g., outside wages and government subsidies. In any event the gross farm income in both Hokkaido and Tohoku is well above the national average. This, however, if taken alone, can be highly misleading. Other elements, not expressed in the above statistics, seriously affect the farmer's livelihood in Tohoku. It is in these that the still backward condition of the region's agriculture is to be found. For instance, extra-farm income is lower in Tohoku than in any other district in Japan except Nankai, where year-around agricultural production is possible. Farm expenditure in Tohoku, moreover, is relatively very high, being almost equal to that of southern Kanto and next to that of Hokkaido. Admittedly these factors are hard to analyze, and official statistics do not cover all of the pertinent elements. Nevertheless the relative picture is reasonably clear. The net income of the Tohoku farmer is low. In spite of recent dramatic increases in rice yields, and in spite of a large total crop and high gross income, Tohoku is still a depressed and backward region ranking ninth in net income among Japan's eleven agricultural districts.

Papers of the Michigan Academy of Science, Arts, and Letters
Vol. XLVII, 1962 (1961 Meeting)

A MAP OF "BURAKU" SETTLEMENTS IN JAPAN

ROBERT B. HALL, Sr.
The University of Michigan

In recent years, the innocuous label "Buraku" has come to be the almost exclusive term applied to Japan's outcast peoples. Essentially it means any hamlet or other small community, but there is no mistaking what is meant by Buraku people.

There are some two to three million of these unfortunates in Japan. The matter of their origin has long been a subject of controversy, and many and varied have been the theories advanced. Most have attempted to prove that, at least the Eta, were of foreign origin. Recently, "progressive" writers have made the claim that the Eta descend from the original Japanese, who were enslaved by "capitalistic exploiters" invading Japan from the continent. Others have claimed a more recent origin, laying their degradation to the feudal system. There is little doubt, if any, but that the Eta and Ie-buni came into existence early in Japanese history and have persisted since. There is every probability that outcastism came with the introduction of Buddhism in Nara time and results from religious taboos on the taking of life, the handling of dead things, and other defiling occupations. Through the centuries, other outcast groups developed, such as the Hinin, Matagi, Kabo, and Sanka. It is ironical that the entire onus of pariahism should, today, be carried by the Eta, and to a lesser extent by the Ie-buni and Mitagi. These were the most fortunate, one might even say the aristocrats of the outcast peoples, during feudal time, as they held economic monopolies and owned property. It was just this that kept them in place and assured their identity. The others, having no fixed abode or insured livelihood merged into the general population, especially in the growing cities.

Any field study of the problem today runs into the difficulties that: (1) legally there are no outcasts, as the system was abolished in 1871 so that their existence can, therefore, not be officially recognized and; (2) ordinary people, in the parts of Japan where outcasts are numerous, are very much afraid of and repulsed by Buraku people.

The geographer's interest in this problem is, of course, primarily that of distribution. Map No. 1 (Fig. 1) was long in the making. It shows the distribution of Eta communities, and no attempt has been made to record individuals. This is not a serious omission as "misery seeks company."

The map shows no communities on the northern and recently settled island of Hokkaido. It shows very few in the "Frontier Zones" of Tohoku (Northern Honshu) and Kyushu, south of the Kuma-kii mountainland. The overwhelming concentration of outcast communities occurs in the area of "Elder Japan"—the lands surrounding the Inland Sea. This is the birthplace of the Japanese Nation, and it is here that most of Japan's population resided when the civilization of Tang China was introduced, including Buddhism. Heavy secondary concentrations are found at the base of the Noto Peninsula (present-day Ishikawa and Toyama prefectures) and in Gumma, Saitama, and Nagano prefectures in central Honshu. These are all areas of known colonization from the Inland Sea area early but after the coming of Buddhism. It is of considerable interest and importance that the map of "Eta," Outcast Communities today (Fig. 1) and the map showing the distribution of Japan's total population in the early tenth century[1] (map No. 2, Fig. 2) present almost identical patterns. Conformity is particularly noticeable in: the Go-Kinai and adjacent provinces, the area immediately around, the old capital cities—Nara, Kyoto, etc.; the shores of the Inland Sea; in ancient Tsukushi-northwest Kyushu; and in western Kanto. Important conformity is also found throughout most of the Chugoku Peninsula—including the old culture centers of Kibi (essentially modern Okayama Prefecture) and the San-in region on the Sea of Japan. Similar conformance occurs at the base of the Noto Peninsula and in interior Shinano Province, now Nagano Prefecture.

There can be little doubt but that the Eta came into being at an early date, had their origin in "Elder Japan," and moved with early large-scale colonization as an integral part of the society.

In the later northward course of Empire—northern Honshu and Hokkaido—those involved seem to have lost their identity in the enforced equality of the frontier and war.

A number of clusters and isolated communities are found, other

[1] After manuscript map by S. Inoue, based primarily on the Engi-shiki.

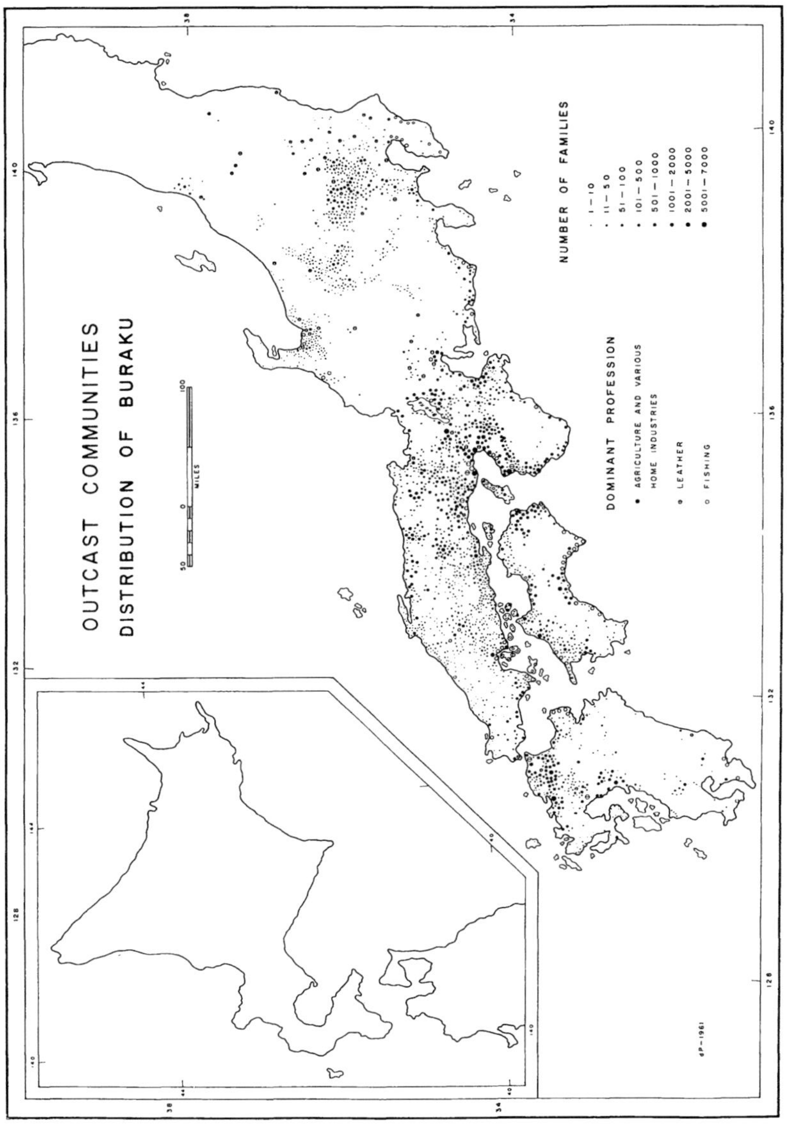
OUTCAST COMMUNITIES
DISTRIBUTION OF BURAKU
50 0 100
MILES
NUMBER OF FAMILIES
1 – 10
11 – 50
51 – 100
101 – 500
501 – 1000
1001 – 2000
2001 – 5000
5001 – 7000
DOMINANT PROFESSION
AGRICULTURE AND VARIOUS HOME INDUSTRIES
LEATHER
FISHING
dP-1961

FIG. 1

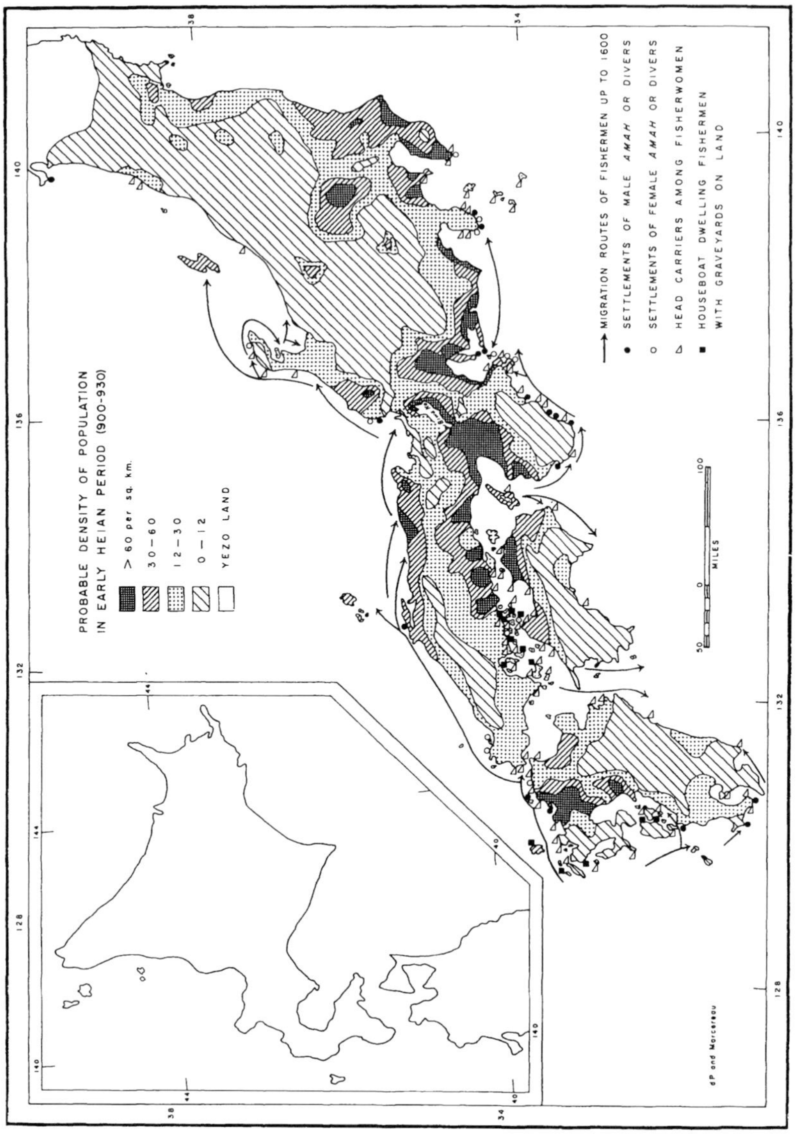
PROBABLE DENSITY OF POPULATION
IN EARLY HEIAN PERIOD (900-930)
> 60 per sq. km.
30 – 60
12 – 30
0 – 12
YEZO LAND
MIGRATION ROUTES OF FISHERMEN UP TO 1600
SETTLEMENTS OF MALE AMAH OR DIVERS
SETTLEMENTS OF FEMALE AMAH OR DIVERS
HEAD CARRIERS AMONG FISHERWOMEN
HOUSEBOAT DWELLING FISHERMEN
WITH GRAVEYARDS ON LAND
50 0 100
MILES
dP and Marcereau

Fig. 2

than in Hokkaido, which for the most part result from the later feudal period. The onus on contact with dead persons and animals and the activities pertaining still persisted, as it does to a lesser extent even today. Checking in detail shows that these widespread and secondary communities are generally situated at or near the sites of one-time castles or other fortifications. It would seem that these Eta groups were imported by the then ruling feudal lords to perform their traditional functions of disposing of the dead and of making those commodities, essential to the culture of the time, which were derived from leather, gut, hair, and bone, such as armor, bowstrings, musical instruments, and footwear. The distribution today of leather workers' communities bears out this hypothesis.

Map No. 2 shows the distribution of Ie-buni or boat-dwelling fishermen. These groups spend their entire life on their boats, except for brief visits to their community graveyards at New Year's and at the time of the O-Bon. These people, too, are confined to Elder Japan. Their range extends from Onomichi in the east to the Ariake Sea and the Goto Islands in the west. These interesting people are today few in number and are rapidly disappearing, due largely to the enforcement of the compulsory education laws. Their outcastism derived from the same source as the Eta, but was never as severe. As Ie-buni, they are still confined to Elder Japan.

Closely related, but even less ostracized are the Amah or diving fishermen—male and female. In localities where the water is deeper and colder the women tend to do the diving and the husband handles the boat and gear. In other localities the reverse is true. Still another related cultural trait is fisherfolk women as head carriers. This is the custom of Ie-buni women as well as women of other, but by no means all, fishing communities.

The present distribution of these several kinds of lowly fisherfolk presents an interesting geographical problem. Map No. 2 shows the pathways of migration of fisherfolk out of Elder Japan. The Ie-buni, with their small houseboats, seemed not to have prospered, as such, outside of the sheltered Inland and Ariake seas.

Off the Sea of Japan coast the Tsushima current flows east and northward and splits at the latitude of the Tsugaru Strait—between Honshu and Hokkaido. One branch flows through the Strait and part of it bends southward and follows the Pacific coast of Honshu to the vicinity of Sendai. Amah and women head-carrying communities were established as far north as the top of the Noto Peninsula before 1600 A.D. The head-carrying trait was extended northward, through the Tsugaru Strait and down the Pacific coast during the centuries following. On the Pacific side the Kuro Siwo or Black or Japan Current gave propulsion to the migrating fishermen as far as the Boso Peninsula. Here this warm current is overridden and forced seaward by the south flowing, cold Oya-shio current. Amah and head-carrying fishing communities were established along this part of Japan's Pacific coast, at an early date. Between the southern tip of the Boso Peninsula and Sendai there is a blank, in so far as early traditional types of fishing communities are concerned. Migration, after 1600, was either by more recent types of fishermen, not burdened by traditional prejudice, or the older type found emancipation by changing their customs or both. There is no ostracization of persons engaged in the modern fishing industry. The shores of Hokkaido have had a similar history.

The Matagi are found only in the Tohoku mountain areas. They are probably the remnants of a much more widely distributed population of hunters and trappers. The probably related Sanka, in the far south, have disappeared. The Matagi persisted because they were protected by the local feudal lords and assured a monopoly of producing medicines from bears, snakes, and other creatures. Today, they find employment as guides for sportsman hunters from the cities to the south.

A study of the location of outcast communities, in detail, shows them to be generally on the outskirts of or nearby the ordinary communities which they served. To them was allocated the more undesirable sites—river banks, swampy areas, and the north slopes of hills and mountains. In the Go-Kinai area many of the ancient Imperial and Noble tombs have, or had until recent times, an Eta community at their foot—telling of their function as caretakers of the dead.

In summary, a study of the distribution of Japan's outcast peoples gives strong support to the suppositions that:

1. Outcastism in Japan developed early and had its origin in the area of Elder Japan.

2. It was firmly established in the areas of mass colonization out of Elder Japan at an early date, but after the acceptance of Buddhism.

3. The system weakened with both time of settlement and the distance from these early centers and generally disappeared on the frontiers.

4. The precise location of most outcast communities, as well as the character of their occupation, gives further proof of the depressed condition of these people through the centuries.

The Japanese Government is seriously trying to correct existing inequalities and there is an awakening of the public conscience. Private organizations are giving their best efforts to the cause of equalization. It will be a slow process and is in many ways reminiscent of our own problem of achieving equality for all of our citizens.

PAPERS OF THE MICHIGAN ACADEMY OF SCIENCE, ARTS, AND LETTERS
VOL. XLVII, 1962 (1961 MEETING)

NEW SOURCES OF RAW MATERIALS FOR JAPANESE INDUSTRY

ROBERT B. HALL, JR.
University of Rochester

ONE of the most remarkable developments in the postwar period has been the sudden reappearance of Japan as one of the industrial giants of the world. The pace of Japan's economic development and growth over the past ten years has been almost unbelievable. While economists are not all in agreement about some details of Japan's economic growth during the 1950's, the period may well be one of unprecedented development for any industrial nation. Most reports point to a 20–25 per cent increase in the industrial index for 1959, bringing the industrial production index to 179.6 (1955 = 100), and 1960 was also a year of record prosperity.[1]

As a result of this miracle of industrial development, Japan is today among the world's leaders in the production of many industrial products, including iron and steel, electrical machinery, automobiles, ships, chemicals, textiles (natural and synthetic), electrical appliances, and optical goods, to mention only a very few. It is again one of the world's major trading nations and has again one of the world's largest and most modern merchant marines, with a total of over six million gross tons.

In addition to the great increase in the production of manufactured goods for domestic consumption and export, Japan has been rapidly developing and expanding its industrial plant, transportation, and port facilities, and is experiencing a major boom in all phases of the construction industry. Some idea of the colossal scope of the construction boom in Japan can be seen in the growth of one single metropolitan area, Tokyo, which grew from six million in 1950 to about ten million in 1960.

With the increase in production and construction has come an unprecedented rise in living standards, including the development of mass markets in Japan for such items as cameras, TV sets (esti-

[1] The Foreign Capital Research Society, *Japanese Industry 1960 %* The Bank of Japan, Tokyo, 1960, p. 4.

mated at one for each four households), sewing machines, and refrigerators.

In order to achieve the great expansion of its industries, Japan has had to import vast quantities of raw materials from foreign sources. The government and industry are coöperating in a program to try to increase domestic production of raw materials and to develop new sources of supply from the meager mineral wealth of the home islands. While there has been an increase in the production of some domestic raw materials, Japan remains today, as before the war, primarily a workshop nation, importing raw materials from all parts of the world and exporting manufactured goods to pay for the imports.

It is, of course, impossible to cover the whole range of raw materials consumed in a nation like Japan, which produces almost all kinds of industrial goods and is continually adding new industries and new products. (A case in point, Japan is now second only to the United States in the production of synthetic fibers and has one of the world's fastest-growing petro-chemical industries.) This study will, therefore, be concerned with Japan's supply of three major raw materials: coking coal, iron materials, and petroleum. All of these are major items in Japan's industrial production, and all are basic to several of the most important industries. In each case Japan produces some of its requirements at home, but is dependent for part or most of its requirements on foreign sources.

Coking coal and iron materials are basic to steel production, and Japan, with a production of over twenty million tons, ranks fifth among the world's steel producers. The present production of steel is about three times that of the 1943 record production of just over seven and a half million tons. If the Japanese iron and steel industry achieves its projected increases in production, the 1970 output will be in the neighborhood of forty million tons. The raw material requirements for the present production and projected production are enormous and cannot possibly be met with domestic supplies.

Coking Coal

Japan's present annual production of coal is about 50 million tons, and this output is sufficient to meet most of the nation's requirements. Most of the production still comes from two areas, Kyushu, with just over half the total production, and Hokkaido, with about one-third of the total production. Unfortunately for the iron and steel industry Japan's coal is not of a high enough grade to take care of the industry's coking coal needs, and most of the high grade coking coal has to be imported. Present imports are running at the rate of about five million tons per year, and the imports of coking coal may be as high as twelve million per year by 1966.[2] Some small increase in the production of domestic lower grade coking coals may be possible, as well as improvement in the blending of coals, but the major increase in supply will have to come from foreign sources.

At present Japan relies on the United States for about three-fourths to four-fifths of the imported coking coal. This is a very different pattern than existed before the war when most of Japan's coking coal was imported from China. For the present, at least, Japan will have to continue to look to the United States and other sources for its coal. While Japan is not at the present trading with China, there is considerable pressure from some groups to reopen trade relations, and to switch to Chinese coking coal, which is obviously much closer to Japan than is that of the United States.

Japan is now diversifying its sources of imported coal and has recently contracted to import small quantities from the Soviet Union, Australia, and Canada, and Japanese steel interests have been studying the possibility of developing a supply of coking coal in Alaska.

That Japan will need more coking coal in the future there is no doubt, and the United States will probably continue to be a major supplier. Japan's steel interests are, however, looking for new sources that will, hopefully, be less costly. There is, however, no doubt but that there will be increasing pressure to try to obtain coal from the much closer resources on mainland China. Japan's steelmakers have imported some coal from mainland China over the past ten years but were not completely satisfied with the results, as the coal was much higher in ash content and sulfur than coal from the United States. The coal industry in China may now have improved the quality of its output, and the Japanese would at least like to find out for themselves whether China's coal would actually be less expensive to import than that from the United States.

Iron Materials

Japan is extremely poor in iron ore resources, and the annual production now amounts to only about one million tons per year, with

[2] *Asahi Evening News*, Tokyo, March 14, 1960, p. 5.

TABLE I*

SUPPLY OF IRON ORE AND OTHERS

Unit: 1,000 Metric Tons

	Domestic Ore	Pyrite Cinder	Iron Sand	Imported Ore
1955	975	1,255	551	5,062
1956	1,051	1,263	789	7,496
1957	1,150	1,511	1,081	8,972
1958	1,140	1,426	941	7,274
1959	1,131	1,633	1,311	10,276

Estimated supply in 1967:
total domestic iron materials, 8,410, imported iron ores, 23,770 to 27,550

* Based on figures from the Japan Iron and Steel Federation and Ministry of International Trade and Industry, as reported in the *Asahi Evening News*, 1, 913 (Tokyo, March 14, 1960) p. 5.

most of the present production coming from Kamaishi, Gumma, and the Kutchan area in Hokkaido. There is obviously little hope that production can be increased much beyond this figure, as the total reserves amount to only about 70 million tons,[3] which is approximately one year's production in the United States.

Japan now imports better than 90 per cent of its iron ore requirements, and virtually all additional needs for iron ore will have to be supplied from outside sources.

There have, however, been two rather important postwar developments in the iron supply situation: the increased use of pyrites and the commercial development of iron sands as a major domestic source of iron. Each of these two resources is now actually more important in total tonnage than domestic iron ore (Tab. I). Through increased use of the pyrite cinder and the iron sands, Japan has managed to supply about 40 per cent of its iron requirements from domestic sources. New techniques of concentration and beneficiation, including better methods of removing titanium and copper, have made the use of these resources possible on a large scale. Pyrites are scattered throughout Japan with the largest mine at Yanahara in Okayama Prefecture. Most of the iron sand is produced in the north in the pre-

[3] Ministry of International Trade and Industry, *Japanese Mining Industry,* Tokyo, 1960, p. 26.

TABLE II

IMPORTS OF IRON ORE BY SOURCE*

1935		1959	
Malaya	40.0%	Malaya	36.1%
China	34.6	India	18.1
Philippines	8.0	Goa	13.4
Korea	6.6	Philippines	12.5
Others	10.8	Canada	6.5
		U.S.A.	5.2
		Others	8.2
	100.0		100.0

* Based on figures from the Japan Iron and Steel Federation and the Ministry of International Trade and Industry, as reported in the *Asahi Evening News*, 1, 913 (Tokyo, March 14, 1960), p 5.

fectures of Aomori, Iwate, and Hokkaido. Through expanded use of the new sources of iron along with domestic ore, Japanese steelmakers hope to have a domestic supply of ore equal to about one-fourth of the requirements, at least into the middle of the 1960's. However, the usable reserves of pyrite and iron sands are limited and amount to about seven and one-half million tons of iron sands and thirty million tons of pyrite.[4]

Present imports of iron ore are running at about ten million tons a year and the amount will more than double during the 1960's (Tab. I). Japan is now importing iron ore from many sources, with the bulk of it coming from South and Southeast Asia. The other suppliers include such distant places as Brazil, Chile, and Peru. The biggest change from the prewar pattern is the disappearance of China as an important supplier (Tab. II). Like the United States, Japan is now looking for new iron resources in almost every part of the world, including Africa. Survey teams have recently been sent to Australia and Rhodesia.

The need for new sources of iron has led the Japanese to invest capital in the Philippines, Malaya, Goa, and now, most important of all, India, which is expected to be Japan's major source in the 1960's. Japan is helping India develop new mines in the northeast and cen-

[4] *Ibid.*, p. 26.

tral sections of the country, and is aiding with the development of transportation facilities and port and harbor improvement. With the development of these new mines, Japan is scheduled to be importing several million more tons of iron each year from India.

Japan is also a major importer of scrap iron, about four million tons annually, with the U. S. as the major supplier. Japan hopes to find new and closer sources of scrap, but this will not be easy to accomplish, given the stage of industrial development of its near neighbors. However, through use of more pig iron and the increasing use of oxygen converters, the steel industry hopes to import less scrap in 1965 than in 1959.[5]

Petroleum

Japan is again one of the world's major markets for crude oil, and the country now has one of the largest refining industries. The 1959 production of petroleum in Japan was only 2,856,149 barrels, and the proven reserves stand at thirty-four million barrels.[6] The 1959 production amounted to less than 2 per cent of Japan's present estimated annual consumption of about 170 million barrels.[7] Petroleum is now Japan's largest import item, and the demand continues to grow at a very high rate.

The most important area of Japanese production is along the northwest coast of Honshu from Niigata to Akita. The area is now experiencing a small boom with the recent report of successful offshore exploration and the development of secondary recovery methods, as well as the extraordinary growth of a small but locally important natural gas industry which provides Japan with a welcome new source of raw materials for its chemical industries.

In the prewar period the United States was Japan's major source of petroleum. Today Japan imports about 85 per cent of its oil from the Middle East. One of the most interesting and significant developments in Japan's postwar quest for raw materials is the development of the Japanese concession in the offshore area of neutral territory between Kuwait and Saudi Arabia. The Japanese company, the Arabian Oil Co., Ltd., holds a forty-four and one-half year concession and

[5] *Asahi Evening News, op. cit.*, p. 5.

[6] *World Oil,* Gulf Publishing Co., August 15, 1960, p. 266.

[7] Carmical, J. H., "Japan Studying Oil Preference," *The New York Times,* December 18, 1960, pp. 1 and 16.

has started production in what now promises to be a very rich field, a continuation of the famous Burgan area. According to *World Oil*, the Arabian Oil Co. plans to produce better than sixty million barrels a year by 1965.[8] A more optimistic estimate suggests the possibility of an annual production of over seventy million barrels by 1962, and the possibility that this will be one of the world's major new oil fields.[9] The company is now constructing facilities for the gathering, storage, and shipment of oil, and is committed to the construction of a refinery some time in the future.

Along with the various projects to increase the amount of raw materials, the Japanese have also been working to lower the costs of the raw materials and to increase efficiency in their use. The steel industry has, for example, using high-grade materials, lowered its coke ratio to the point where it is one of the lowest in the world. This means that more pig iron is produced for each ton of coke consumed. The industry has also developed modern and efficient systems for the handling and processing of raw materials.

One of the most promising approaches to the cost problem is the development of large carriers to haul iron ore to Japanese steel plants. The industry hopes to lower transportation costs and to carry more material per trip with these ships. Several of these ore-carriers are now in use and others are being built. As part of the program to lower freight costs, port facilities are being improved and expanded to handle the larger ships and shipments of raw materials. Super tankers are now being used to transport petroleum to Japanese refineries and the steel industry is working on plans to import coal in special chartered coal carriers.

Conclusions

Japan is today experiencing a period of economic growth beyond anyone's wildest dreams. Figures for industrial production are almost impossible to keep up with at the present time. Where many Japanese used to talk about "poor little Japan," they now contemplate the national honor of surpassing the United Kingdom and West Germany in steel production. Social scientists are beginning to study the problems of leisure time and the "affluent society." This must indeed be a confusing and bewildering time for a people who have ex-

[8] *World Oil, op. cit.*, p. 211.

[9] Carmical, J. H., *op. cit.*, p. 16.

perienced a terrible war, postwar poverty, and endless talk about too many people, too little land, and too few resources.

It is probably impossible to predict, or even to guess, how long this present boom will continue. It shows no signs of a letup now. One thing is certain, however, the present prosperity cannot be maintained without an ever-increasing import of raw materials and expanding markets for Japanese exports.

Because of the rapid changes that are taking place in Japan's economy, it is difficult to predict the future sources of all raw materials. Certain general patterns have emerged, however. First, there will be a continuing dependence on imported raw materials and, unless there is a sudden depression or collapse of the economy, the demands will be much greater in the 1960's than in the 1950's. Second, the Japanese industry is reaching farther and farther afield to satisfy its growing needs. Japan has often been called the "workshop of Asia." It is still a workshop, but more correctly, perhaps, a "world workshop."

This search for material in the Americas, Asia, the Middle East, has its disadvantages—long and expensive transportation hauls, unfamiliar lands, and uncertainty of supply. But it has, at the same time, given Japan a new perspective on itself in relation to the rest of the world, growing self-confidence with its successes in international finance and development, new markets for Japanese exports, and perhaps most important, a flexibility and adaptability to constantly changing patterns of industry and trade.

BIBLIOGRAPHY

Asahi Evening News, Tokyo, March 14, 1960, pp. 5 and 6.

Carmical, J. H., "Japan Studying Oil Preference," *The New York Times,* Section 3, December 18, 1960, pp. 1 and 16.

Ministry of International Trade and Industry, *Japanese Mining Industry,* Tokyo, 1960, p. 26.

The Foreign Capital Research Society, *Japanese Industry, 1960,* c/o The Bank of Japan, Tokyo, 1960, p. 4.

World Oil, 15th Annual International Issue, August 15, 1960, Gulf Publishing Co., Houston, pp. 211 and 266.

Papers of the Michigan Academy of Science, Arts, and Letters
Vol. XLVII, 1962 (1961 Meeting)

TWENTIETH CENTURY SETTLEMENT OF MINDANAO

CANUTE VANDERMEER AND BERNARDO C. AGALOOS*

Pioneer agricultural settlement in the Orient takes three basic forms, each reflecting the inability or reluctance of subsistence farmers to leave the security offered by relatives and friends. The traditional and natural form of settlement beyond the frontier unfolds as a slow, but continuous, extension of dense settlement into a nearby empty area. A compulsory form occurs with the transfer of military personnel or prisoners to distant virgin lands for the purpose of initiating development. An induced form appears in recent times when the government of a country encourages voluntary migration by placing certain lures in distant empty areas to create an attractive frontier in which settlers may expect to improve their living standards with some degree of assuredness. Twentieth century settlement of Mindanao is largely induced—induced with a lure of the benefits derived from roads.

The island of Mindanao has long been a pioneer area in the Philippines, absorbing migrants from the more densely populated regions of the country since the early 1800's. During most of the 1800's, however, only a tiny trickle of people from Luzon, Cebu, and Bohol arrived on the island to occupy parts of the northern coast. Small settlements of Moros existed at the southern tip of the Zamboanga peninsula, along the lower Cotabato River, and in the vicinity of Lake Lanao. Elsewhere, a few small and scattered tribes of people derived a living from hunting, fishing, gathering, and fire-field farming. Large areas of forested but fertile land remained untouched. The rate of population growth for Mindanao during the nineteenth century was probably no higher than that for the Philippines as a whole—about 1.5 per cent per year. A combination of malaria and banditry, the one

* Mr. Agaloos spent several years as a forester with the Philippine Bureau of Forestry on Mindanao and especially in the province of Agusan. Mr. VanderMeer collected a part of the material resulting in this paper while in the Philippines under a fellowship granted by the Ford Foundation. However, the statements and opinions contained herein are those of the writers and are not necessarily those of the Ford Foundation.

inflicted by anopheles mosquitoes and the other perpetrated by Moros, prescribed an abnormally low population growth rate for a large undeveloped area in a country containing regions of great population pressure. During the past twelve years on the other hand, the population of Mindanao has increased at an average rate of almost 6 per cent per year. This paper attempts to demonstrate first that the rapid growth of population in Mindanao during the twentieth century is related closely to road development, and second that the extension of settlement away from the coastline related closely to the extension of roads.

The recent population history of Mindanao may be conveniently divided into four periods each beginning and ending with one of five population censuses. Within these time intervals it is possible to discern three stages in the development of the road net on the island. Therefore, the inter-census periods serve to frame the record both of road development and of population growth.[1] The effect of roads upon the distribution of people on Mindanao appears evident on maps depicting the location of these elements in the provinces of Cotabato and Agusan, both provinces with rich agricultural potential.

Roads and Population Growth

The first census period, from 1903 to 1919, was one of normal population growth and insignificant road construction in Mindanao. Most people lived along the northern coast, utilizing water transportation for access to markets. A sparsely populated interior remained the domain of Lake Lanao Moros and "wild" tribes. The few roads built during these years joined already existing towns with near neighboring villages so that no new land was opened as a result of road con-

[1] The rates of population growth presented herein were computed from population data provided by: Commonwealth of the Philippines, Commission of the Census, *Census of the Philippines, 1939,* Vol. I, Manila: Bureau of Printing, 1940. Republic of the Philippines, Bureau of Census and Statistics, *1948 Census of the Philippines,* Manila: Bureau of Printing, 1951. City of Cebu, Office of Public Information, *Cebu City Hall Reporter,* Vol. III, No. 11, 1960. Letter from Maguio B. Malang, In charge, Agricultural Economics Division, Department of Agriculture and Natural Resources, Cotabato City, February 27, 1961. Letter from P. O. Quintana, Forester, Bureau of Forestry, Department of Agriculture and Natural Resources, Butuan City, Agusan, February 22, 1961.

struction.[2] Slight improvement in the effectiveness of government control over the behavior of Moros and official encouragement of migration caused the annual rate of population growth for Mindanao to be slightly higher than that for the Philippines. Most of the migrants settled along the northern coast where malaria and Moro problems were less prevalent. The provinces of Agusan with malaria and Cotabato with both malaria and Moros had lower rates of population growth than did Mindanao as a whole (Fig. 1).

During the second period, between 1919 and 1939, the government emphasized construction of trunk roads to link the major cities on Mindanao (Fig. 2). Over half the mileage of these new roads followed the path of least resistance along narrow and moderately populated coastal plains, primarily in the north. Late in the period, however, two trunk roads cut across wide expanses of the interior in Bukidnon and Cotabato to connect the cities of Cagayan de Oro, Malaybalay, Cotabato, and Davao.[3] Both the coastal and interior trunk roads allowed people in their vicinities access to larger markets. The roads also facilitated the extension of government services such as police protection and malaria control into areas which previously obtained these services only occasionally. As a result, the average annual rate of population increase for Mindanao rose significantly above that for the nation as a whole. Agusan province obtained a road extending along its entire northern coast and at the same time experienced a population growth rate slightly higher than that for Mindanao. The rate for Cotabato province, however, remained comparatively low because the trunk road crossing the province was constructed late in the period and because the province maintained a reputation for Moro lawlessness (Fig. 1).

[2] Philippine Islands, Census Office, *Census of the Philippine Islands, 1918*, Vol. I, Manila: Bureau of Printing, 1920. Wester, P. J., *Mindanao and the Sulu Archipelago: The Natural Resources and Opportunities for Development*, Bulletin No. 38, Department of Agriculture and Natural Resources, Manila: Bureau of Printing, 1922.

[3] Philippine Islands, Bureau of Public Works, *Official Road Map of the Philippines*, Manila: Bureau of Public Works, no date (1939). Hargis, O. D., and others, *Report of the Mindanao Exploration Commission*, Refugee Economic Corporation, New York, 1939. Commonwealth of the Philippines, Commission of the Census, *Census of the Philippines, 1939*, Vol. 4, Manila: Bureau of Printing, 1940.

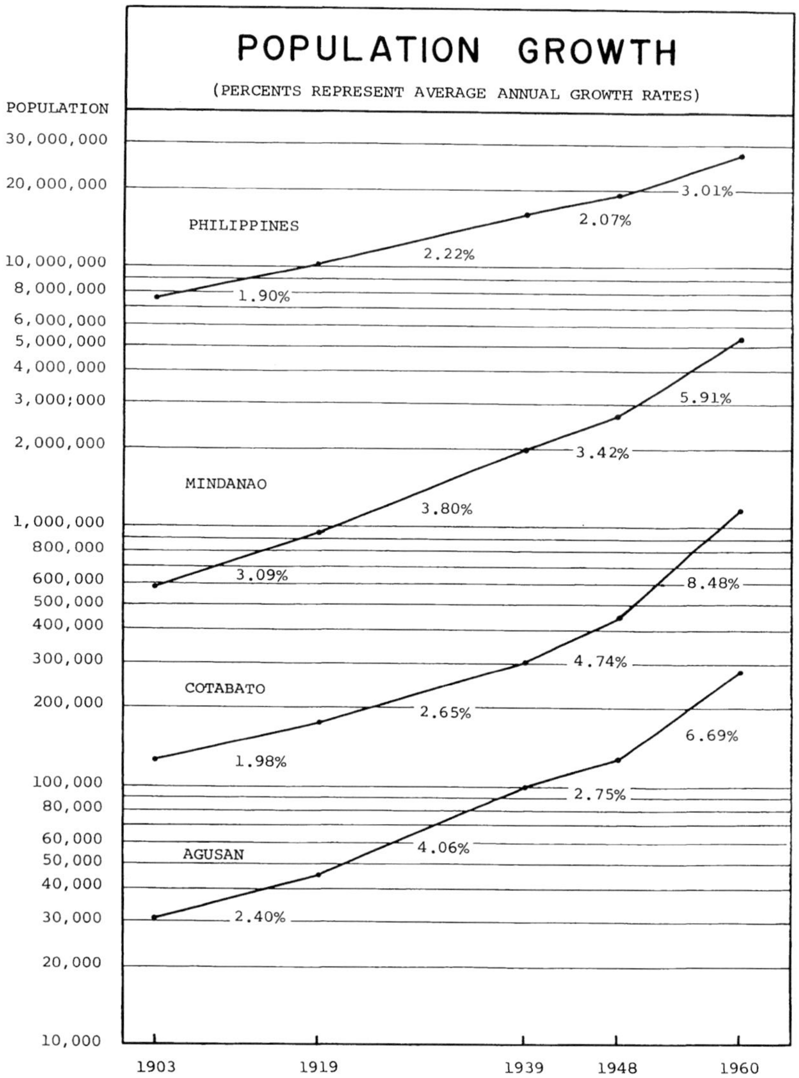

FIG. 1

The period from 1939 to 1948 was marked by a general decrease in the population growth rates for the Philippines and Mindanao, a result of economic and political disruption during World War II. On Mindanao the only major road construction occurred in the prov-

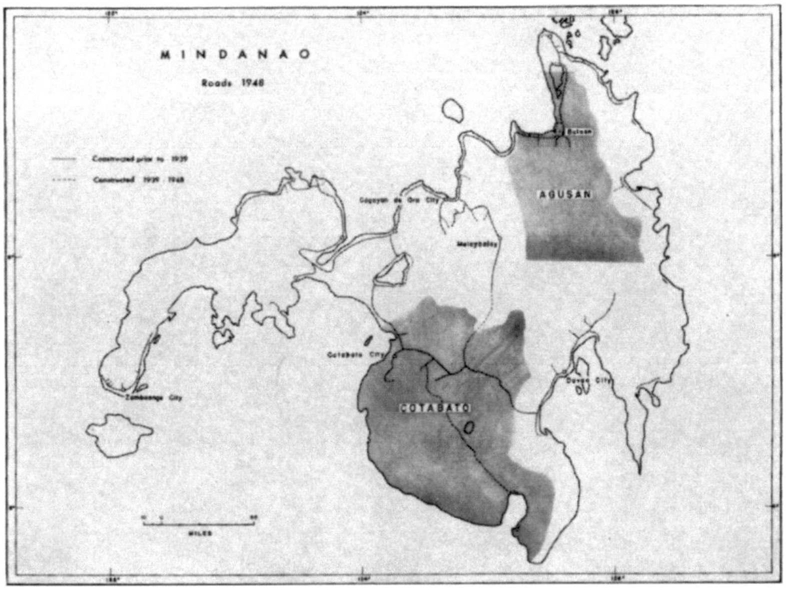
MINDANAO
Roads 1948
Cagayan de Oro City
Malaybalay
AGUSAN
Cotabato City
Zamboanga City
COTABATO
Davao City
MILES

FIG. 2

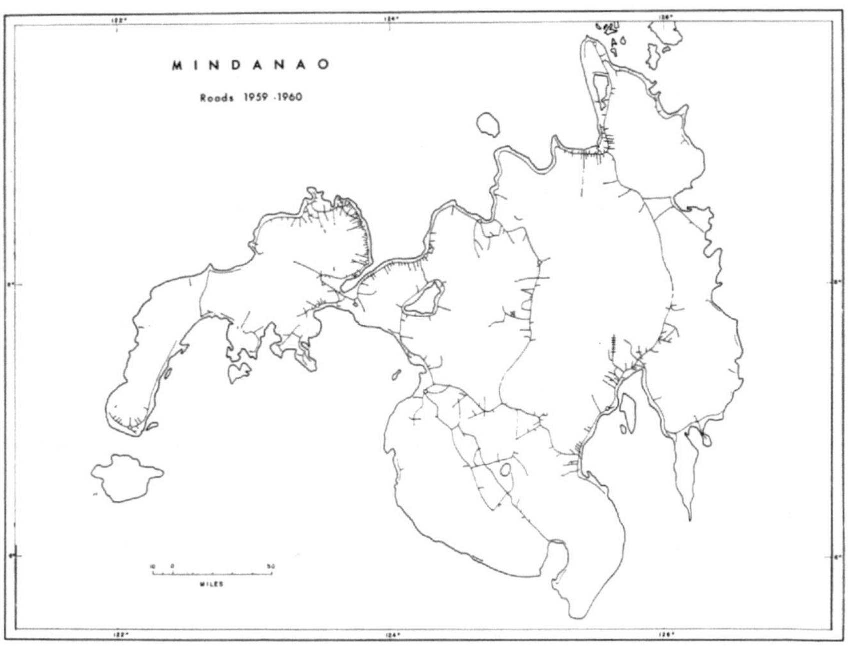
MINDANAO
Roads 1959–1960
MILES

FIG. 3

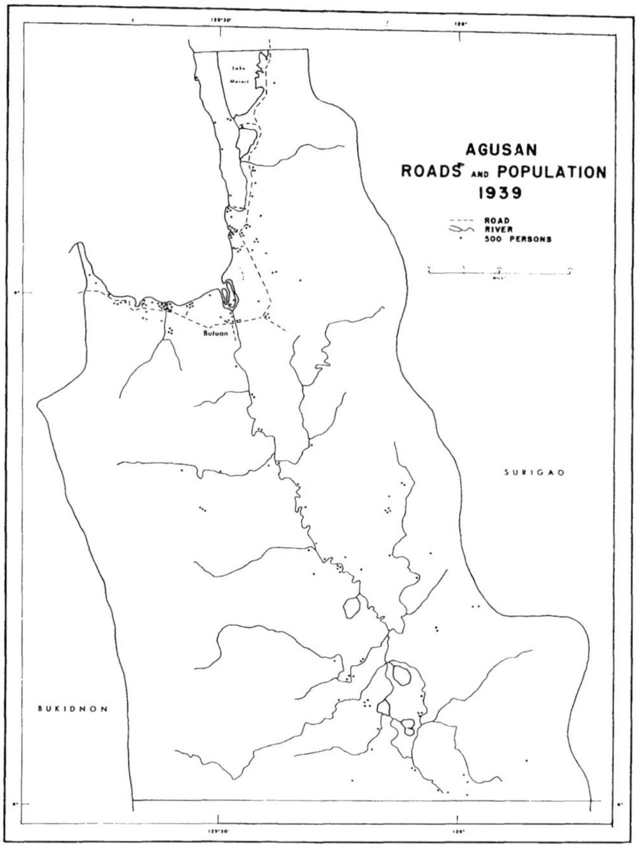

FIG. 4

ince of Cotabato, which in turn experienced rapid population growth.[4] By contrast, Agusan, with few new road miles, experienced a lowering of its growth rate (Figs. 1 and 2).

During the twelve years of the most recent period, from 1948 to 1960, the population of Mindanao doubled with an annual growth rate of almost 6 per cent, whereas the population of the Philippines grew at a rate of 3 per cent per year. The rapid population increase on Mindanao may be attributed primarily but not solely to the con-

[4] U.S. Air Force Aeronautical Chart and Information Center, *World Aeronautical Charts* covering Mindanao, St. Louis: U.S. Air Force, 1948 to 1953.

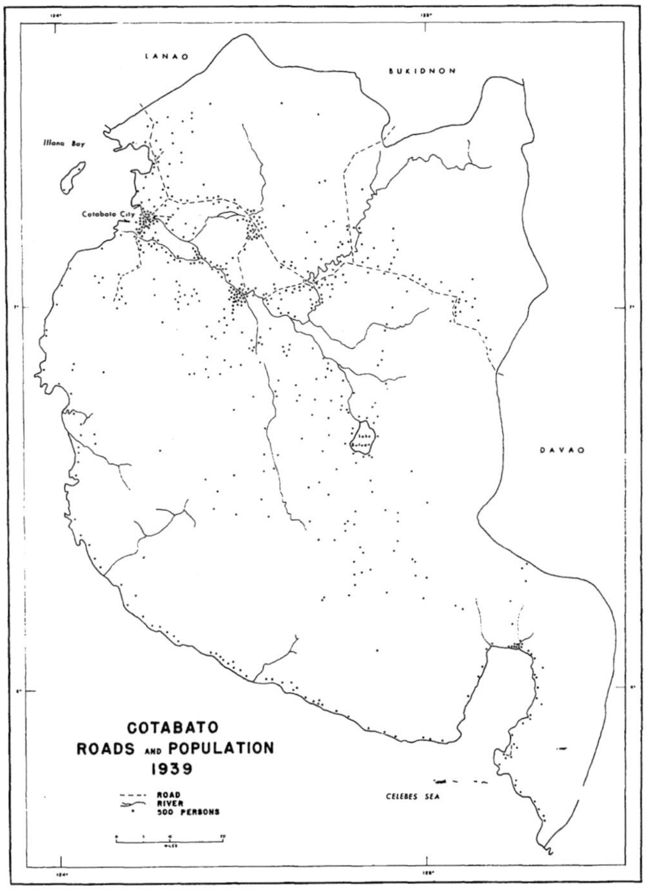

FIG. 5

struction of roads. This period saw emphasis upon the construction of short feeder roads, most of which extended some five to ten miles from existing trunk roads into sparsely populated areas.[5] In addition, a new trunk road was laboriously pushed through the almost unpopulated and extremely fertile but forested Agusan valley to link the cities of Butuan and Davao. Likewise, a second trunk road ap-

[5] Republic of the Philippines, Bureau of Public Highways, *Road Maps* of the individual provinces on Mindanao, Manila: Bureau of Public Highways, 1958 to 1959.

peared in the sparsely populated western part of the rich Cotabato valley (Fig. 3). The construction of these trunk and feeder roads was accompanied in Cotabato by an 8½ per cent average increase of population and in Agusan by a 6.7 per cent annual growth (Fig. 1).

A number of other factors also contributed to the rapid population increase on Mindanao during this last inter-census period. The availability at low cost of war surplus trucks, jeeps, and ships facilitated the movement of food from areas of surplus production such as Mindanao to areas of deficient production such as Cebu. Law enforcement was improved, especially among the Moros in Cotabato.

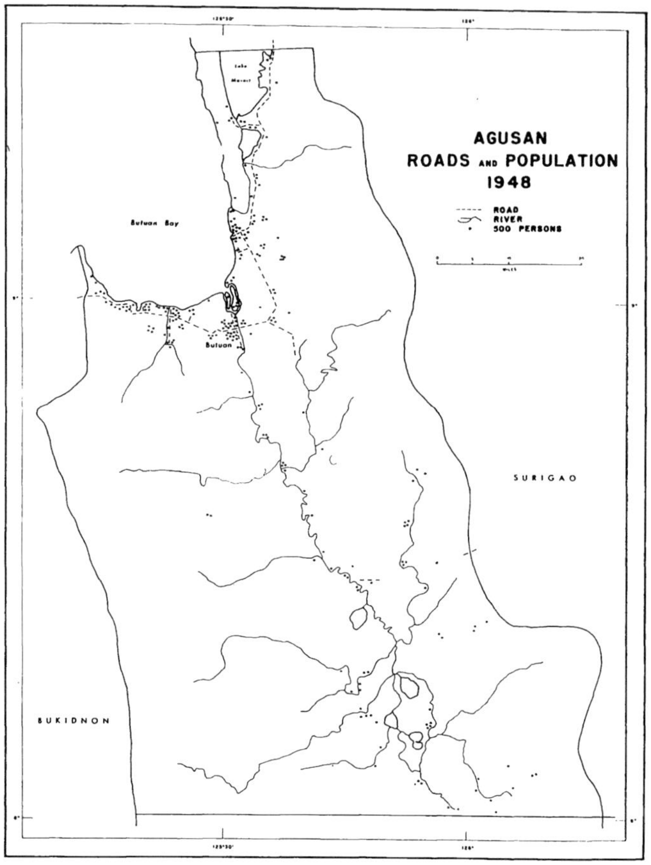

FIG. 6

Malarial control activities spread into hitherto inaccessible interior areas. Finally, government propaganda encouraging migration to and settlement in Mindanao was intensified under the late President Magsaysay. The contributions of each of these four factors were dependent in great part upon the presence of roads, for without roads the trucks and jeeps could not have moved large quantities of surplus corn and rice; nor would the roving police patrols have been so successful in discouraging banditry; nor could the modern techniques of malaria control have been extended into remote villages; nor would

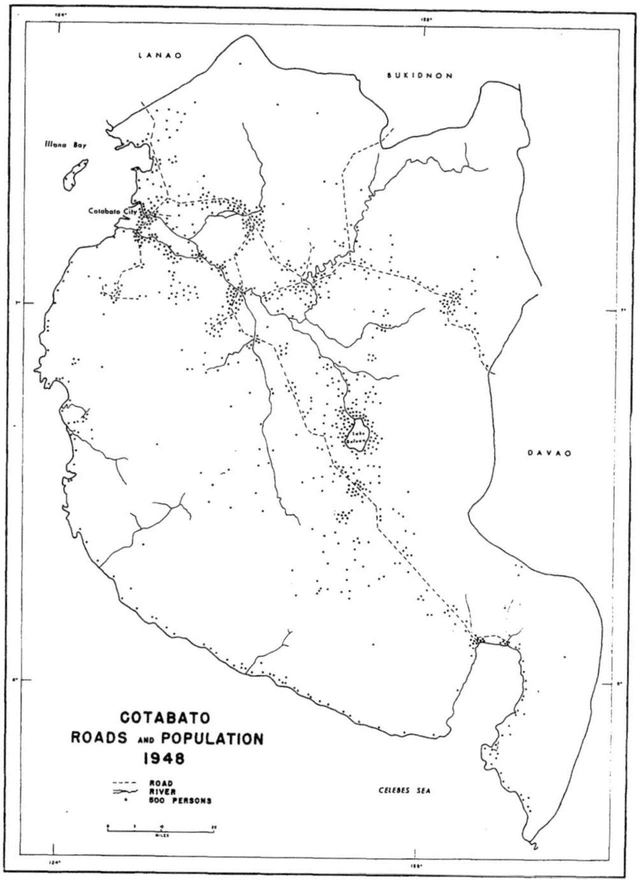

Fig. 7

the propaganda encouraging migration have been so alluring. Less banditry, less malaria, and greater potential for participation in commercial activities offering renumeration above that required for the acquisition of life's necessities; these were the inducements and each was made possible by roads.

Roads and Population Distribution

Maps depicting the distribution of population and the location of roads in Cotabato and Agusan provinces in 1939, 1948, and 1960 illustrate the effect of the road construction upon the extension of set-

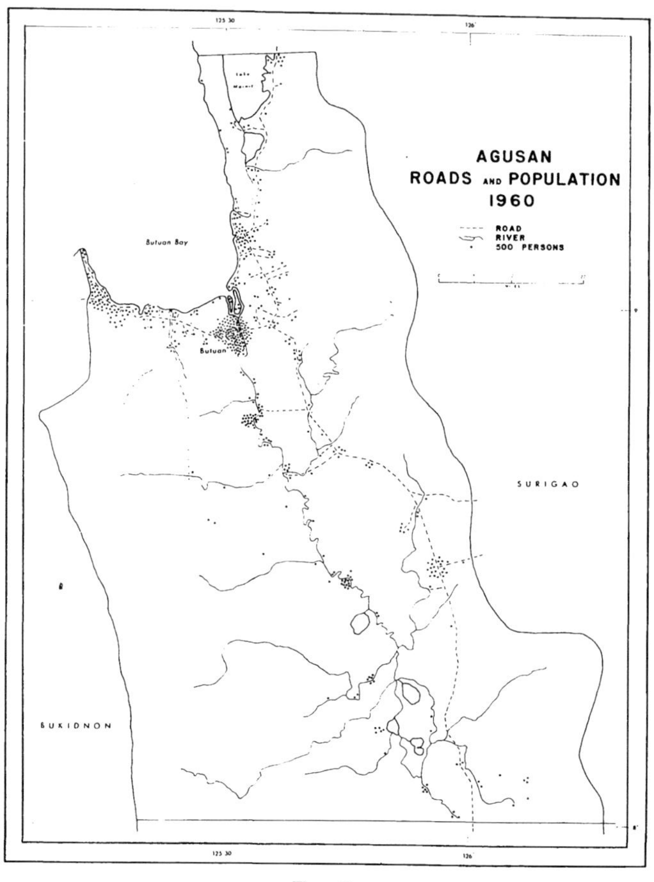

Fig. 8

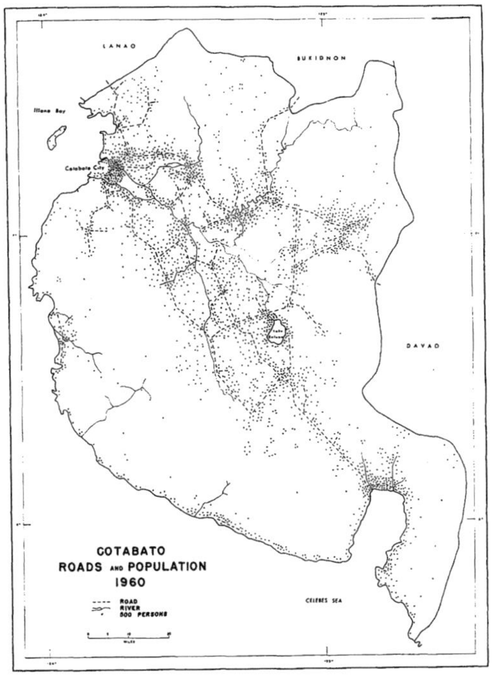

FIG. 9

tlement away from the coastline in Mindanao.[6] In 1903 most people in the two provinces lived near the coast, though some Moro settlements in Cotabato had been established along the lower Cotabato River. By 1919, roads connected each capital city with the nearest important neighboring village some fifteen or twenty miles away. Since both these roads followed the coast, the general patterns of population distribution did not change.

By 1939 however, concentrations of people appeared in the interior of Cotabato, not only along the lower portions of the river but also along the new trunk road which crossed the northern section of the province. About 80 per cent of all the Cotabatans lived within ten miles of a road or coastline. Most of the remainder were scattered thinly through the fertile south-central lowland. Even the rich alluvial soils surrounding Lake Buluan supported a sparse population. In Agusan, the construction of a new road along the already popu-

[6] Population data obtained from the sources indicated in footnote 1.

lated northern coast was accompanied by little or no change in the overall population distribution pattern (Figs. 4 and 5).

In Cotabato the pattern of population distribution changed again by 1948 following the construction of the first trunk road into the Cotabato River valley. A string of pioneer settlements developed in especially favored locations near the road. The alluvial soils of the Lake Buluan area, now within ten miles of a road, supported a large concentration of people. In all, more than 90 per cent of the population of Cotabato lived within ten miles of a road. By way of contrast, Agusan, with no major road construction, again saw no change in the distribution pattern of its population (Figs. 6 and 7).

Both Cotabato and Agusan obtained important new roads during the period from 1948 to 1960 and both likewise experienced important changes in their population distribution patterns. In Agusan, large clusters of people appeared amid the forests of the interior to participate in new or enlarged logging activities and to cultivate the cut-over areas. The lumber milling city of Butuan more than trebled in size. In Cotabato, a second trunk road into the valley attracted many pioneer settlements to its vicinity. At the same time settlements along the older roads expanded. Very few of the 730,000 new people in Cotabato chose to live far from a road. As a result, in 1960 over 95 per cent of the total population resided within ten miles of a road or coastline (Figs. 8 and 9).

The fact that the recent rapid settlement of Mindanao has had to be induced indicates that the Filipino migrant is unwilling to relinquish the benefits of living near relatives and friends in exchange for an economic position just a little better than the one he already enjoys. Despite increasing population pressures at home, he does not migrate on the basis of vague hopes. Before he decides to move he must receive assurances that significant economic benefits will accrue as a result of the move. He must be enticed to move. In Mindanao today, the amenities derived from roads provide the lure. Therefore, road construction has generally preceded rapid agricultural settlement in the frontier, even into fertile areas such as the Cotabato and Agusan river valleys.

PART III

PHILOSOPHY, LITERATURE, AND ART

Papers of the Michigan Academy of Science, Arts, and Letters
Vol. XLVII, 1962 (1961 Meeting)

LITERATURE

POPULISTS, PROGRESSIVES, AND LITERATURE: A RECORD OF FAILURE

RUSSELL B. NYE

Michigan State University

There seems to be general agreement that the period 1880 to 1914 was one of profound change in American life. If we may accept Henry Steele Commager's estimate, these decades form an identifiable "watershed" in American history, separating the old from the new, the past from the present, the remote from the contemporary. Or, if we agree with Henry May, they mark a shift from innocence to complexity, a transition from nineteenth-century adolescence to twentieth-century cultural maturity.[1]

Whatever meaning may be ascribed to these years by historians, they have found (and still find) them rich with suggestiveness, significant for study and speculation. Literary historians, however, have had no such good fortune with them. In literary history, interpretations of the period are likely to be puzzled and uncertain; there is a tendency to move quickly from Whitman to Crane to Hemingway (with a side glance at Theodore Dreiser), passing over the turn-of-the-century decades as the "local color" or "rise of realism" years. The fact is, that these thirty-odd years pose somewhat of a problem to the literary historian.

Why was it that so much happened in American life during these decades, and so little of it got into literature? Why, in the midst of the tremendous conflicts and tempestuous passions that rocked this generation, do its fiction, poetry, drama, and allied literary arts reveal so little of the impact? In other words, where is the *Uncle Tom's Cabin* of populism, the *Moby Dick* of monopoly, the drama of progressivism, or the saga of the Cross of Gold crusade? The record of the literature of the eighties and nineties, insofar as it might have re-

[1] H. S. Commager, *The American Mind* (New York, 1950); Henry F. May, *The End of American Innocence* (New York, 1959).

flected and defined the quality of life around it, is one of failure. It is significant (and ironic as well) that of the three great minds which dominated literature in the period—Henry James, Henry Adams, and Mark Twain—none of them liked what they saw in it, and none of them dealt directly with it. No writer appeared to answer the critic H. H. Boyesen's wish that someone might undertake "the serious task of studying and chronicling our social conditions . . . to stimulate noble and healthful thought on all matters of human concern."[2]

A number of critics, like Boyesen, believed that a new kind of America was emerging which required the creation of a new literature to construe it. William Ordway Partridge hopefully predicted in 1897 the appearance of a "new literature and art and philosophy, all expressing the new spirit pervading American life." Joseph Dana Miller thought that a distinctively American, newly expressive literature would surely spring out of the nineties. Hamlin Garland, observing the vitality of his times, begged for writers, critics, and artists who might represent American civilization in poetry, painting, and fiction as clearly as the new parties in politics represented the new spirit in political life.[3]

These men had good reason for optimism. Certainly few periods in American history seemed to offer more ideas, trends, and events to excite the imagination and to stimulate the literary mind. There was the drama of lumber, the epic of the railroads, the saga of the trusts, the tale of the boss, the machine, and the robber baron. There was even a small, satisfying, victorious war. There were reformers, villains and dragons and crusaders, Grangers and Populists, the great crusade of 1896, the Haymarket affair, Coxey's march, depression and panic, strikes and lockouts, millionaires and hoboes, fear, violence, bitterness, and conflict in farm and city alike. It was the day of Harriman and Hill, Rockefeller and Carnegie, Ingersoll and Bryan, Eugene Debs and Thomas Edison. It was, as Garland called it, "a day of great themes crying to be written."

Nor was there any lack of talent to write them. Henry James, Mark Twain, Stephen Crane, William Dean Howells, Frank Norris,

[2] *Literary and Social Silhouettes* (New York, 1894), 76, 78.

[3] W. O. Partridge, "Art for the Millions," *New Time*, I (August, 1897), 100–102; Joseph D. Miller, "The New Literature," *Arena*, XXVII (March, 1902), 254; Hamlin Garland, "The West in Literature," *Arena*, VI (November, 1892), 669–676.

and Theodore Dreiser were there. Good writers scribbled local color stories by the score—Rose Cooke, Alice French, Sarah Orne Jewett, Mary Murfree, Mary Wilkins Freeman, to name a few of the women. There were clever, talented men such as Frank Stockton, H. C. Bunner, O. Henry, and Richard Harding Davis; expert, polished novelists such as S. Weir Mitchell, Mary Johnston, Henry Blake Fuller, F. Marion Crawford, and the stronger meat of Jack London, Harold Frederic, and Joseph Kirkland. And what did people read? Helen Hunt's *Ramona*, Lew Wallace' *Prince of India*, Frances Hodgson Burnett's *Little Lord Fauntleroy*, F. Marion Crawford's *Don Orsino*, Kate Douglas Wiggin's *Rebecca of Sunnybrook Farm*, Anthony Hope's *Prisoner of Zenda*, Jerome K. Jerome's *Three Men in a Boat*, Charles Major's *When Knighthood Was in Flower*, and so on. In poetry, the most popular poems were James Whitcomb Riley's "Orphan Annie" and Will Carleton's "Over the Hill to the Poorhouse."[4] The popular literary interests of the times, one feels after surveying the lists of best-sellers, seemed to have been imported from another planet. The fact is that with all there was to write about between 1880 and 1914, and all there were capable of writing about it, nothing really happened in literature. The great winds of upheaval and change passed the writers of these decades by; the story of the passing of the old America and the arrival of the new never found its way into the books.

That is, it never found its way into the better ones. There were dozens of third- and fourth-rate novelists who tried to put down on paper what was happening around them, without much success. There are far too many of these to list here, but the C. H. Kerr Company of Chicago, for example, produced a steady stream of typical novels on contemporary issues, among them such items as Margaret Holmes Bates, *Shylock's Daughter* ("a charming story of love and politics dedicated to the People's Party of America"); Edgar Maurice Smith, *Daughter of Humanity* ("a realistic story of love and girls in

[4] These were all best-sellers between 1890 and 1914. Others included books by Gene Stratton Porter, Robert Service, Harold Bell Wright, Zane Grey, Edgar Rice Burroughs' *Tarzan* series, etc. See Frank Luther Mott, *Golden Multitudes* (New York, 1947) appendix. Crawford, the reigning novelist of the period, summed up the taste of the times by decreeing that literature must "deal chiefly with love," that its first object must be "to amuse and interest the reader," and that it must "be of the human heart and truly human. . . ."

a great department store under our present competitive system"); James B. Goode, *The Modern Banker* ("a story of his rapid rise to power and of his dangerous designs"); or Rosa Hudspeth, *Juggernaut of the Moderns* ("an intense, dramatic story of a victim of modern laws in the city").[5] An excellent example of the kind of novel published by the willing but inept was "Anon Moore's" *John Harvey: A Tale of the Twentieth Century,* which the Kerr Company published in 1897. Harvey, a socially conscious millionaire, finds in his Western gold mines a new variety of metal, harder than steel and much lighter. Knowing that the monopolists will attempt to control this metal for their own profit, rather than the public good, Harvey uses the profits from his mines to set up a model commonwealth in Colorado, based on principles of social equality and economic democracy, called "The Nationality." The success of his experiment naturally earns the enmity of "the plutocracy" which rules the United States and the world. Subsequently he settles down to a long and exhausting international war against big business (incidentally blowing up the British Isles with a new explosive) and eventually winning it. He establishes a new government of peace and plenty as an example to the world, which, one feels, will soon follow in the steps of "The Nationality." The men who wrote such novels had much to say about the problems of contemporary life, and a burning sincerity of intent, yet of them all only Hamlin Garland seemed to be able to combine art and politics with some success.

Though Garland later lost himself in a literary blind alley which he called "veritism," in the nineties he observed at firsthand the battle for power going on between the old order and the new and recognized it for what it was—the last stand of the old agrarianism against the new industrialism. The Midwest from which he came was neither the romantic novelist's frontier of adventure nor the local colorist's region of quaint customs. His was the Midwest of Grangerism and Populism, of ten-cent corn and 10 per cent mortgages, of grasshoppers and freight rates, grain prices and drought, railroad kings and bought legislators, foreclosed farms and machine politics. Born in Wisconsin of a farming family, familiar with prairie poverty through experience on Iowa and Dakota farms, Garland knew rural life from the seamy side. When he searched for an explanation of the misery and poverty he saw around him in American life, he found it in the theories of

[5] Reviews in *New Time,* I (July, 1897), 2.

Henry George, the single-tax economist.[6] When Joseph Kirkland, the author of *Zury*, a grim novel of country life, urged him to "write the truth" about what he knew, Garland began in 1887 to write sketches of Midwest farm life, trying first of all to draw an accurate picture of the sweat, debt, and bleakness of the farmer's lot, and secondly to probe deeper into the causes of the farmer's discontent.

Garland's most significant early work appeared in the stories later collected in *Main Traveled Roads* (1891) and *Prairie Folks* (1892). Stories such as "Up the Coolly," "Under the Lion's Paw," "Sim Burns' Wife," "A Branch Road," "A Day's Pleasure," or "Among the Corn Rows" were not only highly skilled literary pieces, but perceptive and careful analyses of the political and economic and social forces which were revolutionizing farm life. In a sense, Garland's tales summarized and refined, in good artistic terms, the great agrarian revolt of the late nineteenth century. "Under the Lion's Paw" is perhaps the best example of Garland's skill. An industrious farmer, Haskins, falls into the clutches of Butler, "a man who believed that land speculation was the surest way of getting rich." Haskins and his family toil for three years on one of Butler's farms, paying interest on their mortgage at 10 per cent, turning the rundown rented land into a prosperous farm by their backbreaking labor. When Haskins has finally saved enough to purchase the land, he finds that all he has done is simply to make the farm more valuable; the rent is doubled, the purchase price is doubled, and Butler has done nothing but collect interest. Haskins, bitter and broken, is "under the lion's paw" of the landlord. Not only is the story a dramatic translation into human terms of Henry George's theory of unearned rent, but it is a nearly perfect balance of message with literary skill, of thesis with art.[7]

Though Garland never again reached the perfection of *Main*

[6] Garland picked up a copy of George's book on a Dakota homestead, writing in his autobiographical *Son of the Middle Border* (New York, 1917), "Up to this time I had never read any book or essay in which our land system had been questioned. . . . I caught some glimpse of the radiant plenty of George's commonwealth. The trumpet call of the closing pages filled me with a desire to battle for the right." Later Garland heard George speak in Boston, joined the Georgist Anti-Poverty League, and lectured on the single tax himself.

[7] The effectiveness of Garland's story is emphasized by comparison, for example, with John Hudspeth's "Bricks Without Straw," *Arena*, VI (September, 1892), 105–136, on almost exactly the same theme.

Traveled Roads, he treated contemporary issues more completely in *Jason Edwards* (1891), which he adapted from a play, "Under the Wheel," written for *Arena* magazine in 1890. His next novel, *A Spoil of Office* (1892), shifted from economic to political issues, tracing in fictional form the course of events which led to the birth of the Populist party, and castigating the graft, cynicism, and corruption of legislature and Congress. The novel was straight propaganda and lacked proportion and art, but his vignettes of current political life—Grange picnics and Farmer's Alliance meetings, debt-ridden farmers and weary farm wives, corrupt congressmen and grafting ward-heelers—provide keen insight into contemporary problems.[8] His next novel, *A Member of the Third House* (1892), was a study of lobbyists and pressure groups, less effective as fiction and without the bite of its predecessor. Gradually Garland, for a number of reasons, abandoned his economic-political novels for criticism, local color, and finally autobiography. He had the equipment necessary to both the reformer and the artist. At first he seemed to be the man best qualified to write the significant novels of the era of protest and change, the man most likely to become the chief artist of the Populist movement. Never quite able to fix his purpose, never quite certain of how to use to the full the talent he undeniably possessed, Garland never again approached the power and art of "Under the Lion's Paw."

The hottest literary controversy of the nineties was stirred up not by a novel but by Edwin Markham's poem, "The Man with the Hoe." Markham, an obscure California poet, wrote it after seeing Millet's picture, "The Gleaners," to symbolize "the toiler, brutalized through long ages of industrial repression." First published by the *San Francisco Examiner* in 1899, the poem was picked up by the Hearst chain and given national distribution. Its theme—the degradation of the common man—and its implicit prophecy of revolution,

> O masters, lords, and rulers in all lands,
> How will the future reckon with the Man?
> How answer his brute question in that hour
> When whirlwinds of rebellion shake all shores?

[8] The difficulty encountered in finding an outlet for controversial writing is illustrated by the exchange in 1893 between Garland and Francis Browne, editor of *The Dial.* In answer to Garland's charge that *The Dial* "couldn't see anything this side of the seventeenth century," Browne replied that "the unseemly spirit of radicalism" was "bad enough in politics; we don't want it in literature." (Mss. correspondence, Newberry Library Collections, Chicago.)

expressed the anger and resentment of the farmer and worker. Hailed by some as an indictment of the privileged and by others as a subversive piece of demagoguery, Markham's short poem became an issue of violent debate almost overnight. If revolution ever came, the New York *Herald* editorialized ominously, "Markham would be its prophet and priest." The Chicago *Evening Post* pronounced it "a dangerous poem," and Collis P. Huntington, the railroad tycoon, offered a prize for the best poetic reply to it. The Reverend Dr. Holland of St. Louis preached nine successive sermons on it, and Bryan made political capital of it. What he meant by the poem, Markham said, was "to let the Hoe-man keep his hoe, but do not grind him down with trusts and combines and infamous monopolies . . ."[9] It was by no means great poetry, but it did give tongue to the troubled times and left a deeper imprint on them than any of the numerous propaganda novels.

The great issues of agrarian unrest and the Populist revolt found no able, artistic chronicler and produced little worth-while literature beyond Garland's scattered stories and Markham's poem. The general problem of democratic processes versus privilege, however, caught the attention of novelists too. Mark Twain's *Gilded Age*, done in 1873 with Charles Dudley Warner, though not a good novel, gave a name to the Grant era and thoroughly demolished its political morality. John W. De Forest's *Honest John Vane* (1875), a portrait of a designing politician, was the first adequate treatment in fiction of the soiled politics of the period. F. Marion Crawford's *An American Politician* (1884) and Paul Leicester Ford's *The Honorable Peter Stirling* (1894), based on Grover Cleveland's career, were less effective. Henry Adams' skeptical study, *Democracy* (1886), C. K. Lush's story, *The Federal Judge* (1897), and H. H. Boyesen's *Mammon of Unrighteousness* (1891) were others in a series of similar novels. For the defense John Hay wrote *The Breadwinners* (1883), a novelized brief for the existing order, and William Allen White, not yet in his progressive phase, attacked reformers and Populists in *The Real Issue* (1896). But the Populists, bosses, and boodlers came and de-

[9] For accounts of the controversy, see E. B. Payne, "The Hoe Man on Trial," *Arena*, XXII (July, 1899), 17–24, and Mark Sullivan, *Our Times* (New York, 1936), II: 236–238. An interesting earlier estimate of Markham is B. O. Flower's in *Arena*, XXVII (April, 1892), 391–411. Also of importance are two interviews with him, *The Public*, I (April 22, 1899), 11–12; and *Arena*, XXVII (April, 1892), 412–414.

parted, and Bryan ran thrice and lost, without producing, in the flood that poured from the presses, a novel or play or poem of major stature dealing with the politics of the period.

After 1900 the "insurgents" and "progressives" (although the names did not arrive until later) in the state legislatures and in Congress furnished the inspiration and plots—people against the boss, honest men against the machine, upright young attorney (or newspaperman, or businessman) against cigar-chewing politician with a love story (usually involving a Montague-Capulet theme) dragged in protestingly by the heels. Most of these novels resembled a fictionalized Lincoln Steffens article, with little to commend them beyond their unanimous indignation at political wickedness. Elliot Flower treated Chicago politics in *The Spoilsman* (1903), which though weak as a novel did a masterful job of analyzing machine government and the sources of civic corruption. J. W. Linn's *Second Generation,* I. K. Friedman's *The Radical,* Frederick Hill's *The Minority,* Joseph Altsheler's *Guthrie of the Times,* Charles Tenney Jackson's *The Midlanders,* Churchill Williams' *J. Devlin, Boss,* H. P. Miller's *The Man Higher Up,* William Sage's *By Right Divine,* Samuel Merwin's *Citadel of Unrest,* Holman Day's *The Ramrodders,* and Alfred Lewis' *The Boss* were all representative of the general pattern. Booth Tarkington's *Gentleman from Indiana* (1899) and his collection of stories, *In the Arena* (1905), drawn from his own experiences in the Indiana state legislature, were no more distinguished than the common run. The best political novel of the period was undoubtedly Brand Whitlock's *The Thirteenth District,* published in 1910. Whitlock, an admirer of Illinois' Governor Altgeld, and himself a four-term reform mayor of Toledo, wrote out of long acquaintance with and deep feelings about practical politics. The novel, a sensitively written story in the vein of nascent realism, dealt with the degradation of democracy in "Macochee" (Urbana), Ohio. A literary craftsman of real skill (as his autobiography, *Forty Years of It,* later proved), Whitlock gave his rather ordinary plot a vital spark of artistry that the other reform-political novels lacked—his was not a tract, nor an exposé, but a carefully constructed and moving story of deeper implications. Years later, after a distinguished career as statesman and diplomat, he returned to the political boss in *Big Matt* (1928), a much less impressive novel.

The story of business, the saga of the millionaire, was not so popular as the muckraking novel of politics, but it attracted more skillful literary practitioners. Ida Tarbell's study of Standard Oil provided a plot pattern for dozens of novels of this variety, much as Steffens' articles did for the political novelists. The question of just how rich a man should be, and how honest he could be while getting rich, became one of great interest to novelists. Furthermore, the creation of Rockefeller's empire of oil or Carnegie's of steel contained elements of the heroic and Homeric; there were dozens like Rockefeller whose careers deserve chroniclers. Samuel Merwin (editor of *Success Magazine*) collaborated with Henry Kitchell Webster on one of the better novels of business, *The Short Line War,* in 1899, a story of the railroads, and later with Webster again in *Calumet K,* dealing with the grain ring in the Northwest. Merwin alone wrote *The Whip Hand* and Webster *The Banker and the Bear,* both studies of the stock market. C. K. Lush's *The Aristocrat* (1901), Francis Lynde's *The Empire Builders* (1907), and Will Payne's *The Automatic Capitalist* (1909) were representative of dozens of similar novels of business.[10]

Better artists than these, however, chose to write of the clash of ethics and politics with a swiftly-expanding business civilization. William Dean Howells' study of *The Rise of Silas Lapham* (1885) was probably the best novel on this theme which the period produced, although the work of Winston Churchill, Robert Herrick, David Graham Phillips, and William Allen White maintained a consistently high level. Churchill's *Coniston* (1906), the story of the attempt of a railroad magnate to dominate a state, was a financial and artistic success. In *Mr. Crewe's Career* he put the millionaire into politics (1908), and in *A Far Country* (1915) made Theodore Roosevelt the implied hero of "progressive political advance." Churchill, however, preferred the historical novel and the problem novel, where

[10] There were apparently no more than a few novels in defense of the business man until Stewart Edward White, Peter B. Kyne, and others appeared after World War I. Gerald Lee's *Inspired Millionaire* (1908) praised the millionaire as "the bellows of great cities, the draught of creative forces and the energies of man," and saw his career as "a kind of colossal naked poetry," but concluded with a proposal for Christianized capitalism. David Parry, who curiously enough was president of the National Association of Manufacturers, explained in *The Scarlet Empire* (1906) that the greatest need of the age was for a kind of capitalistic socialism. The greatest and most influential bulwarks of contemporary business, however, were two men who had nothing to do with literature, Andrew Carnegie and Horatio Alger.

he found his greatest success. Robert Herrick, a professor of English at the University of Chicago, chose as his theme the conflict between wealth and morality, the same question which Andrew Carnegie solved so easily in his wellknown essay, "The Gospel of Wealth," in 1889. *The Common Lot* (1904), *The Memoirs of an American Citizen* (1909), *A Life for a Life* (1910), and others were variations on this theme. But expertly constructed and written as they were, Herrick's novels skimmed the surface of contemporary problems; he saw much of conflict and injustice and error in his society, but he was neither earnest nor angry enough to become the real novelist of insurgency. White's studies of business and labor problems, *A Certain Rich Man* (1909) and *In the Heart of a Fool* (1919) reconsidered the Howells theme of business ethics and, while competent novels, were not considered first class.

The most popular and successful of the reform novelists of the period 1880–1914 was David Graham Phillips of Indiana, who was also a prominent contributor to magazine muckraking, and who served on the New York *World* as an ace reporter. Of his twenty-odd novels, seven dealt directly with current political and economic issues. *The Master Rogue* (1903) was a study of monopoly (a preview of Dreiser's Frank Cowperwood), whereas *The Cost* (1904) described the tie-up of Wall Street with political chicanery. (Its hero was apparently Albert Beveridge, Phillips' college classmate and personal friend.) *The Plum Tree* (1905), which dealt with politics and high finance, was probably the best of his reform novels.[11] *The Deluge* (1905) was a study of trusts, and his series in *Cosmopolitan* in 1906, called "The Treason of the Senate," threw that body into an uproar and led Theodore Roosevelt to coin the phrase "muckraking" to describe what Phillips and others were doing. Phillips laughed off the attack and went on with *The Light-fingered Gentry*, a probe into insurance company rackets, and *The Second Generation*, a study of industrial management. Effective and skillful as Phillips' novels were, they were still minor novels of transitory interest which survive chiefly as footnotes in literary history. Phillips was a first-rate reporter, a good storyteller, and a second-rate artist—he told his story

[11] The title came from a telegram supposedly sent by political boss Matt Quay to his broker, "You buy 5000 Met and I'll shake the plum tree"—a phrase that soon became synonymous with political pickings.

well, but he remained a journalist and never developed into a novelist.[12]

When Frank Norris chose the epic of wheat as his subject, he lifted the novel of business to a higher level of art. The struggle of farmer against railroad, of speculator against producer, of man against the forces of Nature—these struck him as fit material for "a very long, a very serious, and perhaps a very terrible novel" about the West. He planned a trilogy of wheat—one novel about its production, a second on its distribution, and a third on its consumption. The first, *The Octopus* (1901), was based on an actual battle in 1878 between the wheat growers of the San Joaquin valley in California and the Southern Pacific railroad. His second novel, *The Pit* (1903), took the grain to the Chicago market, dominated by financier Curtis Jadwin (modeled on the career of "Crazy" Harper) who attempts to corner the market in wheat and who is swept away to financial destruction in the flood of golden grain. The third novel, *The Wolf*, presumably would have dealt with a famine caused by wheat harvest failure, but Norris' death left it unwritten. His truncated triology, dealing with the struggles of little people and big whose lives and fortunes were ruined by the savage struggles of business, was something more than reform propaganda. He hoped to catch, he said, the temper of life, the spirit of his age, in a "great song which should embrace in itself a whole epoch, a complete era, the voice of an entire people." Only his early death, perhaps, prevented him from realizing his aim, and from becoming the greatest novelist of the era.

As the years before the First World War drew to a close another novelist of equally powerful gifts found, he wrote, "an epic quality in the rise of individuals to merciless and remorseless power through the adaptation of their combative instincts to the peculiar conditions of the American scene." Theodore Dreiser, like Phillips a journalist as well as a novelist, saw meanings in the lives of men such as Rocke-

[12] For estimates of Phillips, see John C. Underwood, *Literature and Insurgency* (New York, 1914), which rates his work far too high; Isaac F. Marcosson, *David Graham Phillips and His Times* (New York, 1932); Kenneth S. Lynn, *The Dream of Success* (Boston, 1955) chapter IV; and Louis Filler, *Crusaders for American Liberalism* (Yellow Springs, Ohio, 1950). The best exposition of Phillips' ideas is his article, "The Menace of Plutocracy," *Arena*, XXXV (March, 1906), 258–264, or his book of essays, *The Reign of Gilt* (1905).

feller and Morgan. Strongly influenced by the thought of Herbert Spencer, Dreiser was convinced that there were but two kinds of people—strong and weak—and that in the careers of the American business tycoons lay the best proofs of the selective principles of Darwinian-Spencerian ethics. He chose as his model Charles T. Yerkes, the Philadelphia and Chicago traction king, and spent months uncovering Yerkes' past in the newspaper files.[13]

Frank Cowperwood, the central figure of Dreiser's novels of business, *The Financier* (1912) and *The Titan* (1914), is a character creation of major stature, and one of the most memorable in the history of the American novel. Brilliant, ruthless, unprincipled, powerful, Cowperwood "weaves his devious way," in Dreiser's phrase, "amid intricacies of law and public opinion and theories as to morals, duty, charity and the like," and the two novels in which he appears are, despite obvious flaws of style and conception, major novels, with Norris' the closest thing to great literature that the period produced. Yet Dreiser, like Norris, never quite lived up to expectations; obsessed more and more by the ironies of fate and the disillusion of naturalism he turned away from the novel of business to *An American Tragedy*, essays, social commentaries, and eventually to the religio-philosophical quest for the meaning of modern life. If Norris died too soon, Dreiser was born too late, for the quality of his mind was suited to the twenties, not to the crusading nineties and the progressive era.

The original question remains, why, with all this activity in literature (and especially in the novel) do the years 1880–1914 show so little of the lasting, the eminent, the superior? Twain's *Gilded Age* (not one of the great); Howells' *Lapham*; Norris' unfinished trilogy; Dreiser's dual study of the millionaire; Garland's stories; Markham's poem—surely this is scant fare for art, particularly when compared to earlier and later epochs. The writers were there, the components ready at hand, the protagonists and antagonists existed in contemporary life—even the rise of realism and naturalism provided literary and philosophical means and methods. Yet out of this crucial period in the development of American life there came no great literature, but only a spate of earnest, competent, second-rate writing.

[13] The best biographical and critical study is F. O. Matthiessen, *Theodore Dreiser* (New York, 1951).

"This silence concerning all the vital things of life," wrote H. H. Boyesen finally,[14]

> I believe to be the most serious defect of present American fiction. The strong forces which are visibly and invisibly at work in our society, fashioning our destinies as a nation, are to a great extent ignored....

There were practical, aesthetic, and historical reasons for this. In the first place, journalism, not literature, was the most vigorous and active intellectual force of the era. A group of powerful, imaginative editors and reporters dominated and shaped much of the thought of the period—editors such as E. W. Scripps, Joseph Pulitzer, Frank Munsey, S. S. McClure, Frank Leslie, W. R. Hearst, John Brisben Walker, B. O. Flower, and Norman Hapgood; and topflight reporters of the caliber of Ray Stannard Baker, Ida Tarbell, Lincoln Steffens, William Hard, Charles Edward Russell, Burton Hendrick, and James Howard Bridge. These brought to journalism a set of techniques and an energy and skill which the novelists, playwrights, and poets did not possess and with which they could not successfully compete. Stephen Crane and to a lesser extent Theodore Dreiser partially absorbed these techniques into literature, but the majority of writers who attempted to get America into their books tended to produce little more than fictionalized reporting, and that not half so well done as the professional journalists did it. The facts, plainly presented and documented, were dramatic enough when dug out and published by a Steffens or a Baker; why write an imaginary story about them? Ida Tarbell's story of Standard Oil, Phillips' "inside story" of the Senate, or Steffens' articles on District Attorney Folk's struggle with Boss Ed Butler of St. Louis were superior to anything even the better novelists produced. Norris, Phillips, and Dreiser, though what they wrote had power and meaning and not a little art, could not really surpass good reporters doing what they did best.

Second, the popularity of regionalism and local-color literature drained off much of the period's literary energy and talent. There were dozens of good writers so caught up in the task of writing about people and places that they neither noticed nor cared about contemporary social, political, economic, and theological problems.[15]

[14] H. H. Boyesen, *op. cit.*, 46.

[15] H. R. Warfel and G. H. Orians, *American Local Color Stories* (New York,

Garland, the nearest approach to a Populist artist of merit, sank deep into local-colorism and deeper into his own kind of it. There must have been many others like him, writers who might have turned their pens to the issues of the times but who found regionalism more interesting, and certainly easier artistically to grasp and control.

Third, whereas slavery had presented the writer with a clear, simple, national set of issues, which Harriet Beecher Stowe was able to put into focus in a clear and simple book, the problems of late nineteenth-century industrial civilization were far too complex and diffuse for any artist to duplicate her accomplishment. Men such as James and Twain (and to some extent Crane) did not attempt to do so, preferring to deal with materials and themes of wider human significance and deeper philosophical import. Norris, who might have done it, died young; Dreiser was not artist enough to succeed.[16]

It is also apparent that the events of the period had greater appeal to the practical than to the imaginative mind—freight rates, direct primaries, tax revisions, honest ballots, anti-trust laws, mortgage rates—and demanded specific answers. Like the conflicts of the Constitutional period, the struggles of the nineties and after lent themselves best to journalism, political expression, and debate, rather than to poetry and fiction. Unlike the slavery question, the economic and political issues of the era did not exist within a group which possessed a strong literary tradition as abolition had existed in New England. There was no Populist Boston, nor any Lowell, Emerson, Thoreau, Whittier, Holmes, or Harriet Stowe to take up the cause of the farmer, the insurgent, and the reformer as the poets of Cambridge once took up the cause of the slave. The temperature of the times stimulated economists, sociologists, and politicians much more than novelists and dramatists. Intellectual activity was likely to be channeled off into Good Government Clubs, Civic Reform Leagues, or party politics; imaginative, creative men such as White, Steffens, and Whitlock who might have been literary artists did not need to

1941), list thirty-three *major* local colorists active between 1880 and 1914. It would be easy to extend the list to a hundred or more.

[16] Upton Sinclair commented on this in "Our Bourgeois Literature," *Collier's,* XXIV (October 8, 1904), 22–24. "There is no call for complexity, and no use for anything hard to understand," he pointed out, since the popularity of regional and romantic literature drives it away. Authors will not consider "the savage and terrible aspects" of life, nor its real issues, but pander to "the positive and genial tastes" of society.

write novels so long as they could express themselves in organizations. Progressivism was more likely to produce a league against something than a poem.

When the period closed in the First World War years, perhaps a half-dozen novels had said about all that there was to say in literature about the great tides of conflict and change that had washed over these three crucial decades which saw the emergence of modern America. The fact is that the ideas, issues, conflicts, and conclusions of the period 1880–1914 expressed themselves in men and events, and not in art.

PAPERS OF THE MICHIGAN ACADEMY OF SCIENCE, ARTS, AND LETTERS
VOL. XLVII, 1962 (1961 MEETING)

SECOND GROWTH IN *WALDEN*

D. GORDON ROHMAN

Michigan State University

F. SCOTT FITZGERALD'S Great Gatsby talked much about the past. He wanted "to recover something, some idea of himself perhaps, that had gone into loving Daisy. His life had been confused and disordered since then, but if he could once return to a certain starting place and go over it all slowly, he could find out what that thing was . . . ," Nick Carraway tells us.[1] But Gatsby never did find out. His quest to recover his "true self" in the past failed.

Thoreau's succeeded. In *Walden,* Thoreau captured his "Platonic conception" through a recovery of the past. But the victory was not cheap. Years of disillusion followed brief joy at the pond. Thoreau became poignantly aware of the precarious and fugitive character of each lived moment. His journal of these years after the pond is his record of the "essential anguish of man—of man in time."[2] It took Thoreau seven years and seven major revisions of his book to seize in art a timeless self from the ruins of time.[3] The result of that quest is the "second growth" of *Walden.*

Thoreau was three things: a man in search of a life, a prophet in search of a revelation, and an artist in search of a form.[4] The man formed the prophet, and the prophet created the problems of the artist. Seven years it took between pond and book (1847–54) for the man, the prophet, and the artist to coincide in art. At the pond, he was essentially the man. Time seemed to Thoreau then a legitimate theater of action; in time a man could reveal his authentic divinity and gain personal immortality. In time, a man could make his own

[1] F. Scott Fitzgerald, *The Great Gatsby* (New York, 1953), p. 84.

[2] Georges Poulet, *Studies in Human Time*, trans. Elliott Coleman (New York: Harper Torchbooks, 1959), p. 10. My paper owes a more general and inspirational debt to Poulet's work than the specific footnotes can reveal.

[3] See J. Lyndon Shanley, *The Making of Walden, with the Text of the First Version*, Chap. II, "The Versions of Walden" (Chicago, 1957), pp. 18–33.

[4] This is a remark made by Mark Schorer about D. H. Lawrence which I have applied because of its surprising aptness to Thoreau. See Schorer's introduction to *Lady Chatterly's Lover* (New York: Grove Press, 1957), p. xxiii.

circumstances. Emerson had kindled and fanned this extravagant hope: there was to be a miraculous salvation from necessity wherein every spirit builds itself a house and beyond its house a world, and beyond its world a heaven. In awe, Thoreau wrote of the new Moses that, "a new heaven opens to him. . . . In his world every man would be a poet, Love would reign, Beauty would take place, Man and Nature would harmonize."[5] Bathed in the currents of Universal Being, this sometime schoolteacher, pencil-maker, magazine writer, and odd jobber, condemned to a house of solicitous sisters, maiden aunts, female boarders, a deaf father, and a gossipy mother, fell in step with the great man who preached that circumstances did not alter cases, who promised boom times of the spirit. "Build, therefore, your own world," Emerson exhorted in *Nature*, identifying idealism with prudence.

Thoreau's personality has always presented something of an enigma to his interpreters. On the one hand, especially in his early life, he often desired to play some part in this world, the stage of his own existence, even if it presented only shows. On the other hand, especially after the pond experience, he longed to pass from appearance to reality, to escape from a world which had become his prison. It is difficult to get the emphasis right. Thoreau was a man who liked to play many parts, who delighted to watch himself playing them, knowing that they were only parts and he himself only a player in life's unsubstantial pageant. Of one thing we can be sure, however, Thoreau, like most Romantics, was self-haunted, unable even if he would, to escape from his own drama. And his drama often turned to tragedy in the years after the pond. It became necessary then for Thoreau to find permanence apart from the hypocritic days; he had to learn the virtues of heroic patience and the discipline of the artist of Kouroo, that parable of the dedicated artist in the conclusion of *Walden* who sheds his mortality in a quest of perfection. Thoreau wrote about his leaving the pond that, "One would think twice before he accepted heaven on such terms. A ticket to Heaven must include tickets to Limbo, Purgatory, and Hell."[6] Thoreau found that

[5] *The Writings of Henry David Thoreau,* 20 vols. (Boston, 1906). For convenience I shall use *J* for *Journal,* volumes VII–XX of the above edition. Since the *Journal* is also numbered I–XIV, I shall use this numbering. The reference in the present note is, *J*, I, 432–433.

[6] *J*, III, 215.

he could not escape his mortality, the inevitable taint upon a finite creature who is always enslaved to time and place.

The purgatory of the years after the pond was isolation in time, a problem Thoreau inherited from the seventeenth century: the individual bereft of God's abiding duration, left anxiously in a universe of discontinuous moments. "Confined in the moment in which it exists, the consciousness watches its successive modes of existence pass, one after another, and escape it. To feel oneself live is to feel oneself leave behind, in every instant, an instant which *was* the very self."[7] The individual was reduced to a mere succession of thoughts—"a parcel of vain strivings tied by a chance bond together," as Thoreau wrote in an early poem. The heart of Thoreau's despair lay in this double fate: continual creation and continual dissolution of consciousness. Herein lies a historical irony.

In the name of consciousness, the Transcendentalists had rejected the eighteenth century clock-maker God, had reasserted the possibility of divinity in each moment. They thought they had discovered in the moment of consciousness what two centuries had denied: duration, an indivisible sense of self and reality. The peaks of Transcendental mysticism and literature testify to moments when this ability to feel incandesces into an emotion of highest intensity and the soul is drunk with its own grandeur: "I become a transparent eyeball; I am nothing; I see all; the currents of the Universal Being circulate through me; I am part or parcel of God," chanted Emerson in *Nature*.

But in between such moments there always came a general diminution when the Transcendentalist found himself in a state very close to nonexistence, torpor, vague reveries, and indefinable uneasiness, when, as Emerson wrote in "Experience," the "whole frame of things preaches indifferency." The problem, Thoreau wrote, was, "Are our serene moments mere foretastes of heaven . . . or simply a transient realization of what might be the whole tenor of our lives?"[8] Temperament, that iron wire upon which the beads of consciousness were strung, apparently was the fate that put all divinity to rout. "Romantic nostalgia appears thus as the forlorn desire of a life that the mind can never give itself fully in any moment and that, notwithstanding, it sees from afar, there, beyond both ends of the mo-

[7] Poulet, p. 16.

[8] *J*, II, 268.

ment, in the elusive realm of duration. . . ."[9] Emerson's solution was to acquiesce: "Since our office is with moments, let us husband them," he wrote in "Experience." The irony was not just that joy was inconstant, but that it apparently was involuntary as well.

Such always has been the curse of quietism or Transcendental "diurnal" philosophy. The hymn to the present, morning, dawn, and awakening that Thoreau sings in *Walden* gives no hint of the dark brooding over this fact in the journal where the mind testifies to nostalgia, decay, and nothingness. "There was a time when the beauty and the music were all within, and I sat and listened to my thoughts, and there was a song in them," Thoreau wrote in the fifties, but by then his cords were loose.[10] Transiency was the curse he labored to exorcise as he lived again and again in the dream of the past. "Ah, that life that I have known! How hard it is to remember what is most memorable! We remember how we itched, not how our hearts beat. I can sometimes recall to mind the quality, the immortality, of my youthful life, but in memory is the only relation to it."[11] He felt that the springs of feeling were drying up; the "flexible sap hardens into heartwood, and there is no further change."[12] His life had turned from poetry to prose. "Does nothing withstand the inevitable march of time?" he cried out. "Why did I not use my eyes when I stood on Pisgah? Now I hear those strains but seldom. . . . Ah, sweet, ineffable reminiscences!"[13] Yearning for fulfillment in the present, Thoreau was, like Gatsby, borne back ceaselessly into the dream of the past.

The great problem for the Romantic artist, as for Thoreau, was to make the dream real again through the exercise of the waking will. In *Walden,* Thoreau succeeded in incarnating himself, the flesh became word. The long quest in memory which the journal records between pond and book brought him finally into a possession of his life in the form of a Keatsian urn or a Yeatsian Byzantium, abstracted and removed from the dissolution and mortality of moments. The solution lay in the use of memory recaptured in art.

By remembering—even by painful remembering—man escapes the purely momentary. "To regret deeply is to live afresh," Thoreau

[9] Poulet, p. 26.
[10] *J,* VI, 294.
[11] *J,* II, 237–238.
[12] *J,* III, 203.
[13] *J,* III, 222.

had written in 1839.[14] And a decade later the truth of the observation was confirmed and restated: "My imagination, my love and reverence and admiration, my sense of the miraculous, is not so excited by any event as by the remembrance of my youth."[15] *Walden's* role in his own life was to present a duration based upon the affective memory. Certain events recalled in their original, qualitative content from the "deep well" of consciousness—events such as those auditory experiences of the chapter on "Sounds" in *Walden*—served to convey a meaning of selfhood which could not be found from the contents of daily experience. "To construct a work of art is to re-construct the world of experience and the self. And thus a concept of the self emerges, through the act of creative recall translated into a process of artistic creation, displaying characteristics of unity and continuity which could not be attributed to the self as given in immediate experience."[16] Through his affective memory, Thoreau linked his present with the dissolving past.

> Methinks my present experience is nothing; my past experience is all in all. I think that no experience which I have to-day comes up to, or is comparable with, the experiences of my boyhood. And not only this is true, but as far back as I can remember I have unconsciously referred to the experiences of a previous state of existence. "For life is a forgetting," etc. Formerly, methought, nature developed as I developed, and grew up with me. My life was ecstasy. In youth, before I lost any of my senses, I can remember that I was all alive, and inhabited my body with inexpressible satisfaction; both its weariness and its refreshment were sweet to me. This earth was the most glorious musical instrument, and I was audience to its strains . . . I can remember how I was astonished.[17]

The wine of this astonishment was the vintage he expressed in *Walden*. As with trees, so with men, Thoreau observed at this time; they commonly have two growths, a spring and a fall growth, the "latter sometimes equalling the former. . . . So is it with men; most have a spring growth only, and never get over this first check to their youthful hopes; but plants of hardier constitution, or perchance planted in a more genial soil, speedily recover themselves, and, though they bear the scar or knot in remembrance of their disappointment,

[14] *J*, I, 95.
[15] *J*, II, 33.
[16] Hans Meyerhoff, *Time in Literature* (Berkeley and Los Angeles: University of California Press, 1955), p. 48.
[17] *J*, II, 306–307.

they push forward again and have a vigorous fall growth which is equivalent to a new spring."[18]

Thoreau resigned himself to await in patience the fortuitous awakenings of the heart, the heroic patience of the artist of Kouroo. *Walden* becomes then both the poetry of memory and the poetry of prophecy. Thoreau's journal in the years after the pond is a record of torments of imminence and dramas of presentiment. The accomplishment of *Walden* was to center a Platonic "I" in the immobile moment of consciousness and bring back to that center, across the temporal distances, both the past and the future, as Thoreau wove the affective moments of his past into a mythic harmony of the seasons. Thus could Thoreau remain secure in the faith of his own divinity, even though he himself remained in the here and now to make pencils and survey woodlots in the enveloping darkness of disease and mortality.

[18] *J*, IV, 227–228.

PAPERS OF THE MICHIGAN ACADEMY OF SCIENCE, ARTS, AND LETTERS
VOL. XLVII, 1962 (1961 MEETING)

ROBERT LOWELL AND THE SENSE OF HISTORY

GLAUCO CAMBON

The University of Michigan

LIKE everything human, poetry is the child of history, since it could hardly take shape in a void; and yet by the same token it rises above its parent, by emerging into the heaven of achieved form. But this form is transparent; it can be a mirror and a judge; poetry judges history by being at the same time *from* it and *above* it. The poet stands in time, yet rises above it, like the philosopher; or at least his presence in time is not entirely timebound, if he is to fulfill his mission as a witness: "Lebe mit deinem Jahrhundert, aber sei nicht sein Geschöpf; leiste deinen Zeitgenossen, aber, was sie bedürfen, nicht, was sie loben" (Live with your century, but don't be its creature; perform for your contemporaries, yes, but what they need, not what they praise). This Schiller quotation from the ninth *Letter on aesthetic education* might be an apt commentary on the work and attitude of Robert Lowell, who is one of the most history-conscious American writers of our time precisely because he has a quarrel with history.

If history is the past, and the past is unchangeable, then Lowell's quarrel with it is futile. But his implicit philosophy is not Crocean; it is closer to Bergson's conception of time past as impinging on time present in the dramatic duration of the experiencing mind. Now his relation to this past is extremely tense, and also complex; to understand it is to see how his poetry comes *towards* our history, and how our history painfully procreated it in its own inverted image.

As an American, Lowell is the heir of the New England past, and this shows clearly in so many poems that, like *Quaker Graveyard in Nantucket, in Memory of Arthur Winslow* and *Mary Winslow*, project family history into public history—an understandable act on his part, if we keep in mind that the Lowells have had a conspicuous stake in the destiny of New England. On the other hand, to inherit New England means to inherit a secession from Europe, from that older past and wider cultural horizon in which Lowell feels involved as a Westerner: "I write the Wonders of the Christian Religion, flying from the Depravations of Europe, to the American Strand . . . " One way to acquire a greater intimacy with Lowell's poetry as a child and judge of history is to read it contrapuntally against the background of Cotton Mather's General Introduction to *Magnalia Christi Americana.* Only that Lowell has reversed the position: what he flees from is not the "depravations of Europe," but the (to him very real) depravations of America itself. His direct inheritance was a promise, but a disinheriting one. The Pilgrim Fathers were indeed "children of Light," but of a blighting light, the light of Lucifer:

> Our fathers wrung their bread from stocks and stones
> And fenced their gardens with the Redman's bones;
> Embarking from the Nether Land of Holland,
> Pilgrims unhouseled by Geneva's night,
> They planted here the Serpent's seeds of light;
> And here the pivoting searchlights probe to shock
> The riotous glass houses built on rock,
> And candles gutter by an empty altar,
> And light is where the landless blood of Cain
> Is burning, burning the unburied grain.

Between the benighted Calvin at Geneva, guilty of having contributed to the breakup of Christian Europe, and the hellfire that sprang from his message, the Mayflower pilgrims were in a bad way when they sailed from the infernal ("nether") land of Holland, for the seeds they brought from betrayed Europe were those of destruction, as the redmen came to know; if they hoped to earn another garden of Eden, death and the snake penetrated it right away, and their very consciences, those searching lights, expose them as "landless blood of Cain," as accursed nomads doomed to sterility, not holy pilgrims sustained by a divine Covenant. Few people would miss here, as in so much else that Lowell wrote, the forceful way in which history begets poetry, and poetry in its turn accuses history, through the alchemy of metaphor: the puns on the Netherlands, on Lucifer, on the searchlights, on the "riotous" houses built on Biblical rock, but too brittle to stand the coming quake, are not clever wordplay, but a structural device bent on eliciting its fearful meaning from the landscape of history, down to our own time. In the same way, the speaker of a later poem (*Mother Marie Therèse*) says of General Montcalm that he lies asleep "on Abraham's bosom," utilizing the geographic name of a famous battlefield near Quebec to bring out transcendent implications; and in still another composition, *The Dead in Europe,* we find the name of Maryland switched from its actual etymology to a religious one that has to do with the Queen of Heaven, not of seventeenth-century England; while the moving epitaph for Santayana, from the 1959 volume *Life Studies,* imaginatively connects the watchful nuns of Santo Stefano Rotondo, where the skeptical philosopher spent the last years of his life, with the Capitol geese that according to a commonplace of legend saved Rome's last stronghold from the nocturnally creeping Gauls. Humor is not absent from this flash of telescoped vision, or, for that matter, from the last book to date with its affectionate family sketches.

But the earlier volumes—*Land of Unlikeness, Lord Weary's Castle, The Mill of the Kavanaughs*—have little or no humor, being attuned to a mood of bitterness which might well make them offensive to whosoever refuses to share Lowell's fiery conversion. Who is, after all, this New Englander who repudiates his spiritual lineage? Isn't he simply a renegade? If so, the poetry will have to be accepted as a merely private expression of troubled genius in a historical context to which it is irrelevant. A personal conversion to the long-abandoned dogmas of Papist Rome,* even if it results in magnificent verse, cannot have any bearing on the realities of Protestant Yankeedom: can you abolish three hundred years of history by a stroke of pen? At this point the historical sense itself will intercede for the "unhouseled" poet by reminding us that his secession from the secessionist forebears actually continues their own attitude, whatever the motivating ideology. Rome is Lowell's Walden, as the impassioned poem on Concord reveals:

> Ten thousand Fords are idle here in search
> Of a tradition. Over these dry sticks—
> The Minute Man, the Irish Catholics,
> The ruined bridge and Walden's fished-out perch—
> The belfry of the Unitarian Church
> Rings out the hanging Jesus. Crucifix,
> How can your whited spindling arms transfix
> Mammon's unbridled industry, the lurch
> For forms to harness Heraclitus' stream!
> This Church is Concord—Concord where Thoreau
> Named all the birds without a gun to probe
> Through darkness to the painted man and bow:
> The death-dance of King Philip and his scream
> Whose echo girdled this imperfect globe.

* This conversion was only temporary, but it left a mark. And it was in itself very "personal" indeed.

The stream of Heraclitean becoming, History in the making, threatens to be our lost chance if we stake everything on "Mammon's industry," for the only forms with which that stream can be "harnessed," i.e., time can be redeemed, are those of meaningful ritual, thought and art, not the feverishly multiplied machines on which we pride ourselves. The "ten thousand Fords" will never carry us to what we most need—the fountainhead of a reliable tradition. What remains of America's promising start, here in the town that fired the Revolutionary War and then the revolutionary thought of Thoreau and Emerson, is just "dry sticks"; the landmarks are pitiful ruins, the local version of Christianity is powerless to counteract the creeping materialism of the "children of light," and what the complacent Muse of history will grant us to hear in response to our devotion aroused by the well-known monuments is not the "shot heard round the world" of Emersonian memory, but the disquieting cry of a remoter war; murdered King Philip claims his long-due revenge, much as Ikkemotubbe summons the Sutpens, the Sartorises, and the Compsons to the earth of which he was defrauded by the white man, in Faulkner's Yoknapatawpha saga.

A perceptive sociological historian like Oscar Handlin emphasizes separation or "apartness" as the fundamental experience of the immigrant to America, therefore of America as such; it is easy to see how Lowell, like other American writers of note, *stands apart* from his native land *within* his native land, and thus takes up the posture of the prophet. It would seem America is one vast community of "sinners in the hands of an angry God"; will it mend its ways and avoid the threatening doom? The conscientious objector cannot tell; he sees what he sees, and he feels it is his mission to awaken, not to soothe. In *Where the Rainbow Ends,* his native Boston swings dangerously between salvation and destruction:

> I saw my city in the Scales, the pans
> Of judgment rising and descending . . .

In *At the Indian Killer's Grave,* the head of the Redskin chieftain announces nemesis to the descendant of his destroyers:

> "Surely, this people is but grass,"
> He whispers, "this will pass;
> But, Sirs, the trollop dances on your skulls
> And breaks the hollow noddle like an egg
> That thought the world an eggshell . . .
> . . . . . . The Judgment is at hand . . .

The Indian killers are no "chosen people":

> ... Your election,
> Hawking above this slime
> For souls as single as their skeletons,
> Flutters and claws in the dead hand of time."

In *As a Plane Tree by the Water,* Boston, and all of America by implication, are equated with Babel:

> Darkness has called to darkness, and disgrace
> Elbows about our windows in this planned
> Babel of Boston where our money talks
> And multiplies the darkness..........
> Our Lady of Babylon, go by, go by,
> I was once the apple of your eye;
> Flies, flies are on the plane tree, on the streets.

Postwar America of the middle and late forties feels secure within its still impregnable defenses, but, the poem concludes, they are just another German "Atlantic wall," likely to crumble as the walls of Jericho; conversion to the Spirit will be the only way to survival. This plea for national soul-searching, which subsequent world events have partly borne out, takes its place with Robinson's *Cassandra* as a powerfully worded reminder to the prosperous republic not to yield to the temptation of *hybris*; but the hymning note of love is not lacking. Christ is envisaged as the liberating force behind the accusing violence of the prophet; Lowell is an apocalyptic believer, like Hopkins, and *Colloquy in Black Rock* embodies to perfection his imaginative thrust from the machine-like rhythm of an earthbound beginning to the transcendent lightning of the crowning line:

> .... my heart,
> The blue kingfisher dives on you in fire.

It is this moral tension, translated into a uniquely taut compactness of style, that enables Lowell to sustain a successful dialogue with formidable literary ancestors like Melville. *The Quaker Graveyard in Nantucket,* one of the modern Bostonian's highest achievements, is precisely such a dialogue. The sailor drowned at sea, Lowell's own relative Warren Winslow, surfaces ominously to stare his surviving fellows-at-arms out of countenance, and can be taken to repeat in the modern context the posthumous gesture of Fedallah in *Moby Dick,* for ill omens are gathering on America's might at sea:

Sailors, who pitch this portent at the sea
Where dreadnaughts shall confess
Its hell-bent deity,
When you are powerless
To sand-bag this Atlantic bulwark, faced
By the earth-shaker, green, unwearied, chaste
In his steel scales: ask for no Orphean lute
To pluck life back. The guns of the steeled fleet
Recoil and then repeat
The hoarse salute.

Poseidon, the "earth-shaker," the scaly seagod, will shatter the Atlantic Wall of America; the Pequod's fate is going to be re-enacted (stanza 2). Life, imagination, spirit, not walls or steel, are the appropriate response to the challenge of present history. One could almost read Toynbee, along with Melville, between the lines. The apocalyptic vision of past and future doom expands to a Melvillian climax when the last stanza develops the whale symbol to encompass the decay of modern civilization culminating in a shipwreck like the Pequod's:

Gobbets of blubber spill to wind and weather,
Sailor, and gulls go round the stoven timbers
Where the morning stars sing out together
And thunder shakes the white surf and dismembers
The red flag hammered in the mast-head...

But the harpooned, hacked, rotting whale is not just a source of corruption; it is life, and can harbor Jonah-Christ in its huge body:

.....Hide,
Our steel, Jonas Messias, in Thy side.

Thus the harpoon of the old Quaker whalemen becomes the spear of Longinus, and the symbolic killing of the whale an act of suicide, since it is a crucifixion of our own culture, without a guaranteed Resurrection; Lowell anticipated Marius Bewley in his Christological interpretation of Moby Dick.

This is the "wisdom" that the dove of Jesus has given to the "exile" (*Where the Rainbow Ends*); but it is, Melville would say, a wisdom that is woe, and the disinherited heir of the Puritans inverts the route of the Mayflower. His *Exile's Return,* counterpointing Malcolm Cowley's book by the same title, shows Europe as a shell-torn battlefield, where, despite the presence of "lily-stands" in the Rhineland and of a "rough cathedral," death lurks; a brief quotation from Dante's *Inferno* lines in Canto III ("lasciate ogni speranza o voi ch'entrate") warns us that these towns are a Hell gate. *The Dead in Europe* tells us, through the imagined chorus of the European victims, what that hell was like:

> After the planes unloaded, we fell down
> Buried together, unmarried men and women;
> Not crown of thorns, not iron, not Lombard crown,
> Not grilled and spindle spires pointing to heaven
> Could save us. Raise us, Mother, we fell down
> Here hugger-mugger in the jellied fire:
> Our sacred earth in our day was our curse.

A daring metamorphosis of the voice expands the significance of the poem in its conclusive stanza, the third, by making the poet himself the speaker and thus implicating America in the spiritual fate of bombed Europe:

> . . . Shall I bear,
> (O Mary!) unmarried man and powder-puppet,
> Witness to the Devil?

The very last lines restore the initial chorus, but as an enlarged "we" that associates Lowell, indeed all of America, to the prayer of the dead:

> . . . . Mary, hear,
> O Mary, marry earth, sea, air and fire;
> Our sacred earth in our day is our curse.

A reshuffle of the elements seems in order to renew life, or perhaps just to end history as such in a Final Judgment; Mary the Marrier will then be a death giver, but also a life giver in the spiritual sense. The transition of the verb tense from the past of Stanzas 1 and 2 to the loaded present of Stanza 3 ("*is* our curse") clinches the effect of the plural-singular-expanded plural sequence as a direct way of prophetically relating past history (recent though this past may here be) to the day which is still being lived by mankind.

The prodigal son goes to Rome, hoping to find there a promise of salvation; and Rome, the cradle of his culture, is violently addressed by him in *Dea Roma* as the locus of sixteen centuries wasted by Christendom "since Maxentius fell"; not only all the roads, but "many sewers" lead to it, yet St. Peter, providential fisherman,

> walks on the waters of a draining Rome
> To bank his catch in the Celestial City.

Evidently a decisive event is approaching, a Second Coming, to seal history; the apocalyptic mood is in tune with the anguish of a Cold War overshadowed by the threat of an atomic cloud. Lowell's sense of history is eschatological and not, properly speaking, "historical," for he speaks like an angry prophet, not certainly like a Hegelian historicist or hopeful humanist. Later poems concerned with Rome, i.e., "Beyond the Alps," "Falling Asleep over the Aeneid," "For George Santayana," develop in a more meditative tone, but still questioning our use of history in the alternative of power, reason, and faith; and the eschatological note is not wanting either.

Because his commitment to spiritual values is so thorough, whatever direction his religious allegiance may have taken since his conversion to Catholicism, Robert Lowell can sustain his poetry at a declamatory pitch that would prove very insidious to a less gifted imagination. There is despair in his faith, and cruelty in his love; but he agrees with his countryman Robert Frost that fire is better than ice. If rebellion is the best way an American has to establish a connection with his tradition, then Lowell's New England ancestry counts for something. The sociologist, the scientific historian has to do with facts and laws; who but the poet can afford to have "a lover's quarrel" with history? This is his way of being available to history-ridden mankind.

Papers of the Michigan Academy of Science, Arts, and Letters
Vol. XLVII, 1962 (1961 Meeting)

DR. WILLIAMS AS EXHORTER: THE MEANING OF AMERICANISM

BERNARD F. ENGEL

Michigan State University

It must be with ironic, even raucous, amusement that William Carlos Williams, long regarded as a dilettante in esoterica, finds himself today saluted as "the America of poets,"[1] the pathfinder whose route will be taken by American poets of the future.[2] As a steady New Jersey obstetrician he must be somewhat bemused by his inclusion in the Beat writers' hall of heroes, and as a lifelong opponent of academicians he must marvel at the use of his techniques as touchstones in recent poetry textbooks.

The turning of American readers to Williams after a generation of obeisance to T. S. Eliot, who is in many ways his opposite and even his enemy, would seem to indicate a revival of interest in the old campaign to found a national literature on native grounds. If Williams' work is to be categorized, the one label which fits both its content and its technique is "American." But though his message is deliberately—at times provocatively—one of nativism, it is not intended to be chauvinistic. His aim, he says, is to advance "a new cultural concept" which is "not to be narrowly designated American though perhaps first adequately conceived here."[3] In discussions of aesthetic theory his rejection of what he sees as foreign ideas and techniques is rigorous, but when in his poetry he can make what he considers significant use of an import he puts it to work without defensive explanation. Thus in his long poem *Paterson* the central observer is a Proteus figure, Book IV is a travesty of a pastoral, and Book V makes use of a unicorn, a quasi-mammal unlikely to appear in anyone's bestiary of northern New Jersey. Williams would armor his countrymen against the exotic not because he thinks the ideas of

[1] Randall Jarrell, *Poetry and the Age* (New York: 1955), p. 218.

[2] Stephen Whicher and Lars Ahnebrink (eds.), *Twelve American Poets* (New York: 1961), p. 115.

[3] Letter to Horace Gregory, 1944, in John C. Thirlwall (ed.), *The Selected Letters of William Carlos Williams* (New York: 1957), p. 228.

other times and places are evil but because he holds that before we can successfully learn from others we must know ourselves.

Viewing himself as a successor to Whitman, Williams accepts as his mission the campaign to develop in America a culture which will allow the individual to fulfill himself. But as a well-read man of the twentieth century Williams acknowledges the colder, more complex climate of his own times. His thinking has been significantly influenced, whether at first or second hand, by Darwin, Freud, Wilhelm Reich, Marx, Beard, Einstein—by all who have brought what he sees as light to our darkened plain. Ever the individualist, and always the artist rather than the philosopher, he is the follower of no man's doctrinal platform. But whether by native inclination or ideational osmosis, he has accepted the general principles of modern socio-psychological libertarianism.

The urgency of his importunate verse and prose arises from desperation, for he sees himself as an analyst whose patient perversely refuses to recognize the existence even of symptoms. Though he has in mind a utopia, he does not and perhaps cannot see either the shape of it or the path toward it:

> . . . Whither? I
> cannot say. . . .
> . . . The how (howl) only
> is at my disposal . . .[4]

His hortatives therefore include more rejections than acceptances. The American at present can take only a tentative first step toward self-realization. He must begin by rejecting all that represses his spirit, whether that frustrating psychological code which Williams labels "Puritan" or that cruel economics which, in the manner of Ezra Pound, he decries as "usury." The American, Williams holds, must resolve the dichotomy which separates head and heart, man and nature, man and mankind. "Puritanism" and "usury" are the strait jackets which keep us schizoid.

This dualism feeds on Americans' failure to understand either their past or their present: they "walk outside their bodies aimlessly" because they "neither know their sources nor the sills of their / disappointments . . . "[5] Efforts to restore either the presumed traditions

[4] *Paterson* (New York: 1951), p. 28. No line numbers are given.

[5] *Ibid.*, p. 14.

of western civilization or a set of specifically American ideals are mistaken and even dangerous, for the past can become meaningful to us only after we have equipped ourselves to understand it. And understanding can come only when through imagination we have developed a comprehension of ourselves in action in our own time and place.

Williams' long opposition to the work of Eliot is grounded in these convictions. He argues that Eliot has turned to sterile philosophies which, however vitalizing in the past, are not viable today. Eliot's career therefore represents a high treason, the abandonment by perhaps the generation's most talented writer of the American poet's obligation to guide his countrymen. By the power of "The Waste Land" Eliot set back by twenty years the furtherance of American poetry;[6] Eliot's turn to theology was "offensive," and his "affectation of authority" makes him "slimy."[7] In *Paterson,* Eliot is the target of several passages of mocking allusion and paraphrase.[8]

Yet Williams thinks the past has its uses. Though it can give us no affirmations, it can show us the roots of our "disappointments," how and where we went wrong. A study of American history is therefore the line of departure from which we must mount the attack on the bastions of American social and psychological error. Williams' story of the American past is that of a moralist. He believes that America has from the beginning represented unrealized potentiality, a chance for the human spirit to achieve a liberation never afforded it in the Old World. His individualistic study of origins, recorded in *In the American Grain,*[9] asserts the source of American moral character to be that "Puritanism" which he condemns. Williams sees the colonial Puritan as having the Elizabethan firmness and courage needed to survive in the New World. But this in itself was surely little enough; and it grew not from breadth of vision but from narrowness, from an inhuman rigidity. The Puritan did not "see" the New World, for he had imported traditional ways of viewing which made him blind to its facts. His strength made for individualism, but narrowness turned this to isolation, to divorce from his fellow men and from nature. The result was that the American became not a self-fulfilling individual but a self-repressing one, a life-denier who

[6] *The Autobiography of William Carlos Williams* (New York: 1951), p. 174.
[7] *Selected Letters,* pp. 224–225.
[8] E.g., pp. 38–40.
[9] (Norfolk, Conn.: 1925).

stultified not only his own soul but also the spirits of his fellow men. Williams will accept no plea that the Puritan perhaps became what he had to become. To make clear his view of Puritan perversity, he presents a contrasting picture of the French Catholic spirit as exemplified by Père Sebastian Rasles, author of *Lettres Édifiantes*. Williams sees Père Rasles as a representative of an expansive morality, one which sought to understand life in the here and now without importing judgments appropriate to some other place and time. Williams insists that the Puritan refused to see the Indian, for example, as anything other than an unformed Puritan and even a devil. He asserts that the Catholic fathers, in contrast, took the Indian on his own terms and were not afraid to touch him, to attempt psychological communion, and to refrain from judgments. Whatever the accuracy of these comments as history, they demonstrate Williams' conviction that the new American who will—must—emerge will draw inspiration not from Miles Standish but from Père Rasles.

Williams' picture of the more recent past and of the present is also that of a moralist. Book I of *Paterson* suggests that the primeval sin, the source of most life-denying dualism, has been greed in the form of a rapacious individualism which misused potentialities because it did not understand or even perceive them except as materials for exploitation. The first prose insert in the poem is an account of what Williams sees as the American's typical savagery toward his natural surroundings, violence accompanied with a disregard for consequences which is all the more horrifying because of its spirit of freebooting gaiety. The incident is a rush for pearls in Notch Brook near Paterson in 1857, a search which resulted in the destruction "with little or no result" of millions of the native mussels: the discoverer of one large pearl ruined it by boiling open the shell.[10] Though the passage is presented without direct editorial comment, its function as an example is made clear by Williams' juxtaposition of it with verse lines ridiculing the merely abstract, all which is not tied to a time and place and an action. These report that Paterson-as-poet twice a month receives "communications from the Pope and Jacques Barzun," the utility of which appears to be only that clerks can ungum the foreign stamps for their children.

Ruin of potentiality is the theme again in a prose passage on life

[10] P. 17.

in colonial New Jersey.[11] This reports that though there was not "beauty" there was "a strangeness and a bold association of wild and cultured life" in early days when the varied flora of the setting was intermingled with the cabins of ironworkers. These idyllic circumstances did not endure, however; the next paragraphs report incidents of frontier and revolutionary violence which perverted the promise, turning the forest into a refuge for roughneck victims of human savagery who became savage in return. Some of these victims were Negro slaves; what other kind of life they might have had is suggested in the immediately following verse passage which reports the vigor of an African chief's nine wives. These lines avoid sentimental picturing of those who live close to nature as necessarily happy, but they do imply that the Negro in his homeland was more truly alive because of his nonexploitative association with nature.

Most vivid of all illustrations of this point in Book I are the passages on the falls of the Passaic River. These deride Alexander Hamilton's wish to use the falls to power a model manufacturing enterprise, and declaim the idiocy of the celebrated daredevil "Sam Patch" who saw the falls only as a scene for foolhardy publicity stunts. No one, that is, attempted any sort of harmony with nature. There is no overt suggestion in *Paterson* as to what harmony would be, for Williams scrupulously limits his function to encouragement of the rejections without which discovery of the new cannot be made.

A desperate, almost a despairing, survey of the contemporary scene in Books II, III, and IV investigates the American's continuing "divorce" from nature and his fellows, the "blocked" state of his communication with them, and the "staleness" and "sterility" which result. The action of the poem ends in Book IV with Paterson overwhelmed by the sea. There is the hint of a possible future in the repeated insistence that the ocean which takes Paterson is "not our home" and in the ending passage's remark that, after the disappearance of the central figure, a man and his dog are seen emerging to the beach. Paterson has died, and he is not to be taken either as a martyr or as a savior to be resurrected. Yet perhaps his death is a metamorphosis: in another guise he or his spirit may return. He appears again in Book V, which is both a continuation and a commentary. Paterson finds that though appearances have changed, Americans

[11] Pp. 21–22.

still exhibit that fundamental dichotomy here symbolized by the opposition between the "virgin and the whore," the "Puritan" inability to comprehend that moral purity may be concomitant with freedom of the spirit.

Given that the American needs a new way of understanding, it follows that he needs a new means of perception. As vigorously as he argues for a new realization of the self, Williams pleads for development of a new poetic "line," an "American" aesthetics which will focus on the particulars of experience. The line which will make possible not merely the expression but the actual discovery of the national plight can be neither a borrowing nor an adaptation, for the words with which men think and speak cannot be separated from their content. The twentieth century American may respect Freneau's observations on metrics, and the struggles of Poe and Whitman to assert a new aesthetics. But the American poet today must look to the world about him, to the speech it uses and to the facts of its experience. Only in these materials can he find both his proper subject, and his means of understanding and expressing it.

The chief technical problem Williams has faced as a poet therefore has been that of dealing with fundamentally abstract concepts while retaining respect for particulars. Like John Dewey, from whom he has said that he took his idea that "there is no universal except in the local,"[12] Williams insists that generalizations can be viable only if tied always to specifics. Throughout *Paterson*—and often in other work as well—he repeats the injunction: "Say it, no ideas but in things." Thus there is in his work none of the persistent romanticizing of, for example, his West Coast contemporary Robinson Jeffers, who sees nature as a source of moral patterns. Nor is there in Williams' work any of Robert Frost's reluctance to give up the tantalizing hope that nature may at some moment provide an inspiration. Williams holds that nature must be perceived almost as a collection of neutral facts, "things" to be made use of, their value depending entirely on what man does with them. Williams' typical presentation of a scene therefore begins with a spare, even stark, notation of details. He insists that valid generalizations can arise only from the facts, never from impositions by the poet. The opening line of *Paterson* asserts

[12] *Selected Letters*, p. 224. This is argument-by-authority, of course, for Williams' convictions are apparent in works written long before publication of Dewey's *Art and Experience* (1934), the book Williams is referring to.

that "rigor of beauty is the quest," and to Williams rigor demands abstention from editorializing intrusions. His verse therefore relies heavily upon juxtapositions and subtle repetitions, for his chief device for communicating values is the ordering of his facts. His metrics too must reject the past. An "American" measure will have not what Williams sees as the outworn accentual-syllabic tradition suited to the language of another time, but a felt rhythm akin to that of what he conceives to be twentieth century American speech; in argumentive moments he does not hesitate to cite Einstein's theory of relativity as justification for his practices.[13] An "American" poetry will be one in which the poet functions as a maker not of interpretations but of presentations, a scrupulous reporter of things as they are. The mind must follow not inherited or environmental preconceptions but the accurately observed, concretely visualized facts of experience.

Williams' nativism is a result and function of both his ideas and his techniques. It is oriented to the future, expansive enough to incorporate ideas taken from science without being bowled over by them. In an age of "two cultures," it argues that poetry is a vehicle of the imagination which may subsume both. If it is the truth about ourselves that will make us free, it is an imagination grounded in the facts of experience which will provide the pathway to that truth. Meanwhile, Williams seems to say, let American poets get on with the job. They will find that if they are to use the exotic instead of surrendering to it, they must first develop the capacities native to themselves as Americans.

BIBLIOGRAPHY

(Works referred to, and a few other useful ones.)

DONOGHUE, DENIS. "For a Redeeming Language." *Twentieth Century*, CLXIII, June, 1958, pp. 532–542.

HARTUNG, C. V. "Poetry of Experience." *University of Kansas City Review*, XXV, October, 1958, pp. 65–9.

JARRELL, RANDALL. *Poetry and the Age*. New York, 1955.

KOCH, VIVIENNE. *William Carlos Williams*. Norfolk, Conn., 1950.

QUINN, SISTER M. BERNETTA. *The Metamorphic Tradition in Modern Poetry*. New Brunswick, N.J., 1955.

SOUTHWORTH, JAMES G., *More Modern American Poets*. Oxford, 1954.

THIRLWALL, JOHN C., ed., *The Selected Letters of William Carlos Williams*. New York, 1957.

---

[13] E.g., *Selected Letters*, p. 332; and *Paterson*, p. 65.

WHICHER, STEPHEN, AND LARS AHNEBRINK, eds. *Twelve American Poets.* New York, 1961.
WILLIAMS, WILLIAM CARLOS. *The Autobiography of William Carlos Williams.* New York, 1951.
——. *In the American Grain.* Norfolk, Conn., 1925.
——. *I Wanted to Write a Poem.* Boston, 1958. Ed. by Edith Heal.
——. *Paterson,* Books I–IV. New York, 1951.
——. *Ibid,* Book V. New York, 1958.

Papers of the Michigan Academy of Science, Arts, and Letters
Vol. XLVII, 1962 (1961 Meeting)

TECHNIQUES IN THE TWENTIETH-CENTURY NOVEL FOR RELATING THE PARTICULAR TO THE UNIVERSAL: *SET THIS HOUSE ON FIRE*

ALICE R. BENSEN

Eastern Michigan University

Upon its publication in June 1960, William Styron's novel *Set This House on Fire* received very uneven critical comment. Much of the discussion turned upon the novel's presumed bid for "greatness." Since this novel is perhaps the most outstanding of those published by younger men during the past year, it offers a good point of departure for a consideration of twentieth-century techniques for relating the particular to the universal.

That Styron intended this novel to be received as a major work is obvious from its theme, its length, its scope, its style, and its explicit relation to other works of literature. It is clear that Styron intended to achieve a degree of universality.

The presentations in a novel may be individual, plural, or universal—categories which are not, of course, mutually exclusive. For example, it has long been axiomatic that by "truly" knowing one human being one knows all men. The "truly" implies the immanence of universality. "Man," the universal, from Greek drama on has often been represented in a single protagonist. Here, then, are two modes of presenting an individual. The author may concern himself with uniqueness: the individual as the unique locus where certain circumstances and events intersect (as, for example, Taine's "la race, le milieu et le moment"), or as a plant that has resulted from some *saltus*. Or, on the other hand, he may seek chiefly, as Tennyson with his cranny-flower, to see the universal as it declares itself even in this particular.

Concern with the plural—with classes—a concern once associated chiefly with the authors of classical "characters," is in our day the approach of the sociological novelist, who, in his partial role of scientist, is attempting to understand the problems of a whole class at once. The class is a limited part of universal man, and the individual is important only in what he shares with the class.

The term "universal" brings up problems of application. Northrop Frye, in his study of William Blake,[1] points out Blake's insistence that if a universal is equated with a least common denominator, it sinks to something valueless. We are not interested in a novel whose chief matter is the motions of involuntary muscles, however human. From this level we rise—if we are using moral criteria—to vices and virtues. Selfishness is, perhaps, universal; some one of the virtues might likewise be offered. But when we think of Oedipus or Hamlet as being "universal" we do not have in mind such characteristics as they might share with Tartuffe or Lady Bountiful. We are rather following the line of thought traced, again, by Frye: "A man, a dog and a tree are all alive; but the man is the most alive; and it is in man that we should look for the image, or form, of universal life." And "all great visions in art lead up to visions of the unfallen world."[2] Whatever the terminology we use, we are aware that some passages of literature present, through whatever incantatory means, glimpses of the struggle toward the wholeness of the "Man" that is in "man"—his attempt to know and to achieve the "unfallen world." Such literature may be ranked, according to one's system of criticism, as the "highest" or merely as a category like any other. Whatever its rank, Styron was attempting in *Set This House on Fire* to produce literature of this sort.

Gene Baro, in his review of this novel, declares: "It is as if a novel with a sociological bias were somehow fused to a would-be Greek Tragedy."[3] This description might be taken as roughly summarizing the history of the twentieth-century novel. It has preserved the gains that accrued to it from Naturalism but has been impatient with the restrictions of that school. Naturalism opened out to the novelist many fields of observation, but it restricted him to recording behavior and to theorizing in terms of scientific laws. Consequently, he found himself limited to presenting his characters at only two distances: as individuals—seen objectively—and as typical members of a class.

But early in the century Dorothy Richardson and her fellow explorers began to look more deeply into themselves and their char-

[1] *Fearful Symmetry*, Princeton, 1947, p. 35.

[2] *Ibid.*, p. 40.

[3] *New York Herald Tribune Book Reviews*, June 5, 1960, p. 1.

acters. By introspection they were able to present the unique flow of thought and feeling in themselves or their imagined characters. They presented these nascent cerebral and emotional *données* directly by "stream-of-consciousness" transcription and also metaphorically by expressionistic techniques.

In addition to revindicating uniqueness, many novelists have revindicated universality of the sort implied in Baro's reference to Greek tragedy. They have transfigured their characters and situations by a variety of techniques—some of which have been resumed from the pre-Naturalistic past, and some of which are largely new. Several of these are used in Styron's novel of last summer.

Styron has given his novel a title taken from a passage in another work of literature. Contrasting with such titles as individual names (*Esther Waters, Sister Carrie, My Antonia, Son Excellence Eugène Rougon, Germinie Lacerteux*), or names of substances or places or seasons (*Wheat, Lourdes, Germinal*), literary titles give an extra dimension of some sort, though not necessarily universality. Huxley, for example, achieves a point of wit or irony in the titles *After Many a Summer* and *Brave New World.* Quite different is Forster's use of Shelley's *The Longest Journey,* Hemingway's use of Donne's *For Whom the Bell Tolls,* and Faulkner's Biblical *Absalom, Absalom!* These titles indicate an aspect of the life of universal man. To this group belongs Styron's title—again from Donne—*Set This House on Fire.* In Donne's context it refers to one of the acts of chastisement with which God, before hiding his face from his creature, has awakened him to grace; Styron uses it in this sense and also, following the total meaning of Donne's passage, as an exhortation, a prayer for reawakening.

Like many other writers of this century Styron calls forth a powerful work of literature of the past to assist in the orchestration of his values and to project, as by an incantation, the implications of his characters beyond their immediate being. As Joyce conjures up the *Odyssey,* Thomas Mann *Faust,* and Faulkner the New Testament, so Styron does *Oedipus at Colonus.*

In 1953, in the first issue of the *Paris Review,* writing something of a manifesto for this group of young writers, Styron declared:

> A critic nowadays will set up straw-men, saying that Mailer had Ahab in mind when he created Sergeant Croft, that Jim Jones thought of Hamlet when he came up with his bedevilled Private Prewitt, stating further, however, that

> neither of these young men have created figures worthy of Melville or Shakespeare; they do this, or they leap to the opposite pole and cry out that no one writing today even *tries* to create figures of the tragic stature of Lear. For a writer, God forbid either course.... The writer's duty is to keep on writing, creating memorable Pvt. Prewitts and Sgt. Crofts, and to hell with Ahab.[4]

Perhaps, cold sober and back in Carolina, Styron's unschooled hero, Cass Kinsolving, might say "To hell with Oedipus." (Except that he does not, and, in fact, is not averse to giving his life an extra dimension by quoting from *Lear.*) But in southern Italy, blind drunk, and in an extremity of distress, he chooses for uttering his misery the words of the blinded, exiled king. Furthermore, some six specific references to the play occur scattered at intervals throughout the book, forming a sort of burden, so that the reader is led to look past the new novel to the time-honored drama and see further parallels to Cass's situation.

It would appear that in our century—an age both of iconoclasm and of a desperate search for myths—there is set up not infrequently a sort of reciprocal relation between a contemporary work and an older work. In spite of some contrary trends, we seem to wear sufficiently freely the bonds of a Work of Art for a Work of Art's Sake to find it a normal project to contemplate Joyce's *Ulysses* and the *Odyssey* simultaneously and to observe how each affects the other. Similarly, contemplating both Styron's novel and *Oedipus at Colonus* in their entirety, one recognizes many parallels; and a timelessness sets in, in which Creon and the Thebans take on the menace of the phony, as represented in the contemporary piece by the perverted playboy Mason Flagg and the movie stars, the companion Antigone merges with the courageous fragile wife Poppy, and the chorus is—in its inquiring moments—the naïf sophisticate Peter Leverett (it has been pointed out that his name means "Peter Rabbit")[5] and—in its monitory moments—the policeman Luigi.

Like Oedipus, Cass had received a prophecy that he might find "a resting place" where he "might round out" his "bitter life"; he had assumed that this was Italy, but his sojourn there was followed by a translation and he finally found peace in Carolina. He too was visited by "portents"—"earthquake, thunder"; his hallucinations took these

[4] "Letter to an Editor," *The Paris Review,* I (Spring 1953), 12–13.

[5] Robert Gorham Davis, "Styron and the Students," *Critique,* III (Summer 1960), p. 42.

and similar forms. His "mind was a boiling cauldron; nothing so sweet As death . . . could have been given" him. He shared the prayer: "Grant me, then, goddesses, passage from life at last . . ." His relation to his fragile, faithful wife, Poppy, was much the same as that of Oedipus to his daughter Antigone; he needed "this girl To lead me, her frailty to put my weight on"; she "has wandered ever with me . . ." In his relation with Mason Flagg, Cass is, like Oedipus, threatened by one who wants to derive "strength" from him: ". . . you shall be much solicited by our people Before your death . . . for their welfare. . . . The riddler has it that their strength's in you—"; Cass's Creon, Flagg, and his people would "settle you near" them "and so Have you at hand . . . Just at the border, . . . where you'll not be free." Flagg, like Creon, would "take" Cass for Flagg's, not Cass's, own good; his thought runs like Creon's: "I see you ravaged as you are, a stranger Everywhere, never at rest, With only a girl to serve you in your need." Oedipus' description of Creon fits Flagg exactly: "You brazen rascal! Playing your rascal's tricks In righteous speeches, as you always would!" And "What repartee! I know no honest man Able to speak so well under all conditions." Like Oedipus, Cass resolves, impotently: "They shall never hold me in their power!" But, as Polyneices points out to Oedipus: "Are we not beggars . . . and exiles . . .? We live by paying court to other men." Like Oedipus, Cass rages helplessly after Flagg's rape of Francesca: "Creon, here, has abducted my two children." Cass actually quotes the passage: "Not to be born beats all philosophy. The second best is to have seen the light And then to go back quickly whence we came." Poppy, pleading with Flagg to stop humiliating Cass, parallels Antigone's "I beg Mercy for him, the beaten man! . . . Be kinder than you seem!" Cass became accustomed to thinking of himself, in his drunkenness and inertia, as relatively blameless: "And yet, how was I evil in myself? I had been wronged, I retaliated." And he scorns the inexperienced Americans in the PX: "The only ones Fit to be fellow sufferers of mine Are those with such experience as I have." Like Oedipus, though cognizant of his guilt he still feels himself essentially decent and the possessor of a holy endowment: "For I come here as one endowed with grace"; and he makes his demand: ". . . give me this shelter." His first desperate desire to give himself over for punishment after his murder of Flagg is calmed by Luigi, who thinks his "rage excessive," his "punishment too great for what [he] had done." Luigi, like the chorus, thinks:

"... you are surely worth our pity." One might see a parallel between the lengthy cold shower that Cass takes before his final self-freeing actions and the ceremonial bath that Oedipus takes in the grove before his translation. Peter, persisting in his inquiries, might say with the chorus: "What evil things have slept since long ago It is not sweet to awaken; And yet I long to be told . . . I only want to hear it truly told." Antigone's declaration could be uttered by Peter: "We know only In all that we have seen and done Bewildering mystery." And Luigi's philosophy is voiced by the chorus' final words: "Now let the weeping cease; Let no one mourn again. These things are in the hands of God." Cass is content at last in his true Colonus, Charleston: "He shall not seek another home, For this, in all the earth and air, Is most secure and loveliest."

Approximately half the book is composed of Peter's account of Cass's narration of his boyhood experiences, his wartime psychosis, the resentment against his epoch and his personal lot that resulted in the inert, drunken years of his postwar life in Paris and Italy. Sometimes this is given in the third person and sometimes by Cass himself in the first person. The account is in the form of total recall, and, as other critics have pointed out, amounts to a monologue. Used as it is here, this device makes for universality. The surrealism of total recall in itself raises Cass's account above the realistic setting of his act—a Charleston riverbank. The language, though Cass when reported objectively speaks with an habitual rich Southern rhetoric, goes beyond this into an extremely evocative metaphorical style, creating a vast tapestry of the fullness of life into which Cass's more folksy utterances fit. The result is that Cass, though never becoming in any degree an allegorical figure or a symbol, is nevertheless understood so thoroughly and in such a rich context that he signifies to us more than a single individual or a mere class.

Abraham Rothberg judges that Styron's novel suffers from being cast in detective-story form, with facts being withheld from the reader by the narrator and with a complicated schedule of flashbacks. He thinks the story would be better if told "straightaway," with Cass as the sole narrator.[6] There is undoubtedly some awkwardness and factitiousness in the telling. But a valuable result seems to accrue to the actual scheme. If Cass told his story "straightaway," it would

[6] Review in *New Leader,* July 4–11, 1960, pp. 24–26.

have, though at a different pitch, something of the emotional structure of Masefield's *The Everlasting Mercy*. The actual form of this novel is both centripetal and centrifugal. By the time we have read the first seventh of the book we have met directly or mediately all the real agonists and most of the bystanders, and recognized and viewed —if through a glass darkly—their plights. From the rim of the circle —or "gyre"—that they compose, we move both inward and outward. We come to understand the motivations and relationships of these people more clearly. And we are led to see along rays, as it were, going out infinitely from the center of their circle, these problems—with their ambiguous causes and victims—filling all the world.

Several critics judge that Styron does not succeed in realizing any of his women characters. Rothberg points out that the reality of the women never approaches that of the men. His stricture needs, perhaps, to be restated, and with a different emphasis. This novel, like *The Long March,* is about men and their world. Women enter it only as they are perceived by these particular men. The only "important" women in the novel are Poppy and Francesca. Celia is introduced only to show up Flagg's crudity, while Rosemarie fulfills expositional functions. Robert Gorham Davis complains that, "like so many heroines of apparently serious American novels, [Francesca] is a dream figure out of medieval romances or boyish fantasies."[7] But this comment is inadequate. Francesca is presented objectively only three times and then extremely briefly. During the arrest scene and during her brief encounter with Peter as she flees from Flagg, she is sufficiently complex to be realistic. And surely in her flashing account to Cass of her final vengeful defense against Flagg she is triumphantly actualized. The rest of the time she appears only through the mind of Cass, and in his distressed, bleared meditation she does indeed play a somewhat dreamlike role. But it is a role that presents to the imagination the poetic potentialities of Woman: spiritual and physical loveliness, devotion, a gay resourcefulness, warmth, and endurance. Poppy, too, is recognizably realized in a series of vivid, though brief, vignettes. It is true, however, that she is often shown only as an adjunct to Cass's sense of guilt, and in such passages she is somewhat abstracted. But Poppy trails clouds of glory. As Francesca is associated with calling, Poppy is with singing—a disembodied song

[7] *Op. cit.*, p. 44.

among the lemon groves. An incantatory ambiance surrounds Poppy; in her sturdy fragility are united actuality and the unfallen world.

A ritual-conscious generation, we are grateful to works that symbolize the passage from disequilibrium to equilibrium and the anticipation of new free action, whether the path is by recognition of guilt and atonement, recognition of the absurd and acceptance of its implications, or recognition and acceptance of the minuscule range of human potentialities. We cherish such books from the past, like *Crime and Punishment*; and of this century such as *The Longest Journey*, *The Magic Mountain*, *Mrs. Dalloway*, *Finnegans Wake*, *Absalom, Absalom!*, Camus' *L'Étranger*, and even such fragile pieces as those of Marguerite Duras. It is clear that *Set This House on Fire* involves an existential ritual. Paying tribute to the absurd with a consistent tincture of irony, the novel opens with a statement of confusion and horror and, after many agonizing episodes of evasion, concludes with the symbolic destruction of the phony, the achievement of a fair degree of self-knowledge, and the commitment to "being" as against "nothingness."

The irony that is never lost sight of suggests one more device making for universality. Both Cass and Peter are conscious of the humor and wryness of their situations, actions, and emotions. The diction, the rhythms—the total rhetoric of almost any paragraph uttered by either of them—indicates a constant awareness of a multiplicity of frames of reference. These do not cancel each other out; their interaction is more like that of the voices in a fugue. We recognize the strength of each factor while seeing each as a part of the ineluctable totality.

Papers of the Michigan Academy of Science, Arts, and Letters
Vol. XLVII, 1962 (1961 Meeting)

GROUPING OF PILGRIMS IN THE GENERAL PROLOGUE TO *THE CANTERBURY TALES*

JOHN REIDY

The University of Michigan

While the individual pilgrims in the *Prologue* have all received much attention, little has been paid to their groupings. O. F. Emerson suggested six groups, but did not go beyond a superficial treatment,[1] and Kemp Malone points out a number of the pilgrims such as the Parson and his brother, who were together prior to joining the "wel nine and twenty."[2] J. Swart has given a more subtle analysis,[3] but my treatment is so different that I can do no more here than make passing reference to him. The groupings I am proposing correspond closely to Professor Emerson's but I hope to go beyond external features and show criteria for the groupings which are related to Chaucer's literary purposes and to the general structure of the *Prologue.*

The Knight, Squire, and Yeoman, members of the same household, form an obvious group with which anyone recounting the Pilgrimage would naturally begin. Chaucer apologizes for not setting all the pilgrims in their degree,[4] but even the shortest wit would recall the Knight first, and the others follow naturally by association. Yet apart from this verisimilitude there is a further reason for beginning with these.

In the *Canterbury Tales* a court poet was proposing to present dramatically real people of all classes. Though he may never have actually read any of this work to the king and court, yet it seems a fair assumption, especially in view of the subsequent half-comic

[1] Oliver Farrar Emerson, "Some Notes on Chaucer and Some Conjectures," *PQ*, II (1923), p. 90.

[2] Kemp Malone, "Style and Structure in the Prologue to the *Canterbury Tales*," *ELH*, XIII (1946), 38–45.

[3] J. Swart, "The Construction of Chaucer's *General Prologue*," *Neophilogus*, XXXVIII (1954), 127–136.

[4] *Canterbury Tales* A.743–746. All quotations are from *The Works of Geoffrey Chaucer*, ed. F. N. Robinson (Boston, 2d ed., 1957).

apologies for "vileynye,"[5] that he had such an audience in mind. Some of the more grotesquely vulgar, brutal or vicious characters might,—though not necessarily[6]—be displeasing to them as a refined audience; certainly as a ruling class audience the pointed social satire might well be still more displeasing. Our present familiarity with the *General Prologue* makes us take it for granted. But at its first appearance a poet so conscious of his audience pretty certainly realized the need for tact when, for the first time in a long career, he was presenting to his refined hearers such vivacious disorderly persons who would come so unhandsomely betwixt the wind and their nobility.

Having announced, therefore, that the pilgrims are "sondry folk," he takes care not to offend anyone's susceptibilities at first. A courtly audience could heartily approve of the first two characters. The Knight is "parfit," and while older men might smile a little at the Squire's youthful exuberance in dress or in love, nobody would wish away this ribband in the cap of youth. Mr. Gaylord has indeed suggested in a paper read before this academy two years ago that the disgraceful nature of the war in which the Squire had campaigned implies that Chaucer was hinting dissatisfaction with the chivalric system.[7] This may be so, but none of the disgrace comes on the Squire; he remains an ideal young man. Mr. Gaylord's thesis I take as supporting evidence that Chaucer was paying special attention to his audience's susceptibilities, for the hint of dissatisfaction is a very guarded one; elsewhere in the *Prologue* Chaucer, when willing to wound, is not afraid to strike, yet here he will "just hint a fault, and hesitate dislike."

The literary function that I am proposing for the first group, then, is that of gaining the initial goodwill of the audience towards a new kind of poem. Though there can be no actual proof of this, the suggestion gains plausibility from a consideration of the Yeoman, the third member of the group. The Knight and Squire are sufficiently individualized and credible, though models, but the model Yeoman

[5] A.725–742.

[6] See, for example, the lord and lady in the *Summoner's Tale* (D. 2200 ff.), and the reception of the *Miller's Tale* (A.3855–3858); but the gentlefolk did object to what they thought the Pardoner might produce (C. 323–326).

[7] Alan Gaylord, "A. 85-88: Chaucer's Squire and the Glorious Campaign," *PMASAL*, XLV (1960), 341–?61.

is completely stereotyped. He is not a man, just a Yeoman. Such a typical Yeoman, in fact, that when in the *Friar's Tale* the devil disguises himself as a Yeoman, Robinson notes that the description of him is strikingly like that of the Yeoman in the *Prologue*.[8] Surely Chaucer could do better than this! But in this group Chaucer is maintaining an aristocratic point of view, and so his presentation of the first member of a lower class reflects this attitude. An aristocrat would be interested in a Knight's individuality, or a Squire's, but hardly in a Yeoman's. The Yeoman is skilled in woodcraft, his arrows are well cared-for, his dagger is sharp; he is a model from his superior's point of view. He resembles the Colonel's ideal of a soldier, smart, clean, impersonal, and no trouble; with his private feelings the Colonel is not concerned.

The unity of this first group, therefore, lies in the consistent point of view, wholly in keeping with the predispositions of the audience. It is possible that not many of them would in fact be upset by the realism and low life in the *Prologue,* or even by the clear statement of a great deal that was amiss in England—though this would be likely to make a conservative ruling class uncomfortable—yet still it is good policy for an author to make his audience comfortable at first, to present people they could easily sympathize with and be interested in, and, in general, to lead them gently towards the satire and novelty to come. Such a policy requires the treatment of the Yeoman which we actually find.

In the second group, the Prioress, Monk, and Friar, a satirical tone first appears. To ease the transition Chaucer makes the satire on the Prioress very gentle, and also exactly the kind most acceptable to a court audience. For after a few lines which are only amusing in retrospect,[9] we have the account of the Prioress's table manners, and of the pains she took to imitate court behavior and to be dignified. Here is a member of a lower class attempting to appear gentle, but succeeding through overemphasis in being merely genteel. To such an attempt Chaucer's audience would be most sensitive, and by it their laughter most readily provoked. The likelihood that the nuns of St. Leonard's, Bromley, made a practice of imitating their betters at Barking has long been recognized,[10] and although Marie P. Hamilton has shown

[8] Note on D. 1380 ff.

[9] A.118–124.

[10] See E. P. Kuhl, "Chaucer's Prioress," *PQ,* II (1923), 306–309.

that St. Leonard's was not so poor compared to Barking, nor so middle-class as has been assumed,[11] yet still it was not so rich and important, and the likelihood remains of a certain snobbery in their mutual relations.

This laugh, or smile, at the Prioress forms a bridge from the unsatirical, "straight" point of view of the first group to that critical, ironic one which subsequently prevails. And as a bridge it is firmly anchored in the account of the Prioress. For the satirical treatment which unifies the second group is based on the varying attitudes of these three ecclesiastical persons toward their vocations. Now as to the Prioress's regard for her vocation as a nun, her infringements of rule are minor and external. Even if we disallow Sister Mary Madaleva's plausible extenuations of these flaws,[12] yet still the small dogs, the fair forehead, and the gold brooch indicate only a minor degree of worldliness, and it is just such a minor degree of worldliness that is so perfectly expressed by her exaggeration of prescribed neatness into "counterfeiting cheer of court."

The satire, gentle for the Prioress, becomes sterner with the Monk. He ignores his vocation and devotes himself to field sports, the table, and the business of the monastic estates. He admits his neglect openly, dismissing the rule as out of date. He condemns himself out of his own mouth, and there is a fine ironical cast to the reasoning: "How shall the world be served?" Serving the world should be none of his concern. Yet though a bad Monk he does not seem a bad man; for he appears sufficiently serious-minded and respectable, and in this he differs from the Friar. With the Friar the satire becomes most severe, and we can see the progress of its development through this group. The Prioress fails externally, the Monk ignores his vocation, but the Friar abuses his, perverting it for his own pleasure and greed. As a Friar he has theological learning,[13] moral status, and sacerdotal powers, and in his cynical application of confession he employs all three for money-making. In the portrait of the Friar the satirical animus reaches a height not attained again until we get to the last two characters, the Summoner and the Pardoner.

[11] "The Convent of Chaucer's Prioress and her Priests," *Philologica: The Malone Anniversary Studies* (Baltimore, 1949), 179–190.

[12] *A Lost Language* (New York, 1951), pp. 27–60.

[13] He was a "maister" (A.261), presumably of Divinity, like the Friar in the *Summoner's Tale* (D. 2184 ff. and Robinson's note).

For each of these three satirical portraits Chaucer provides an appropriate satirical norm, with respect to which the character is judged. For the Prioress the insinuation is subtle, almost imperceptible, fitting the satire. Leaving aside the possible ambiguity of the "Amor vincit omnia" on her brooch, there are the ambiguous meanings, religious and secular, of *reverence, conscience,* and *charitable,* and it is borne in upon us as we read that though these could be used in describing the religious aspects of a character, in the description of the Prioress the secular meanings are foremost. When Chaucer sums up her disposition before describing her appearance, by saying "And al was conscience and tendre herte,"[14] everything he has mentioned insists on the natural, not the religious meaning. The occurrence of these words all within three lines,[15] and the repetition of *conscience* at the beginning and end of a paragraph,[16] remind us both of her vocation and of how she falls short of it. With the Monk, as the satire is more obvious, so is the norm. The rule of St. Benedict and some common monastic shibboleths are explicitly introduced and as explicitly rejected. As to the Friar, so flagrant are his transgressions that there is almost no need to establish an explicit norm, yet Chaucer does so, and in doing so emphasizes his satire. Miss McKisack in her recent work points out how St. Francis so impressed the imagination of the Middle Ages that any shortcomings of the Friars were embarrassingly evident by contrast.[17] Presumably a contemporary of Chaucer would make the comparison between the Friar and St. Francis without a reminder. However, Chaucer rubs in the contrast, informing us that his Friar knew the taverns well, "And everich hostiler and tappestere/ Bet than a lazar or a beggestere /For unto swich a worthy man as he/ Acorded nat as by his facultee/ To have with sike lazars acqueyntaunce."[18] Few would forget St. Francis among the lepers, and one of the well-known stories of St. Francis is that of the leper to whom he turned back to embrace.[19]

[14] A.150.

[15] A.141–143. See *Middle English Dictionary,* ed. H. Kurath and S. M. Kuhn: *charitable* adj., senses 1(a) and (c); *conscience* n., senses 4 and 2.

[16] A. 142–150.

[17] May McKisack, *The Fourteenth Century, 1307–1399* (Oxford, 1959), p. 309.

[18] A.241–245.

[19] See Paul Sabatier, *Life of St. Francis of Assisi,* trans. Louise S. Houghton (New York, 1928), p. 26.

The third group is of three laymen, the Merchant, Man of Law and Franklin, and the Clerk. The transition link from the Friar to the Merchant, emphasized by the linking rhyme, is their common interest in money-making. The Friar sells his sacerdotal powers as the Merchant sells his shields in exchange. Professor Emerson labels these four a "professional" group,[20] but they have a closer connection. The principle of unity here is not an abstraction—point of view, or attitude to a vocation—but a person, the Clerk, on whom the group pivots.

Consider first the Merchant, with his impressive exterior; apparently not merely a merchant but an important financier, yet, especially in contrast to the Friar, a great bore. He is a type too common in our Chambers of Commerce and even in our Business Schools, talking solemnly of the obvious. His political views are restricted to the piracy problem, his other conversation full of "bargaynes" and "chevyssaunce," "Sownynge alwey th'encrees of his wynnyng."[21] He has biset his wit well, but on an unworthy object. No greater contrast to the Clerk can be imagined, for the Clerk has devoted himself entirely to the treasures of the mind, completely neglecting money. In external appearance, in speech,[22] in intellectual interest, and—apparently—in moral probity, he is exactly opposed to the Merchant. The Merchant is secretive; the Clerk freely shares his treasure.[23] The nobility of the one's pursuit emphasizes the baseness of the other's. The Clerk has chosen the better part, while the Merchant, even in this life, seems to have lost the good of intellect. Yet if the Merchant's whole way of life is judged by the Clerk's, so also is the Man of Law's. The Merchant is wholly devoted to gain, the Clerk to learning, to knowledge as its own end, and for a higher end. The Sergeant too has gained knowledge, but in his case it is what Matthew Arnold might call an instrumental knowledge. The Sergeant uses it for gain, to his clients and to himself. Moreover, like the Merchant, he has some pretenses: his "reverence" is only seeming.[24]

The Clerk, in his concentration on knowledge, constitutes an ideal by which to judge the Merchant's aims, and in his use of knowledge

[20] *Op. cit.*, p. 90.
[21] A.275.
[22] "Sownynge in moral vertu." A. 307.
[23] A.308.
[24] A.312–313.

an ideal by which to judge the Lawyer's. The connection with the Franklin is not so clear, yet the Franklin, being in company with the Lawyer, demands a place in this group. He belongs to a busy class of country gentlemen,[25] yet his administrative and judicial activities are dismissed in three lines, exceptionally few,[26] and Chaucer concentrates to an exceptional degree on his interest in, and love of, food, in fact his gluttony. He bisets his wit on the pleasures of the table, and it is the abstemious sober Clerk who is there to judge him. Since he is not juxtaposed immediately with the Clerk, Chaucer makes the contrast the more blatant so that it will be observed.

By now Chaucer has disposed of the pilgrims who in real life might be familiar to his audience—the knightly family, moderately important Church dignitaries, and men important in the government of the realm, financially, legally, administratively—and is ready to turn to those performing more humble work. To these he provides a bridge, namely the five guildsmen, our fourth group. Like those who follow in the fifth group, they drive a trade, yet they are wealthy enough to be able to participate in London civic politics, which at times could rise to the level of national politics.[27] This group of five is something of a puzzle; they tell no tales, their characters are not developed. The function I suggest for them of forming a bridge between the socially important pilgrims and those of little social importance seems in fact to be their only function.

The six who follow, the Cook, Shipman, Doctor, Wife of Bath, Parson, and Plowman, are heterogeneous. Emerson calls them (except the Cook) a "provincial group."[28] The Cook forms an obvious link with the guildsmen, but the only principle of unity here is that all are engaged in some practical, nonadministrative type of work—cookery, navigation, doctoring, weaving, parish work, and agricultural labor. I see no trace here of the more subtle bonds which I have suggested for the earlier groups. Perhaps it is appropriate that at the lower end of the social scale, furthest from the king, the theoretical principle of unity, the elements of the *Prologue* seem most randomly arranged. This is perhaps too fanciful. Yet a larger order is asserted,

[25] K. L. Wood-Legh, "The Franklin," *RES,* IV (1928), p. 148.

[26] Only the Wife of Bath and the Miller have fewer lines devoted to their occupation (A.447–448, 562–563).

[27] As in 1388. See McKisack, pp. 454–459.

[28] *Op. cit.,* p. 90.

for this section ends with the model Parson and Plowman, who with the model Knight symbolize the three estates of medieval social theory.[29]

Such a bow to the concept of the three estates might have been an effective ending to the list of pilgrims, but there are five more, the Reeve, Miller, Summoner, Pardoner, and Manciple, clearly set off by their being listed together,[30] a rogue's gallery contrasting sharply with the Parson and Plowman, and especially since the break comes in the middle of a couplet. Their common characteristic is their swindling, as Swart points out, maintaining that nobody described earlier could be called an undetected criminal.[31] Accepting this statement with reserve in view of the Shipman and the Friar, we must agree that this is a concentration of villainy. And it is interesting that, excepting the Manciple, this group has provided Professor Curry with his richest field in studying Chaucer's use of medieval science for describing character.[32] Up to now significant physiognomical and medical details have been rare and scattered—the Wife of Bath's teeth, the Cook's "Mormal," the Friar's neck, and possibly the Monk's eyes[33]—but now they are concentrated. Chaucer's tone is more violent and even angry, less delicately ironic than before. Compared with the handling of the Friar the tone becomes quite brutal, and the scientific evidence of diseased bodies and warped minds helps to generate and maintain this tone.

This special tone, special method of character description, and special marking of them off formally as a group, all seem to indicate that their position in the sequence, after the round of society has apparently been completed and summed up by the representatives of the three estates, is no accident, but that Chaucer was indicating that their various kinds of swindling set them in some way outside society as parasites upon it.

As the flat impersonal Yeoman helped us establish the tone of the

[29] Cf. W. H. Clawson, "The Framework of *The Canterbury Tales*," in *Chaucer, Modern Essays in Criticism*, ed. Edward Wagenknecht (New York, 1959), p. 13.

[30] A.542–544.

[31] *Op. cit.*, p. 129. He speaks of the Shipman's "impulsiveness."

[32] Walter C. Curry, *Chaucer and the Medieval Sciences* (New York, 1926), pp. 37–47, 54–90.

[33] On shining, starting eyes, see *The Works of John Metham*, ed. Hardin Craig (EETS 132, London, 1916), p. 130.

first group, so here in the sixth does the flat and impersonal Manciple. The other four have striking gifts of strength, personality, or mental power—the Reeve least of all, but even he is feared.[34] The Manciple is characterized only by his swindling; this is his *metier*, his gift, his fair grace from God, his one talent which paradoxically bears interest through concealment. His lack of interesting personal qualities is artistically necessary not only to focus our attention on the point of swindling, but because he represents this antisocial force. The others are conspicuous enough, he is not. He is the essential parasite, anonymous, inconspicuous, ruinous. It is appropriate that the great legal masters, representing the impersonal force of law, should be preyed on by so impersonal a manciple.

Among this parasitic, antisocial group, Chaucer casually places himself. Exponents of biographical criticism may consider in which of his capacities he felt qualified for such society—as poet or as civil servant.

I have endeavoured to show that in the *General Prologue* there is evidence of certain groupings; that certain fairly obvious groups have a unifying principle more subtle and significant than appears on the surface; and that Chaucer arranged these groups for the literary purposes of manipulating his audience and for setting up different satirical standards, as well as for dealing with the pilgrims roughly in accordance with their status or function. These groups are linked by different bridges of association, except for the last group, where the abruptness of the break and its relation to the general structure of the rest constitute a means of controlling the audience's attitude towards it. The order in which we meet the pilgrims, though apparently merely the haphazard sequence of a casual memory, is really quite carefully planned.

[34] A.605.

Papers of the Michigan Academy of Science, Arts, and Letters
Vol. XLVII, 1962 (1961 Meeting)

CHAUCER'S SUMMONER: AN EXAMPLE OF THE ASSIMILATION LAG IN SCHOLARSHIP

THOMAS JAY GARBÁTY

The University of Michigan

It is a peculiar fact, and one worth noting, that scholars no sooner give life to a publication than they enter it into the journals, thereby to bury it. The journals, instead of transmitting knowledge, too often become a coffin for ideas which are only exhumed years later, and sometimes do not come to light at all. This is a serious problem of modern scholarship, and one that is interdisciplinary. I would call it the "assimilation lag," the pause, often long and with serious results, between the publication of a correct idea and its general acceptance as a teaching tool or guide to the deeper understanding and enjoyment of literature, or a science. In this paper I wish briefly to outline such a lag in a problem of Chaucer research.

The Summoner in the *Canterbury Tales* was the official of the medieval church whose chief purpose was to discover immorality and fornication in the parish and bring the culprit into a church court. That this office had already become corrupted in the late fourteenth century seems an obvious fact. Chaucer himself, in the Friar's Tale (D 1323–1326), describes such a Summoner as having doings with the low women of the town, who told him their secrets and with whom he was on an intimate footing. When the poet wrote "Ful prively a fynche eek koude he pulle," as early as 1909 Kittredge remarked [*MLN* VII (1909–1910) 476] that Chaucer described the Summoner's own method of fornication. The man looked evil, he scared away children, and he was evil. Haselmayer [*Speculum* II, (1937) 56–57] attributes this dreadful portrait of the Summoner to fourteenth century literary convention, which made the official into a villainous bogeyman to frighten rather precocious children. There seems no question but that the Summoner was, for Chaucer, already a literary type.

To the great joy of historical scholars, the Summoner also had a terrible skin disease: narrow eyes, depilated eyebrows and beard, highly flushed (cherubinny) face with fiery knobs and white, pustular whelks. Here was bait indeed, and the researchers sprang to the attack. To Walter Clyde Curry [*Chaucer and the Medieval Sciences* (New York, 1960), p. 38] we owe the traditional analysis of the Summoner's sickness. Curry, after reading widely in medieval medical texts, concluded, and I believe rightly, that the disease was called in the Middle Ages *alopecia,* a form of leprosy. Arnoldus de Villa stated that a complete depilation of the eyebrows and beard was definitely a sign of leprosy. John of Gaddesden (whom Chaucer's Physician mentioned in the *Prologue*) referred to the husky voice (identified by Curry as the "stif burdoun" which the Summoner carries to the Pardoner's song), falling of hair, and roundness of the eyes as leprous signs. Excess of salt in the system (saucefleem face), too much wine and general debauchery also were direct causes of this feared disease. According to Curry, it seems most understandable that Chaucer, reading the medieval texts, might choose to portray the Summoner as a leper. This theory stands today; it is to be questioned, and yet Curry in the original analysis was right.

Ever since Manly, literary historians have had the uncomfortable feeling in their work with Chaucer's pilgrims that the man they happened to be studying as a literary type may, perhaps, really have been alive at one time. There has been some critical opinion that this was the lot of the Summoner, since Chaucer was unduly harsh in his portrayal, more so than in his description of the other villains. Perhaps this was due to some personal feeling of antipathy toward a particular individual. In my own research I could not afford to overlook this aspect of the case, and my resolution was strengthened by a little note in the A.M.A. *Archives of Dermatology* giving the Summoner's symptoms and asking for a diagnosis [Marc Weiner, "A Case for Diagnosis, by Geoffrey Chaucer," LXXX (1959) 87–88]. There was the question. The Summoner's illness was described as leprosy in the Middle Ages; what was it called today?

With the help of medical advice and a modern text [Stokes, Beerman, Ingraham, Jr., *Modern Clinical Syphilology* (Philadelphia, 1945)], I must conclude that the Summoner was suffering from *Lues,* or a secondary stage of syphilis, with paretic neurosyphilitic involvement. All modern and medieval facts, symptoms, and signs, point to this doleful conclusion. The signs of the secondary stage of *Lues* are syphilitic alopecia of the eyebrows and patchy hair loss, any widely scattered nonblistering eruption, hoarseness and aphonia, ptosis (the dropped eyelid), neurasthenia, and conduct slump. All of these the medieval diagnosticians had recognized with acuity. The knobs on the Summoner's face are either papular or pustular *Lues*, and the syphilids, in their crusted state, have a greasiness which is remarkably reminiscent of the greasy skin of medieval lepers. In reference to the "scalled browes" and "piled beard" syphilologists warn that "every clinician should learn to watch the eyebrows. The clue to a secondary syphilis, a lepra, . . . may be found there." [Stokes, et al., p. 532] The "stif burdoun" which Curry describes as hoarseness is quite similar to the simple and erosive laryngitis so common to relapse symptoms in *Lues*. The white whelkes have always been mentioned as specific leprous signs because of the depigmentation of the skin, but syphilitic leukoderma causes the same reaction and is very common in patients with papular syphilids.

Chaucer mentions that the Summoner uses "quyk-silver" on his face and ointment that would cleanse and bite, which Curry describes as arsenic. It is known that mercury and the arsenicals are the oldest drugs to which syphilitics have resorted for centuries. The process was that of inunction, the rubbing of the ointment into the face, which caused a sudden and terrifying flare-up of all syphilitic lesions and a redness and discharge not before present. It is known as the Jarisch-Herxheimer effect, and the earliest patients complained that the cure was worse than the disease. Also, alcohol is the one drug which physicians consider most dangerous to patients under treatment for *Lues*, and we know that the Summoner liked much of it. Pustular infections, furuncles, and carbuncles are stimulated, and the skin is flushed, perhaps to the point of being "fyr-red."

In the discussion of paresis and neurosyphilis we leave the factual comparison of symptoms and range into the field of conjecture, and I might give less validity to this. Alopecia and leukoderma are included here, and also what physicians call the conduct slump, slackening of sexual morals, at times maniacal excitement, persecutory ideas which may subside, only to take the form of a well-directed attack later on a single object or person, as for instance the Summoner's tale, the attack on the Friar. "Upon this Frere his herte was so wood/ That lyk an aspen leef he quook for ire" (D 1666–1667). We lack a Wasserman and a spinal tap. Were the man living today he would be recognized as an advanced syphilitic.

There are some facts, however, which make me believe that the

Summoner was an artistic creation of Chaucer. The poet uses stock phrases to describe him, as Curry has already found. Beyond that, the phrase "Wel loved he garleek, oynons, and eek lekes" was also a cliché. Robert Kaske found a similar phrase in the Book of Numbers [*MLN*, LXXIV, 481–484], and I have found John of Burgundy use it as a warning against the plague, and John Arderne as a warning against hemorrhoids. If then, the Summoner was fictitious and Chaucer did give him a disease called leprosy in the texts, the obvious question would be why leprosy? Was the *cause* important?

It is of great interest that all the medieval medical texts make specific mention that leprosy was caused by infection through coitus. Trevisa's translation of the Bartholomaeus, *De proprietatibus rerum,* mentions that leprosy "commeth of fleshlye lyking be a woman soone after that a leprous man hath laye by her." John of Gaddesden remarks that "He who has lain with a woman with whom a leprous man has had intercourse feels pricks between flesh and skin . . . with pustulation of the face, then of the body." Lanfrank, in his *Science of Cirurgie* states of leprous men that "þei wilneþ mycle to comne wiþ wommen," and that afterwards pustules will grow in their tongue, a rather suspicious sign of the primary lesion. Guy de Chauliac, undoubtedly the most important medical writer of his time, tells that leprosy is got from "filth of gendring" and in a later time the men become "wily bigyling and wode." We are confronted with the undeniable fact, then, that leprosy in the Middle Ages was a venereal disease.

Since this is so, and the Summoner obviously is suffering from such a disease, we can be certain that this is not modern leprosy for two reasons: (1) Leprosy, in spite of all historical and Biblical opinion is of an extremely low order of infectiousness, compared with such diseases as measles and smallpox for instance. Contact must be very long and intimate before Hansen's bacillus is transmitted. Patients are usually infected when they are children. But a noteworthy instance where long and intimate contact only extremely rarely results in infection is in the conjugal relationship. "Very few husbands have infected their wives, or wives their husbands." [Ralph Hopkins, "Heredity in Leprosy" in Moulton ed., *Tuberculosis and Leprosy, the Mycobacterial Diseases,* 1938, p. 115] Yet aside from this fact, there is another reason. (2) Presuming the *Tales* are fiction and a literary creation, we know that Chaucer would never have purposely included a known leper among his pilgrims, because this would have been a social impossibility. Fear of leprosy ran to phobic proportions in the Middle Ages. A look, a whisper from a leper infected, and the wretches had to use a clapper at their approach, had to take their food with a stick, and finally, had to isolate themselves from the community. By order of a law strictly enforced, lepers whenever possible were segregated in lazarettos and leprosaria. In the thirteenth century there were nineteen thousand of these in Europe. From 1350–1400 London joined in this fight against the disease and isolated its sick to the extent that by the middle of the fifteenth century leprosy was almost completely stamped out in Europe. [Clay, *The Medieval Hospitals of England* (London, 1909), p. 28]

The problem of confusion of syphilis with leprosy is an acute one today among physicians who are not specially trained in the specific differentiation. *Lues*, the most protean of diseases, is known as the "Great Imitator," and it wears the mask of leprosy more often than of any other sickness. "The differential problems are most difficult in the earlier stages of the disease [leprosy] when the hoarseness, alopecia of the eyebrows, papules and nodules in the face . . . are remarkably suggestive of syphilis." [Stokes, et al. p. 702] If there is confusion today, one wonders how great the mixup in the jungle of dermatology was in the Middle Ages? One might say that the stringent laws mentioned above and the fact that the Summoner's presence would have caused great concern among the Pilgrims are evidence against the authenticity of the *Tales*. Chaucer, in concentrating on the moral causes of a disease, may have forgotten the social connotations. If so, it was not the first time the poet nodded over his "rekenynges" to the detriment of realism albeit the benefit of literature.

The problem has always been that the plague of syphilis was first recognized at the epidemic of Naples among the French army in December 1494, and historians have felt, since shortly before the outbreak Columbus returned from his second voyage, that perhaps the disease was brought over from America. There are weak points in this theory. Why was there not an outbreak in Spain where Columbus landed or at the Azores? Also, the descriptions of the French pox list the pustular and nodular facial symptoms of the secondary stage, which on the average appear only in the second year of infection. Inoculation, therefore, must have been pre-Columbian. Many scholars hold that the large pox was known among the Arabs during the

Crusades. There certainly is almost incontestable diagnostic evidence among the medieval texts even to the primary lesion, though a diagnosis of chancroid or "soft chancre" may perhaps not be precluded here: John of Arderne in his *Treatises of Fistula in Ano* describes a remedy for "þe cancre of a mannez ȝerde."

The conclusions would be then that the Summoner did not have leprosy, but a venereal disease, with symptoms which can be closely identified with secondary syphilis. To know this enhances the whole great irony of the picture. The idea of leprosy is a useless device; the idea of a syphilitic Summoner, the watchdog of morality, who shows the signs of his lechery in his face, is a huge joke, in the true Chaucerian sense. Not to know the full connotation of the disease, is to see Chaucer through a glass darkly.

And here is the lag, for it was Curry himself who first hinted at syphilis in a short and parenthetical sentence in which he commented on a possible medieval confusion. He never meant to have a literal interpretation follow his research. Ever and again scholars have mentioned, very vaguely, the syphilis idea, and Professor Albert C. Baugh wrote me recently that he believed the Cook and the Summoner were both suffering from the disease. But Manly accepted leprosy and the standard anthologies for high school and college students footnote the disease as "the symptoms and suggested treatment indicate a form of leprosy," "afflicted with a disease not unlike leprosy," "the Summoner was suffering from a disease akin to leprosy," "the Summoner possibly has leprosy," "Whelks (leprous) sores." Not one book mentions the venereal origin of the disease. What a loss, albeit perhaps a small one, in the appreciation of Chaucer's satire!

The assimilation lag can be seen further on an important point of Chaucer's life. In 1955, in the *Recueil de Travaux offert à M. Clovis Brunel* [*Société de l'école des Chartes*, Suzanne Honoré-Duvergé, "Chaucer en Espagne?" 9–13] was published a document that gave Chaucer safe conduct into Castile in 1366, putting light on a period of the poet's life of which hitherto we have known nothing. The document had been previously published in France in 1890, and again in 1936. Yet Robinson's revision of Chaucer in 1957 still mentions the erroneous theory of Chaucer's campaign in Ireland with Prince Lionel, and so do all the anthologies printed in the last two years. This gap is still open.

But the arts are not alone in suffering from the lag. It is much more dangerous in the sciences. What may not be commonly known, and which was reprinted in the *Alpine Journal* in 1952 [Notes, LVIII, pp. 536–38] is the fact that John Tyndall, mountain climber and physicist, discovered penicillin in 1875, somewhat more than fifty years before Sir Alexander Fleming in 1928. In an essay prepared for the Royal Society Meeting in 1876 and published in 1881 [*Essays on the Floating Matter of the Air*, (London: Longmans, Green, and Co.) 1881], Tyndall mentioned the following: "The mutton in the study gathered over it a thick blanket of Penicillium. On the 13th (December, 1875) it had assumed a light brown color, as if by a faint admixture of clay; but the infusion became transparent. The clay here was the slime of dormant or dead bacteria, the cause of their quiescence being the blanket of penicillium . . . In every case where the mold was thick and coherent the Bacteria died, or became dormant and fell to the bottom as sediment. . ." Had the bacteriologists noticed this publication, had there been no lag, lives might have been spared. The problem in the arts may not be fatal, but it is still serious. The assimilation lag will grow. New journals are appearing; the tide of publication is rising. The answer must be found soon, else much that is good in scholarship will be overlooked with the bad. The solution lies, of course, in publications like the *Abstracts of English Studies* on a wider scale. Regional organizations under the *MLA* can fill the gap, and there is already forthcoming a content index for articles in the medieval field. Perhaps the Renaissance, Victorian, and Shakespeare societies can publish yearly abstracts of their own. The fact remains that not all published articles are good, and that no matter how much an institution produces, a series of poor articles will affect its reputation worse than none at all. Some screening will have to be done, and it may have to come from the schools themselves, else we shall find ourselves looking for needles in a haystack, and soon we shall not be looking at all.

Papers of the Michigan Academy of Science, Arts, and Letters
Vol. XLVII, 1962 (1961 Meeting)

THE UNCONQUERED TALE OF THE PRIORESS

ALAN T. GAYLORD
The University of Michigan

Although Chaucer's Prioress has long been a favorite of the scholarly and not-so-scholarly commentators, her portrait in the *General Prologue* has received considerably more attention than her Tale. Presumably this is because her story of the "litel clergeon" and the Blessed Virgin invites appreciation rather than discussion, and offers few of the problems that have been discovered in the portrait. If the Tale has seemed simple and pious, the portrait has been taken as complex and ambiguous, exhibiting a kind of mildly flawed humanity with which we seem particularly comfortable these days; more so, at any rate, than with an account of supernatural intervention and devotional intolerance. While the Prioress herself has often been approved as a pleasant representative of the later Middle Ages, one whom the contemporary may view with a mixture of familiarity and condescension, her Tale has been quite a different matter. Certainly, it sounds good like a Prioress' Tale should—but it is not read very much, either. The portrait has generally stood alone.

Yet I would argue that the *General Prologue* is not a static gallery of "characters" in the later Overburyean sense; its original purpose was to present the *dramatis personae* of an action which Chaucer at one time conceived in elaborate scale, a pilgrimage where differentiated characters would act and react through storytelling. The action is not complete, certainly, and Chaucer seems to have developed some of his ideas as he worked, but as a general hypothesis, Kittredge's suggestion that "the Pilgrims do not exist for the sake of the stories, but *vice versa*"[1] is both sensible and persuasive. In the case of the Prioress, however, such a hypothesis runs immediately into difficulty. Does her portrait introduce an expectation which her Tale fulfills? Since most scholars have agreed that there

[1] George Lyman Kittredge, *Chaucer and His Poetry* (Cambridge, Mass., 1951 [1915]), p. 155.

is some satire in the portrait, is the reader to find that her Tale is handled satirically, too? Here it is impossible to find agreement.

F. N. Robinson, speaking of the Tale's "perfect adaptation to the teller," maintains that Chaucer "is as far from showing disrespect for the story as for the devout lady who tells it," and that it serves admirably "to complete the portrait of the Prioress in the Prologue."[2] His choice of the verb, "complete," further intimates his theory concerning the relationship between teller and tale. We may take his views as representing the conservative, or "soft" reading of the Tale which, to be sure, the majority of the critics, including such apologists as Sister M. Madaleva[3] and Sister Brigetta McCarthy,[4] have elected. But there is a "hard" reading, too. E. T. Donaldson calls the Tale "a strange mixture of delicacy and horror," and sums up his views by saying: "[Chaucer] made the Prioress' Tale in some ways as pretty as her own brooch, but it is the failure of her character, not his, which makes the poem so imperfect an expression of the motto *Amor Vincit Omnia*."[5] Paull F. Baum has a short examination, agreeing in substance with Donaldson, which suggests that the Tale "matches the reverent raillery of our first introduction to the Prioress."[6] And R. J. Schoeck has recently gone the farthest with a "hard" interpretation by maintaining that "the widely circulated ritual murder legend is held up for implicit condemnation as vicious and hypocritical."[7] This is surely as far as one can get from Robinson's opinions cited above, or from Robert Kilburn Root's, that the Tale "breathes the spirit of earnest, heartfelt religion."[8] In one sense, then, the Prioress' Tale remains unconquered, and with it the question of its dramatic propriety.

[2] In his edition of *The Works of Geoffrey Chaucer* (Boston, 2d ed., 1957), p. 11.

[3] "Chaucer's Nuns," in *A Lost Language and Other Essays on Chaucer* (New York, 1951), pp. 29–60.

[4] "Chaucer's Pilgrim-Prioress," *The Benedictine Review,* VI(January, 1951), 38–40.

[5] In his *Chaucer's Poetry An Anthology for the Modern Reader* (New York, 1958), pp. 932, 934.

[6] *Chaucer: A Critical Appreciation* (Durham, N. C., 1958), p. 79.

[7] "Chaucer's Prioress: Mercy and Tender Heart," in *Chaucer Criticism* The Canterbury Tales, ed. Schoeck and Jerome Taylor (Notre Dame, Ind., 1960), p. 246.

[8] *The Poetry of Chaucer, A Guide to Its Study and Appreciation* (New York, rev. ed., 1950 [1934]), p. 190 f.

No matter which way one chooses to turn, whether towards the "soft" or the "hard" interpretations of the Tale, he still faces problems. If the Tale is a shining representative of a Miracle of Our Lady, it is a signal instance of elaborate work in a portrait being laid aside or overlooked when the tales begin, thus suggesting that the *dramatis personae* theory of the *General Prologue* can only be taken so far.[9] If the Tale is a subtle instance of satirical undercutting not only of a character but a whole genre, it is practically a unique instance of a poet undeniably Christian working out a devotional atmosphere in a poem which is contradicted almost on its own terms. The implications could be far-reaching. The Tale would then have to be read as in part an indictment of certain areas of Christian tradition, and in part an attack on "sentimentality," and we are not sure Chaucer's independence of his social and religious milieu extended that far. In short, the way one avoids difficulties with the Prioress' Tale is by denying that her portrait is satirical at all. Otherwise, difficulties can increase and multiply.

But for me, denying that her portrait is satirical would involve the greatest possible difficulty in proof. I find it to be highly satirical; and, in fact, I go farther than most in my reading of it, for I cannot see what is very reverent about it any more than I can find all the love and affectionate acceptance of her foibles which so many commentators have discovered in Chaucer's attitude. To demonstrate this, to pursue the question further, and to prepare for a discussion of the Prioress' Tale, it is first necessary to re-examine her portrait, attempting to establish as accurately as possible the attitudes and the techniques that can be found there; otherwise, there will be nothing very definite to match against the Tale. Such a procedure, I recognize, is tantamount to begging part of the question. It has not been demonstrated that Chaucer necessarily intended much of a link between every portrait and its tale, and I know that a case could be made for considering this Tale in isolation. But I will assume for the moment that it is worth while to precede work on the Tale by consulting the portrait, chiefly because the main thesis of this paper is that a very significant matching between the two is possible, one which confirms the "hard" reading by revealing a delicate but nonetheless devastating satire operating in both places and in roughly

[9] Which, indeed, Bertrand H. Bronson has recently suggested. See *In Search of Chaucer* (Toronto, 1960), pp. 66–72.

parallel manner. Thus to examine the portrait is to define further the questions this paper attempts to answer.

It is true that there are no explicitly condemnatory or pejorative words in the portrait, and that the diction appears even enthusiastic: "ful symple," "ful weel," "ful semely," "ful faire," "ful plesaunt," "ful fetys," are no more caustic than they are particularly reverent. But with Chaucer, the arrangement of his material is just as important a way of speaking as individual phrases; and what he will do in this case with context is demonstrated most brilliantly in the first four lines:

> Ther was also a Nonne, a PRIORESSE,
> That of hir smyling was ful symple and coy;
> Hire gretteste ooth was but by Seinte Loy;
> And she was cleped madame Eglentyne.[10]

Here Chaucer gives with one hand and takes away with the other. It is a technique of the whole portrait, which is very carefully shaped after such a manner. First, her important status: she is not just a nun, she is a prioress, and must be a model to all other nuns, a mother to her children, a shepherd to her flock. Speaking of the abbot of a monastery, the office that corresponded to the Prioress' (or, technically, to an Abbess'), the Benedictine Rule says:

> Above all things let him not slight nor make little of the souls committed to his care, heeding more fleeting, worldly and frivolous things; but let him remember always that he has undertaken the government of souls, of which he shall also have to give an account.[11]

There are only two people among the Canterbury pilgrims that have charge over the spiritual welfare of others: the Prioress and the Parson.[12] Concerning the Parson, Chaucer emphasizes one fact:

> 496 This noble ensample to his sheep he yaf,
> That first he wroghte, and afterward he taughte. (p. 22)

Almost the entire portrait is given over to an elaboration of this idea.

[10] Robinson's edition, p. 18. Subsequent page numbers will appear in parentheses after each quote.

[11] *The Rule of St. Benedict*, trans. Cardinal Gasquet (London, 1936), p. 14.

[12] Perhaps we should add the Monk, who is "kepere of the celle," and "a manly man, to been an abbot able," though he is not one; and for that matter, all the ecclesiastics are somewhat eligible for inclusion because of their "pastoral" functions. But the Prioress and the Parson are most clearly in charge of a flock.

The Prioress must be evaluated in the same light. It will not do to speak of her primarily as a woman and a nun. Chaucer's assigning her the role of Prioress was a result only of his own design, and we must not overlook its importance. The establishment of her status makes more effective the reversal of the following lines, for if they elaborate upon the kind of nun and Prioress she is, they do so in an unusual way. We begin to learn not about her "government of souls," but about "more fleeting, worldly and frivolous things."

In "that of hir smylyng was ful symple and coy," attention is riveted upon an image by means of the long vowel in "smylyng" and the primary stress it receives in the line. Madame Eglentyne is a smiling prioress. There is nothing wrong with that, one presumes. She is the only smiler in the pilgrimage, although Friar Huberd's eyes twinkle. But this is hardly the point. The real fillip of the line is the "symple and coy," and John Livingston Lowes has shown why, in a short piece on poetic diction which is a classic of Chaucerian scholarship.[13] The "smylyng" and the "symple and coy" go together, although here I differ from Lowes, who evidently imagined the Prioress as a young woman for whom the faintly erotic overtones of the latter phrase would be appropriate. For it is unlikely that the "mother" Prioress would be young, and rather probable she would be at least middle-aged.[14] The "smylyng" in fact qualifies the "sym-

[13] "Simple and Coy. A Note on Fourteenth Century Poetic Diction," *Anglia,* XXXIII(1910), 440–451.

[14] It is a question of opinion, rather than fact, and of relative likelihoods. Chaucer does not mention the Knight's age either, but his listing of the Knight's campaigns gives as much of a hint as would a modern writer who might tell us that a hero had been "a campaigner in World War I." In the case of the Prioress, the question would be whether or not there was a "typical" age for a Prioress that would figure in the mental picture Chaucer's audience would be forming. Here I turn to Eileen Power, *Medieval English Nunneries* (Cambridge, 1922). The reader is at liberty to draw his own conclusions from Chapter II, "The Head of the House," pp. 42–95. By Benedictine rules, the Prioress had to be at least twenty-one. The post was elective, though confirmed by the Bishop, and ordinarily of life tenure. This would make it normally unlikely that, when an elderly Prioress died or resigned, a young newcomer would take over instead of an older and well-known nun (though such did happen in certain instances). Thus a Prioress might be any age after twenty-one, but "statistically" likely to be over thirty—which I conclude after looking through the records. The Council of Trent, incidentally, established the *minimum* age at forty. Since it seems to have been "unusual" for a Prioress to be young (she had a number of heavy responsibilities, after all), I would expect Chaucer to have

ple and coy"; Madame Eglentyne resembles her namesakes in amorous courtly verse only from the grin up, so to speak. But the possible graciousness of her smile is associated in this line not with the charity of Christ but with the "cheere of court."

The third line of the portrait seems gratuitous only if we ignore Chaucer's meaningful disposition of his material. It looks like an approving observation: her greatest oath was *but* on St. Loy, and in a naughty world which daily dismembered Christ with its swearing, such restraint seems fully ladylike. But the line is comparable to the Monk's asking "How shal the world be served?" (line 187, p. 19). It is not the Monk's business to look after the world's business, any more than it is the Prioress' business to swear at all.[15] Furthermore,

been explicit if he wished her to be taken as younger than ordinary. The lines about her femininity can be just as amusing when we consider her middle-aged as when she is taken as a sweet young thing, as I argue below.

[15] Consistent with the predictable weakness of historical scholarship for diverting byways is the little bibliography that has grown up over the matter of St. Loy, or St. Eligius, and the Prioress' "oath." We have now a great deal of information on this saint, most particularly in the early and still classic article of Lowes, "The Prioress's Oath," *Romanic Review,* V (1914), 368–385, and the recent article of John M. Steadman, "'Hir Gretteste Ooth': The Prioress, St. Eligius, and St. Godebertha," *Neophilologus,* XLIII (January, 1959), 49–57 (which also notes earlier articles not cited here). Discussion of the appropriateness of this choice is a model of some of the larger debate on appropriateness of Tale to portrait. But the discussion is too often inclined to substitute the question "what might the Prioress know about St. Loy?" for the more valid, if more complex, "what did Chaucer know and expect his audience to know about St. Loy?" Surely much of the information collected becomes irrelevant in light of this latter question. For example, Steadman underlines again the fact that St. Eligius was once cited as having refused to take an oath, and suggests that "in swearing by a non-juror, she achieved the ultimate refinement—the hypothetical vanishing-point—in swearing" (p. 57). It is an attractive point, but does it not take Chaucer too seriously? Steadman advances it because "the primary interpretation of this passage should hinge not on why St. Loy is the Prioress' favorite saint, but on why Chaucer calls this her *greatest* oath" (p. 56). The easiest interpretation still seems to me to be the least complicated: the ironical mention of a *peccatum* thrown up to public view by the very act of ostensibly passing over it lightly. That the Prioress *thinks* she is being ladylike, courtly, very French, is also obvious, though it seems unnecessary to collect data about St. Loy that goes beyond, say, the *Golden Legend,* which is where the *Second Nun's Tale* comes from. Perhaps her swearing may be admitted to be ladylike, but it is not very Prioress-like . . . and she shares her saint with the carters! (See *The Friar's Tale,* line 1564, p. 92, and Lowes' article on the oath, pp. 382–385; Lowes does not explore the intriguing

the tone of this third line is in keeping with that of the second, and continues the falling away from the expected direction the first line had suggested. The opening of the portrait is then rounded off by the fourth line which fulfills the atmosphere of the court but not the cloister. Her name "is exactly what a little girl would like to call her favorite doll," says Sister Madaleva,[16] and one hardly needs to say more, except to observe that thinking of the Prioress as an older person gives added tang to all of the language used to describe her in terms of youth and stylized loveliness from the French Romances.

In fact, if the Prioress is not a young lady, some of the portrait becomes decidedly acidulous. When Chaucer returns to the person of the Prioress,

> Hir nose tretys, hir eyen greye as glas,
> Hir mouth ful smal, and thereto softe and reed,

he does in essence the same thing with these traits that he does with the very arrangement of his words throughout: for although each feature is conventionally part of the medieval list of charms, beauty is essentially a matter of proportions, of parts in relation to a whole. And when Chaucer finishes his tactful catalogue, he has not particularly helped matters:

> But sikerly she hadde a fair forheed;
> It was almoost a spanne brood, I trowe;
> For, hardily, she was nat undergrowe. (p. 18)

I find persuasive the suggestion of Gordon H. Harper that in this last line, "Chaucer discloses the real point; the Prioress is very large of forehead, and her bodily proportions are quite in keeping.... she was very large, and her manners sat the more amusingly upon her broad spinster's brow."[17] In his reading, and in mine, Chaucer is seen to have prepared this "surprise" for the reader, one which requires a reaching back and adjusting of the several images already recorded. The Prioress may conceivably be some Amazonian beauty,

---

paradox of having the same saint provide both a "très aristocratique" and a "très democratique" allusion.)

[16] "Chaucer's Nuns," p. 35.

[17] "Chaucer's Big Prioress," *PQ,* XII(July, 1933), p. 309. To call her "spinster" seems a little ungentlemanly to me, not to mention the label of "bulbous" which Mr. Harper affixes to her!

but she is not the fragile and dainty creature that Chaucer's diction has ostensibly prepared us for.

This technique of giving with one hand and taking away with the other is remarkably of a piece. Chaucer continues with it after the first four lines of the portrait, speaking in the same sentence of her "semely" singing of the service . . . which she "entuned in hir nose," and of her "faire" and "fetis" French . . . with the Stratford accent.[18] I find it impossible to take these items and this technique as anything but a dig, nor can I see how they are particularly reverent.

The "symple and coy," the "Seinte Loy," the "Eglentyne," and the French all lead naturally in one sense, unnaturally in another, to her table manners; for the English learned their "curteisie" from the French, and the Prioress is à la mode. Two standards of courtesy have already been intimated in the *General Prologue,* if I am correct in seeing contrasts between the description of the Knight and the Squire,[19] and correct in identifying those two standards as essentially those of morals and those of manners.[20] The Prioress' mode is unmistakably that of the Squire; the line about "in curteisie was set ful muchel hir lest" is sandwiched between a discussion of her manners. As the Squire is described partly in terms used in the *Romance of the Rose,* so is she. I believe it was Lowes who first noted that the description of her table manners may be associated with the un-

[18] On these two items, as on the oath by St. Loy, there is again bibliography and controversy among the "soft" and the "hard." The French is rather clearly an instance, as Harper says, "of speaking French like an Englishman," and one would expect the service-singing to work in the same way. However, nothing of real authority has been done yet on the subject. Muriel Bowden in her chapter on the Prioress in *A Commentary on the* General Prologue *to the* Canterbury Tales (New York, 1957 [1948]) simply observes, "Chanting the service has always demanded a nasal quality to avoid strain on the vocal chords" (p. 102); and Robinson notes, "This mode of nasal intonation is traditional with the recitative portions of the church service" (p. 654, n. to line 123). But for a counterstatement, see M. J. Brennan, "Speaking of the Prioress," *MLQ,* X(1949), 451–457.

[19] There is a preliminary exploration in my "A85–88: Chaucer's Squire and the Glorious Campaign," *PMASAL,* XLV(1960), 341–360.

[20] Cf. Nevill Coghill, "Education in Courtesy," in his *Geoffrey Chaucer,* No. 79 in *Writers and Their Work,* ed. Dobrée (London, 1956), pp. 20–27. See also Chap. VIII, "The Literature of Nobility," pp. 212–232, in my unpublished Princeton University dissertation, *Seed of Felicity* (Princeton, 1958: University Microfilms, 1960).

godly instructions of La Vieille.[21] Yet the association has not been adequately explained. Surely Chaucer cannot have imagined his audience or even his reader to have noticed these allusions to a short passage in a vast work of 21,780 lines; we must rather imagine that he was using the *Romance* as a kind of source book for a formula which would express the "breeding" of one kind of courtesy in the courtliest of terms. Chaucer is not trying to associate the Prioress with La Vieille, the old whore of the earlier romances; at least it seems to me that he has plenty to do in his satire without needing to throw the basic virtue or chastity of the Prioress in the shade. He does not need to intimate that she is a bad woman; it is enough to intimate that she is not a good Prioress.

But why are meticulous table manners to be taken as a matter for reproach? Since it was the Prioress who would entertain visitors to any nunnery and feed them at her private table, should we not rejoice that she was "wel ytaught" at meat? Once again, of course, it is the context that is important. The Prioress' "condicioun" is being described in terms of good table manners and courtly decorum. There is little need to elaborate upon this subject, for many others have covered it well. Chaucer moves from the descriptive details of French romances to the characteristic traits of courtly "courtesy," and at the center of the portrait is explicit about the model that the Prioress is imitating. Whether her attempts to be "estatlich of manere," and to be "holden digne of reverence" arise because she is a member of a lower class attempting to appear gentle, or because she is a member of the nobility who has found it hard to put away her expectations of a debonaire way of life does not really matter in assessing her essential role as Prioress; although I am inclined to the latter view on the basis of Eileen Power's work on the social classes in the nunneries.[22] The point is, to be impolitely flat about it, that she imitates

[21] As in lines 13,385–13,474 of the *Roman*. See Lowes, "Simple and Coy," p. 441. I do not think anyone has pointed out another intriguing item: preceding this passage, La Vieille gives instruction in how a lady should *smile*! (Lines 13,352–13,366; ed. Langlois, *SATF*, IV, 16 f.)

[22] The argument is scarcely less tenuous than the one traced out in note 14 above. See Power, Chap. I, "The Novice," especially such statements as the following: "the medieval nunneries were recruited almost entirely from among the upper classes. They were essentially aristocratic institutions, the refuge of the gently born" (p. 4). But daughters from rich merchant families did enter convents in Chaucer's day. Power takes Chaucer's Prioress as the complete "lady"

not the court of the New Jerusalem but of Babylon. As Prioress she is a member by rights of Christ's nobility; but in more ways than the sheer mechanical arrangement of the procession of pilgrims, she follows the Squire and could dance in his garden.

I have spoken thus far of Chaucer's description of the person of the Prioress and of her manners. It remains to speak of her spirit, or inner personality, or, as Chaucer puts it, "of hire conscience." Once again, there is little that could be said that has not already been said better. The mousetrap emotions are silly sentimentality; the dogs are illegal, their bread extravagant, and their beatings improperly mourned. The defenses of these feelings as sweetness of spirit are surely misguided: not that one should treat the lower creatures of God with anything less than tenderness and pity—St. Francis would surely have extended his kindness even to Brother Mouse—but, once again, that Chaucer is meaningful through what he does not say, and in the way he arranges his material. Between "But, for to speken of hire conscience" and "al was conscience and tendre herte" there is nothing but the shallowness of bathos. A most useful anatomy of the Prioress' "conscience" has been provided by John M. Steadman, who says in part:

> Chaucer has not only pointedly omitted any reference to the true objects of charity—God and neighbor. He has also deliberately cited as the recipients of her charity objects demonstrably false. Such irony is all the more telling, since in the portrait of the Plowman he defines the true objects of charity with exactness and precision.[23]

(p. 8). The Barking-St. Leonard's contrast may have been suggested to the audience of Richard's court (see Robinson's notes, p. 654), although the idea is too close to the Manly theory of a "local Christmas revue" to persuade me; but even so, the contrast would be between rich and poor rather than between the elite and the "pushers." That Chaucer's audience, however, would laugh at the Prioress because she was "aping her betters" strikes me as an incomplete accounting of Chaucer's satire, making her status as Prioress an accidental rather than an essential part of the irony. But even the possibility of a Barking-St. Leonard allusion and contrast seems pointless if we accept the reasoning of Marie Padgett Hamilton, in her "The Convent of Chaucer's Prioress and Her Priests," in *Philologica: The Malone Anniversary Studies,* ed. Thomas A. Kirby (Baltimore, Md., 1949), pp. 179–190; for she suggests that the convent was neither as little, as poor, nor as low-class as some scholars have thought. See esp. pp. 183–190; her evidence is not incontrovertible, but is fully as persuasive as and generally more factual than many of the previous accounts given of the two convents.

[23] "The Prioress' Dogs and Benedictine Discipline," *MP,* LIV(August, 1956), 3.

And all of this not simply of a Nun, but of a Prioress. "If gold ruste, what shal iren do?" After such description, after the return to physiognomy and fashion, and after what I have described as the "surprise" of her full figure, the final business of the "ambiguous coral" rosary[24] and the brooch with its Latin motto can only seem like the strongest irony. I say strong irony because I cannot come to these items as they are placed without feeling that the portrait closes with a quiet but nonetheless firm act of devastation. Lowes' note on this is often quoted: *"Which of the two loves does 'amor' mean to the Prioress?* I do not know, but I think she thought she meant love celestial."[25] But I fear this is to be led slightly out of the way. Could one ever seriously wonder if the Prioress "thought" that *Amor* contained even a trace of earthly love? "This were an inpossible!" I am more inclined to say that the brooch reminds us that the Prioress is good about representing charity on jewelry but not so good about defining it. In terms of the portrait, with both its omissions and commissions, the Prioress has not proved she truly understands what celestial love means.

The portrait has been called "gentle and understanding." The terms seem to me most applicable in a sense originally unintended. The portrait may be "gentle" in the way Dryden implied: "there is . . . a vast difference betwixt the slovenly butchering of a man, and the fineness of a stroke that separates the head from the body and leaves it standing in its place." And it may be "understanding," I think, in the way Dante understood Paolo and Francesca. It would be both crude and rude to call the Prioress a "sinner," in our flat modern way of using the word, and it is certainly true that she is not "guilty" in the way Zimri or Francesca was. That she has virtually nothing in common with ecclesiastics like the Pardoner or the Summoner is beyond all doubt. Yet to say this is not to conclude that Chaucer loved her for what she was, or even found her particularly delightful. Chaucer's Parson can provide a commonplace distinction which will be relevant here:

For sothe, synne is in two maneres; outher it is venial, or deedly synne. Soothly, whan man loveth any creature moore than Jhesu Crist oure Crea-

[24] See Francis Manley, "Chaucer's Rosary and Donne's Bracelet: Ambiguous Coral," *MLN*, LXXIV(May, 1959), 385–388. The "ambiguity" is analogous to that in the word, *amor*.

[25] *Convention and Revolt in Poetry* (Boston, 1922), p. 66.

tour, thanne is it deedly synne. And venial synne is it, if man love Jhesu Crist lasse than hym oghte. (p. 238)

The distinctions are made in terms of love. "When we seek to know how good a man is," said Augustine, "we ask not what he believes, or what he hopes, but what he loves."[26] By these tokens, Chaucer's portrait of the Prioress has shown us that she is not properly "conquered" by the proper kind of love.

To sum up: Chaucer concentrates on the "manere" of the Prioress and describes her almost exclusively in terms of externalities. Although he uses superlatives to tell how well she does things, he nowhere speaks directly of her character in superlative terms. Contrast the Knight: "And though that he were worthy, he was wys . . . He was a verray, parfit gentil knyght"; or the Parson: "riche he was of hooly thoght and werk . . . A bettre preest I trowe that nowher noon ys." He emphasizes her scrupulous attention to etiquette, her genteel affectation, her attention to the adornment of her person, including a "youthfulness" perhaps not quite appropriate to her age and position, and her sentimentality. The appurtenances of her status are associated more with an aesthetic than an ascetic point of view, and discussion of the practice of her piety and the extension of her charity are otherwise entirely absent from the portrait. Such a reading is, as I said it would be, rather "hard" on the Prioress. But I do not think in terms of some person who should be defended from calumny. I think, rather, of a *persona* whose very creation and details of being constitute a criticism of life or a part of life.

What, then, of the Prioress' Tale? In the "soft" view, it should be found to complement, to balance, even to correct, the impressions of the portrait, as the somewhat worldly lady redeems herself with a tale of the lovely lady of another world. But in the "hard" view, it should be found to use the technique of "giving with one hand what it takes away with the other," and ultimately to be "improperly conquered by love." This latter view, of course, is what I intend to argue, having been led originally to such a conclusion by discovering that the correspondences between Tale and teller extend beyond the

[26] *Enchiridion,* XXXI.117; ed. Sizoo, p. 102. And cf. the handling of "the theme of love, earthly and celestial" in Arthur W. Hoffman, "Chaucer's Prologue to Pilgrimage: The Two Voices," in *Chaucer Modern Essays in Criticism,* ed. Wagenknecht (New York, 1959), pp. 30–45, esp. 43 f. [reprinted from *ELH,* XXI(1954), 1–16].

general items of religious appropriateness to mirror the personality of the Prioress as her portrait in the *General Prologue* reveals her. That mirroring, it seems to me, is subtle enough to be easily overlooked, but not necessarily more subtle than the handling of the portrait. It may seem much more subtle, but this, I think, is because so much more attention has been given to the portrait than the Tale.

When we first examine the Tale we must admit there seems little to laugh about. There is nothing to scandalize or shock an audience, nothing that will even come close to smirching the Prioress' "dignitee." She begins with a prayer to God and the Blessed Virgin, couched in terms of great humility and solemnity, and announces her theme in the language of Psalm 8: "by the mouth of children thy bountee/ Parfourned is" (lines 457–8, p. 161). And the story begins with a child, a "litel clergeon" of seven years who lives in a "greet citee" of Asia, and who is remarkable for his great and intuitive piety. Like the Prioress, he is devoted to the Virgin, and is ravished when he first hears sung the antiphon in her honor, *Alma redemptoris mater*. Out of the mouth of this "litel child" comes praise: although he does not understand the Latin, he begins to learn the words and then has their meaning explained to him by an older "felawe." At some possible expense to his regular studies, he learns the whole antiphon and then sings it every day, on the way to school and home.

As the Prioress continues her story, the tone of devotion, a mixture of affection for the boy and wonder at the miracle, is not diminished. We learn that the little "clergeon" must pass through the Jewish quarter, and that the Devil stirs up the Jews to resent such "despit" as his innocent singing offers them. They hire an assassin, who cuts the boy's throat and casts him into a pit, or, more accurately, a privy. When the anguished mother sets out to look for her son who has not returned from school, she does not find him until, guided by the grace of Christ, she comes near the place in the ghetto where he is lying and hears his voice, miraculously singing the *Alma redemptoris mater*. The Christian authorities recover the boy, punish the Jews, and bear him to a convent for a Requiem Mass. Miracles continue: when the "clergeon" is sprinkled with holy water, he sings once more, and reveals that the Blessed Virgin had come to him before his death, and bade him continue singing until a grain which she placed on his tongue should be removed. The abbot of the convent removes the grain, the boy gives up the ghost, and all are prostrated with weeping and with praise of Mary. The Tale ends with an invocation to Hugh of Lincoln, also martyred by the Jews, that he pray God to multiply his mercy on the whole company "for reverence of his mooder Marie."

Where is the satire here, where the sour note, the cutting edge? The modern reader is likely to feel most uneasy over the way the Prioress handles the Jews, and Schoeck builds up his "hard" interpretation from the basic anti-Semitism that is present in the Tale, arguing that Chaucer is not anti-Semitic but that the Prioress is, and advancing his suggestion that her Tale is held up for "implicit condemnation as vicious and hypocritical." But the argument is more complicated than he makes it. In spite of the moving defenses of the Jews which he can cite from the Middle Ages, there can be no question that many listeners would find nothing in the Prioress' Tale to disapprove. He does not deny this, of course, but does suggest that the cosmopolitan court of Richard II might have been more tolerant than the general public and hence more sensitive to blind anti-Semitism.[27] His theory is supposititious; until we can have external evidence pointing one way or another, we are best advised to stay as close to the Tale and its movements as possible, and argue from the inside out.

In order to understand Chaucer's emphases and meaningful arrangement of his material, we must first know just what in the Tale is distinctively his. The briefest of investigations will show that his invention is concerned hardly at all with the outline of this Miracle of the Blessed Virgin. The studies of the Tale's sources demonstrate that all the elements of the plot were already available many times over,[28] and they make it look quite probable that Chaucer had one particular version propped up in front of him as he wrote. At this stage it would be unwarranted to claim that much tailoring of a tale could be made, other than deciding that the basic material was appropriate for a certain kind of character. This would be a most gen-

[27] "Chaucer's Prioress," p. 252 f.

[28] See Carleton Brown, "The Prioress's Tale," in *Sources and Analogues of Chaucer's Canterbury Tales,* ed. W. F. Bryan and Germaine Dempster (New York, 1958), pp. 447–485. The story was at least 200 years old by Chaucer's day. Brown's conclusion that the Tale belongs to "Group C" among the families of mss., and that Chaucer "based his narrative upon a single source" is on p. 460.

eral kind of fitting, not allowing for much internal subtlety. But there is a second stage of investigation, and it concerns the special way Chaucer's style embraces the Miracle. Here there is a good deal of individuality—involved with the *elocutio* more than the *inventio* or *dispositio*—and here it is where we discover that Chaucer has managed to shape the narrative so that it is stamped with the kind of character I found in the Prioress' portrait, reflecting her attitude towards the material. The signs of such a shaping may be divided into the more or less "neutral," i.e., unsatiric, correspondences with her personality, and the definitely less neutral, ultimately satirical ones. I consider the "neutral" signs first.

The concerns of the Tale are, first of all, devotional and inspirational, involving an act of faith and an act of praise, centering in the worship of Mary, and exalting both purity and chastity. All good Christians theoretically should respond to these things, but especially nuns. Sister Madaleva's incontrovertible experience supports such a view.[29] Second, the language of the Tale often suggests or directly translates passages from the Divine Office, or the "Little Office" of the Church, particularly where the Blessed Virgin is invoked.[30] The Prioress does recite her office, we know, through her nose and in a most "semely" fashion. Third, the Tale exhibits great love and "pitee" for its protagonist, entirely fitting for a woman of "conscience and tendre herte," especially if one wishes to argue that her "pitee" could include other objects beside "smale houndes" and mice in traps. The atmosphere melts with motherly ruth; tears are frequent, and flow from many eyes. Fourth, the Tale is as "dainty" as the Prioress might wish to be herself. It is in immaculate rime royal, and shows an absolute mastery of the form. Every word counts; I cannot find a single place where Chaucer pads to fill up a line, or manipulates the syntax to get a rime, or repeats himself needlessly. The invocation alone is a shimmering illustration of complex thoughts made to appear simple by the art that conceals art. The uses of enjambment and of caesura are worthy of independent study. Neither the *Troïlus* nor the *Clerk's Tale* achieves very often or ever sustains what the poetry of the Prior-

[29] Cf. her "Chaucer's Nuns," pp. 55, 56.

[30] See Sister Madaleva's article and the unpublished University of Washington thesis of John Paul McManus, *The Little Office of the Blessed Virgin Mary and the 'Invocacio Ad Mariam' of the "Second Nun's Prologue"* (Seattle, 1953).

ess achieves. This alone may seem to militate against reading her Tale as a satire. Would Chaucer spend so much time on a Tale if its complex purpose were ultimately destructive and if he did not give full allegiance to its contents and its attitude? The question cannot really be answered at this level; I raise it, and pass on, granting that "ultimately destructive" is an unnecessarily harsh phrase to describe possible satire.[31] Fifth, the Tale does not get at all compromised with earthly love and does not seem to demonstrate anything of which the Prioress should be ashamed. This is not a neutral point, in fact, but one in positive favor of the Prioress and against the "hard" interpreters. Her Tale does not dwell improperly on the glitter or pomp of this world; it seems not at all "courtly," which might surprise those who, after reading the Prioress' portrait, would look forward to hearing something from her to match her name and her "cheere of court."[32] It has nothing at all to do with food and table manners, and its immaculateness is not a matter of clean linen. It begins in humility and ends with a prayer for mercy. All of these last points give support to those who claim that the Prioress' Tale corrects, cancels, or ignores the elements of satire in her portrait.

It is a difficult task to move now to the second category of correspondences in the Tale to the Prioress' character, and I have deliberately made it as hard as possible for myself. But I am convinced there is something to be said here, perhaps in terms of shading rather than main stroke, or, to return to an earlier metaphor, in terms of taking away with one hand what the other had given. At issue is the assumed definition of satire. My reading of the portrait, after all, is not in terms of much that could be called "funny." And again, much if not most medieval satire was intended more for instruction than delight and might wield the surgeon's knife as often as it would wear the comedian's sock. The more appropriate term, then, may be the

[31] I earlier suggested that to find satire in the Tale might lead one to conclude that Chaucer was satirizing the whole genre of Miracles, but this does not necessarily follow if one observes the distinction between *choice* of Tale and *handling* of Tale: the discussion below concentrates chiefly on the revelation of the Prioress' attitude towards her material, and not Chaucer's attitude towards supernatural events as ordinarily reported.

[32] Baum, *Chaucer: A Critical Appreciation*, discusses the whole of Fragment VII as "The Surprise Group" and handles this point in considering the function of the Prioress' Tale. See esp. pp. 82–84. The whole section is surely offered by Baum "bitwixen ernest and game"?

familiar term, "irony." I only suggest that the Prioress' portrait is organized in terms of a "second look" which could be called ironic, that the careful manipulation of external things is intended to point towards a quality of inner life which is never explicitly described. In this sense I think I find a good deal of irony, and hence a kind of satire, in the Prioress' Tale which directly illuminates that inner life.

I put first the question of anti-Semitism, returning to it after previously qualifying Schoeck's approach to the matter. Surely we would all like to think that Chaucer was aware of what Wordsworth called the "fierce bigotry" of such lines as these:

558 Oure firste foo, the serpent Sathanas,
That hath in Jues herte his waspes nest . . . (p. 162)
. . . . . .
570 This cursed Jew hym hente, and heeld hym faste,
And kitte his throte, and in a pit hym caste. (p. 162)
. . . . . .
O cursed folk of Herodes al newe,
575 What may youre yvel entente yow availle? (p. 162)
. . . . . .
578 The blood out crieth on youre cursed dede. (p. 162)
. . . . . .
With torment and with shameful deeth echon
This provost dooth thise Jewes for to sterve
630 That of this mordre wiste, and that anon.
He nolde no swich cursednesse observe,
"Yvele shal have that yvele wol deserve";
Therefore with wilde hors he dide hem drawe,
And after that he heng hem by the lawe. (p. 163)
. . . . . .
O yonge Hugh of Lyncoln, slayn also
685 With cursed Jewes, as it is notable,
For it is but a litel while ago. . . . (p. 164)

I say we would like to think that Chaucer was fully aware of the quality of anti-Semitism displayed here, but I do not see that completely satisfactory proof can be adduced. Certainly, the sources and analogues of this Tale speak also of the cursed Jews, of their inhuman plans, and their cruel deeds; furthermore, many of them recount with relish the harsh justice the Jews bring upon themselves. Chaucer not only reproduces these elements but elaborates upon them. But if we are not satisfied in concluding that the poet was the anti-Semite, we may argue hypothetically that he is intensifying the harshness of the Prioress' attitude deliberately.[33] And although it is not in the realm of proof, it seems eminently reasonable to think that Chaucer realized that when the Provost says "Yvele shal have that yvele wol deserve," he is aware of how completely such a sentiment accords with the Old Law rather than the New. Elsewhere he shows himself exceptionally sensitive to the need for justice to be tempered with mercy, as even the pagan Theseus shows.[34] Schoeck argues along these lines (p. 253f.) and is quite persuasive. But the very fact that the Jews had been officially expelled from England since 1290, leads me to wonder whether Chaucer would have thought of anti-Semitism as much of an issue, or written a tale with an eye of particular compassion towards the Jews. I doubt it. The Jews in the Prioress' Tale are not characterized at all, do not appeal to our sympathy or conscience; they are simply the catalytic agents that release the energy of a miracle. The important thing about the Jews is what the Prioress thinks about them; although this is admittedly a slim distinction, it leads me to surer ground, and to my second point.[35]

For the second point of correspondence is that there is no real mercy in this Tale. It is really a tale of power, not of love. The Prioress' sympathy for the "litel clergeon" is not mercy. The portrait lacks any mention of the conventional "works of mercy," and the Tale lacks evidence of their spirit. I would argue that Chaucer emphasizes the quick reactions of the Prioress to the Jews to show how for her it was not even a question of charity or "tendre herte" when they were involved. Schoeck (p. 254) rightly notes the irony of her final prayer

[33] It is particularly interesting to see that several families of manuscripts with versions related to Chaucer's source handle the Jews differently: the miracle astonishes the Jews, softens their hearts, and many if not all of the witnesses "repent" and are converted. (This is a frequent course for the Jews to take in other legends I have seen.) It is of course impossible to know whether or not Chaucer knew of these versions with the happier ending and deliberately rejected them in picking the "right" Tale for his Prioress.

[34] Cf. *The Knight's Tale,* lines 1733–1781, p. 34. The blend he demonstrates is of *justitia* and *misericordia.*

[35] I must admit, however, that the Prioress is up on her anti-Semitic literature, being careful to bring in a reference to the "ritual murder" of "yonge Hugh of Lyncoln," speaking of it as "but a litel while ago" though the murder allegedly took place over a hundred years earlier. At least Chaucer does not have her go to work on this Tale, which would otherwise refresh the reader with a crucifixion, disembowelment, and other grisly deeds by the "cursed Jewes."

which requests a mercy she herself is conspicuously not giving.[36] I have called it a tale of power, of the miraculous power of the Virgin to save and sustain her favorites under the most trying conditions, but one may substitute for "power" the word "brutality." For there is a brutality here, a brutality of attitude, action, and imagery. There is a continual violence in all of the Prioress' references to the Jews and to the murder, beginning with

> a Jewerye,
> Sustened by a lord of that contree
> For foule usure and lucre of vileyneye,
> Hateful to Crist and to his compaignye. (p. 161)

The murder is described in starkest terms:

> This cursed Jew hym hente, and heeld hym faste,
> And kitte his throte, and in a pit hym caste.
>
> I seye that in a wardrobe they hym threwe
> Where as thise Jewes purgen hire entraille. (p. 162)

The bestiality of the deed and of the description of the hiding place is meant, of course, to emphasize the inhumanity of the "Hebrayk peple," but the brutality of the language tends to belie the total softness and sweetness of the lady who swears by Saint Loy.[37] And the quality of her imagery at the moment of discovery of the murdered "innocent" is ghoulishly aesthetic: at the bottom of the filthy privy where the Jews purge their entrails the boy gives his praise *de profundis*:

[36] Cf. *The Parson's Tale, s. v.* "Remedium contra peccatum Invidie," concerning the love of one's enemies: "... the lovynge of oure enemy hath confounded the venym of the devel./ For right as the devel is disconfited by humylitee, right so is he wounded to the deeth by love of oure enemy" (p. 244); and *s. v.* "remedium contra peccatum Ire," concerning "Debonairetee" and "Pacience" or "Suffrance": "The philosophre seith that pacience is thilke vertu that suffreth debonairely alle the outrages of adversitee and every wikked word./ This vertu maketh a man lyk to God, and maketh hym Goddes owene deere child, as seith Crist.... And therfore seith the wise man, "If thow wolt venquysse thyn enemy, lerne to suffre" (p. 249).

[37] The most eloquent commentary that I can think of is provided wordlessly by the Reverend Walter W. Skeat's student edition of *Chaucer The Prioresses Tale, Sire Thopas, etc.* (Oxford, 7th ed., rev., 1897), p. 13. A series of dots between lines 570 and 576 of the Prioress' Tale show that the gentle Lady, in the editor's estimation, was too much for the gentle reader and had to be quietly suppressed!

This gemme of chastite, this emeraude,
And eek of martirdom the ruby bright,
Ther he with throte ykorven lay upright,
He *Alma redemptoris* gan to synge
So loude that al the place gan to rynge. (p. 163)

Excrement, cut throat, chastity, martyrdom, and the clear and surprising antiphon: jewels and jakes in direct juxtaposition. This is the effect not of original source but of Chaucer's craft, and that it can be a strong effect the "hard" interpreters abundantly testify.

Often psychologically related in a very intimate way with the second point above, is a third: there is a peculiar quality of "tendre herte" in the Tale which may be called sentimentality. Deciding where sentiment stops and sentimentality begins is always a delicate matter, and I have mentioned at the beginning of this paper the problems involved in taking this as any kind of critique of sentimentality. Still, I think we are brought to that position. Brutality and sentimentality in combination are unfortunately common in certain types of character, and I think Chaucer shows them here. It certainly is not difficult to see the sentimentality in the portrait, although Chaucer does not prepare us there, except by expectation and omission of detail, perhaps, for the strain of brutality in the Tale. Examine, for example, the following lines:

And whan this abbot hadde this wonder seyn,
His salte teeris trikled doun as reyn,
And gruf he fil al plat upon the grounde,
And stille he lay as he had ben ybounde.

The covent eek lay on the pavement
Wepynge, and herying Cristes mooder deere.... (p. 163)

I find the very sound of "and gruf he fil al plat upon the grounde" suggestive of a calculated effect of the slightly overdone. We may note also the small scale of the Tale's diction; that is, the Prioress repeatedly dwells on the youth of the "litel clergeon," his innocence and his diminutive size.[38] "Litel" and "innocent" are used over and over, in addition to "sely," "yong," "tendre," and "smale." The amount of emotion lavished on him is in correspondence with the extravagant reactions of the Prioress to poor dumb beasts. For in

[38] Schoeck makes the same point (n. 1, p. 257), and adds "this is consonant with the Prioress's excessive and false charity over her 'little dogs.' "

terms of Christian faith, after all, the "litel clergeon" 's story should not be considered a tragedy, although its ending in Chaucer's version has been so described. The unhappiest person should be the widow who is left childless, but she is not mentioned at all beyond one effective stanza where she is anxiously searching for her son. I would add at this point, that to call these reactions of the Prioress sentimentality is not to be forced to conclude that Chaucer is thus criticizing the basic attitudes of the Christianity he knew, although there is not time to go into the matter adequately. The mention of the Second Nun must suffice here.

The Tale that the Second Nun tells, who has no corresponding portrait in the *General Prologue* (the same is true of the Nun's Priest), is in a separate fragment, and is often placed after Fragment VII, which contains the Prioress' Tale. Did Chaucer intend for the tale of the martyrdom of St. Cecelia to be compared with that of the "litel clergeon"? I am not prepared to speculate on this matter yet, but I do find the two nuns' tales to make a fascinating comparison. For the construction and content of the Second Nun's Tale offers such a strong contrast to the Prioress' that their very difference fits into my theory. I will not attempt to claim that the literary craftsmanship of the Second Nun's Tale is superior to the perfection of the Prioress', but I will claim that the former is "better" in all other ways, and I strongly suspect Chaucer would feel it represented his "best" efforts in the field.[39]

The Prioress' Tale is about a "litel clergeon" of seven years, while the Second Nun's Tale is about a noble Roman girl, Cecilie, who marries a young man named Valerian. The "litel clergeon" is pious by instinct and through nurture; Cecilie "bar [Christ's] gospel in hir mynde" (line 123, p. 208), and both receives and conveys instruction. There is no conversion in the Prioress' Tale; in the Second Nun's, Cecilie converts both her husband and his brother to the Christian faith, entirely by reason and persuasion, and the ministers of Almachius, the prefect, are also converted "at hir wise loore" (l. 414, p. 212). There is no love within the Prioress' Tale, except the Prioress' towards the "clergeon" and the widow's towards her son; there is love

[39] The criterion of dramatic propriety being excepted, of course, since the Second Nun is not a developed figure. But does not the very earliness of the Tale indicate the value Chaucer may have attached to it by including it in *The Canterbury Tales*?

in the Second Nun's Tale not only between Cecilie and her husband, but between "Seint Urban" and the other Christians and among the whole Christian community. The faith of the "litel clergeon" is not tested, he is surprised and assassinated; but the faith of the Christians, among them predominantly Cecilie, but including the converted officer Maximus, is sorely tried over a period of time and adversity. The narration of the Prioress is full of sentimentality, while—and I count this most significant—what I would identify as sentimentality is absent from that of the Second Nun. The "litel clergeon" is an "innocent" and reveals a devotion which the Blessed Virgin rewards; Cecilie is pure and spotless, but her virtue is considerably exercised, and reveals a courage of the toughest fiber. The "litel clergeon" never really has contact with the Jews except at the moment of his death; Cecilie stands and debates with Almachius with firmness, vigor, freedom, and even humor. There is a kind of muscular joy about her faith which is entirely lacking in the other Tale, and all the miracles in her Tale are symbols of a quality of character she has proved through her faith *and* her works, and the kind of victory she has won. Her Tale involves mercy, knowledge, grace, wisdom, faith, hope, and love. The only of these qualities much in evidence in the Prioress' Tale was faith, and even that was of a different nature from Cecilie's. But if all that this contrast illustrates is the fundamental differences between Saints' Legends and Miracles of the Virgin, then I would tentatively suggest that Chaucer preferred the former, and more positively maintain that, for all the reasons I have mentioned, he felt the latter to be most appropriate for his Prioress.

My fourth and final point flows out of these observations. Although it sounds harsh, I think we may say that the Tale is a case of arrested development. That is, the Prioress not only identifies herself with the "litel clergeon," she keeps the whole Tale at the emotional and intellectual level of a child. The correspondence here would be with the affected "girlishness" of this Mother Superior that is revealed in the *General Prologue*. In her invocation she concludes:

> My konnyng is so wayk, o blisful Queene,
> For to declare thy grete worthynesse
> That I ne may the weighte nat susteene;
> But as a child of twelf month oold, or lesse,
> That kan unnethes any word expresse,
> Right so fare I, and therfore I yow preye,
> Gydeth my song that I shal of yow seye. (p. 161)

The form of the poem and the style, of course, scarcely accord with untutored lore. Furthermore, our study of sources reveals one highly significant change. Chaucer has moved the age of the "litel clergeon" back to seven years, instead of the ten the earlier versions almost always give. He thus places the boy at a stage where he is theologically to be considered an "innocent," where he can neither read nor write, and where he is on the threshold of the age of consent, the age when his will may be spoken of as entirely "free." But the Prioress makes it clear that his piety is a matter of his nurture: "thus hath this wydwe hir litel sone ytaught.... For sely child wol alday soone leere." He first sings the *Alma redemptoris mater* without understanding it, although it is true that it is later explained to him. Yet even here, the explanation given is

"This song, I have herd seye,
Was maked of our blisful Lady free,
Hire to salue, and eek hire for to preye
To been oure help and socour whan we deye" (p. 162);

while the reply of the "litel clergeon" is simply:

"And is this song maked in reverence
Of Cristes mooder?" seyde this innocent,
"Now, certes, I wol do my diligence
To konne it al er Cristemasse be went." (p. 162)

If the "litel clergeon" sings the antiphon for its petition, he does not say so. He sings it for reverence, which is in accordance with the Prioress' announced theme: "by the mouth of children thy bounte/ Parfourned is." Part of the whole point of this is the almost automatic efficacy of such piety, even when not accompanied by full understanding. The direction of the will is more basic than the condition of the intellect. The "litel clergeon" is an innocent which, by philosophical definition, must mean that although he is pure he cannot be considered "virtuous."

I cannot prove that the Prioress' predilection for this kind of tale is necessarily a failing in her, although I think it is, particularly if we consider it as coming not just from a nun but a Prioress. Her Tale appears to be peculiarly unconquered by the *amor* that is charity, by mercy, by emotional control, and by the mature intellect. The attitudes and techniques of the Tale have, of course, some relevance to the words of Christ: "Amen, I say to you, unless you be converted, and become as little children, you shall not enter into the kingdom of heaven."[40] But we may recall the Apostle Paul's very applicable warning as a contrast. He writes: "Brethren, do not become children in sense: but in malice be children, and in sense be perfect" (*I Cor.* 14:20).

Thus, if there is validity in my reading of the Prioress' Tale, we must conclude that not only has its interpretation been unconquered by the critics, but that, in ironic contrast to the motto on her brooch, both the portrait and the Tale of the Prioress are unconquered in a very special way: they are unconquered by the fullest kind of love that Chaucer knows.

[40] *Matthew* 18:3; from the Douai version.

PAPERS OF THE MICHIGAN ACADEMY OF SCIENCE, ARTS, AND LETTERS
VOL. XLVII, 1962 (1961 MEETING)

BACON'S HUMANISM

EDMUND H. CREETH
The University of Michigan

To some fellow humanists in this scientific age, my theme may appear self-contradictory or infinitely subtle, as though one were to broach, say, Milton's Bawdy. Douglas Bush can cite Bacon as a threadbare illustration of antithesis between humanistic thought and scientific progress. He "sacrifices that scale of divine and humane values which the best minds of antiquity, the Middle Ages, and the Renaissance had striven to make prevail."[1] As to faith, his Church is a branch of the Civil Service. Basil Willey's celebrated study of Bacon, dissector of the scholastic synthesis, leads to similar conclusions.[2] These scholars, the Christian humanism defined by the one differing largely in emphasis and variety of classical sources from the scholasticism mourned by the other, concur in judging Francis Bacon a great writer irrelevant to the vital needs of our time. A prophet is without honor in the land he prophesied.

Yet Bacon may also be represented as more truly a Renaissance humanist, in a classical and philosophical sense, than the Moses, or the Solomon, of the present New Atlantis. In this view, as Rudolph Metz has put it, he stands between two ages but looks back, not really perceiving how science will develop, spiritually closer to Vives and Montaigne than to Descartes and Locke; he recommends knowledge, not for the crass materialism that has in fact come about, but for the creative advancement of civilization.[3]

Without trying, in brief space, to cope with all the questions raised by this disagreement, I wish to propose Bacon's first philosophical work, published in 1605 when he was forty-four, as a document in English humanism. In Book I Bacon addresses to the British Solomon an oration in defense of learning in general, without em-

[1] *The Renaissance and English Humanism* (Toronto, 1939), p. 93.

[2] "The Rejection of Scholasticism," "Bacon and the Rehabilitation of Nature," *The Seventeenth Century Background* (London, 1942), pp. 1–40.

[3] "Bacon's Part in the Intellectual Movement of His Time," *Seventeenth Century Studies presented to Sir Herbert Grierson* (Oxford, 1938), pp. 21–32.

phasis on natural philosophy. At the apex of that Book, having defined "the last or furthest end of knowledge" as being "for the glory of the Creator and the relief of man's estate," Bacon urges that "contemplation and action . . . be more nearly and straitly conjoined and united together than they have been; a conjunction like unto that of the two highest planets, Saturn the planet of rest and contemplation, and Jupiter the planet of civil society and action."[4] The antecedents for this revival of the classical interaction of the active and contemplative lives are Florentine, and Bacon's main line of argument in Book I, having to do with the virtuous effects of learning in the ruling class, is almost quaintly traditional. Bacon closes the peroration on St. Matthew (xi. 10): "*Justificata est sapientia a filiis suis.*" From the encyclopedic Book II and from the general title, *The Proficience and Advancement of Learning Divine and Humane,* it is clear that his definition of *sapientia* is Ciceronian: the knowledge of things divine and human and their causes.[5] The meaning he would attach to these words is not less traditional than his overriding Renaissance ideal of marrying contemplation and action for human benefit. So far from being secular, Bacon's piety subordinates moral and natural philosophy beneath their places in traditional Christian humanism in the Renaissance. He will not allow moral philosophy equal to divinity. His corresponding limiting of natural philosophy to the realm of secondary causes and abstract forms is what makes possible its unprecedented expansion in the *Great Instauration,* no doubt to the neglect or destruction of traditional values to which he holds in the *Advancement.*

Bacon's philosophical conservatism, of which only remnants survive in the *Novum Organum,* is medieval. Augustine, as Eugene Rice has shown, takes the ancient notion of wisdom as a *rerum humanarum divinarumque scientia,* divides it, and distinguishes the eternal objects of wisdom from the temporal objects of rational knowledge.[6] Bacon's distinction between divine and human learning is thus akin to that of Augustine between *sapientia* and *scientia.*

[4] *The Works of Francis Bacon,* ed. James Spedding (New York, 1869), VI, 134–135. Geoffrey Bullough, "Bacon and the Defense of Learning," Grierson *Studies,* pp. 1–20, calls Book I an appeal to James for favor and relates it to other defenses of learning.

[5] *De Officiis,* II, ii, 5.

[6] *The Renaissance Idea of Wisdom* (Cambridge, Mass., 1958), pp. 4–13.

Again, Aquinas in his *Summa Theologica* (II, ii, Q. 1, art. 5) argues that the objects of faith cannot be those of science, that "faith and science are not about the same things," for the objects of science must in some sense be "seen" whereas those of belief cannot be. In saying this Bacon would not mean precisely what Aquinas means, but their views are hardly antithetical. In the *Advancement* and the utopian *New Atlantis*, which is nearer in spirit to it than to the *Organum*, faith, civil knowledge, and natural science, though distinct, exist in harmony and balance. Natural science was not inevitably hostile to the humanistic side of Bacon's thought but, in fact, was once part of it, and his interest in nature study long antedates even the *Advancement of Learning*. But it must be admitted that later in life the techniques for interrogating nature preoccupied his mind to the exclusion of other "human things," if not to the exclusion of faith, and there is little of the peace of God in the *Novum Organum*, even if, as James is said to have commented, it passeth understanding. Let there be laboratories, Verulam said in effect, and there were laboratories. While he was still in public life he set up and in extreme moments himself worshiped the Idol of the Laboratory. Hence much of Bacon's mature thought is neglected and distorted when we take the *Advancement* to be merely preliminary to the *Great Instauration*, as the arrangement in the standard edition of Bacon by James Spedding encourages us to do. Discussions of Bacon's philosophy drawn from both the early and the late work should discriminate carefully between them.

Against "the disgraces which learning receiveth from politiques" Bacon, in the preliminary part to Book I of the *Advancement*, restates the sort of defense employed by humanists like Castiglione and Sidney; Sir Thomas Elyot, three quarters of a century before Bacon wrote, had found particularly urgent need to counter the pestiferous English opinion that learning makes a man unfit to govern. It is "a matter of doubtful consequence, if states be managed by empiric statesmen, not well mingled with men grounded in learning," Bacon asserts, and contrariwise hardly ever was a disastrous government in the hands of learned governors. Rome did not reach heights of government until it ascended to heights in the learned arts of poetry, historiography, antiquarianism, and oratory.[7] This

[7] *Works,* VI, 97–105.

humanistic idea that good leaders are educated is taken up again and forms the major phase of the "human proofs" of the dignity of learning in Book I. The ardor with which Bacon pursues the argument precludes our taking it to be lip-service or flattery of the new monarch. Though modified by Bacon's characteristic reference to the practical, the notion retains essential traditional elements: political Platonism, patriotism, and sense of the grandeur that was Rome. The social benefit when precepts, laws, religion are touched with "eloquence and persuasion of books, of sermons, of harangues . . ." appears the more manifestly

> when kings themselves, or persons of authority under them, or other governors in commonwealths and popular estates, are endued with learning. For although he might be thought partial to his own profession, that said *Then should people and estates be happy, when either kings were philosophers, or philosophers kings*; yet so much is verified by experience, that under learned princes and governors there have been ever the best times.[8]

To illustrate, Bacon recounts the rules of six Roman princes, in which we see "the blessed effects of learning in sovereignty, painted forth in the greatest table of the world," and closes the section by comparing the government of Elizabeth. Bacon illustrates with enthusiasm the concurrence of learning with military virtue by discussing Alexander, Caesar, and, remarkably, Xenophon. Bacon leads up to his peroration by considering knowledge as a source of moral and private virtue and perspective, of power and fortune, and of the finest pleasure.

So much for the human proofs of the dignity of letters. Much of the less carefully wrought second book is more novel, but toward the relationship of both moral and natural philosophy to the holy faith Bacon's attitude may be described as a piously conservative Christian humanism. Neither branch of philosophy should aspire to teach points of theology. Their sphere is worldly. Though he does not yet broach the new method for natural science, Bacon devotes a sizable section to distinguishing physics and metaphysics and redefining the latter so as to divorce it from religion. Under human philosophy, he criticizes logic and presages the Idols. But he devotes the largest share of attention to nonscientific matters such as history, rhetoric,

[8] *Works,* VI, 146. See Fritz Caspari, *Humanism and the Social Order in Tudor England* (Chicago, 1954) for detailed discussion of this line of argument before Bacon.

virtue, duty, and civil knowledge. He thinks of Jacobean England as the scene of the third revival of learning, to match those in Greece and Rome.[9]

"And if it be said that the cure of men's minds belongeth to sacred Divinity," Bacon asserts in connection with moral philosophy, "it is most true: but yet Moral Philosophy may be preferred unto her as a wise servant and humble handmaid." She ought "to give a constant attention to the doctrines of Divinity, and yet so as it may yield of herself (within due limits) many sound and profitable directions." Aristotle and Pliny the Younger sensed the divinity of virtue. "But these be heathen and profane passages, having but a shadow of that divine state of mind which religion and the holy faith doth conduct men unto. . . . " It is in fact the holy faith, so plainly exalting "the good which is communicative," which "doth judge and determine most of the controversies wherein Moral Philosophy is conversant" and which in particular "decideth the question touching the preferment of the contemplative or active life, and decideth it against Aristotle. For all the reasons which he bringeth for the contemplative are private, and respecting the pleasure and dignity of a man's self, (in which respects no question the contemplative life hath the pre-eminence). . . . But men must know, that in this theatre of man's life it is reserved only for God and Angels to be lookers on."[10] Bacon always disapproved wisdom for a man's self.

This return to a reflective life ultimately active, after the medieval preference for the contemplative, is familiar in both Florentine and Tudor humanism. Coluccio Salutati and Leonardo Bruni and, in his own way, Sir Thomas Elyot, all motivated partly by patriotism, argue for application of the fruits of contemplation to active civic life. The Florentines take their inspiration largely from Cicero, who included human things in his definition of wisdom and in *De Finibus* (II, xiii, 40) quoted Aristotle to the effect that man is born both to know and to act. In conscious imitation of Socrates, they scorn the study of natural science, as Bacon of course does not. Bacon, in fact, feels it necessary to protest almost directly after his

[9] *Works,* VI, 392.

[10] *Works,* VI, 330, 344; 314. Bacon's idea that philosophy functions as the handmaid (*ancilla*) of sacred doctrine is developed in Aquinas, *Sum. Th.* I, Q. 1, art. 6.

plea for the uniting of action and contemplation: "Neither is my meaning, as was spoken of Socrates, to call philosophy down from heaven to converse upon the earth; that is, to leave natural philosophy aside, and to apply knowledge *only* to manners and policy."[11] But the reason given by Salutati and Bruni for rejecting natural philosophy is that, since it seeks to rise to an understanding of heaven, it is ultimately *contemplative*. Bacon's novelty is that while denying "science" validity as a means of ascent to divine truths he yet grants it validity in its human and sublunary sphere, which for him is that of moral philosophy also. He veritably calls *natural* philosophy down from heaven to converse upon the earth. His inclusion of it in the socially applicable fusion of action and contemplation does not in itself distinguish him essentially from Bruni, who "insisted that Aristotle's speculative wisdom be perfected by an active life."[12] His eventual preoccupation with natural science was made possible by his disengaging it from contemplative ends and directing it horizontally, so to speak, into the world of secondary causes. His original motive for doing so was not that otherwise scientific studies were contemplative but that they were presumptuous.

Arguments to the effect that Bacon rejects the solid values of earlier learning sometimes make use of the preliminary defenses of learning in the *Advancement* itself, although they ignore the main body of Book I, the divine and human proofs, presumably taking them to be lip-service. The defense against politiques is also ignored, but the defense of natural science against religious zeal, the censure of the methods of the schoolmen, and the criticism of Ciceronianism are taken as evidence of Bacon's hostility to solid values of the past.

Basil Willey stresses the point that, "at the outset of his great work," Bacon feels it necessary "to show that whatever the forbidden knowledge may be"—it consists for Bacon in prying into God's secrets with the intent to give a law unto ourselves—"it is not natural science. Bacon's argument is of great importance, because it furnished the scientists of the following two centuries with a technique for reconciling science with religion." Willey then cites Bacon's primary evidence—that

[11] *Works,* VI, 135. My italics.

[12] Rice, *op. cit.*, p. 56. This paragraph is much indebted to Rice's chapter "Active and Contemplative Ideals."

it was not the pure knowledge of nature and universality, a knowledge by the light whereof man did give names unto other creatures in Paradise, as they were brought before him, according to their proprieties, which gave the occasion to the fall; but it was the proud knowledge of good and evil....

—and proceeds directly to comment: "To Bacon the logic-spinning of the schoolmen was a kind of forbidden knowledge; it was a presumptuous attempt to read the secret purposes of God...."[13]

True, Bacon presently declares that a principal vice of learning, which indeed makes it liable to religious objections, lies in presuming, "by the contemplation of nature to attain to the mysteries of God," and that he ends the section with the caution that men "do not unwisely mingle or confound these learnings [that from God's works and that from God's word] together." His more detailed description of this fault shows, however, that he does not refer to scholasticism but to a branch of Renaissance neo-platonism:

for if any man shall think by view and inquiry into these sensible and material things to attain that light whereby he may reveal unto himself the nature or will of God, then indeed is he spoiled by vain philosophy....And hence it is true that it hath proceeded that diverse great learned men have been heretical, whilst they have sought to fly up to the secrets of the Deity by the waxen wings of the senses."[14]

The philosopher behind this heresy is Plato, not Aristotle, and the indictment more readily calls to mind Sadoleto, Conrad Celtis, and Elyot than Aquinas. Bacon has in mind the sort of men of his own age whom Sir Thomas More had mentioned, without pious condemnation, in his well-known letter to Oxford: "There are even some who make the knowledge of things natural a road to heavenly contemplation, and so pass from philosophy and the liberal arts... to theology."[15]

We have seen that Aquinas too distinguishes the objects of sense and faith. Moreover, as Spedding notes, Bacon's first proof of the purity of knowledge of the world, that before the fall Adam had knowledge to name the other creatures "according to their proprieties," is itself scholastic. On the question "whether the first man had knowledge of all things" (*scientiam omnium*), Aquinas argues, against various objections, that (Gen. ii) Adam named the animals. "*Nomina*

[13] Willey, *op. cit.*, pp. 35–36.

[14] *Works*, VI, 95–96.

[15] E. F. Rogers, *The Correspondence of Sir Thomas More* (Princeton, 1947), letter 60. Quoted in Caspari, *op. cit.*, p. 71.

autem debent naturis rerum congruere. Ergo Adam scivit naturas omnium animalium: et pari ratione, habuit omnium aliorum scientiam." (*Sum. Th.*, I. Q. 94, art. 3) Aquinas does not raise the objection which Bacon had to face, that knowledge of the world is evil. The three objections he refutes have rather to do with the fact that in the present state of things such knowledge is cumulative. The third, specifically, is that man's present condition is intended partly so that he may advance in knowledge. Aquinas counters that prelapsarian man would have learned through experience what he already knew speculatively. Hence Bacon's position that the pursuit of knowledge can help man partially to regain the earthly paradise seems implicit in Aquinas. It is more nearly true that Bacon reasserts a scholastic argument in face of the jealousy of Renaissance divines than that he advances natural science as part of a rejection of scholasticism. Willey's study of Bacon and the rejection of scholasticism depends too largely upon passages in which Bacon is not thinking of the schoolmen at all. It also creates the misleading impression that Book I of the *Advancement* is primarily devoted to natural science.

The last section of Bacon's preliminary defense opens with the notorious censure of sixteenth-century *copia* and of the schoolmen. Be it noted however that Bacon's underlying purpose is "by a censure and separation of the errors, to make a justification of that which is good and sound, and to deliver that from the aspersion of the other."

In Bacon's view, as the Reformation called up the Renaissance, it was Luther who "was enforced to awake all antiquity and to call former times to his succors to make a party against the present time; so that the ancient authors, both in divinity and in humanity, which had long time slept in libraries, began generally to be read and revolved." This commendable revival of learning required "a more exquisite travail in the languages original wherein those authors did write, for the better understanding of those authors and the better advantage of pressing and applying their words." Thus far Bacon expresses no disapproval. What he censures is the "affectionate study of eloquence and copie of speech" which, for several reasons, then began to flourish and which "grew speedily to an excess," to the point that "men began to hunt more after words than matter." "Then did Car of Cambridge, and Ascham, . . . almost deify Cicero and Demosthenes, and allure all young men that were studious into that delicate and polished kind of learning."[16] In reference to this passage, Bush calls Bacon unjust for not recognizing that to most humanists good Latin stood "for a whole orthodoxy, cultural, ethical, political, and religious." Yet as we have seen Bacon does not mean to attack good Latin, and he is able to name Erasmus as a predecessor in scoffing at this vain excess of rhetoric. It is important that we do not, in calling this passage an attack on Ciceronianism, suppose out of prejudices respecting Bacon as scientist, that he does not himself revere Cicero and admire rhetoric. Whenever he himself cites Cicero, as in the essay on atheism,[17] it is enthusiastically as a partisan and authority. Of rhetoric, Bacon immediately takes pains not to be misunderstood: "notwithstanding it is a thing not hastily to be condemned, to clothe and adorn the obscurity even of philosophy itself with sensible and plausible elocution." Taking up the topic again in Book II, Bacon writes much as Bush has written of *him* that "it was great injustice in Plato, though springing out of a just hatred of the rhetoricians of his time, to esteem of Rhetoric but as a voluptuary art. . . . " Bacon does not quite share Cicero's sense that *eloquentia* is the outward sign of a noble nature, yet he goes on to say that much as Plato "said elegantly, *That virtue, if she could be seen, would move great love and affection*," the next best thing, this being impossible, is through rhetoric "to shew her to the Imagination in lively representation."[18]

One of the ascribed reasons why the New Learning had led to excess in rhetoric was "the enmity and opposition that the propounders of those (primitive but seeming new) opinions had against the schoolmen; who were generally of the contrary part, and whose writings were altogether in a differing style and form; taking liberty to coin and frame new terms of art to express their own sense and to avoid circuit of speech, without regard to the pureness, pleasantness, and (as I may call it) lawfulness of the phrase or word." Here Bacon's sympathy is with "lawful" Latin at the expense of the language of the schools. But in the controversy between humanism and

[16] *Works*, VI, 119.

[17] *Works*, XII, 134–135.

[18] *Works*, VI, 298. Cf. Cicero, *De Officiis*, I, v, 15, quoted by Bush, *op. cit.*, p. 59, in definition of Renaissance humanism.

scholasticism to which he alludes Bacon accepts what is sound on each side and condemns the excesses of each, closing his censure of the Ciceronians with the comment that under them the learning of the schoolmen grew "to be utterly despised as barbarous. In sum, the whole bent and inclination of those times was rather toward copie than weight."

Bacon, then, would not seem to despise scholasticism. Yet the discussion to which he now turns, of a kind of degenerate learning which "did chiefly reign amongst the schoolmen," is conventionally cited to show his rejection of this branch of medieval thought. The distemper consists, not of vain words, but of vain matter. Bacon calls up the authority of St. Paul against sciences marked by novelty of terms and "strictness of positions, which of necessity doth induce oppositions, and so questions and altercations." The schoolmen, having read too narrowly, chiefly in their dictator Aristotle, and knowing too little of natural and human history, are the notable offenders. Thus they spin out "those laborious webs of learning which are extant in their books." The method rests not so much on the sound basis of "evidence of truth proved by arguments, authorities, similitudes, examples, as upon particular confutations and solutions of every scruple, cavillation, and objection." It is one of the ironies of the history of ideas that Bacon in fact accuses the schoolmen of fragmenting knowledge: "*Questionum minutiis scientiarum frangunt solidatem.*" Each part of knowledge should support the others; the strength of the old man's faggot lay in the bond. As Bacon elsewhere phrased it, "let this be a rule, that all partitions of knowledge be accepted rather for lines and veins, than for sections and separations; and that the continuance and entireness of knowledge be preserved."

In this his most extended discussion of the schoolmen, Bacon is far from complaining that they fused faith and reason, and his criticism constitutes no fundamental rejection of scholastic thought. Before you descend to their distinctions, he asserts, "the generalities of the schoolmen are . . . good and profitable." It is only that, in the general censure, they take corruption from their particular fault of contentiousness. They leave God's word and God's works alike to spin cobwebs; it is not that they confound the two. The respect for their "generalities" underlying Bacon's criticism of their "digladiation about subtilities" is significant, if we are to assert with Bush that Renaissance humanism is fundamentally medieval and Christian.[19]

In the *Novum Organum,* the arrogant heart of the *Great Instauration,* Bacon glibly cites the *Advancement* of fifteen years ago as proof that he is no enemy of humanistic studies and need not waste words asserting the fact now. He apologetically designates that work, revised and done into Latin, to be a part, however imperfect, of the *Instauration.* But in his exasperation at the inability of traditional disciplines to penetrate the complexities of external nature, Bacon changes his position in the second of his two principal philosophical works to one unquestionably hostile to many of the finer things of life. The humanist carried away by "admiration" to think of Alexander, "not as Alexander the Great, but as Aristotle's scholar," now thinks of Aristotle only as scientist and finds need to pen a *new* Organon. The writer who, in the manner of Sidney, composed a peroration on the immortality conferred by books now proudly humbles himself to the study of facts "unincumbered with literature and book-learning." Natural philosophy itself tends to become the Prima Philosophia which, in the *Advancement,* united the branches of natural, divine, and human philosophy, so that in an incredible aphorism (Book I, CXXVII) Verulam declares that he means all knowledges, including ethics and politics, to be reinstituted according to the inductive method developed for studying such phenomena as heat and cold, light, and vegetation. Instead of distinguishing the knowledges as lines or veins, Bacon here foretells the current dismal entropy.

It is not a question of his sacrificing scales of values descended from antiquity, but of his allowing them needlessly to be swept into insignificance by the passion to interrogate nature. It is not so much that he divorces some medieval marriage of faith and the world as that he preserves a medieval division between them while wanting, like ourselves, to give full play to each. In the *Advancement* and its narrative offshoot the *New Atlantis* Bacon brings natural philosophy down into the theater of man's life. The completed *New Atlantis* (had not science drawn him away from that work) would have given scientific research its distinct place in an ideal com-

[19] Even in the *Novum Organum,* though he includes scholasticism among the Idols of the Theatre, Bacon finds it perhaps the least objectionable of the systems mentioned.

monwealth. Bacon's mature humanism, more than the scientific preoccupations of his old age and retirement, gives him relevance to a society (and an Academy) in which the two cultures are mutually exclusive. "Let there be . . . ," he wrote in a moment of anticipation of Sir Charles Snow,

> (and may it be for the benefit of both) two streams and two dispensations of knowledge; and in like manner two tribes or kindreds of students in philosophy —tribes not hostile or alien to each other, but bound together by mutual services;—let there in short be one method for the cultivation, another for the invention, of knowledge.[20]

---

[20] *Works*, VIII, 63, from the Preface to the *Novum Organum*. This work as a whole, however, takes a contemptuous attitude toward those who do not seek knowledge in the study of things.

Papers of the Michigan Academy of Science, Arts, and Letters
Vol. XLVII, 1962 (1961 Meeting)

DRAMATIC CONVENTION AND THE PSYCHOLOGICAL STUDY OF CHARACTER IN *OTHELLO*

LEO F. McNAMARA

The University of Michigan

A psychological criticism of literature, that is, a criticism that draws upon contemporary psychological theory and knowledge, is much practiced today. Some of us look favorably upon this enterprise and expect from it an aid to our understanding. Why is it that all too frequently we find it instead an assault upon our intelligence and rather a strain on our credulity? One reason may be that some critics of a psychologizing bent evidently measure the validity and effectiveness of their criticism by the number and intensity of the shocks it administers to the traditional understanding of a work, by the extent to which it ignores or gratuitously repudiates the well-established scholarship associated with the work, and by the degree of novelty in the interpretation it offers. But if we are to have a responsible, helpful psychological criticism—and there is no reason why we may not—it will be one that undertakes to explain a work in such a fashion that traditional ideas about the work and the scholarship supporting those ideas are not ignored but accounted for and illuminated. This criticism will take as its criterion not, as so often at present, a vacuous novelty of interpretation; it will seek confirmation of its results in their concordance with the findings of scholarship and the common understanding of serious readers.

It seems reasonable to hold that the extent to which a psychological analysis is appropriate and the mode of analysis chosen depend upon the kind of literature studied. Not all works are amenable to the same kind of psychological analysis, nor are all contemporary psychological systems equally adaptable to the criticism of a given work. These are general propositions that, if granted, would help bring about a more responsible psychological criticism. Perhaps I can illustrate some of the ramifications of these propositions by taking up in this paper one particular problem: the question of the psychological study of character in Shakespeare's *Othello* and the relation of this study to the dramatic conventions that partly define the characters in that play. In what follows there will perhaps be more questions raised than solved. I do not intend an exercise in psychological criticism, but an exploration of some of the problems that such a criticism must face up to.

In choosing *Othello* for particular comment I wish to reply to John V. Hagopian's "Psychology and the Coherent Form of Shakespeare's *Othello*," read before this section of the Academy in 1959.[1] In his paper Professor Hagopian employs principles and terminology derived from the Freudian and Gestalt psychologies in an attempt to explain the action of *Othello* as "a single, unique whole human action involving the interrelationships of complex characters acting upon each other with clear (if not always conscious) motivation."[2] The result is an essay that to my mind renders Shakespeare's work unrecognizable. The tragedy of *Othello* becomes reduced to the proportions of a rather shabby domestic melodrama, superior in no discernible way to the crude tale in Cinthio, Shakespeare's source.

Briefly to summarize Professor Hagopian's paper, he begins by rejecting what he terms "the reduction [of *Othello*] to moral allegory so popular among traditional Shakespeare scholars,"[3] that is, the equation of Desdemona with absolute innocence and purity, Othello with great and simple virtue flawed only by the belief that men are what they seem, and Iago with "motiveless malignancy," the principle of evil incarnate. Certain excesses in the work engage his attention: (1) Iago's villainy, disproportionate to his original desire for revenge; (2) Desdemona's attraction to Othello and later, her extraordinary interest in Cassio's reinstatement; (3) Othello's readiness to believe the absurd, impossible insinuations of Iago on the basis of insufficient evidence. Proposing to explain these supposed excesses through the application of Freudian conceptions, Professor Hagopian finds that the text supplies "plentiful evidence" for asserting sexual inadequacy in Othello. This inadequacy leads to frustration in Desdemona, a passionate woman whose sexual expectations are thus disappointed. A consequent subconscious yearning for Cassio explains her extraordinary interest in his reinstatement in Othello's favor. Othello's own "agonized awareness" of his

[1] *PMASAL,* XLV (1960), 373–380.

[2] Hagopian, p. 374.

[3] Hagopian, p. 373.

failure to satisfy her makes him ready to believe Iago's insinuations about Desdemona and Cassio. Iago himself is interpreted as "a simple honest soldier who has been disillusioned [by Othello's failure to promote him]" and "a man with an uncanny, shrewd insight into the characters around him."[4] His revenge powerfully exceeds its original limits because it works upon Othello's sexual inadequacy and so sets off a psychological "chain reaction" made possible by the motives and dispositions outlined above.

It will be readily seen that Professor Hagopian's whole case rests upon reading certain passages in the play so as to construe sexual inadequacy in Othello. These passages, it is scarcely necessary to say, are capable of receiving and usually receive constructions quite different and more probable. We might take one instance to indicate the tenor of Professor Hagopian's interpretation. In Act I, Scene iii, Othello is informed by the Senate that he must go immediately to the defence of Cyprus. He replies, "With all my heart" (1.278). Othello's readiness to comply with the command of the Senate is usually taken as the response of one willing to put aside personal satisfaction to do "the state some service." Professor Hagopian, however, finds it "significant that this bridegroom of only a few hours should so eagerly welcome a postponement of the consummation of his marriage."[5] But this comment is hardly illuminating, since the question immediately poses itself: why, then, is Othello a bridegroom? Why create the need for postponement in the first place? I have no doubt that some rather unusual and charming answers can be produced, but at the expense of a further flight into Freudian speculation and away from the plain sense of our text. Since I do not view the Freudian element in Professor Hagopian's paper as its chief fault (it is a symptom of a more basic error) I shall not stay with this issue. Suffice it to say that the whole notion of sexual inadequacy is implicitly contradicted by the author himself when to bolster a different part of his argument he cites this passage, Othello addressing Desdemona on Cyprus:

> Come, my dear love,
> The purchase made, the fruits are to ensue;
> That profit's yet to come 'tween me and you.
> (II, iii, 8–10)

[4] Hagopian, p. 376.
[5] Hagopian, pp. 376–377.

Perhaps the really disquieting thing about Professor Hagopian's view of the play is not so much its lack of agreement with the text as its quite amazing agreement with Iago. Othello's supposed inadequacy, frustration in Desdemona, a consequent yearning for another, younger man: Professor Hagopian's explanation of the action has been anticipated in all its essentials in the explanation Iago offers to his gull Roderigo.[6] There is no denying that Iago is a shrewd psychologist, but it would be rather ingenuous in us to accept as a satisfactory view of the characters and the action what amounts to the account fobbed off on Roderigo. In this reading, who's he then that says Iago plays the villain? We end up with an honest Iago, however, by failing to distinguish between what he seems and what he is, a distinction obscured in the romantic criticism exemplified by Professor Hagopian's paper.

For this paper is certainly in the romantic tradition of Shakespeare criticism. The psychological analysis of character is not, after all, something new in Shakespeare study: it is at least as old as Maurice Morgann's *An Essay on the Dramatic Character of Sir John Falstaff*, published in 1777. There is really very little that is novel about psychoanalytic criticism as applied to Shakespeare: the underlying assumptions are pretty much the same as those that inform the kind of criticism that began with Morgann's curious essay and reached its high-water mark in Bradley. 'How many children had Lady Macbeth?' becomes 'How many complexes had Hamlet?' The changes in method and terminology are superficial in comparison with the similarity in doctrine concerning the nature of dramatic character. Romantic criticism insists on treating characters in any drama as if they were real persons, no matter how flagrantly this runs counter to the nature of the drama in any given instance. This mistaken notion is basic to Professor Hagopian's paper.

I have suggested that it is instructive to view Professor Hagopian's paper in relation to Morgann's curious essay, the granddaddy of all the romantic psychological studies of Shakespeare. Doctor Johnson made a shrewd estimate of things to come when Boswell asked him his opinion of Morgann's essay. "Why, sir, we shall have the man come forth again; and as he has proved Falstaff to be no

[6] This explanation is offered by Iago in his several interviews with Roderigo: *Othello* I, iii, 302–388; II, i, 215–294. Textual references are to G. B. Harrison, ed., *Shakespeare: The Complete Works* (New York, 1952).

coward, he may prove Iago to be a very good character."[7] Now Professor Hagopian has not gone so far toward vindicating Iago as have some who used to excite the spleen of E. E. Stoll,[8] but he does offer an interpretation that apparently makes an honest man of Iago once again. And the tenor of his paper certainly mitigates the evil of Iago while it reduces the stature of Desdemona and Othello. The whole scale of the tragedy is reduced.

But the enormity of Iago's evil, the innocence of Desdemona, the flawed nobility of Othello are not illusions to be explained away; they are the facts from which we must start. If these facts evaporate in the course of a psychological criticism, so much the worse for the criticism.

"The trouble with Shakespearian criticism," E. E. Stoll writes, " . . . is that it has been prompted and guided by the spirit of literalism . . . The characters have been taken for separable copies of reality."[9] What Stoll is here attacking is the romantic view of character, the view that is implicitly assumed in Professor Hagopian's paper. His use of psychoanalytic principles is incidental to this basic conception; the essential point is that characters are treated as if they were real persons, and it is here that his interpretation is fundamentally at fault.

The literalist or, if you like, realist convention of character has of course been prominent in modern drama as in modern literature generally. That this particular sort of characterization has not always been dominant in English drama is immediately evident to anyone with the slightest familiarity with medieval and Tudor drama. What is not so immediately obvious, apparently, even after a half-century of scholarship, is how deeply and pervasively the drama of Shakespeare is influenced by the conventions of the Moralities and Interludes. Romantic criticism of the plays fostered some odd notions about character in Shakespeare, and we are still strongly affected by these notions. Some scholars and critics still

[7] *Boswell's Life of Johnson*, ed. G. B. Hill (Oxford, 1887), IV, 191n.

[8] " . . . some of the critics have been as much taken in by him [Iago] as the people on the stage; a sensible American even called him a prince of good fellows, and thought he would not, if it kept out of his way, had trod upon a worm; and every now and then, in the conjuring game of criticism, something of this monstrous misconception is revived." Edgar Elmer Stoll, *Art and Artifice in Shakespeare* (Cambridge, 1934), p. 38.

[9] Stoll, *Art and Artifice,* p. 48.

think and write as if the appearance of Shakespeare were accompanied by a total change from the allegorical to the realist conventions. But if anything is clearly discernible in the best Shakespearian scholarship of the past four or five decades it is the growing recognition that the assumptions of a literalist criticism do not apply to Shakespeare—the realist measure simply will not do.

This denial of the sufficiency of literalist notions, though it does not imply a retreat into simple allegorical interpretation, entails a view of Shakespearian drama as one in transition from a predominantly allegorical to a predominantly realist mode of character presentation and hence a drama greatly shaped by the older mode and not justly appreciated by a criticism more adapted to the newer: adapted to the beautiful plays of Chekhov, say, the works of Eugene O'Neill, or (to descend to the present) the fables and fantasies of Tennessee Williams or the dreary sentimentalities of William Inge. The view of character found in Freudian and some other (but not all) contemporary psychologies applies much more aptly to these than to Shakespearian drama.

To return to *Othello*: what does a psychological criticism have to account for here? Taking the three major characters, it has to account for a Desdemona whom we are plainly asked to accept as wholly good and pure and true; anything else is to take the Iago-view and change the play utterly. It has to account for an Othello whose nobility and fall are alike evident, and whose representative character (his relation to the Everyman and Mankind of the Moralities) is ably demonstrated by Professor Edmund H. Creeth in a paper[10] exemplifying the best qualities of contemporary Shakespearian scholarship. And it has to account for a Iago whose kinship with the pre-Shakespearian Vice figure has long been noted by scholars such as Stoll[11] and Farnham[12] and has recently been fully and most convincingly documented in Bernard Spivack's *Shakespeare and the Allegory of Evil.*[13] It has to account for a tragedy in

[10] E. H. Creeth, "Moral and Tragic Recognition: The Uniqueness of *Othello, Macbeth,* and *King Lear,*" *PMASAL,* XLV (1960), 381–394.

[11] Edgar Elmer Stoll, *Shakespeare and Other Masters* (Harvard, 1940), pp. 230–280 *passim.*

[12] Willard Farnham, *The Medieval Heritage of Elizabethan Tragedy* (Univ. of California, 1936), p. 451.

[13] Bernard Spivack, *Shakespeare and the Allegory of Evil* (Columbia Univ. Press, 1958).

which, as Professor Creeth puts it, "the spiritual career of Othello . . . follow[s] in full the general design for a protagonist's experience which we find in the Macro plays and their Tudor parallels" and in which "Othello himself re-enacts in specific and concrete terms the career devised to stage allegorically the life of typical man and transmitted as a dramatic convention, along with the Vice, to Tudor playwrights".[14]

If we recognize the conventions operating in *Othello,* they tell us much about the alleged "excesses" which have disturbed some readers and which Professor Hagopian's essay undertakes to account for. Desdemona's "extraordinary" interest in Cassio, for example, is seen as an expression of extraordinary kindness instead of a manifestation of a subconscious yen for the man. Othello's succumbing to the temptation of Iago is seen not as the delusion of a man *peculiarly* credulous and gullible (a persistent critical error), but as the re-enactment of this phase of the struggle with Evil, shown by Creeth to be the enduring structural design inherited from the Moralities. Professor Hagopian's complaint that Othello readily believes his ancient on insufficient evidence is oddly reminiscent of Rhymer's contention that Shakespeare should have provided a garter instead of a handkerchief. In the dramatic convention, the temptation must succeed. Iago's evidence will *always* seem to us to be insufficient, for the simple reason that it cannot be sufficient: what Iago wishes Othello to believe about Desdemona is not, in fact, true.

The remaining excess alleged by Professor Hagopian touches upon the vexed question of Iago's motives. He finds a disproportion between Iago's villainy and his "originally vague (but apparently limited), desire for revenge on Othello".[15] Why limited? This is presumably a reference to Iago's initial explanation to Roderigo in the first scene of the play that he desires revenge because Cassio and not he has been made lieutenant. But if we remember that Iago is here replying to the impatient suspicions of his gull, there is no reason why we should accept this as Iago's original, limited desire for revenge, and every reason why we should suspect it. The first time we have a genuine opportunity for unequivocal statement is

[14] Creeth, pp. 385, 392–393.
[15] Hagopian, pp. 374–375.

in Iago's first soliloquy (I, iii, 389–410) and there we have not limit but an intimation of chaos:

> I have't. It is engender'd. Hell and night
> Must bring this monstrous birth to the world's light.

The question of Iago's motives has been prominent since Coleridge's striking pronouncement about the villain's "motiveless malignity." A chief virtue of Spivack's *Shakespeare and the Allegory of Evil* is his demonstration that much of the puzzlement over Iago's motives stems from the assumption, once again, of erroneous literalist notions; Iago is supposed to have motives as an actual human being would have them. When we realize to how great an extent the shaping of the character of Iago depended upon the rich and potent Vice tradition, the literalist measure becomes meaningless. Recognition of the Vice convention powerfully illuminates many parts of the play: the boisterous opening scene before the house of Brabantio, for example, and the drinking-scene on Cyprus as well as the comic colloquy of Iago, Desdemona, and Emilia (a version of the traditional quarrel of soldiers and housewives) as they await Othello's arrival on Cyprus (Act II, sc. i) all draw upon the comic element in the Vice tradition; the frequency of diabolic imagery[16] that culminates in Othello's stabbing of the "devil" Iago, Iago's taunting response—"I bleed, sir; but not kill'd"—and his final mysterious silence, this imagery derives from the Vice's diabolic aspect; Iago's role as false adviser and tempter is the supreme working of a part played by the Vice in scores of pre-Shakespearian dramas. Do not misunderstand. This is not source-hunting, but an attempt to convey some suggestion, in brief compass, of the large contribution of convention to the character of Iago. A full presentation of this thesis will be found in Spivack's work, cited previously.

One final aspect of the convention deserves to be noted because a misunderstanding of it has often led to false inferences. Professor

[16] S. L. Bethel, "Shakespeare's Imagery: The Diabolic Images in *Othello*," in *Shakespeare Survey 5*, ed. Allardyce Nicoll (Cambridge, 1952), pp. 62–80. Bethel shows that the incidence of images of the diabolic is proportionately greater in this than in any other of Shakespeare's plays. He also shows an interesting decrease in the frequency of Iago's using of diabolic terms with a corresponding increase of these terms in the speech of Othello, as the ancient progresses in dominating the mind and actions of his captain.

Hagopian as have others before him sets great store by Iago's reputation for honesty, and he asks, when this honest service is bypassed and Cassio made lieutenant, "should we be surprised that the moral basis of his universe collapses and he turns to villainy"?[17] Another reminder of the romantic bias, this. Have we left the girlhood of Shakespeare's heroines only to meditate the boyhood of his villains? This intimation of a prior Iago is totally unwarranted. What do we know, from the text, of any "collapse" in a presumed moral basis, or of a "turning to" villainy? Nothing. What we do know is that Iago does have a reputation for honesty, that he is called "honest Iago" so frequently that the epithet is almost a part of the name, and that he is not, in the only sense that really signifies in the play, honest. The false-naming convention is apparent here: Iago is known as "honest" in precisely the same way that Vyce in *Horestes* goes by the name Courage; Hypocrisy in *Lusty Juventus* is known as Friendship; Avarice, the Vice of the play *Respublica* simulates Policie and the gallants Insolence, Oppression, and Adulation assume the names Authoritie, Reformation, and Honestie; Fancy in Skelton's *Magnificence* presents himself as Largesse—the list of instances could go on indefinitely. The pertinent fact about Iago's reputation for honesty (a convention parallel to the simpler naming device in the allegories) is that he is enabled to accomplish his role as false counselor and tempter. It tells us nothing about his supposed motives.

I have tried to show some of the things in *Othello* that a psychological criticism must account for, and some of the ways in which a psychological approach that involves improper assumptions about the nature of character fails to answer to our knowledge of the play. At the outset I promised more problems than solutions for psychological criticism, so I do not intend to develop such a criticism at this point. I will only suggest that it seems to me that, among contemporary psychologies, that of Jung presents some interesting possibilities for a study of *Othello*. Maud Bodkin brings up the question of how far we can "identify Iago as a projected image of forces present in Othello." She remarks that the temptation of Iago and the whole drama are built not only upon falsehoods, "but also on partial truths of human nature which the romantic vision

[17] Hagopian, p. 376.

ignores".[18] In this view the temptation scene becomes a scene manifesting a conflict within Othello, as the dimly perceived terrible half-truths in the back of his mind struggle with and finally overthrow the truths by which he has lived. In the psychological criticism suggested by Maud Bodkin, a criticism that restores Othello to his place of central importance in the play and makes an appropriate assumption about the nature of the character Iago, I think we find a striking confirmation, arrived at by a completely independent line of inquiry, of what we learn from a study of dramatic convention. This mutual concordance of scholarship and criticism is, I believe, the criterion of validity for the psychological criticism of *Othello*.

[18] Maud Bodkin, *Archetypal Patterns in Poetry* (Oxford, 1948), p. 221. See also her Appendix II, "Psychological Criticism and Dramatic Conventions," for a statement with which this paper is in substantial agreement.

PAPERS OF THE MICHIGAN ACADEMY OF SCIENCE, ARTS, AND LETTERS
VOL. XLVII, 1962 (1961 MEETING)

TOM JONES AND JACOB DE LA VALLÉE

LYALL H. POWERS

The University of Michigan

THE role of Marivaux in Anglo-French literary relations has been generally recognized: his connection with Richardson has received some close attention, but the influence of his novels on those of Fielding has been comparatively neglected.[1] The similarity between *Le Paysan parvenu* (1735) and *Joseph Andrews* has indeed been noted—after Fielding's encouragement in the opening chapter of Book III of that novel, where he mentions "the History of Marianne and le Paisan Parvenu." But the influence of Marivaux's novel on *Tom Jones* has barely been noticed. G-E Parfitt has observed that "Dans *Tom Jones* et *Joseph Andrews* on trouve les traits qui peuvent bien venir du *Paysan Parvenu,* car, des deux romans de Marivaux, celui-ce est évidemment le plus rapproché de l'esprit et du sujet des romans de Fielding."[2] And he adds a brief general comparison of Tom and Jacob de la Vallée. The striking similarity between these two heroes, which Parfitt merely touches on, encourages closer examination and invites us to consider Fielding's use of Marivaux's novel. We recall the passage in the introductory chapter of Book XIII of *Tom Jones*: "Come, thou that hast inspired thy Aristophanes, thy Lucian, thy Cervantes, thy Rabelais, thy Molière, thy Shakespeare, thy Swift, thy Marivaux, fill my page with humour; till mankind learn the good-nature to laugh only at the folly of others, and the humility to grieve at their own."[3]

There are first, of course, certain obvious differences between these two novels of the picaresque tradition: *Le Paysan parvenu* is

[1] See, for example, Joseph Texte, *Jean-Jacques Rousseau et les origines du cosmopolitisme littéraire* (Paris, 1895), chap. III; and the unpublished dissertation of Sidney E. Glenn, "Some French Influences on Henry Fielding," Urbana, Illinois, 1932. See also F. C. Green, *Minuet,* London, 1935, and Alan D. McKillop, *Samuel Richardson,* Chapel Hill, 1936.

[2] *L'Influence française dans les oeuvres de Fielding et dans le théâtre contemporain de ses comédies* (Paris, 1928), p. 120.

[3] *Tom Jones,* Modern Library; all quotations are from this edition, and subsequent references to book and chapter are given in the text.

first-person narration, *Tom Jones* is third-person—and the narrative is regularly interrupted by generous comments in the author's own person; Jacob is nearly twenty when his career begins, while we know Tom almost before he is born, and it takes Fielding one-sixth of his novel to bring Tom to the age of Jacob. One important reason for Fielding's slowness in this is that he must set out all the various strands of his intricate plot before he can well start Tom on his way. The plot itself is another obvious difference between the novels: that of *Le Paysan* is rather severely simple, while that of *Tom Jones* is ingeniously complex.[4]

Perhaps the initial similarity to be noted is the generally humanistic attitude of Marivaux and Fielding in their handling of the question of good and evil as it affects their respective heroes. Fielding's benevolist treatment of Tom Jones is much like Marivaux's conception of Jacob as an *honnête homme.* In both cases the basis of judgment seems to be a humanistic ethic rather than religion. Indeed, as Oscar Haac has recently pointed out, Marivaux's "conception of *honnêteté* is like a religion which advocates the ideals of humility, kindness, and human understanding"—"a lay substitute for the Christian ideal of the gentleman."[5] In refusing the place offered him by his benefactor, M. de Fécour, in favor of the ailing Dorville, Jacob de la Vallée says, virtually, I am doing unto another as I would he do unto me.[6] Both heroes are guided through their adventures by a sort of instinctive goodness. Jacob regularly insists that his actions are motivated by natural promptings. In speaking of the mistress of his first conquest, Geneviève, Jacob explains his interested reaction in this way: (I quote from the 1735 English translation)

> The *Doeux Yeux* of a Man of the *Beau Monde,* have nothing of Novelty for a polite Woman; . . .
>
> But this was not my Case; for my Looks had nothing of Gallantry in them, nor had they any Notion of a Compliment. I was a Peasant, young and Handsome withall, and the Homage I paid her Charms, proceeded from nothing but the pure Pleasure I took in seeing them. My Rusticity was void of dissimula-

[4] See R. S. Crane, "The Plot of *Tom Jones,*" *Journal of General Education,* IV (1949–50), 112–130.

[5] "Marivaux and the *Honnête Homme,*" *Romanic Review,* L (Dec. 1959), 266.

[6] See *Le Paysan parvenu,* Classiques Garniers, p. 220.

tion, and was only the greatest Flatterer by its not knowing how to flatter at all.[7]

Fielding describes what is essentially the same characteristic in his Tom Jones this way: "Mr. Jones had somewhat about him, which, though I think writers are not thoroughly agreed in its name, doth certainly inhabit some human breasts; whose use is not so properly to distinguish right from wrong, as to prompt and incite them to the former, and to restrain and withhold them from the latter" (IV, vi). Indeed, both Jacob and Tom participate to some extent in the ideal of the noble savage: that is to say, they are both young men of the natural countryside who are initially uncorrupted by the taints of modern civilization, and particularly free of hypocrisy. Marivaux insists more often on this characteristic of his hero than does Fielding, but the characteristic is not the less apparent in Jones for all that.

The instinctive tendency to do good guides the respective heroes through the list of adventures that comprise their respective novels, and this benevolent tendency is rewarded (suitably, as the reader is made to feel) at the conclusion of each novel. Both novels are, of course, success stories: in both, the hero passes through a series of adventures that brings him gradually nearer to his goal. In both instances the reader feels all along that the hero deserves a fair reward, but he is likewise pleased to see that as the hero progresses through the chain of experiences he is gradually rendered more fit to enjoy that reward. And it is in the style of presentation of that series of improving experiences that the novel of Fielding seems to indicate most clearly its debt to *Le Paysan parvenu.*

The method of development employed by Marivaux is summed up well enough for our purposes by an early editor of *Le Paysan parvenu*, Duviquet:

En nous montrant un paysan marchant à la fortune, grâce à la bonne opinion que sa tournure inspire aux femmes, Marivaux a voulu faire passer son héros par toutes les modifications de ce qu'on appelle *amour*.[8]

---

[7] *Le Paysan Parvenu: or, the Fortunate Peasant. Being Memoirs of the Life of Mr.* ———. Translated from the French of M. de Marivaux (London, 1735), p. 16.

[8] *Le Paysan parvenu, Oeuvres Complètes de Marivaux* (Paris, 1825–27), VIII, 45, note 1.

The history of Jacob is a series of encounters with women who are attracted by his masculinity, and who contribute to his final success. Tom Jones is equally a favorite with the ladies, and his history is likewise punctuated by the intervention of various women who are attracted to him. If he does not pass through precisely the same graduated scale of amorous adventures that Jacob does, Tom Jones does experience three social levels of his world through their respective representatives, Molly Seagrim, Mrs. Waters, and Lady Bellaston, a series which is similar to Jacob's correspondents—Geneviève, Mme de Fécour, Mme de Ferval. Tom's career is, of course, something more than an *éducation sentimentale*, yet this aspect of his development is interestingly similar to Jacob's; and in any case, the heroes' stages of development are in both novels marked off by these amatory affairs. In comparing this aspect of the two novels we shall see most clearly the probable echoes of the French one in the English.

As the heroes pass along the series of "affairs" there is the constant question of their fidelity: as Jacob generously shares his favors with the various women he encounters, there is the shadow of Mlle Habert always in the background; and as Jones entertains the women he meets, there is at his back the shadow of Sophia. There is an important difference here, of course, which we shall consider more fully in a moment: Jacob de la Vallée is actually married to la Habert, while Tom Jones is not even engaged to Sophia—this is a difference which cannot but mitigate our condemnation of Tom's "infidelity." Nevertheless, in both novels it is the female figure at the heroes' back that makes them suffer pangs of conscience as they pursue their various escapades.

The initial liaison in both novels is entered before the hero is quite aware of the woman to whom he is to owe responsibility. Jacob's first real conquest is the unfortunate Geneviève, who wants to help him make his fortune in a way he cannot accept. Tom's first love is the earthy Molly Seagrim. These two girls are of about the same social standing, a good step beneath their successors in the life of the respective heroes. For both heroes the next woman encountered is she to whom they are to feel responsible during their subsequent careers with the obliging ladies. Beyond this, Mlle Habert and Sophia Western have little of importance in common. Their situations, however, are similar enough, as we see in comparing the roles of Mlle Habert *aînée* and Mrs. Western: these two furies, in their stout opposition to the respective heroes, contribute considerably to the discomfort of Mlle Habert *cadette* and of Sophia.

Both heroes, then, have been supplied with similar goads for their conscience as they set out on their other conquests. Jacob immediately wins the hearts of Mme d'Alain and her daughter, Agathe. Nothing of a serious nature can develop in that situation, for Mlle Habert is too near at hand. But these two serve to emphasize the idea of Jacob's attractiveness. As Mlle Habert *aînée* attempts to make trouble for Jacob by having him summoned before M. le Président, Jacob again gives evidence of his attractiveness for the fair sex as he at once wins to his side both la Présidente and Mme de Ferval. The latter plays an important role in the novel as she is the woman with whom Jacob is to become most deeply involved, and who comes closest to causing him serious trouble: while he is keeping a rendezvous with her Jacob is interrupted by the Chevalier, who remembers him from the country and who might possibly expose him to society. Mme de Ferval appears in the least favorable light of any of the women who figure in Jacob's illicit affairs: she is actually a friend of the woman whom she is helping Jacob to deceive. Her value to Jacob is that she gives him money. Through la Ferval Jacob is enabled to meet and impress the lusty Mme de Fécour. This appealing creature—*à la gorge furieuse*—is perhaps the most likable of the women in Jacob's series of affairs. She is in many ways his female counterpart. Mme de Fécour is, with Jacob himself, the character who most clearly points ahead to the style of Fielding. She is a sort of embodiment of the attitude to love which is characteristic of the best of Fielding. Here is Jacob's sketch of her:

> Directly I saw a pretty fat Woman enter the Room, of a moderate Stature, but with one of the most furious Bosoms I had ever beheld; she seem'd otherwise a Woman without Ceremony, and at first sight a lover of Pleasure....
>
> Mrs. Fecour was about three or four Years younger than Mrs. Ferval, I believe she had been handsome in her Youth; but what was now most observable in her Countenance, was a frank Cordial Air, which made it very agreeable to look at.
>
> ...to which you may add a hale robust Look, a certain amiable Freshness of Complexion....
>
> ...it was one of her peculiar Qualities not so much as to know what Affectation meant.
>
> For example, if she lik'd you, this Bosom which I spoke of, was display'd

to you a thousand different Ways, not so much with an Intention to tempt your Heart, as so to let you know that you had touch'd her's [*sic*]; it was her manner of declaring Love (pp. 233–234).

Her association with Jacob consists rather in the promise of amorous joy than in the actual realization of it, but she takes her place in the series just the same.

The two women Jones first encounters after he has set out on his career are Mrs. Waters and Lady Bellaston. The resemblance of these two to Mme de Fécour is perhaps not obvious, yet Mrs. Waters' attitude to Tom is much like la Fécour's to Jacob: her amorous dalliance with Tom for the pure natural joy of it is quite what we should expect from *la dame à la gorge furieuse* (and it is to be remembered that the *gorge assez furieuse* of Mrs. Waters has its own rather important little role in the opening pages of the ninth book of *Tom Jones*). And Lady Bellaston shares many traits with her French predecessor, Mme de Ferval. Like la Ferval she is a woman of a certain age, perhaps just past her prime but still very attractive and a widow—or at least a woman who no longer has a husband. Lady Bellaston's relationship to the woman to whom the hero feels he is being untrue is even closer than the corresponding relationship of Mme de Ferval to Mlle Habert. And as la Ferval had supplied Jacob with gifts of money, so does Lady Bellaston supply Tom Jones. It is with her that Tom's dalliance leads him the farthest—he is reduced to proposing marriage in order to free himself, and this proposal almost proves to be his ultimate undoing, later on—and it is his affair with Lady Bellaston that comes closest to preventing him from gaining his ultimate reward, Sophia Western.

Tom Jones's relationship with Mrs. Miller, his landlady in London, and with her daughter Nancy recalls Jacob's situation with Mme d'Alain and her daughter. The attitude of Mrs. Miller is rather more sober than that of Mme d'Alain, but the initial attitude of Nancy is not much different from that of Agathe. But as far as functional value is concerned, the role of the Miller family in Fielding's novel more nearly follows that of the Dorvilles in Marivaux's. Jacob's generous *honnêteté* in his behavior to the Dorvilles gains their favor, but, furthermore, his association with them leads to his meeting the highly placed Comte de Dorsan and, apparently, to his ultimate success. Fielding's development is similar: Tom Jones's generosity to the Millers—and especially to Nancy—wins him their admiration, and they are instrumental in effecting his reconciliation with Squire Allworthy and gaining his ultimate reward, Sophia.

[There are other, incidental similarities, such as the lengthy digressions in both novels: in *Le Paysan parvenu* the story of *le plaideur* told en route to Versailles, and Mme de Dorville's story of her encounter with the wolf; in *Tom Jones* we have the story of the Quaker, the long episode of the Man of the Hill, and the history of Mrs. Fitzpatrick. In both novels, all of these digressions serve as oblique commentary on the main fable: all have to do with unfortunate or loveless unions. There is, too, a reminiscence of the friendship of Jacob and Dorsan in Tom's friendship with Nightingale (the former associated with the Dorvilles, the latter associated with the Millers).]

But we cannot pursue the parallels beyond Jacob's taking up with Dorsan, for this is as far as Marivaux leads his hero on the road to success. At the end of the fifth part Jacob seems comfortably set with his new friend and protector, yet he has not achieved that conclusive success the reader had hoped for and, indeed, been led to expect. Yet Fielding might well have perceived, as any sensitive reader can perceive, the direction in which the motives of the first five parts of Marivaux's novel were directing Jacob. His amorous education in manners has led Jacob a good distance, and in his reaction to the delightful Mme de Dorville we see a nobler refinement of sensibility. By the middle of the fifth part we are warned that Mme de la Vallée is not to live much longer; and, as scholars have pointed out, the continuation of Marivaux's novel through three additional parts seems to have followed with some faithfulness the expectations created in the first five: it seems logical that Jacob, safely widowed, should ultimately gain the hand of someone like the charming Mme de Vambures.

Whether, as it would appear, Marivaux's intention was precisely of that sort or not, the main lines of Jacob's career are established, and upon them Fielding seems to have relied to some considerable extent in fashioning the history of Tom Jones. Furthermore, the dramatic situations, the characters themselves, and the general attitude of Fielding's novel are strongly reminiscent of Marivaux's. In the treatment of the series of amatory adventures, both authors have attempted to excuse somewhat their respective heroes: not only do both distinguish nicely between the sentiment or passion that attracts the hero to the various "other" women, and the sentiment of true love; they both make it clear that in every case it is the "other" woman who is the aggressor and the hero the passive (though not unwilling) partner, so to speak. It is precisely here, however, in the handling of the question of infidelity, that the marked difference between the two novels is most noticeable—here that the originality and power of Henry Fielding most clearly emerge. The important differences in treatment are these: (1) that Jacob is a married man while Tom is a bachelor; (2) that Jacob's extramarital affairs are, in the narrow sense of the term, virtually chaste while Tom's affairs are manifestly not; and (3) that the figure to whom Jacob owes responsibility and who is therefore his prick of conscience (Mlle Habert) and the figure who would represent his ultimate reward (not yet created in the first five books) are combined by Fielding into the single character Sophia Western.

Now it would seem, at first glance, that Marivaux's arrangement gives him the advantage: Jacob's comparative chastity in his affairs makes the reader less liable to condemn him for libertinage; and furthermore, the satisfaction of Jacob's and the reader's expectations is but slightly postponed, for Jacob dutifully returns after his adventures to be engulfed by the eager limbs of the uxorious Habert. And for his dutiful attention to la Habert during his *éducation sentimentale*, we would agree that Jacob at last deserves to be rewarded with some true-love. But at second glance it becomes apparent that the advantage is really Fielding's: by keeping his hero a bachelor Fielding has to some considerable extent removed the stain of infidelity from Tom Jones; Tom of course feels guilty about his affairs, feels he is being unfaithful to his Sophie, but his infidelity (as he feels it) is in part a result of his despair, and at least Tom is legally free. In any case, Tom has a rather acceptable quasi-chivalric explanation for his behavior. Fielding writes (XIII, vii): "Jones had never less inclination to an amour than at present [the masquerade at which he hoped for a last interview with his Sophie]; but gallantry to the ladies was among his principles of honour, and he held it as much incumbent on him to accept a challenge to love as if it had been a challenge to fight." (This point is well argued in an essay by William Empson, who feels that in Tom's love affairs we have a good example of what he calls Fielding's doctrine of mutuality of impulse —"evidently the secret message which Fielding boasts of in *Tom Jones*."[9]) It is true that Jacob, like Tom, is regularly the passive

[9] "Tom Jones," *Kenyon Review*, XX (Spring, 1958), 217–249, esp. 224–226; quotation is from p. 226.

partner; yet we feel strongly that Jacob is "on the make"—ready to take advantage of amorous ladies for the joy of it, yes, but principally for the advancement they may afford him.

But most important of all, surely, is Fielding's decision to unite in Sophia Western the functions of Marivaux's Mlle Habert and whoever would have been Jacob's final reward. As William Blake would certainly object to Jacob's *dutiful* attention to his wife, so to some extent must we. Jacob's regular return to Mlle Habert, his dutifulness, is unmistakably tinged with reluctance:[10] his fidelity is not inspired by true love, and it is quite apparent that his ultimate goal consistently lies beyond. The arrangement of Tom's situation is much more effective: his conscience is pricked by the figure of his Sophie, his duty to her is self-imposed and inspired by the sincerest love. How much better that the figure to whom he feels his duty is owing and the figure who is his ultimate goal and reward are one and the same!

I should conclude, then, that Fielding apparently saw in Marivaux's *Le Paysan parvenu* a reasonable and useful basic pattern for the instructive story of a young man's learning wisdom or prudence—much as he later found in the *Aeneid* a useful pattern for his *Amelia*. The main lines of Tom Jones's career follow rather closely those of Jacob de la Vallée's, but Fielding was interested, finally, in something more than a sort of *éducation sentimentale*: Tom indeed has his reward in the lovely Sophia, but he has gained something more even than this exceedingly well-named lady—he has gained that necessary *sophia* we all wished him to have, that wise prudence that makes him a full man. He has not, perhaps, gained that *hagia sophia* won by the good Captain Booth of the later *Amelia*; but Tom's wisdom, his prudence, is of an eminently practical sort and his education has achieved a kind of perfection.

[10] See for example Jacob's own expression of his attitude:

"Revenons. Je m'en retourne enfin chez moi; je vais retrouver madame de La Vallée qui m'aimait tant. . . .

"Je crois pourtant que je l'aurais aimée davantage si je n'avais été que son amant (j'appelle aimer d'amour); mais quand on a d'aussi grandes obligations à une femme, en vérité, ce n'est pas avec de l'amour qu'un bon coeur les acquitte; il se pénètre de sentimens plus sérieux; il sent de l'amitié et *de la reconnaissance*; aussi en étais-je plein, et je pense que *l'amour en souffrait un peu*" (p. 259; my italics).

PAPERS OF THE MICHIGAN ACADEMY OF SCIENCE, ARTS, AND LETTERS
VOL. XLVII, 1962 (1961 MEETING)

UNCLE TOBY'S PIPE AND WHISTLE

FREDERICK P. KROEGER
Flint Community Junior College

SHAKESPEARE'S Falstaff has, through the ages, collected a host of admirers who have always enjoyed him and who have claimed him as their favorite character. In a similar way, many readers of Laurence Sterne's *Tristram Shandy* have become "Tobyolaters."[1] Uncle Toby is a character who is long remembered for his gentleness, his genuine sweetness, and, of course, for his Hobby Horse.

In the life and opinions of Walter Shandy "my uncle Toby" is like a rock upon which "my father's" rhetoric breaks, and only occasionally flows smoothly around. The manipulation of uncle Toby's pipe or his lips is usually a signal to the reader that pyrotechnics, to change the metaphor, will soon appear from "my father's" quarter. At the risk of appearing to be a "Tobyolater" I should like to try to show how Sterne has used uncle Toby's pipe, and *Lillabullero* to fill out his character, and most particularly to show the final, and in my opinion most brilliant application of these devices in the affair of the Widow Wadman. Before I do this, however, I feel I must discuss the device of the Hobby Horse, and its place in the characterization. A Hobby Horse is a man's ruling passion.[2] My uncle Toby's ruling passion was an unceasing interest in battles, sieges, and weapons of war. All of his thoughts were on campaigns and fortifications, and he fought battles over again either in his mind or in a miniature battlefield that his servant Corporal Trim had constructed in the back yard.

In the digression on character drawing by the use of a Hobby Horse Sterne says: "There is no instrument so fit to draw such a thing

[1] I am indebted to George Saintsbury for this term.

[2] "When a man gives himself up to the government of a ruling passion,—or, in other words, when his Hobby Horse grows headstrong,—fairwel cool reason and fair discretion." Bk. II, Ch. V. All references are based on the division into nine volumes or books of *The Life and Opinions of Tristram Shandy Gentleman* by Laurence Sterne. The Everyman Edition No. 617, (1927) is the most convenient and easily available edition with this division.

as a man's character as that which I have pitched upon. . . . —by long journeys and much friction, it so happens, that the body of the rider is at length filled as full of Hobby-Horsical matter as it can hold; —so that if you are able to give but a clear description of the one [the Hobby Horse], you may form a pretty exact notion of the genius and character of the other."[3] It must be admitted that this Shandean technique is a good one. Sterne uses it often to characterize other characters in the book as well as the narrator himself, but it is a technique that produces only a flat character. Now in my opinion uncle Toby though a somewhat shallow character, is certainly not a flat one; he is not just an old soldier, furiously riding an old soldier's Hobby Horse. Something has been added very subtly to make him more a live personality. The pipe and the whistle have been used not "to have given you the great outlines of my uncle Toby's most whimsical character; . . . not the great contours of it; . . . but some familiar strokes and faint designations of it. . . ."[4] These "faint designations" occur throughout the book, but one of the two most important ones occurs in connection with the discussion of the Hobby Horse. He develops his ruling passion because, "my uncle Toby by nature was not eloquent."[5] He had to have a map of the area of Namur, where he received his wound, so that he could point to the scarp, the covered way, the counterscarp, and all the other areas where complicated maneuvers took place. " 'Twas not by ideas,—by Heaven; his life was put in jeopardy by words."[6] He shows his character by gestures and not by words. Stern tells us what he is like, and then allows Toby to act in a manner that will demonstrate the truth of the author's statements about him. He is a military man of courage, but with childlike reasoning ability and a peaceful and placid nature. He has all the finer sentiments of humanity: gentleness, mildness, courtesy, good temper, and an unwillingness to hurt anything or anybody. And finally, there is another trait most important for the author's purposes:

My uncle Toby Shandy . . . was a gentleman, who, with the virtues which usually constitute the character of a man of honor and rectitude,—possessed one in a very eminent degree, which is seldom if ever put into the catalogue; and that was a most extreme and unparallel'd modesty of nature; . . . modesty in the

---

[3] Bk. I, Ch. IV.
[4] Bk. I, Ch. XXII.
[5] Bk. II, Ch. V.
[6] Bk. II, Ch. V.

truest sense of it . . . not in regard to words, for he was so unhappy as to have very little choice in them,—but to things;—and this kind of modesty so possessed him, and it arose to such a height in him, as almost to equal, if such a thing could be, even the modesty of a woman:[7]

This sense of modesty in "my Uncle Toby" is always shocked when "my father" mentions the affair that "my Aunt Dinah" had with the coachman, and the first mention of *Lillabullero* occurs in connection with this story. When "my father" answers Toby's cry, "For God's sake . . . and for my sake, and for all our sakes, my dear brother Shandy,—do let the story of our aunt and her ashes lie in peace," with a philosophical argument, he is completely at a loss for words. "My Uncle Toby would never offer to answer . . . by any other kind of argument, than that of whistling half a dozen bars of *Lillabullero*."[8] Thus he expresses himself by an action rather than words, although I wish to suggest later that in certain places words which Toby is too polite to utter are there also. In this instance Sterne tells us that this was the usual channel through which he gave vent to his passions when anything surprised or shocked him, or when anything seemed very absurd to him.[9] This whistle is an intrinsic part of his character, and though the Hobby Horse and *Lillabullero* stand out as quite obvious devices, I feel that a study of the situations in which uncle Toby whistles will show some things which aren't quite so obvious. Ever since Bishop Percy published his *Reliques* critics have been saying, as did the good Bishop, that the whistling of *Lilla Berllero* was used wonderfully by Sterne to characterize an old soldier. The second reference we have to it in *Tristram Shandy* is clearly an indication that uncle Toby is an old soldier. Stern tells us so in so many words:

For tho' it might have suited my uncle Toby's character as a soldier excellently well . . . to whistle the *Lillabullero* . . .[10]

This is fine as far as it goes, but there is a lot more to the whistling than this. Aside from this one place in the book, the whistling is used to indicate other things. More frequently, *Lillabullero* is a sign that my uncle Toby has mounted his Hobby Horse and is galloping as fast

[7] Bk. I, Ch. XXI.
[8] Bk. I, Ch. XXI.
[9] *Ibid.*
[10] Bk. II, Ch. II.

as he can, away from the discussion around him. "There is no treating the subject [philosophy] whilst my uncle is whistling Lillabullero."[11] When a philosophical argument begins Toby closes his mind to it, for he seldom understands it. He assumes the character of a soldier in his mind, and begins his daydream, though outwardly he is the placid, pipe-smoking, country gentleman. At times, when this transition takes place he whistles his favorite song, indicating that he has begun his flight.

At the dinner given to discuss the possibility of changing Tristram's name, Toby is both bored and shocked. Kysarcius makes the statement that the mother is not kin to the child. This statement shocks Toby for a moment, and he instinctively purses his lips for an exclamatory whistle. At the same instant both he and "my father" realize that they are going to hear a long argument to prove this point:

> My father instantly clapped his hand upon my uncle Toby's mouth, under colour of whispering in his ear;—the truth was, he was alarmed for Lillabullero—and having a great desire to hear more of so curious an argument—he begged my uncle Toby, for Heaven's sake, not to disappoint him in it.—My uncle Toby gave a nod—resumed his pipe, and contenting himself with whistling *Lillabullero* inwardly.[12]

My father hears the argument and Toby takes refuge in his Hobby Horse.

Uncle Toby whistles when something he considers to be absurd has been said and perhaps that is another reason for the instinctive pursing of the lips at Kysarcius' remark. I have suggested before that, occasionally, behind the whistle, there are words that Toby is too polite to utter. I believe that many of Sterne's more discerning contemporary readers supplied these words and got quite a bit more fun out of Toby by so doing. The first line of *Lillabullero* is "Ho broder Teague dost hear de decree?" By supplying these words as an exclamation at the end of the next quotation, quite an amusing effect may be obtained.

> Brother *Toby*, quoth my father, I do believe thee to be as honest a man, and with as good and as upright a heart as ever God created;—nor is it thy fault, if all the children which have been, may, can, shall, will, or ought to be begotten, come with their heads foremost into the world:—but believe me, dear *Toby*, the

[11] Bk. IX, Ch. XVII.
[12] Bk. IV, Ch. XXIX.

accidents which unavoidably waylay them, not only in the article of our begetting 'em—though these, in my opinion, are well worth considering,—but the dangers and difficulties our children are beset with, after they are got forth into the world, are enow—little need is there to expose them to unnecessary ones in their passage to it.—Are these dangers, quoth my uncle Toby, laying his hand upon my father's knee, and looking up seriously in his face for an answer,—are these dangers greater now o' days, brother, than in times past? Brother *Toby*, answered my father, if a child was but fairly begot, and born alive, and healthy, and the mother did well after it,—our forefathers never looked farther.—My uncle *Toby* instantly withdrew his hand from off my father's knee, reclined his body gently back in his chair, raised his head till he could just see the cornice of the room, and directing the buccinatory muscles along his cheeks, and the orbicular muscles around his lips to do their duty—he whistled *Lillabullero*.[13]

On the other hand, we might interpret the whistle as uncle Toby saying to himself, "Well if thats the case it is the soldier's life for me. This is much too shocking a state of affairs for me to participate in and I shall be off." I think either interpretation will stand, but if the words of *Lillabullero* were immediate in one's memory I think the first suggestion I have made to be the better one.

Lillabullero was an anti-Catholic song, and uncle Toby whistles it as loud as he can on that night that "I" was born when "my father" provides Dr. Slop with an excommunication of the Church of Rome to read out loud so that Obadiah will be properly cursed. The absurdity of the situation is immediately realized by Toby for he has heard the excommunication before, and as it is a fearful one, being read only because Dr. Slop has cut his thumb, it is indeed inappropriate. Furthermore, "imagine to yourself a little squat, uncourtly figure of a Doctor *Slop*, of about four feet and a half perpendicular height, with a breadth of back, and a sesquipedality of belly, which might have done honor to a sergeant in the horse guards," and a Papist, reading an excommunication written by Ernulphus the bishop.[14]

"Ho broder Teague dost hear de decree?"[15]

---

[13] Bk. III, Ch. VI.

[14] Bk. II, Ch. IX.

[15] "Now, now de hereticks all go down
By Christ and Shaint Patrick, de nations our own.
Lerro lerro, lilli burlero, lilli burlero bullen-a-la

Dare was an old prophecy found in a bog
Ireland shall be ruled by an ass and a dog."

The whistling of *Lillabullero* has become a sign that here indeed is an absurdity. The whistling is not always used in this way, but this seems to be its major usage. When the author has made his long trip to drop a tear on the grave of Amadis and Amanda, the great lovers, only to find that it is not there, he wishes uncle Toby were there to whistle *Lillabullero*. Toby recognizes the absurdity of many situations where none of the other characters see it. I wonder if he was such a "confused, pudding headed, muddle-headed fellow" after all.[16] Sterne says he was no fool, and I am tempted, in the light of *Lillabullero,* to believe it.

In Book III we find uncle Toby whistling again. Sterne says " 'Twas no inconsistent part of my uncle Toby's character—that he feared God, and reverenced religion." Toby has just told his brother that God is the one who determines the length of a man's nose. He says, " 'Tis he . . . who makes us all, and frames and puts us together in such forms and proportions . . . as is agreeable to his infinite wisdom." " 'Tis a pious account, cried my father, but not philosophical —there is more religion in it than sound science."[17] At this Toby begins whistling *Lillabullero* with more zeal than usual.

Again I think the words to the tune apply, but I should like to suggest another interpretation as well. Since this is an anti-Catholic song, could it not be that Toby in his innocence is trying to intercede with God for his brother? I suggest that it is quite in character for Toby, with his naïve ideas to think he can placate God by whistling *Lillabullero*. He whistles it with "zeal," and considering the connotations of this word in the eighteenth century I think the contemporary reader understood the whistle in some such sense.

The last time uncle Toby whistles in *Tristram Shandy* is during the affair of the Widow Wadman. Here all the effects of the whistle, and all of the causes for the whistle seem to apply. Just as the device of the pipe is used with the greatest brilliance in this episode, so too is the whistle.

Toby has at last decided to declare his love for the widow. He has arrayed himself in his regimentals and sallied forth. He has delayed his entrance by stopping in the path to hear the complete story of Trim's brother Tom's amours. At the end of the tale he has marched off in the opposite direction only to recall his mission and

[16] Bk. II, Ch. II.
[17] Bk. III, Ch. XLI.

return, marching straight up to Mrs. Wadman's door. Here he hesitates for the last time while Trim lifts the knocker, but before he can call a halt to Trim's action the knocker falls and all is lost. "My uncle Toby perceiving that all hopes of a conference were knock'd on the head by it——whistled Lillabullero."[18] Toby, broadly speaking, is whistling to keep his courage up, yet in this last scene I think that every aspect previously discussed concerning why he whistles may be seen.

In the first place Toby is dressed as Captain Shandy, so we see him in his soldier character whistling his soldier song as he goes into "battle." Secondly, we see Toby wishing with all his might that he were somewhere else, even if it were on the actual battlefield. So *Lillabullero* indicates a desire to flee from an unpleasant situation, a situation where he will have to use words, a situation also that shocks his natural modesty. He wishes he were on his Hobby Horse riding fast away. And, finally, the words of the song apply with their reference to the beginning of a new government. Also, I think the history of the song, as it was probably known in the eighteenth century, adds a wonderful element of humor and absurdity to the situation.

Bishop Percy says, without any authority but his own, that the words "Lilli burlero, and bullen-a-la" were words of distinction used by the Irish Papists during the massacre of the Protestants in 1641.[19] Now if this were general common knowledge among Sterne's readers would it not add a great deal to the humor of the situation?[20]

The *Reliques* was published in 1765, the same year that books VII and VIII were published. It seems to me reasonable to believe that if Sterne's readers were not already acquainted with *Lilla-*

[18] Bk. IX, Ch. XVI.

[19] Thomas Percy, *Reliques of Ancient English Poetry*. The 5th edition in three volumes (London, 1812), Vol. II, p. 405.

[20] Sterne changed publishers in late 1761 and Becket took Dodsley's place. It is curious that Books V and VI printed for Becket still retain the italicized proper names, and *Lillabullero* is also italicized as it was in the previous Dodsley publication. In Books VII, VIII, and IX, published by Becket, names are not italicized and neither is *Lillabullero*. I had hoped to show that Sterne meant to refer definitely to the words "Lilla-bullero," and not to the song title *Lillabullero* by not italicizing here, but I am afraid it cannot be done, both because Lillo berlero is not italicized in the Amandus and Amanda episode in Bk. VII, Ch. XL, and because of the lack of italicizing all the way through these last books.

bullero by this time, that they would look it up in the *Reliques* to refresh their memories. This interesting bit of history would then be familiar to them if it was not before.

In this final scene Sterne has utilized all the knowledge the reader has gained about uncle Toby and *Lillabullero* to get the most out of it. He will do the same with the pipe.

Uncle Toby's pipe is mentioned 29 times in twenty chapters, and I don't believe in any one instance do we remember anything about it. It is a stage prop used to illustrate or reinforce statements Sterne has made about uncle Toby's character. Although Sterne has said he will use no instrument to draw his character, I think it possible to say that he has certainly used the pipe for "familiar strokes and faint designations of it."

As one would expect, the best example of this character-drawing technique occurs while Sterne is slowly sketching in Toby's character. "My father," asks Toby after an hour and one half's silence during which Toby has been smoking his social pipe, "in mute contemplation of a new pair of black plush-breeches which he had got on:—What can they be doing brother? . . . we can scarce hear ourselves talk."

"I think, replied my uncle Toby, taking his pipe from his mouth, and striking the head of it two or three times on his left thumb," and here Sterne leaves him, pipe in hand, "in order to give you the outlines of his character."[21] Thus the first impression we have of Toby is one of a pipe-smoker.

In the history of literature we find many properties used to indicate character. The most obvious is a sword to indicate a soldier, and in the same way, Toby's pipe is used to indicate his character. He is a pipe-smoker. This immediately creates certain images for the reader. Sterne does not fail to make these images concrete; for example, he pictures Toby smoking by the fire or smoking and meditating as he walks about his battlefield. Since the author has provided his character with a pipe he takes care to mention it every time Toby uses it. One characteristic of Sterne's style is his minute observation and description. "He loved to transfigure and interpret the obvious."[22] This is well illustrated by the fact that if uncle Toby is smoking, and wishes to make a statement, Sterne tells us he takes

[21] Bk. I, Ch. XXI.

[22] Walter Sichel, *Sterne, A Study*, (London, 1910), p. 187, and see *Tristram Shandy*, Bk. II, Ch. XVII, and Bk. III, Ch. II, and XXIX.

his pipe out of his mouth, as in the quotation above. Similarly he must remove his pipe to shake his head.

> "... my uncle Toby took his pipe out of his mouth and shook his head."[23]

These references maintain our impression of Toby as a pipe-smoker for I think we may consider them examples of the statement by Sterne, "I must sustain my character properly."[24]

To return to my first example, Sterne has sustained Toby for eleven chapters, sitting there with his pipe in his hand ready to speak until Book II, Chapter VI. In these eleven chapters Sterne has told us a great deal about Toby, including the description of his Hobby Horse and how he came by it. I suggest that the technique used here is something like the technique of the playwright using a revolving stage.[25] In this case Sterne has set up a cozy scene of uncle Toby and "my father" and then has revolved the set so that the narrator has the center of the stage. As he describes Toby in his long digressions the reader in his imagination can glance over to the motionless figure of Toby sitting silently there, and then again give his attention to the narrator. When Sterne's narrator is through speaking he says, "At present the scene must drop—and change for the parlour fire-side."[26] The scene revolves back and Toby says, "I think—it would not be amiss brother, if we rang the bell."[27]

It has been my contention that one of Toby's most important traits is that he is inarticulate. What better way to reinforce this notion than to have him in suspended animation with his pipe in his hand ready to speak after one hour and one-half of silence, and then let him say thirteen humdrum words? And all this disregards the wonderful humor of the technique.

Since Toby speaks little, natural actions with the pipe help to indicate other character traits also:

> Then brother Shandy, answered my uncle Toby raising himself off the chair and laying down his pipe to take hold of my father's other hand,—I humbly beg I may recommend poor Le Fever's son to you.[28]

---

[23] Bk. V, Ch. XXXVII.

[24] Bk. II, Ch. II.

[25] Sterne uses stage terminology all through the book as he changes from scene to scene, and even calls the book a drama.

[26] Bk. II, Ch. V.

[27] Bk. II, Ch. VI.

[28] Bk. VI, Ch. V.

Here in the act of laying down the pipe to take his brother's hand may be seen gentleness, kindness, and humility. Further we see "Toby laying the end of his pipe upon my father's hand in a kindly way of interruption."[29]

Kindness, politeness, and caution and deliberate action are indicated here:

> My uncle Toby was no chronologer—so not caring to advance one step but upon safe ground, he laid down his pipe deliberately upon the table . . . and taking my mother most kindly by the hand . . . he led her out.[30]

Politeness and good manners seem to be indicated by the fact that he does not smoke while walking with his sister-in-law.

We see him as a soldier when we find "Toby, waving his tobacco-pipe, as he would have done his sword at the head of a regiment."[31] Toby usually has his pipe in his hand, so Sterne has very cleverly utilized it to show his soldier character. This appears as a perfectly natural action in the context and seems to show the author's complete foreknowledge of the character. He knows him so well that he naturally has Toby wave his pipe as if it were a sword.

Toby on his battlefield is riding his Hobby Horse, and waving his pipe in the example above. The pipe is often used to show Toby in some phase of his Hobby Horsical activity, for

> Now it happened then, as indeed it had often done before, that my uncle *Toby's* fancy, during the time of my father's explanation of *Prignitz* to him—having nothing to stay it there, had taken a short flight to the bowling-green!—his body might as well have taken a turn there too—so that with all the semblance of a deep school-man intent upon the *medius terminus*—my uncle Toby was in fact as ignorant of the whole lecture, and all its pros and cons, as if my father had been translating *Hafen Slawkenbergius* from the *Latin* tongue into the *Cherokee*. But the word *siege*, like a talismanic power, in my father's metaphor, wafting back my uncle *Toby's* fancy, quick as a note could follow the touch—he open'd his ears—and my father observing that he took his pipe out of his mouth, and shuffled his chair nearer the table, as with a desire to profit—my father with great pleasure began his sentence again.[32]

He thinks about battles and sieges most of the time when he is not actually re-enacting them on his "rood and one half of the ground."

[29] Bk. V, Ch. III.
[30] Bk. V, Ch. XIV.
[31] Bk. V, Ch. XXXIII
[32] Bk. III, Ch. XLI.

Similar occasions occur in Book IV, Chapter XXIX, and Book V, Chapter III.

While the passage above shows uncle Toby's Hobby Horsical activity, it also shows another character trait. Sterne, in this passage, shows uncle Toby in a state of polite boredom and at the same time fooling "my father" by indicating false interest by removing his pipe, and contemplating it with "wonderful seriousness." He is capable of fooling "my father" with this gesture, but he dissembles interest because he is too kind and too polite to willingly disturb "my father" while he is indulging in his favorite pastime. When his interest really wanes, or is put to an abrupt halt he immediately replaces his pipe in his mouth and continues his daydream.

> Do not, I beseech thee, interrupt me at this crisis.—My uncle Toby put his pipe into his mouth.[33]

Furthermore, in this instance we find uncle Toby taking solace in his pipe. In another instance uncle Toby, being ridiculed by his brother, makes no defence. He takes refuge in his pipe, and smokes so hard to cover his discomfort that the smoke sets his brother to coughing. Here we see Toby sensitive, yet too gentle to argue with his brother. Also, I feel, we see Toby's eternal frustration, for the two brothers can never understand each other. As soon as Toby sees his smoking has caused his brother to cough he comforts him, and even goes so far as to dry his eyes. This is one of the extremes of sentimentalism Sterne indulges in, yet it shows Toby to be very tenderhearted.[34] For proof of the above statements let us look at another instance of similar nature:

> 'Twas well my father's passions lasted not long; for so long as they did last, they led him a busy life on't; . . . as the unexpected strokes his science met with from the quaint simplicity of my uncle Toby's questions (it was as if ten dozen hornets had stung him behind).
>
> 'Twas all one to my uncle Toby—he smoked his pipe on with unvaried composure—his heart never intended offence to his brother—and as his head could seldom find out where the sting of it lay—he always gave my father the credit of cooling by himself.—He was five minutes and thirty-five seconds about it . . .[35]

---

[33] Bk. V, Ch. II. See also, Bk. III, Ch. XLI.
[34] Bk. III, Ch. XXV.
[35] Bk. III, Ch. XLI.

The pipe may be used as a pipe of peace between the two brothers, giving comfort to them both.

> The best friends, in the world may differ sometimes.—True,—brother Toby, quoth my father shaking hands with him—we'll fill our pipes brother.[36]

Differences of opinion, however, were not too frequent between the brothers because Toby has a great amount of patience.

> My uncle Toby would give my father all possible fair play in this attempt: to interest him in philosophy and with infinite patience would sit smoking his pipe for whole hours together. . . .[37]

The pipe contributes a great deal to our knowledge of Toby in the story of Le Fever (Book VI, Chapters VI-XIII). First; upon hearing of the sick Lieutenant, Toby is so upset that he cannot finish his dinner. Trim brings him his pipe to comfort him. Then he sends Trim to help the poor man, and while he is gone he smokes two more pipes as he thinks what he can do for the sick man. Later Toby is even more concerned about Le Fever, and so again takes solace in his pipe and a glass of sack.[38]

Before I discuss the part the pipe plays in the courtship of Mrs. Wadman I should like to quote three passages to sum up uncle Toby's character. Hazlitt says, "My uncle Toby is one of the finest compliments ever paid to human nature. He is the most unoffending of God's creatures . . ."[39]

> In the sharpest exacerbations of his wound (like that on his groin) he never dropt one fretful or discontented word—he blamed neither heaven nor earth—or thought or spoke an injurious thing of any body, or any part of it; he sat solitary and pensive with his pipe—looking at his lame leg—then whiffing out a sentimental heigh ho! which mixing with the smoke, incommoded no one mortal.[40]

---

[36] Bk. II, Ch. XVII.

[37] Bk. III, Ch. XXXIX.

[38] Sterne used uncle Toby's pipe to record the passage of time in quite an orthodox way since "a pipe" can be either a measure of time or distance. In the Story of Le Fever we find an example of this in Bk. VI, end of Ch. VI, and beginning of Ch. VII.

"It was not till my uncle Toby had knocked the ashes out of his third pipe, that Corporal Trim returned from the inn . . ."

[39] Walther Sichel, *op. cit.*, p. 200.

[40] Bk. VIII, Ch. XXVII.

> My uncle Toby entering the room with marks of infinite benevolence and forgiveness in his looks, my father's eloquence rekindled against the passion—. . . —as soon as my uncle Toby was seated by the fire, and had filled his pipe, my father broke out in this manner.[41]

In these quotations we are given a fairly complete mental picture of Toby's outward appearance and inner feelings. The reader, however, may feel that he is too good, too forgiving, and too kind. Sterne has kept him from being a saint, fortunately, for in one place he shows Toby as a real human struggling with his desires and finally overcome by them.

> My uncle Toby took the ivory pipe out of the corporal's hand, looked at it for half a minute, and returned it.
>
> In less than two minutes, my uncle Toby took the pipe from the corporal again, and raised it half way to his mouth—then hastily gave it back a second time.
>
> The corporal redoubled the attack,—my uncle Toby smiled,—then looked grave, —then smiled for a moment,—then looked serious for a long time;—Give me hold of the ivory pipe, Trim, said my uncle Toby—my uncle Toby put it to his lips,—drew it back directly,—gave a peep over the horn-beam hedge;—never did my uncle Toby's mouth water so much for a pipe in his life. —My uncle Toby retired into the sentry-box with the pipe in his hand.—
>
> —Dear uncle Toby! don't go into the sentry-box with the pipe,—there's no trusting a man's self with such a thing in such a corner.[42]

This amusing scene humanizes Toby. Saints can be boring, and here we see him not quite a saint, and he is better for it.

I think that now we can see what an important part uncle Toby's pipe played in Sterne's characterization technique. It is true that Sterne tells us in so many words that Toby is benevolent, kind, gentle, pensive, given to day-dreaming, etc., but he makes us believe that he is telling the truth about his character by making Toby's actions with the pipe fit the descriptive words. We might say that Sterne, the playwright, has sketched a character, and Toby, the actor, has played it out before our eyes.

At the end of Chapter V in Book II Sterne tells us that he will tell more about uncle Toby's activities on his battlefield later, and in Chapter VII of the same book he first mentions Widow Wadman. I think it is fortunate for us that Sterne's idea of structure allowed

[41] Bk. IX, Ch. XXXIII.
[42] Bk. VI, Ch. XXVIII.

him to put off these accounts until the last books of *Tristram Shandy*, and I think it is also fortunate that through the accident of no more being written *Tristram Shandy* concludes with these long-promised events. I believe that Sterne deliberately held back as long as he could, for he says in Book II:

How my uncle Toby and Corporal Trim managed this matter, [the construction of the miniature battlefield]—with the history of their campaigns, *which were no way barren of events,*—may make no uninteresting under-plot in the epistasis and working up of this drama.[43]

I suggest that this story would not be half so rich if the reader were not perfectly acquainted with uncle Toby, his whistle, and his pipe.

The reader knows how addicted to tobacco our hero is, and how he is almost never without his pipe. He also knows that when Toby removes his pipe from his mouth he is interested. Sterne has given us Toby's pipe for a gauge of his emotions. Consequently, when we find in Book VIII, Chapter XVI, that the Widow Wadman is actually able to take his pipe out of his hand in order to get her fingers close to Toby's, we may feel that Mrs. Wadman is progressing very well in her campaign.

Madam Wadman attacks Toby in his sentry box. First she looks at the map of the battle, which immediately arouses Toby's passions and he begins an explanation of the battle, removing his pipe from his mouth to use it as a pointer.

When the attack was advanced to this point;—the world will naturally enter into the reasons of Mrs. *Wadman's* next stroke of generalship—which was, to take my uncle Toby's tobacco-pipe out of his hand as soon as she possibly could;[44]

She deprives him of his main weapon and leaves him defenseless, for he would be too polite, if he were not already completely bemused, to use *Lillabullero*.

As the widow continues to work her wiles upon the innocent victim we see Toby completely forget his pipe. She has claimed his whole attention.

I see him yonder with his pipe pendulous in his hand, and the ashes falling out

[43] Bk. II, Ch. V. Italics mine.

[44] Bk. VIII, Ch. XVI.

of it—looking—and looking—then rubbing his eyes—and looking again, with twice the good-nature that ever Galileo looked for a spot in the sun.[45]

The next time we do not see the pipe in action, but we have a comment by Walter Shandy which is amusing in the light of what we know about Toby and his pipe. Toby tells Mrs. Wadman he loves her, and then without another word settles back in silence.

> My father was always in raptures with this system of my uncle Toby's, as he falsely called it, and would often say, that could his brother Toby to his process have added but a pipe of tobacco—he had wherewithal to have found his way, if there was faith in a Spanish proverb, towards the hearts of half the women upon the globe.[46]

It has been said of Sterne, "Give him a situation, a thought which appeals strongly either to his imagination or to his humanitarian instincts . . . and no man knows so well how to lead up to it; how to make the most of it; how, by cunning arrangement . . . to show it off to the best possible advantage."[47]

The denouement of the story of the courtship of Mrs. Wadman is a fine example of the quotation above. Chapter XXXI, Book IX, is worked out with the pipe being used with the perfection of art. The culmination of this use is found in those memorable lines, that end the chapter, "My uncle Toby laid down his pipe as gently upon the fender, as if it had been spun from the unravellings of a spider's web."

Here we see Sterne's artistry at its height, for before the sentence just quoted he makes sure that the reader is aware of the pipe by mentioning it four times. It is worth notice that this is the only chapter in which this occurs, and it is a very short one. The reader sees uncle Toby picking up his pipe to meditate on the most fitting way to describe Mrs. Wadman's "Humanity," laying it down to count on his fingers all her perfections, making a sign with it for Trim to take a chair so that he may write down these perfections, and pointing with it to show the proper place on the sheet to begin the list.

Using this method Sterne has prepared his reader for that final laying down of the pipe.

[45] Bk. VIII, Ch. XXIV.

[46] Bk. IX, "The Eighteenth Chapter".

[47] *Cambridge History of English Literature,* Vol. X, p. 61.

One of Mrs. Wadman's perfections is her great sympathy and compassion. He claims this for her because she so often inquires about his wound. Trim tells him that she is indeed interested in it, but had the wound been on his knee cap, as was Trim's, she would not be half so interested for:

> The knee is such a distance from the main body—whereas the groin, your honor knows, is upon the very *curtain* of the *place*.[48]

At last Toby knows. He gives a long low whistle and lays "down his pipe as gently upon the fender, as if it had been spun from the unravellings of a spider's web—." Both the story and the play are completed. With *Lillabullero* perhaps expiring in this last suspiration, the actor, using his pipe to subtly reinforce his actions, has played his part to perfection, for even the final action is characteristically gentle.

Sterne has utilized all the knowledge the reader has gained about uncle Toby and the pipe just as he did in this last book with *Lillabullero*. This seems to indicate a quite natural character development throughout his years of writing. There is nothing startling in this, but it is interesting to see how he has done it.

Critics have suggested that Walter and Toby combined would be much like Sterne's father. I do not know whether this is true or not, but I suggest that by Book IX these two characters are so fully developed that there is no possibility of combining them into one man. I cannot help but feel that Sterne knew he had a good thing in Toby's pipe, and *Lillabullero,* and so used them to more fully develop the character. The masterly use of these devices in the last book seems to bear out my opinion.

The Hobby Horse, the whistle, and the pipe are the devices used to characterize uncle Toby. Yet to say that he is an old soldier who rides a veteran's hobby horse, smokes a pipe, and occasionally whistles *Lillabullero* would be ridiculous. Still Sterne never gives elaborate descriptions of his physical actions as he does, for example, of "my father" or Corporal Trim. He never lets him say much, not even as much as Dr. Slop or Yorick; Toby is an inarticulate man.[49]

[48] Bk. IX, Ch. XXXI.

[49] Except for Bk. II, Ch. XII, where the speech given to Toby is appropriate in its gentleness, but almost too profound in its simplicity. Or is Toby wiser than we think?

He uses only such trite words as gentle, kind, innocent, brave, modest, and, of course, his own coinage *Hobby-Horsical* to describe him, yet he lives in our memories probably more than any other character in the book. It is the pipe and the whistle that convince us that simple descriptive words are truths about his character, though the little actions associated with the pipe and whistle are almost all we ever see Toby doing beyond sitting. It takes genius in an author to maintain a convincing character with so little, and to finally use these devices so brilliantly in Books VIII and IX, the last books he produced of *The Life and Opinions of Tristram Shandy*.

Papers of the Michigan Academy of Science, Arts, and Letters
Vol. XLVII, 1962 (1961 Meeting)

THE MIDDLE-IRISH VERSION OF THE *THEBAID* OF STATIUS

ROBERT T. MEYER

Catholic University of America

The student of mediaeval literature working in Middle-Irish finds himself in a no-man's land, half-way between the well-plotted Old Irish of the Glosses, the *Glossen-irisch* of Thurneysen, and the language of the modern period, with no full grammar and as yet no complete dictionary with which to work. Some of his texts will be transcriptions of older materials or arbitrary redactions of earlier texts. Most of this is still unedited and while it has been described as a scholar's paradise, there are still a good many pitfalls. But when one comes upon the various adaptations of the Latin classics, he finds himself upon a more solid base. He can to some extent control his materials. The rehandling of earlier works is of course a commonplace of the literary history of the Middle Ages.[1] The twelfth century was a great period for the translation of Latin authors in mediaeval Irish literature. It was the period of the translation of Lucan's *Pharsalia,*[2] Vergil's *Aeneid,* and the tale of Troy, the *Togail Troí,* a paraphrase of Dares Phrygius' *De Excidio Troiae.*

The *matière de Rome la grant* included all classical antiquity, and the Middle Ages were dependent upon various Latin authors for the Greek stories as well. The saga of Thebes then was conveniently available in the *Thebaid* of Statius,[3] which must have been extremely popular to judge from the numerous manuscripts still extant. An age which did not make the neat distinction of the moderns between silver and golden Latinity found a good deal of pleasure and information too in Statius, and he was a curriculum author as well.[4] Juvenal tells us that when Statius promised a public reading the

[1] Thomas Chestre, *Sir Launfal,* edited by A. J. Bliss [London, 1960], p. 1.

[2] Cf. R. T. Meyer, "The Middle-Irish Version of the *Pharsalia* of Lucan," *PMASAL,* XLIV [Ann Arbor, 1959], pp. 355–363.

[3] E. R. Curtius, *Europäische Literatur und Lateinisches Mittelalter* [Berne, 1948], p. 89.

[4] *Ibid.,* pp. 56 f.; 264.

Romans flocked to hear his pleasing voice and his beloved *Thebais* (uocem iucundam et carmen amicae / Thebaidos . . .).[5] In more recent times Goethe, greatest of the moderns, goes on record as saying:

> Statius is a very laudable poet and worthy of diligent study. I am not offended by the copious effusions of his genius, but in him I admire the art which every poet should have, of clear comprehension and precise description.... The things which he outlines in words seem present to our very eyes.[6]

Today Statius has fallen into disrepute along with Lucan, his fellow-writer, and a recent examination of college catalogues shows no school offering a course in either of these "last of the Romans." The late H. E. Butler enumerated the reasons which render the Theban legend unfit for epic poetry.[7] Aside from a lack of unity the story is unpleasant from beginning to end. Horror adds upon horror, crime upon crime, and the only softening feature is the figure of Antigone coming in at the very end. These very features, however, found a ready audience among the Irish, and the many episodes so closely modeled on Vergil's use of earlier legends provided the Irish with a long epic with many *remscéla,* but with this difference, that they do not lead up to a final catastrophe as in the Battle of Thessaly so well described in Lucan.

Besides this, the subject matter of the *Thebaid* could be readily fitted into the native Irish catalogue of story. Native tradition had classified stories by types, and we have two old lists of sagas where these types are enumerated:

> Destructions, Cattle-raids, Courtships, Battles, Cave-stories, Voyages, Tragedies, Adventures, Banquets, Sieges, Plunderings, etc.[8]

In this native framework of tales, then, there was room for the *Togail na Tebe,* the Siege of Thebes as it is called. Native sagas such as the *Togail Bruidne da Derga,* The Storming of Derga's Hostel, are representative of this literary genre, and the story of Troy is likewise the *Togail Troí.*

[5] *Iuvenalis Saturae* VII. 82–87, ed.W. V. Clausen [Oxonii, 1959], pp. 99 f.

[6] Quoted in Moses Hadas, *Ancilla to Classical Reading* [New York, 1954], p. 357.

[7] H. E. Butler, *Post-Augustan Poetry from Seneca to Juvenal* [Oxford, 1909], p. 208 f.

[8] R. Thurneysen, "Die Sagenlisten," in his *Die irische Helden- und Königsage bis zum siebzehnten Jahrhundert* [Halle, a.S., 1921], pp. 21–24.

There are two complete manuscripts of this tale. One is the famous Egerton MS. 1781 in the British Museum, a fifteenth century vellum. It is a genuine *bibliotheca,* a library consisting of romantic tales, lives of foreign (non-Irish) saints, chiefly from Latin sources, other texts such as the Legend of the Invention of the Holy Cross, the *Historia Caroli Magni,* an Irish version of the Pseudo-Turpin chronicle, etc.[9] The other, the Edinburgh manuscript, is in the National Library of Scotland, formerly known as Advocates' Library, and bears the shelf list number MS. VIII, Kilbride, No. 4. There are but 26 leaves, folio size, in this manuscript; a few are lacking. It dates from the early fifteenth century and was written by two scribes.[10] There is also a fragment in the Library of Trinity College, Dublin (MS. H.2.7, now No. 1298), consisting of but four leaves which bear the date 1479. This manuscript once belonged to Edward Lhwyd, the famous philologist and antiquary, who wrote the first comparative Celtic grammar (*Archaeologia Britannica,* [Oxford, 1707]).[11] While the two manuscripts, Egerton and Edinburgh, have a common origin without a doubt, the relationship of the Dublin fragment to these has not yet been established. Professor Donald Mackinnon began an edition of the *Togail na Tebe* with a translation in the *Celtic Review,*[12] which ceased publication with No. 40 in 1916, two issues after his death. In the meanwhile George Calder of the Department of Celtic at Glasgow University began to prepare an edition quite independently, which was published by Cambridge University Press in 1922.[13]

The two manuscripts have their own external points of interest

[9] Robin Flower, *Catalogue of Irish Manuscripts in the British Museum.* Vol. II [London, 1926], pp. 526–545 describes in great detail the contents of this fascinating book.

[10] Donald Mackinnon, *A Descriptive Catalogue of Gaelic Manuscripts in the Advocates' Library, Edinburgh . . .* [Edinburgh, 1912], pp. 195–198.

[11] T. K. Abbott and E. J. Gwynn, *Catalogue of the Irish Manuscripts in the Library of Trinity College, Dublin . . .* [Dublin, 1921], pp. 80; 337. John O'Donovan said it was a fragment of the *Togail Troí* and Abbott followed this suggestion in his catalogue of 1900. Dr. Gwynn corrected this in his version.

[12] *The Celtic Review,* vol. 7 [1911], pp. 106–121; 204–219; 318–335; vol. 8 [1912–13], pp. 8–21; 100–111; 218–233; 296–313; vol. 9 [1913–14], pp. 16–33; 112–127; 210–225; 292–309; vol. 10 [1914–16], pp. 24–39; 106–115; 198–211; 312–322.

[13] *Togail na Tebe, The Thebaid of Statius: The Irish text edited from two MSS. with introduction, translation, vocabulary, and notes* by George Calder . . . [Cambridge, 1922], xxiii + 431pp.

to any student of late mediaeval literature. Edinburgh has interesting glosses and marginalia. Some bare spaces in the vellum are filled with *penn gles* (Lat. *probatio pennae*), where the scribe tried his pen after he had sharpened it or had mixed up a new batch of ink. The term *penn gles* is written on folio 19. One scribe sighs for a refreshing drink of barley-whey as a rest from his labor of copying.[14] Another writes on the blank end of a leaf:

> This blank space at the end of the leaf has no need but to be filled up. p. 16.

Other marginalia prove that the scribes were interested in the progress of the story they were copying:

> Oh God and O Mary! Great is that spiteful madness that was upon Lycurgus as he mourned his hapless infant, compelling the Greeks to destroy their gold and their treasures. p. 29.

There follows an invocation to our Blessed Lady:

> Oh Son of Mary, O Mary, help me!! p. 36.

Again back to the story proper:

> Ill-fated indeed was that watch that the Thebans set upon the Greeks, seeing that they themselves had no awakening out of sleep till they were killed. p. 45.[15]

If the Edinburgh manuscript gave us an insight into the mind of the scribe working in his monastic carrel, the Egerton manuscript carries an interesting colophon telling something of the circumstances of the copying of that text. The scribe traveled about from house to house to copy manuscripts and enjoyed the food and hospitality of various patrons of learning. He does not sigh for his daily ration of barley-whey nor does he complain of the tiresome labor over his vellums. The man of learning went on a free circuit—*saorchuairt*—wherever there were great books to be copied and good living to be had. He also gives us notes on the history of the times, whose houses

[14] *The Rule of the Céli Dé*, 40, says "Further, when they are thirsty, a *bochtan* of whey or buttermilk, with water added, may be taken. This is to be drunk in sips." *The Rule of Tallaght*, edited by Edward Gwynn [Dublin, 1927], (*Hermathena*, no. XLIV, Second supplemental volume), pp. 74–75.

[15] See C. Plummer, "On the colophons and marginalia of Irish Scribes," *British Academy Proc.*, XII [1926], pp. 11–44 for many other interesting examples.

he stayed at, who was bishop of that diocese, what battles were being fought, who was slain in a local skirmish, etc.

The *Togail na Tebe* proper does not begin with the *Thebaid* itself but gives us some background to the saga proper through an account of the foundation of Thebes by Cadmus; the story of Oedipus was also added by the redactor. It then goes on to tell us that about this time there was a great, well-born, eminent poet among the Franks whose name was Statius (*Stait don filid Frangcach sochinelach*), who thought it well to chronicle all these happenings. Now Statius was commonly thought by mediaeval writers to be of French origin because of a confusion with another Statius, a grammarian who had dwelt at Toulouse. Dante and others claimed him for that place. So also a passage in the *Togail Troí* from the Stowe manuscript:

> *Conid amlaidh sin indisis Sdait in fili socenelach do Franccaib cetimrum luingi Argo.*
>
> [So does Statius the eminent poet (*fili*) of the Franks tell the tale of the first sailing of the ship Argo.][16]

Now this does not mean that the *Togail na Tebe* was from a French version of Statius, some such source as the *Roman de Thèbes.* Had such been true the form of the name would have been *Stais,* just as Chaucer speaks of *Stace.* The Middle-Irish version of Statius is as different from the Old French as that version is from its Latin original. The Irish redaction of Statius was a free reworking of the Latin, as we saw in the case of Lucan's *Pharsalia.* The Irish must first of all make it into a prose epic, for such was the native tradition. All the great Irish native masterpieces are written in prose from the *Táin Bó Cúalnge* to the histories of Keating in the seventeenth century. But the long Irish epics were broken up into divisions called *remscéla,* "prefatory tales," the assumption being that they led up to a climax. Such actually was the case in the *Cath Catharda* of Lucan. In some instances the divisions in the *Togail na Tebe* agree with the book divisions in the Latin original. So for some of the subtitles of the Theban saga in the Irish redaction:

> *finit do scel an mundtuirc,* 827
> (finit to the tale of the necklace)

[16] Kuno Meyer, "Ueber einige Quellenangaben der *Togail Troí,*" *Zeitschrift für Celtische Philologie,* vol. 10 [1915], pp. 358–360.

Imtusa Mairt conuigi sin, 1087
(concerning Mars thus far)
Tochastol na nGrec anso bodeasta, 1377
(the muster of the Greeks now hereafter)
Sochraidi sloig Arcaidi andsin, 1586
(the allies of the host of Arcadia there)
cetcath na Tebi sin, 2982
(that is the first battle of Thebes)

Sometimes such a division closes the preceding section and introduces the next:

> *Cained Tid connigi sin ⁊ lamchomairt Polineiceis moir mic Edip andsin,* 3396–97.
> (The lamentation (*keening*) for Tydeus thus far, and the handsmiting of great Polynices, son of Oedipus, there)

Corresponding with the end of Book VIII, the Irish has:

conaid confad Thid conigi sin, 3388
(the fury of Tydeus thus far.)

The redactor must place the story in its own proper setting. In his school days he had learnt that "the four things to be asked of every composition must be asked of this composition, viz., place, and person and time, and cause of invention."[17] So the story of Cadmus must needs be told, and how Jove stole his daughter and gave her the continent of Europe, which was the third part of the whole world. We hearken back in this threefold division to material as old certainly as Isidore of Seville and Orosius, the historian who also gave us some geography. This is a well-established tradition in the beginning of any mediaeval epic, to name only the *Waltharius* of Ekkehard or the *Alexandreis* of Gautier de Chatillon. And as the story of Oedipus is told and how he had unwittingly killed his own father and espoused his mother —

> . . . and knew nothing of that fact till the attention of Queen Iocasta fell upon the king's naked feet, for thus they were with a hole through them, the Queen asked:
> "What has pierced thy feet?"
> "Not hard," [*ni ansa*] said he — – 112–114

we are back once more in the world of Irish story-telling, the cate-

[17] Cf. the *incipit* of the *Aislinge Meic Conglinne, the Vision of MacConglinne, a Middle-Irish wonder-tale* edited with a translation, notes, and a glossary by Kuno Meyer . . . [London, 1892], pp. 2–3.

chetical beginning of many a saga, as in the famous Deirdre story: "Why was the exile of the Sons of Uisliu?" "Not hard," etc. . . .[18]

The actual beginning of our tale, however, is a flat historical statement: "There was a certain, noble, revered, and most honorable king, etc. . . . . " 1–2.

This was by far the more common beginning of the older Irish stories such as the *Scéla Mucce meic Dathó* (The Story of Mac Datho's Pig) or the *Tochmarc Étaine* (The Wooing of Étain).

As in the case of the Irish *Aeneid,* the redactor gives us the background, does not begin as either Statius or Vergil did *in medias res.* Once we get to the matter of the tale proper we are not to expect anything quite so bald as the Bohn translations of the classics, or anything quite so literary as the productions of the Loeb Classical Library. There was a certain element in which the Irish redactor and his audience would have no interest, e.g., the long complicated genealogies of the royal family of Thebes, reaching back as it did into prehistory much older than the Trojan heroes. He put the scenes and action into his own day as far as possible. This was nothing new, however: the Old French romances of antiquity did likewise, and their manuscripts in some instances are decorated with richly colored miniatures with Trojan heroes decked out as mediaeval knights playing chess near a pavilion where fair ladies listen to the reading of some mediaeval romance or to the musical composition of some troubadour. And so we find Irish expressions substituted for a literal translation of the Latin, as when Aëtion, "skilled in treating wounds" in the original (*Theb.* X.734), becomes the "chief leech" (*prím-lega,* 4180). Typical Irish similes are very effective: "as white as the linen of a smock after a full cleansing" (480) describes the pallor on the maidens' faces in *Theb.* I.537–38. Statius tells us that Oeclides ran swifter than the East wind (*volat ocior Euro*): the Irish informs us that "his speed and quickness were as swift as a grey cold blast of wind in a rough winter season" (2402–3). The "long blue arrow" is mentioned (1962, 2312, 2347) which occurs again in native compositions as the *Cath Ruis na Ríg* and the *Battle of Magh Rath.* Again the "lime-white shield" or the "lime-white bones" of dead heroes is a commonplace; they are always enlargements upon the Latin original of Statius. Polynices became "blanched like a white clover

[18] Cid dia·mboí loṅges mac n-Usnig? Ni ansa. *Longes mac n-Uislenn, the Exile of the Sons of Uisliu,* edited by Vernam Hull [New York, 1949] [M.L.A. Monograph Series, vol. XVI], p. 43.

flower, or like a fully bleached smock, and his muscles as hard as a handstaff of a kiln" (3365–67). The great warrior among his men appeared like a whale among "little speckled salmon" (3520). Phegeus is a "tower of battle and combat" (996); Harmonia's necklace a "blush to gloom, a boiling of rage" (767–8). Tydeus carries a "little spit of a toy javelin" (*birán bunsaigi baisi*), the very words used of the weapons of the little boy Cuchulain in the great *Táin Bó Cúalnge*. What appears to be the observation of *geis* or tabu occurs at the very beginning of the tale, when the redactor is giving us background before translating the Latin original. Prophets and wizards said that Laius would die should he look upon his children (84 f.). And again, when Oedipus and Laius finally met, "each of them was demanding from the other his name and neither of them would give the other his name" (105 f.).[19]

When Oedipus attends the banquet (*Theb.* VIII.241–6) the Irish redactor tells us that he was brought in to "drink and be merry" and "it is related that his hair had not been rinsed and his head had not been washed since he was blinded," recalling the bath and head-washing that was given to guests in the epics before their entertainment.[20] Funeral games are made into fairs (oenach) in at least eight different instances throughout the epic. Cloaks of Tyrian purple set round with gems in Statius (*Theb.* VI.63) are enlarged upon in the Irish version:

> bright beautiful purple garments of the land of the Syrians were spread over him outside his exterior, with many an ornament of beautiful golden thread and of round carbuncle gems. (2213–15).

This is not unlike descriptions of mantles in the native sagas, "set round with precious carbuncles." We are, for the time being, in the Celtic world of shape-shifting and magic. The soothsayers and magicians of Statius are "wizards and true men of skill." (4218) When the *Thebaid* (VII.771) tells us that Apollo revealed his god-head to the people, the Irish goes on to say that he "dropped his wizard shape." (2916) Mercury carried a wand (*caduceus*) of such quali-

[19] John Revell Reinhard, *The Survival of Geis in Mediaeval Romance* [Halle a.S., 1933], pp. 129, 159.

[20] Cf. E. Windisch, "*Táin Bó Cúailnge* nach der Handschrift Egerton 1782," *Zeitschrift für Celtische Philologie,* vol. 9 [1913], p. 128, 9–10. This is imitated in the *Merugud Uilix maic Leirtis,* 112, ed. R. T. Meyer [Dublin, 1958] [Mediaeval and Modern Irish Series, vol. 17], p. 4. Cf. note *ad loc.* p. 21.

ties that one end of it would wake all the world's dead and the other would kill all men alive. (590 f.). Mackinnon remarked that the translator's mind was more concentrated on the *slacan druidheachd,* the "druidic beetle" of the Gaelic Tales than on Statius' text when rendering this passage.[21] Other personalities of antiquity are given Irish names and characteristics. Bellona, the goddess of battle, is Badbha, the "red lipped and furious" (*badbha belderga bruthmara,* 3018) of Statius, *Theb.* VIII.348; the "Badb of battles" (*in badbha catha,* 1365) from Statius, IV.6. But again it might be Statius' Tisiphone, XI.58 where she is the "hideous and complaining" (*an badb granna geranach,* 4313). The king is always high-king (*ard-rí*), as in the Old Irish epics; so Adrastus is high-king of the Greeks (3000 f.), Eteocles of the Thebans (3026). Mars as god of battle, the Bellipotens of *Theb.* VIII.384, is called in Irish the "cruel and mighty Mars the god of war" (*in croda chumachtach Mairt .i. dea in chatha,* 3052 f.).

Native tradition again comes to the fore in the mode of justice employed by the three judges of Hades, Minos, Aeacus, and Rhadamanthus (1662–65):

> They have an urn of justice with white and black stones therein, and when the white stone used to come up first, the decision was true; but when the black stone used to come up the judgment was untrue.

Now the *Old Irish Ordeals* mentions the three stones:

> the bucket filled with bogstuff and coal and any other black thing and three stones were put into it, a white, a black, and a speckled stone. Then one would put his hand therein, and if the truth were with him, he would bring out the white stone. If he were false he would bring out the black stone. If he were half guilty, he would bring out the speckled stone.[22]

The Irish redactor may be harking back to something very old in early Celtic tradition when he enlarges upon the "still dry spear" (*etiamnum cuspide sicca*) of *Theb.* VIII.383. He speaks of the

> spear in his hand which used to drip and flow profuse streams of blood from point to haft, presaging the death and doom of heroes in each of the two armies. (1353–56)

[21] *The Celtic Review,* vol. 7 [1911], p. 327, fn. 3.

[22] *The Irish Ordeals,* 20, "Tre-lia Mothair," ed. Whitley Stokes, *Irische Texte,* III, Pt. 1 [Leipzig, 1891], p. 191, trans. pp. 209 f. See also the remarks by H. Meroney, "Studies in Early Irish Satire," *Journal of Celtic Studies,* I [1950] pp. 225 f.

Somehow it reminds one of the bleeding lance in the mediaeval Welsh tales of *Culhwch* and *Olwen* and the *Peredur* in the so-called Mabinogion.

The twelfth century Irish looked at some of the practices of pagan antiquity with an eye askance. So the redactor would naturally skip over genealogies of the gods, with which he was not familiar anyhow. He excuses a certain rite because these were the pagan customs, dictated by "the false, lying pagan fables of these people," (*amal fhorglit na faibli guacha gendtligi,* 785). Again he speaks of "the gods you worship" or "the gods they worshipped in those days" (423, 1024, 2851). When Statius mentions the soul of a warrior going to Hades, the Irish redactor substitutes the Christian hell (*iffern*). *Thebaid* X.854 speaks of the impending last destruction (*suprema lues*), which the Irish renders as the "troubled Day of Judgment" (*an lá buaidirthi brátha,* 4232). Apollo tells his servant to depart to the joy of the Elysian fields (*Theb.* VII.775 f.), a passage completely misunderstood in the Irish version, which bids him to "arise and go into the awful pains of hell as has been appointed thee" (2920 f.), a possible Scriptural reminiscence. Another occurs in the passage where they embalm the infant son of Lycurgus with spices from the East (*Theb.* VI.61):

> cinnamon, balsam, myrrh, and frankincense . . .
> (.i. sinomum 7 ballsam 7 mirr 7 tuis . . . 2212 f.)

a possible reminiscence of *Eccl.* 24.20:

> sicut *cinnamomum et balsamum* aromatizans odorem dedi :
> quasi *myrrha* electa dedi suavitatem odoris . . .

This would have been familiar to any literate person at this time as it occurs in the Office for our Blessed Lady in the Lesson for Matins and again at Nones. *Cant.* 4.6 would have suggested myrrh and frankincense:

> vadam ad montem *myrrhae* et collem *thuris.*

That the redactor had at hand a well-glossed text of the Latin *Thebaid* with something like the learned scholia of Lactantius Placidus, the famous sixth century commentator, is not at all unlikely. When he departed from the Latin text to explain classical references he must have depended upon such aid. The Irish version (255–62) tells the story of Sciron, the infamous robber who enjoined strangers to kiss his feet; if they refused they were cast headlong into the sea. There is nothing of this in Statius, but the whole story is told in the commentary of Lactantius Placidus in a scholion to *Theb.* I.333.[23] Antigone is mentioned (2645 f.) as sister of Eteocles and Polynices, following the scholia of Lactantius on *Theb.* VII.244:

> ANTIGONE Oedipi filia, Polynicis soror.

In *Togail na Tebe* (2647–50) Antigone observes the host from a tower apart:

> For this was a custom they observed that hosts and multitudes should not see their maidens, till they were married. And visible to her there was the march of the hosts away over the plain, but she was not visible to them.

where Lact. Placid. had commented upon *Theb.* VII.243:

> NONDVM CONCESSA VIDERI pudicarum virginum morem descripsit, quia numquam uirgines facile uidebantur a populo.

The story of Sthenoboea, who wished to seduce her stepson Bellerophon, is not told in Statius, but the Middle Irish account, 1688–95) tells the whole story from Lact. Placid., where it occurs as a long scholia on *Theb.* IV.589. So too the story of the blinding of Tiresias the seer (1624–36), told in the scholion to *Theb.* II.95 and not at IV.408, where the translator has just been working. It is possible that the Statius manuscript used by the translator carried a reference to the earlier occurrence. That Semele was mother of Bacchus and daughter of Cadmus the Middle Irish redactor tells us when translating *Theb.* IV.656 ff., but there is no mention of Semele here in Statius. However, Lact. Placid. adds this gem to explain Statius' *materna ad moenia*:

> MATERNA AD MOENIA, id est ad Thebas, quia exinde fuit Cadmi filia Semele, mater Liberi.

Very well, but the Irish account goes on to give the details of the birth of Bacchus, which is not to be found here but at the very beginning, where the words *quod saeve Ivnonis opus* (I.12) draw a long scholion. At the very end the Irish says that Thersander son of Polynices assumed the sovereignty of Thebes and that he was one of the twenty-seven kings present with the Greeks at the siege of Troy

[23] *Lactantii Placidi qui dicitur Commentarios in Statii Thebaida et Commentarium in Achilleida,* recensuit Ricardus Jahnke [Lipsiae, 1898], p. 39.

(4913–15). This is not from Statius, but Dictys tells us that Thersander of Thebes came to Troy with fifty ships.[24]

Something should be said of the stylistic qualities of this redaction. Whereas the ancient Irish epics were written in prose with some poetry intermixed for the purple passages, prophecies, prayers, oaths, etc., by this time there is no longer any verse to be found in works of this type. Alliteration is freely used, however, the general idea being to have qualifying adjectives alliterate with the name of the person or place they modify. The serpent is at first *nemnach,* "poisonous," to agree with *nathrach,* but less than twenty lines further it is "venomous, hostile, with crests and stings" (*nathrach neimneach naimdidi com m[b]eandaib 7 birinib*, 200). The Fury came up from hell: "she came through the fierce, idle, and fiendish rabble," (*tanic tresna daescurshluagaib disciuri dinaime deamnacha*, 190). Statius speaks of the "well known road to Thebes" (*notum iter ad Thebas*, I.101) which is rendered *sligid suaichnid saineamail,* "the well-known and excellent road" (198). Polynices the Theban is qualified as the "violent, stubborn, stalwart Theban ..." (*in trenfher tend talchai Tiauanda* ... 1406 f.); but Tydeus as soldier (*milid*) is the "handsome, high-spirited soldier ..." (*miled maisech mormenmnach*, 1437). Statius speaks of the "heights of Cirrha" (ardua Cirrhae, II.63), upon which the Irish improves: "through the very smooth easily traversed roads of Mount Cirrha" *(sligthib soreidi soimtheachta slebe Cirra*, 612). But the alliterative effect can best be seen in many different patterns of which I can offer but one (lines 182–86):

Is andsin ro eirig in Fhuir
[Then arose the Fury,]
demnach dasachtach
[fiendish, mad,]
cona trillsib do
[with her locks]
nathrachaib nemnacha ima ceand
[of venomous serpents about her head,]
ac siangail 7 ic sibsanaig
[a-puffing and a-croaking,]
7 ic feadgaire 7 ac foluamnigud
[and a-whistling and a-fluttering,]

[24] *Dictys Cretensis Ephemeridos Belli Troiani,* I.17, edidit Ferdinandus Meister [Lipsiae, 1872], p. 13 *ad fin.*

7 ac sugud int shrotha
[and a-sucking the stream,]
tibrechtaich teindtigi . . .
[welling and fiery . . .]

Others would be the simile of arrows falling like hail in a storm (1982–86) and a battle ending as a storm ends (1987–89), corresponding to *Theb.* V.394–406, or the sacking of Thebes (4113–15), *Theb.* X.522 ff. But here the translator simply followed the contemporary style in the treatment of native themes. So in such Middle Irish compilations as the *Battle of Magh Rath* or *Magh Lena* the same sort of bombast appears. At first it comes as a novelty, but like the ever-occurring *cid tra acht, imthusa, immorro,* or *is andsin,* to mark paragraph or chapter divisions, it becomes very dull indeed.[25] If anything the Middle Irish redaction is more horrendous, if I may use the word, than its Latin original. One is apt more than once in the course of his reading to re-echo the ejaculations of the scribe of the Edinburgh manuscript. Statius would be glad to know that his *Thebaid* lived on in the poetry of Dante, Chaucer, the unknown author of the *Roman de Thèbes,* and in this Irish redaction, for it was his devout hope that the work would survive him, yet not rival the divine *Aeneid.* Both prayers were answered and the scribe of the Egerton manuscript adds a note on the copying of his text and asks the blessing of God on his soul.

[25] These are all given by C. Plummer as examples of the turgidity of style to be found in late Middle-Irish compositions (*Miscellanea Hagiographica Hibernica* [Bruxelles, 1925] (Subsidia Hagiographica, 15), p. 97).

Papers of the Michigan Academy of Science, Arts, and Letters
Vol. XLVII, 1962 (1961 Meeting)

ART

THE 1961 SCENE: MICHIGAN ACADEMY EXHIBITION

CHARLES E. MEYER

This year's show was small, consisting of only seventy pieces, mostly painting. There were four ceramic entries, three pieces of sculpture, one enamel, and one woven entry. In view of the large number of practicing artists in our state, the proportion of artists exhibiting with the Academy has become disappointingly low.

In spite of the limited size of the show it was beautifully installed in the new gallery area of Wayne State University. The show was most enhanced by the light, airy, display space. It also demonstrated a range of directions that was most refreshing to see at a time in which abstract expressionism seems to dominate as a movement.

The choice of prizes indicates as well the broad ranging appreciation of the jurors, Mr. Fred Maurice of the Kalamazoo Art Center and Dr. Stuart Hodge of the Flint Institute of Arts.

The first prize in oils went to Raymond Gloeckler for *The Promoters,* a surrealistic fantasy of misshapen, humorously pathetic figures, painted with masterly craftsmanship (Fig. 1).

The first prize in watercolors was won by Elizabeth L. Dulmage for *Glass Spectrum* (Fig. 2). This is a transparent usage of the medium in the most honest manner. Utilizing this basic transparency the artist has created a rich blue, red, green, and yellow blend of transparent planes. The effect is one of looking through varicolored bottles that have the richness of medieval stained glass.

The first prize for prints was taken by Mr. William T. Woodward for *Hibernation* (Fig. 3). This print is an extremely rich combination of rusty oranges and browns and weathered green. The simplicity of the block print technique's directness is emphasized by utilization of the wood grain as textural accent, and the clean distribution of large forms held together by sharp lines.

The prize for the best representational painting was awarded for *Cornfield in Winter,* a watercolor by Rodolphe La Riviere (Fig. 4). This painting well deserved this particular award, but it should be noted that the abstract devices of general compositional disposition,

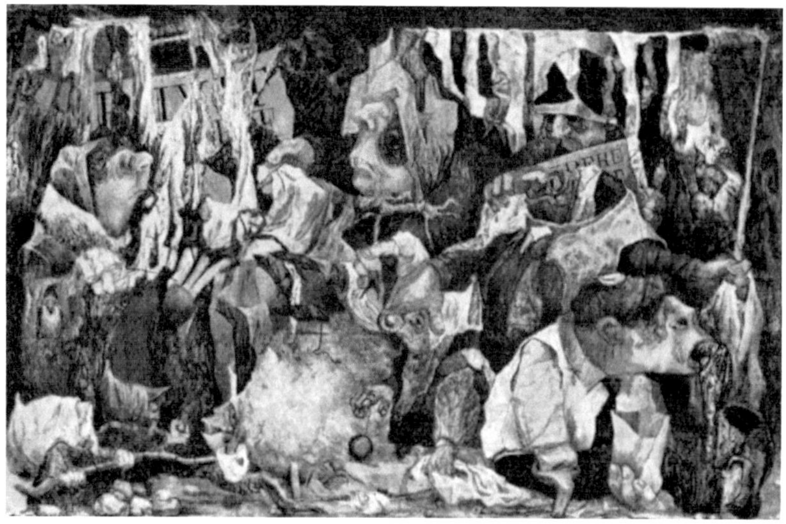

FIG. 1
Raymond Gloeckler, *The Promoters.*
Oil, First Prize, 1961

emphasis upon the varied textural surfaces employed mark a high degree of sensitivity to medium and form often lost in representational painting.

The purchase prize for this year was given for *Barn Window,* an oil by Aimé Martinat (Fig. 5). This painting, like that of La Riviere, is traditional in emphasis upon an almost minute attention to detail; it too, however, demonstrates the artist's capacity as a composer of abstract forms, and as a fine colorist. The richness of the paint surfaces and the tastefully rich selection of color demonstrate the best sort of attention to medium that La Riviere's work also shows.

Ranging further into the most developed abstract expressionist form of presentation is the oil *Composition 25-60* by Dr. Allen Leepa (Fig. 6). This painting was awarded an honorable mention in oils and represents the best of this direction. It is a vital, rhythmic, vibrant play of brilliant pastel yellows, oranges, and blues. The attention to

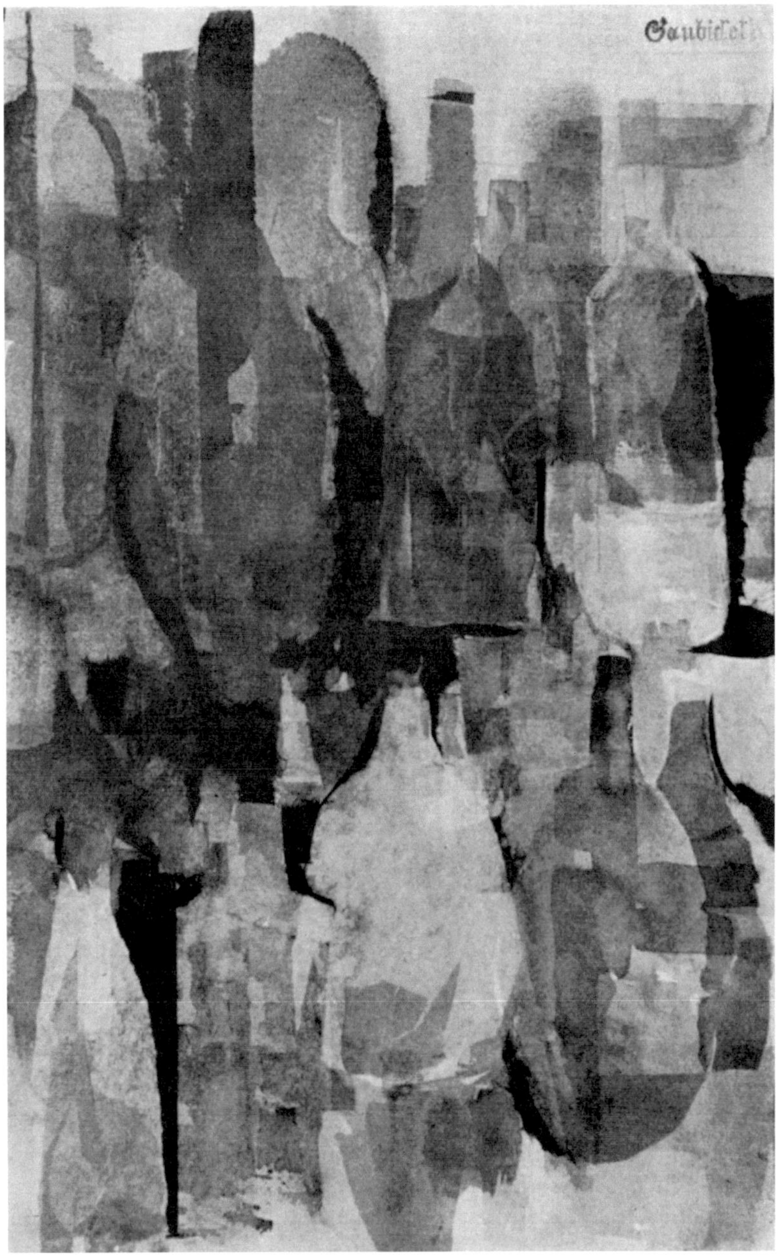

Fig. 2

Elizabeth L. Dulmage, *Glass Spectrum*.
Watercolor, First Prize, 1961

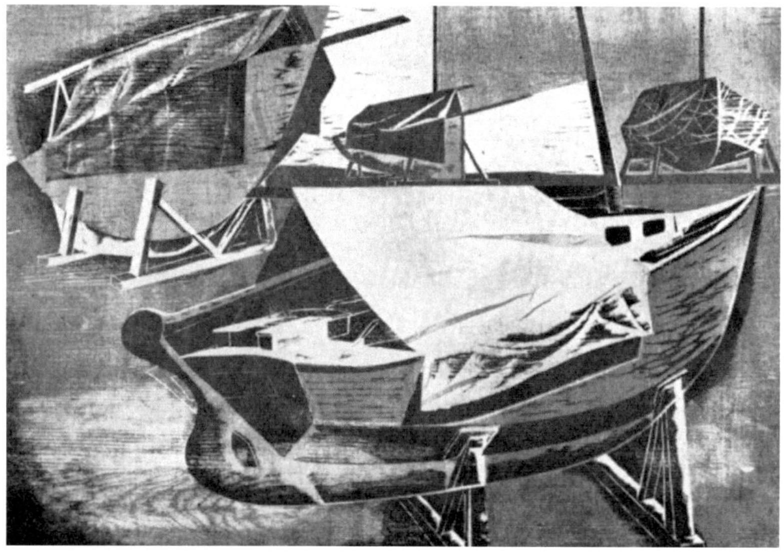

Fig. 3
William T. Woodward, *Hibernation*.
Print, First Prize, 1961

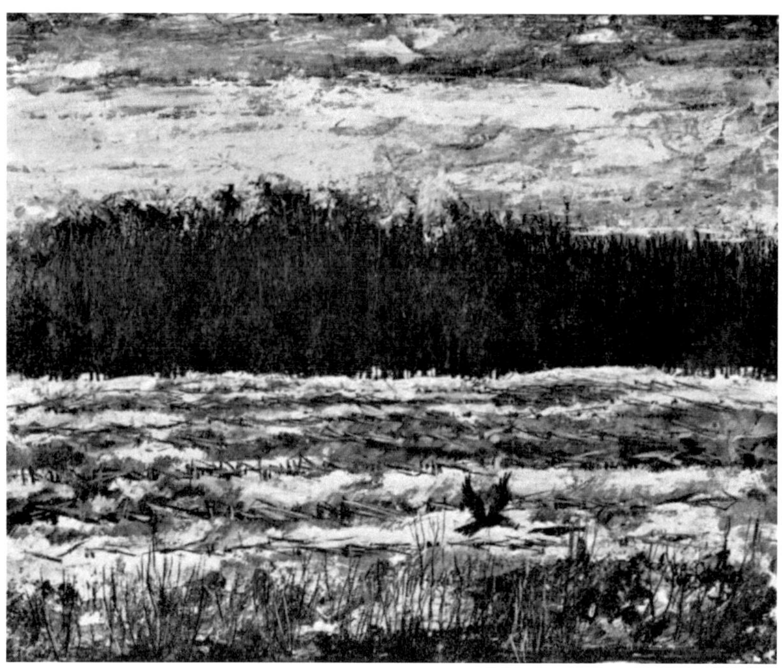

Fig. 4
Rodolphe La Riviere, *Cornfield in Winter*.
Watercolor, Prize for Best Representational Painting, 1961

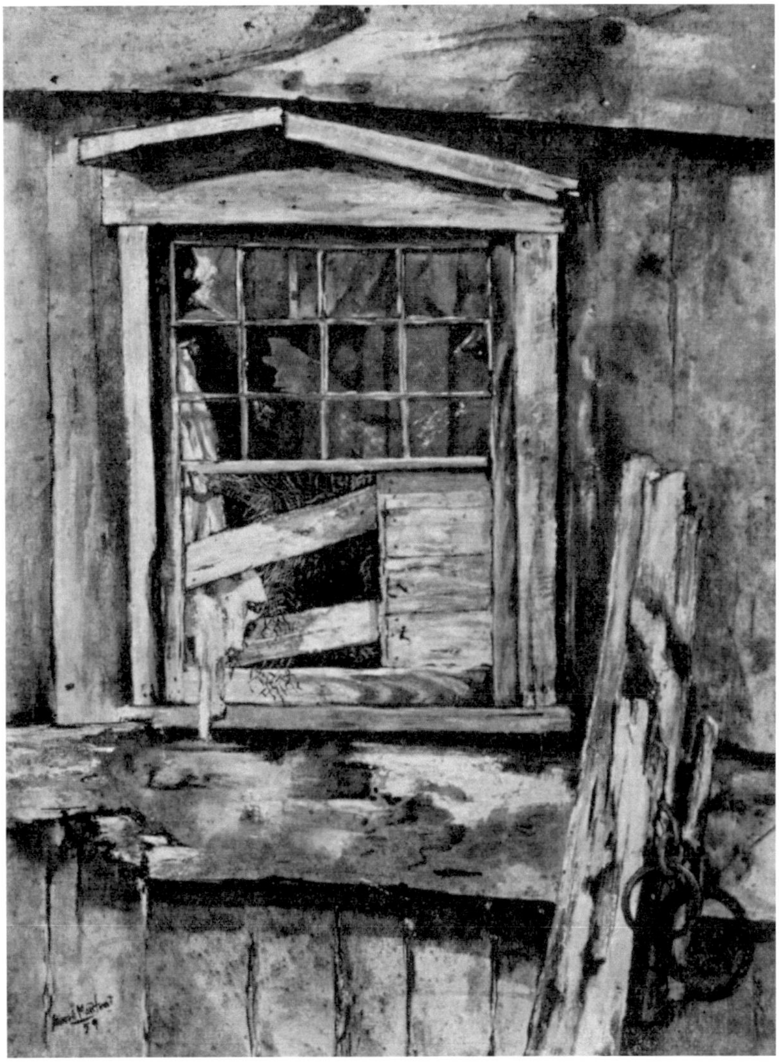

FIG. 5

Aimé Martinat, *Barn Window*.
Oil, Purchase Prize, 1961

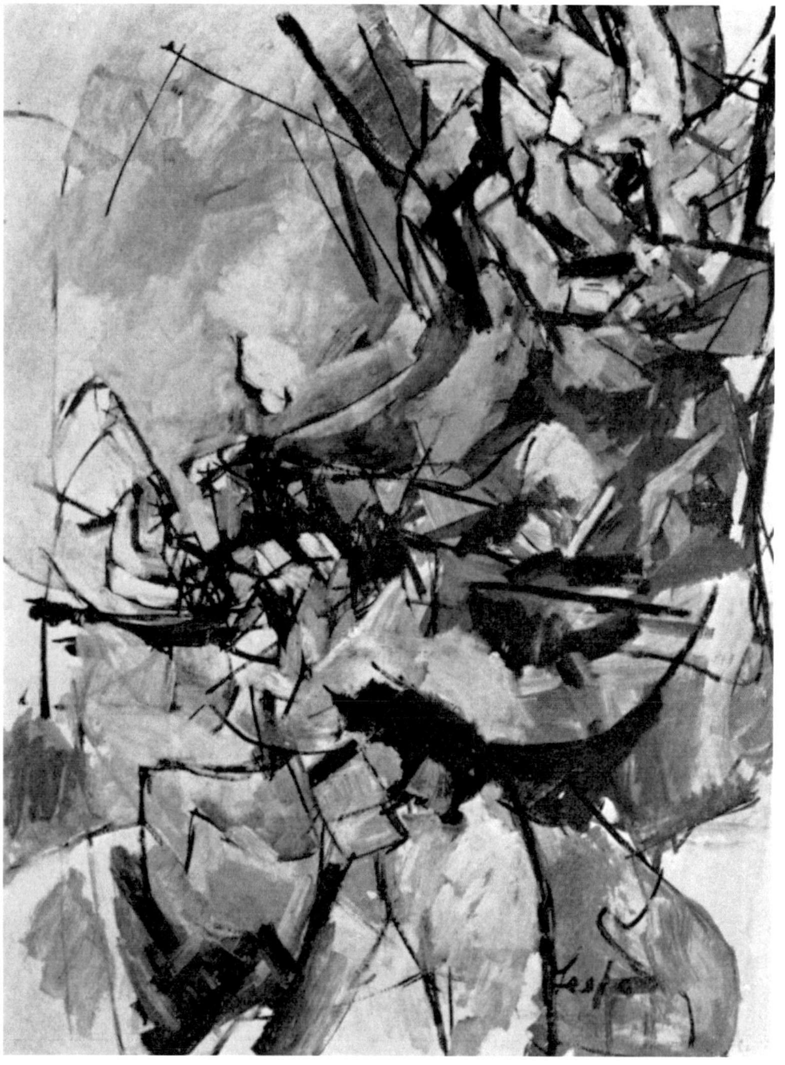

Fig. 6

Allen Leepa, *Composition 25-60.*
Oil, Honorable Mention, 1961

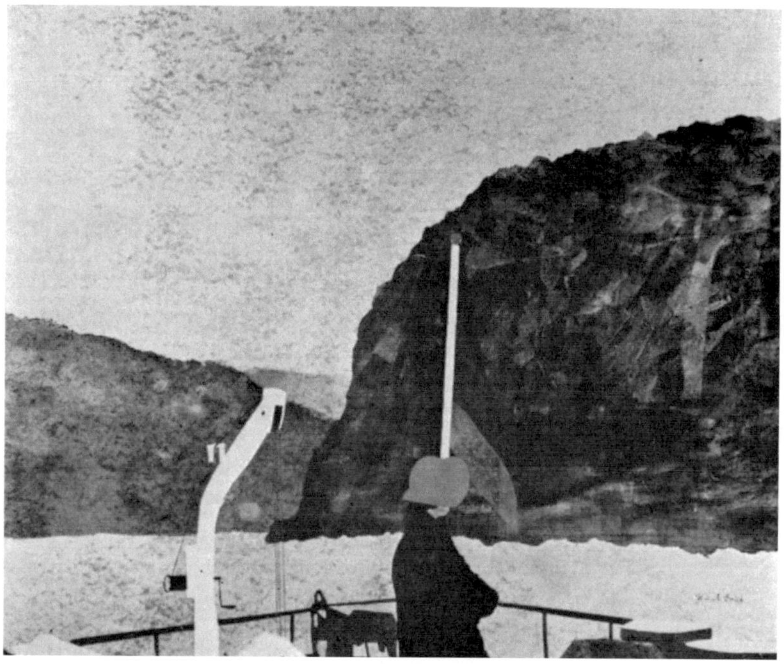

FIG. 7
William Bostick, *Die Lorelei.*
Prize for Unusual Medium, 1961

and joy in the medium itself that was noted above in a different context is here as well, and the intense vitality that breathes through is as real as that of a representational painting.

Another prize for the most unusual medium went to Mr. William Bostick for *Die Lorelei* (Fig. 7). This is a construction neatly composed around a bright spot of red in the woman's head. In terms of the other paintings above, it marks a position between representationalism and total abstraction.

Related to this position is another honorable mention watercolor by Ernest Scanes called *Arkansas Cotton* (Fig. 8).

Worthy of special mention is Louise J. Nobili's oil *Where the Sun Is Oblique* (Fig. 9). This marks something new in the Academy in that

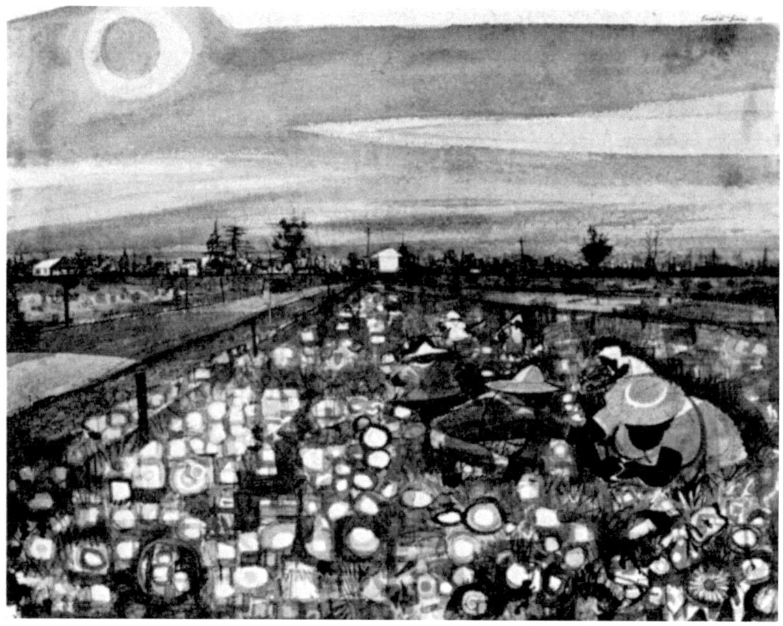

Fig. 8
Ernest Scanes, *Arkansas Cotton.*
Watercolor, Honorable Mention, 1961

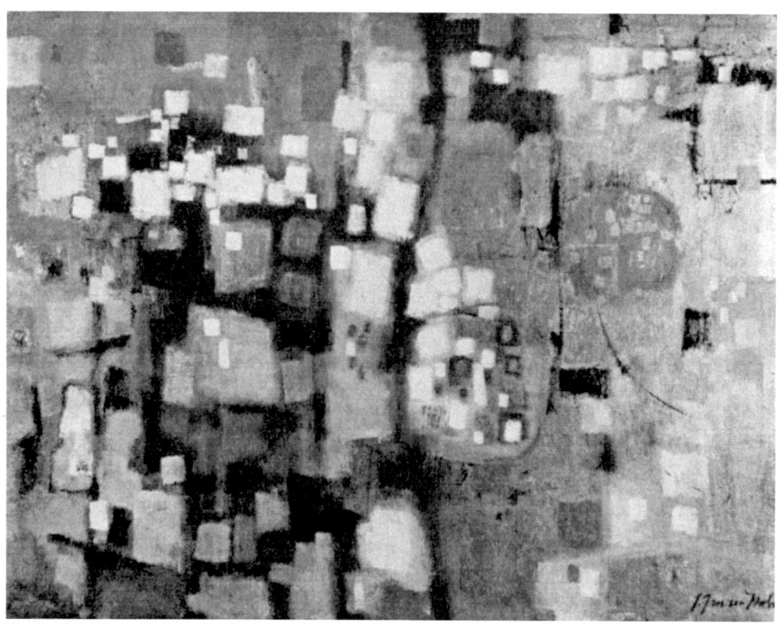

Fig. 9
Louise J. Nobili, *Where the Sun Is Oblique.*
Oil, Special Mention, 1961.

Mrs. Nobili's previous entries have been superb watercolors. This indicates that she is a superb oil painter as well. The painting is constructed of soft, fuzzy, pale golds, grays, and black. The total quality communicates a moonstruck effect that is most pleasing.

The range of the total show then is most wide, and the quality indicates that some of the better Michigan artists have remained Academy members. Let us hope that the Academy will attract others, however, for its number of artists is at present small and includes only a small percentage of the truly good artists in Michigan. There is much room for improvement.

The aims of the Academy have been clearly defined in its latest publications and they are those of many artists who do not hold memberships. Perhaps broader publication of these aims is necessary to attract our fellow artists. Wider solicitation for entries to the annual show, with emphasis upon the attractive prizes to be won, might be worth developing. As wide ranging as the exhibit has been, it cannot be said to have been completely representative in the last three or four years. Let us strive to move friends and colleagues to participate.